NMR STUDIES OF TRANSLATIONAL MOTION

Translational motion in solution, either diffusion or fluid flow, is at the heart of chemical and biochemical reactivity. Nuclear magnetic resonance (NMR) provides a powerful non-invasive technique for studying the phenomena using magnetic field gradient methods. Describing the physical basis of measurement techniques, with particular emphasis on diffusion, balancing theory with experimental observations and assuming little mathematical knowledge, this is a strong, yet accessible, introduction to the field. A detailed discussion of magnetic field gradient methods applied to magnetic resonance imaging (MRI) is included, alongside extensive referencing throughout, providing a timely, definitive book to the subject, ideal for researchers in the fields of physics, chemistry and biology.

William S. Price is Professor and Chair of Nanotechnology in the School of Biomedical and Health Sciences and the School of Medicine, University of Western Sydney. He directs the Biomedical Magnetic Resonance Facility and leads the Nanoscale Organisation and Dynamics Group, University of Western Sydney. Professor Price is an expert in the theory, development and application of NMR spectroscopy, focussing on NMR diffusion measurements and MRI techniques for studying molecular association and molecular dynamics.

Cambridge Molecular Science

As we enter the twenty-first century, chemistry has positioned itself as the central science. Its subject matter, atoms and the bonds between them, is now central to many of the life sciences on the one hand, as biological chemistry brings the subject to the atomic level, and to condensed matter and molecular physics on the other. Developments in quantum chemistry and in statistical mechanics have also created a fruitful overlap with mathematics and theoretical physics. Consequently, boundaries between chemistry and other traditional sciences are fading and the term *Molecular Science* now describes this vibrant area of research.

Molecular science has made giant strides in recent years. Bolstered both by instrumental and by theoretical developments, it covers the temporal scale down to femtoseconds, a time scale sufficient to define atomic dynamics with precision, and the spatial scale down to a small fraction of an Angstrom. This has led to a very sophisticated level of understanding of the properties of small molecule systems, but there has also been a remarkable series of developments in more complex systems. These include protein engineering; surfaces and interfaces; polymers; colloids; and biophysical chemistry. This series provides a vehicle for the publication of advanced textbooks and monographs introducing and reviewing these exciting developments.

NMR STUDIES OF TRANSLATIONAL MOTION

WILLIAM S. PRICE
Nanoscale Organisation and Dynamics Group
University of Western Sydney, Australia

CAMBRIDGE
UNIVERSITY PRESS

CAMBRIDGE
UNIVERSITY PRESS

Shaftesbury Road, Cambridge CB2 8EA, United Kingdom

One Liberty Plaza, 20th Floor, New York, NY 10006, USA

477 Williamstown Road, Port Melbourne, VIC 3207, Australia

314–321, 3rd Floor, Plot 3, Splendor Forum, Jasola District Centre, New Delhi – 110025, India

103 Penang Road, #05–06/07, Visioncrest Commercial, Singapore 238467

Cambridge University Press is part of Cambridge University Press & Assessment,
a department of the University of Cambridge.

We share the University's mission to contribute to society through the pursuit of
education, learning and research at the highest international levels of excellence.

www.cambridge.org
Information on this title: www.cambridge.org/9780521806961

First published 2009

A catalogue record for this publication is available from the British Library

Library of Congress Cataloging-in-Publication data
Price, William S., 1964–
NMR studies of translational motion / William S. Price.
p. cm.
1. Translational motion – Measurement. 2. Reactivity (Chemistry).
3. Nuclear magnetic resonance. 4. Diffusion and 5. Magnetic resonance imaging. I. Title.
QC762.6.T73P75 2008
541′.39–dc22
2008027008

ISBN 978-0-521-80696-1 Hardback

Contents

Preface

Translational motion in solution (e.g., diffusion, flow or advection) plays a central role in science. Self-diffusion can be rightfully considered as being the most fundamental form of transport at the molecular level and, consequently, it lies at the heart of many chemical reactions and can even govern the kinetics. Diffusion, due to its very ubiquity, is encountered in a myriad of scientific studies ranging from diseases to separation science and nanotechnology. Further, the translational motion of a species not only reflects intrinsic properties of the species itself (e.g., hydrodynamics), but can also shed light on the surrounding environment (e.g., intermolecular dynamics or motional restriction). Consequently, being able to study and ultimately understand the translational motion of molecules and molecular systems in their native environment is of inestimable scientific value.

Measuring translational motion at the molecular level presents special difficulties since labelling (e.g., radiotracers) or the introduction of thermodynamic gradients (which leads to mutual diffusion and consequently irreversible thermodynamics) in the measurement process can have deleterious effects on the outcome. Also, in many instances it is of interest to measure the diffusion of species at quite high concentrations. Fortunately, nuclear magnetic resonance (NMR) provides a means of unparalleled utility and convenience for performing non-invasive measurements of translational motion. Of particular significance is that, in general, the species of interest inherently contain NMR-sensitive nuclei and thus sample preparation generally requires nothing more than placing the sample into the NMR spectrometer.

NMR is an inherently quantum mechanical subject, yet translational motion of molecules lies between the microscopic and the macroscopic and is most conveniently described with classical physics. Abiding by Ockham's razor, this is the path that will be taken here. This is not an NMR textbook per se, and for the fundamentals of NMR numerous excellent texts are currently available.[1–6] A tabulation of many of the commonly used NMR acronyms can be found elsewhere.[7]

The last two decades have seen the advent of widespread commercial availability of MRI imagers and magnetic gradient probes and gradient drivers for NMR spectrometers. Magnetic gradients now pervade almost all areas of NMR ranging from coherence selection, gradient shimming, selective excitation and signal suppression[8–21] in high-resolution NMR to inputting spatial dependencies into NMR imaging (also known as MRI, and NMR microscopy) and NMR diffusion measurements. The use of gradient NMR allows diffusion to be added to the list of standard NMR observables.

This book grew out of two widely cited pedagogical papers that I wrote sometime ago on measuring diffusion with NMR[22,23] using magnetic field (i.e., B_0) gradient methods. The overall aim of the book is rather modest: to present a clear overview of this field with particular emphasis on trying to impart a clear physical picture – ideally such that the literature may become more accessible to readers interested in expanding their research into this field. Specifically, this book endeavours first to explain the physical basis of magnetic gradient-based NMR measurements of translational motion – especially diffusion measurements involving pulsed magnetic field gradients. Flow- and imaging-based measurements are closely related technically and theoretically and so some coverage of these kindred fields is presented. Secondly, to provide a clear correlation between experimental methodology and theoretical data analysis and the sorts of applications that are feasible. I have cited rather more references than is typical in a monograph to give the reader more chance of understanding the present text and also the opportunity to read further. I also liberally cite some textbooks that I found particularly lucid – especially in mathematical areas since many of the readers of this book may not come from a strong mathematics/physics background and to this end I also include the derivations of some of the more important equations. Although in general I have tried to maintain historical lineage, in some cases I have cited a more accessible reference for a concept than the original reference. As is so often the case, what started as my attempt to clearly explain an area of science became an exercise in trying to make myself really understand the area. This area of NMR application, theory and methodology has grown exponentially in the past three decades. Thus, it is neither possible nor reasonable to attempt anything approaching a comprehensive coverage; further the ordering of the material was made particularly difficult by many of the studies being justifiably included under more than one of the subheadings.

The outline of this book is as follows: Chapter 1 serves as an introduction to translational motion with special attention to diffusion and begins with some introductory remarks on the relationship between diffusion and chemical and biological phenomena, followed by a detailed exposition on its physical basis and how it can be mathematically modelled – these concepts will be heavily drawn upon in later chapters. This basis also allows the final section, where some of the more common techniques for measuring diffusion are briefly reviewed, to be placed in perspective as it allows the strengths and limitations of the different methods of diffusion measurements to be more clearly understood. Chapter 1 is rather solid and parts can be skipped on a first read or by those only interested in the NMR techniques and not the fundamental aspects of diffusion and its modelling.

Chapter 2 introduces the basic concepts behind magnetic gradient–based NMR diffusion measurements in freely diffusing systems and in particular the workings of basic pulsed gradient spin-echoes (PGSE), the most common (analytical and approximate) mathematical procedures for correlating the experimental variables, diffusion (and flow) with the observed NMR signal. Measurements of samples containing mixtures of species and aggregating species, but where the exchange is slow on the measurement timescale, are also considered.

Chapter 3 builds upon Chapter 2 and extends the coverage to the cases of simple porous systems, that is where a species is confined to a single highly symmetrical pore of some description. Whereas analytical relations could be given for the relationship between experimental variables and observed signal in the case of free diffusion, such analysis even for simple highly symmetrical pores rapidly becomes mathematically intractable and thus some consideration is given to the validity of the commonly used mathematical approximations and the various numerical approaches that are sometimes used.

Chapter 4 continues on from Chapter 3 and considers phenomena such as exchange, anisotropic and flowing systems and diffusion in polymers. Diffusion measurements in internal magnetic fields are also given brief mention.

Chapter 5 provides an overview of the additional instrumentation for an NMR spectrometer needed to conduct diffusion, flow and, by extension to a system capable of generating magnetic field gradients in three orthogonal directions, imaging-based measurements.

Chapter 6 details fundamental experimental considerations such as the selection of delays, gradient parameters and solvent suppression technique in diffusion measurements and how the applied gradient strength might be calibrated. This chapter ends with a section on how to transform and analyse the PGSE data resulting from simple freely diffusing samples, complex mixtures and samples involving restricted diffusion.

Chapter 7 concerns most of the issues that thwart attempts to get high-quality data such as background gradients, eddy currents induced by the rapidly pulsed magnetic field gradients and imperfectly generated gradient pulses. The symptoms and suggestions for obviating or at least meliorating their effects are considered.

Chapter 8 provides an overview of some of the more sophisticated sequences for measuring diffusion, flow and related phenomena.

Chapter 9 gives an overview of the inclusion of conventional nuclear magnetic resonance imaging (aka MRI) methodology with diffusion and flow measurements.

Whereas in the previous chapters the systems being studied might be microscopically heterogeneous, there was an implicit assumption that they were macroscopically homogeneous. However, in many cases (biological tissues such as brain being a prominent example) the sample is also heterogeneous on a length scale that can be accessed by MRI methodology. The combination of MRI techniques with gradient-based diffusion and flow-measuring techniques provides powerful tools for characterising natural and synthetic materials.

Chapter 10 gives a brief overview of the use of radio frequency (i.e., B_1) field gradients to measure translational motion.

Chapter 11 surveys some of the applications to which gradient-based measurements of translational motion have been put.

References

1. R. R. Ernst, G. Bodenhausen, and A. Wokaun, *Principles of Magnetic Resonance in One and Two Dimensions*. (London: Clarendon Press, 1987).
2. J. W. Hennel and J. Klinowski, *Fundamentals of Nuclear Magnetic Resonance*. (Essex: Longman Scientific & Technical, 1993).
3. D. Canet, *Nuclear Magnetic Resonance Concepts and Methods*. (New York: Wiley, 1996).
4. D. M. Grant and R. K. Harris (eds), *Encyclopedia of Nuclear Magnetic Resonance*. (New York: Wiley, 1996).
5. M. Levitt, *Spin Dynamics – Basic Principles of NMR Spectroscopy*, 2nd edn. (New York: Wiley, 2008).
6. J. Keeler, *Understanding NMR Spectroscopy*. (New York: Wiley, 2005).
7. D. D. Traficante, Abbreviations and Acronyms Used in Magnetic Resonance. *Concepts Magn. Reson.* **10** (1998), 59–62.
8. P. C. M. Van Zijl and C. T. W. Moonen, Complete Water Suppression for Solutions of Large Molecules Based on Diffusional Differences Between Solute and Solvent (DRYCLEAN). *J. Magn. Reson.* **87** (1990), 18–25.
9. D. M. Doddrell, Application of Pulsed Field Gradients in High-Resolution NMR Spectroscopy. *J. Chin. Chem. Soc. (Taipei)* **38** (1991), 107–17.
10. P. C. M. Van Zijl and C. T. W. Moonen, Solvent Suppression Strategies for In Vivo Magnetic Resonance Spectroscopy. *NMR Basic Princ. Prog.* **26** (1992), 67–108.
11. J. Keeler, R. T. Clowes, A. L. Davis, and E. D. Laue, Pulsed-Field Gradients: Theory and Practice. *Methods Enzymol.* **239** (1994), 145–207.
12. D. Canet and M. Décorps, Applications of Field Gradients in NMR. In *Dynamics of Solutions and Fluid Mixtures*, ed. J.-J. Delpuech. (New York: Wiley, 1995), pp. 309–43.
13. L. Mitschang, H. Ponstingl, D. Grindrod, and H. Oschkinat, Geometrical Representation of Coherence Transfer Selection by Pulsed Field Gradients in High-Resolution Nuclear Magnetic Resonance. *J. Chem. Phys.* **102** (1995), 3089–98.
14. J. R. Tolman and J. H. Prestegard, Homonuclear Correlation Experiments Using Pulsed-Field Gradients. *Concepts Magn. Reson.* **7** (1995), 247–62.
15. J.-M. Zhu and I. C. P. Smith, Selection of Coherence Transfer Pathways by Pulsed-Field Gradients in NMR Spectroscopy. *Concepts Magn. Reson.* **7** (1995), 281–91.

16. R. E. Hurd, Field Gradients & Their Application. In *Encyclopedia of Nuclear Magnetic Resonance*, ed. D. M. Grant and R. K. Harris. vol. 3. (New York: Wiley, 1996), pp. 1990–2005.
17. T. Parella, High-Quality 1D Spectra by Implementing Pulsed-Field Gradients as the Coherence Pathway Selection Procedure. *Magn. Reson. Chem.* **34** (1996), 329–47.
18. W. S. Price, Gradient NMR. In *Annual Reports on NMR Spectroscopy*, ed. G. A. Webb. vol. 32. (London: Academic Press, 1996), pp. 51–142.
19. S. Berger, NMR Techniques Employing Selective Radiofrequency Pulses in Combination with Pulsed Field Gradients. *Prog. NMR Spectrosc.* **30** (1997), 137–56.
20. W. S. Price, Water Signal Suppression in NMR Spectroscopy. In *Annual Reports on NMR Spectroscopy*, ed. G. A. Webb. vol. 38. (London: Academic Press, 1999), pp. 289–354.
21. M. Weiger, T. Speck, and M. Fey, Gradient Shimming with Spectrum Optimisation. *J. Magn. Reson.* **182** (2006), 38–48.
22. W. S. Price, Pulsed Field Gradient NMR as a Tool for Studying Translational Diffusion, Part I. Basic Theory. *Concepts Magn. Reson.* **9** (1997), 299–336.
23. W. S. Price, Pulsed Field Gradient NMR as a Tool for Studying Translational Diffusion, Part II. Experimental Aspects. *Concepts Magn. Reson.* **10** (1998), 197–237.

Acknowledgements

Many people have contributed to this book either directly or indirectly. I thank the authors whose work I have included. Particular thanks to my collaborators, colleagues and mentors over the years, including Dr Yuichi Aihara, Prof. Yoji Arata, Assoc. Prof. Janice Aldrich-Wright, Dr Alexander Barzykin, Prof. Paul Callaghan, Dr Reynaldo Castillo, Dr Bob Chapman, Dr Gary Dennis, Prof. István Furó, Prof. Kikuko Hayamizu, Prof. Lian-Pin Hwang, Prof. Masatsune Kainosho, Prof. Jorg Kärger, Prof. Philip Kuchel, Assoc. Prof. Andrew Shalliker, Prof. Olle Söderman, Dr Tim Stait-Gardner, Prof. Peter Stilbs, Dr Allan Torres and Prof. Sergey Traytak. The efforts of two of my PhD students, Mr Gang Zheng and Mr Nirbhay Yadav, for many hours spent proofreading are gratefully acknowledged. All errors that are contained in this book are, however, attributable only to me.

I also thank my editor, Dr Michelle Carey, who probably never thought that the book would ever be finished, and her editorial staff including Sarah Matthews.

Most of all I thank my wife, Hasna, and my two children, Edward and Roxanne, for their patience and support while I wrote this book. And to my late wonderful parents who always gave me every support and encouragement in life and also in pursuing an academic career.

The NSW State Government is acknowledged for financial support through a *Bio*First Award.

Abbreviations and Symbols

A	cross-sectional area of a sample tube
a	characteristic distance (e.g., radius of a pore or half the inter-planar separation)
ADC	apparent diffusion coefficient (*see also* D^{app})
a_M	relaxation length
$a_{\min}$	uncertainty in initial and final position during a gradient pulse
b	gradient or diffusion weighting factor
$\mathbf{b}$	gradient or diffusion weighting vector
$\mathbf{B}_0$	static magnetic field
$\mathbf{B}_1$	radio frequency (rf) field
BPP	bipolar gradient pulses
BSA	bovine serum albumin
$C(Z, \Delta)$	diffusion envelope
CORE	component resolved spectroscopy
COSY	correlation spectroscopy
CPMG	Carr–Purcell–Meiboom–Gill sequence
CTPG	constant time, pulse, and gradient and amplitude diffusion experiment
D	self-diffusion coefficient
d	embedding dimension
D_i	individual self-diffusion coefficient
D_i^0	self-diffusion coefficient of the ith oligomeric species at infinite dilution
$\langle D\rangle_P$	population-weighted averaged diffusion coefficient
$\langle D\rangle_W$	mass averaged diffusion coefficient
$\langle D\rangle_W^C$	mass averaged diffusion coefficient including obstruction effects

$\mathbf{D}$	self-diffusion tensor
$\mathbf{D}^{\mathrm{lab}}$	diffusion tensor in the laboratory axes frame (x', y', z')
$\mathbf{D}_{\omega}$	diffusion coefficient spectrum or tensor
D_{κ}	fractional diffusion coefficient
D_{ω}	frequency-dependent diffusion coefficient
D^{∞}	effective or long-time self-diffusion coefficient
D^{*}	asymptotic dispersion coefficient
$D(t)$	(measuring time-dependent) diffusion coefficient obtained in the SGP limit
$\mathbf{D}(t)$	time-dependent diffusion tensor
D^{0}	infinite dilution (or short time) self-diffusion coefficient
D_{anion}	anion diffusion coefficient
D_{cation}	cation diffusion coefficient
d_{f}	dimension of the fractal space
d_{w}	random walk dimension
D^{app}, $D^{\mathrm{app}}(t)$ or $D^{\mathrm{app}}(\Delta)$	'apparent diffusion coefficient' obtained by simplistically applying the free diffusion solution to the analysis of a more complicated system
D^{eff}	synonymous with D^{∞}
D^{*}_{eff}	effective time-dependent dispersion coefficient
D_{b}	bound ligand diffusion coefficient
D_{f}	free ligand diffusion coefficient
D_{i}	'distinct' diffusion coefficient
D/D^{0}	relative diffusion coefficient
$\mathbf{D}_{\mathrm{M}}$	mutual diffusion tensor
D_{M}	mutual diffusion coefficient
DOSY	diffusion ordered spectroscopy
$\mathbf{D}^{\mathrm{pr}}$	self-diffusion tensor in the principal axes frame (x, y, z)
DQ	double quantum
d_{s}	spectral dimension
DCNMR	*see* electrophoretic NMR
DDCOSY	diffusion–diffusion correlation spectroscopy
DRCOSY	diffusion–relaxation correlation spectroscopy
DDIF	diffusion decay in the internal field
DTI	diffusion tensor imaging
DWI	diffusion-weighted imaging
$\Delta\bar{P}(\mathbf{R}, \Delta)_{\frac{1}{2}}$	average propagator width at half-height
E	elliptic integral of the second kind
E, $E(\mathbf{q}, \Delta)$	spin-echo attenuation (normally synonymous with E_{Diff})

E_{dc}	pulsed electric field
E_{Diff}	spin-echo attenuation due to diffusion
$E(\mathrm{g}_1)$	spin-echo attenuation due to B_1 gradients (normally synonymous with E_{Diff})
$E(\mathbf{q}, \infty)$	long–time diffusive attenuation
E_{phase}	signal attenuation due to the phase-twist or residual phase-twist
E_{poly}	spin-echo attenuation in a polydisperse system
E_{Relax}	spin-echo attenuation due to relaxation
ENMR	electrophoretic NMR
F	Perrin factor
$\mathrm{FT}\{L(Z)\}$	reciprocal lattice
f	friction coefficient
F_1	indirectly detected dimension in a 2D NMR experiment
F_2	directly detected dimension in a 2D NMR experiment
f_{s}	free volume contributed by the solvent
FID	free induction decay
FT	Fourier transform
$\mathbf{g}$	magnetic field gradient (applied) – normally spatially homogeneous (i.e., constant)
$\mathbf{g}_0$	background or constant magnetic gradient
g^{app}	initial guess of gradient strength
g_{e}	encoding gradient
g_{eff}	effective gradient
$\mathbf{g}_{\mathrm{int}}$	magnetic gradient arising from internal susceptibility differences (*see* background (magnetic field) gradients)
g_{r}	read gradient
$\mathrm{G}_{x,y,z}$	gradient used for spatial localisation in MRI
GPD	Gaussian phase distribution
$\mathbf{H}$	mean square displacement tensor
H	Hurst exponent
h	reduced permeability (= Ma/D)
HMQC	heteronuclear multiple-quantum coherence
HRMAS	high-resolution magic angle spinning
HSA	human serum albumin
$\mathfrak{I}$	tortuosity
$\mathbf{i}, \mathbf{j}, \mathbf{k}$	unit coordinate vectors
IMFG	internal magnetic field gradient (*see* background (magnetic field) gradients)
$\mathbf{J}_A(t)$	flux associated with species A at time t

$J_n(x)$	Bessel function of the first kind of order n
$j_n(x)$	spherical Bessel function of order n
K	modified Bessel function of the third kind
K	elliptic integral of the first kind
K	mobility factor
$\mathbf{k}$ (or k)	dephasing strength of gradients when used in imaging (*see* **k**, k-space)
k	gradient or diffusion weighting factor, more commonly written as b
k_1	first-order rate constant
k_{+1}	(forward) first-order rate constant
k_{-1}	(reverse) first-order rate constant
K_d	dissociation constant
K_e	equilibrium constant
l	step length or size, length of sample tube, length of receiver coil
$L(Z)$	lattice correlation function
$\mathcal{L}$	Laplace transform
LED	longitudinal eddy current delay
M	relaxivity (or permeability)
M_0	thermal equilibrium magnetisation
MAGROFI	magnetisation rotating frame imaging
MGSE	Modulated Gradient Spectroscopy *see* OGSE
MAS	magic angle spinning
MMME	multiple modulation multiple-echo
MOSY	mobility ordered spectroscopy
MRI	(nuclear) magnetic resonance imaging
MSD, $\langle R^2 \rangle$	mean square displacement
MW	molecular weight
M_n	number-average molecular weight
M_w	weight-average molecular weight
LED	longitudinal eddy current delay
$\mathbf{n}$	outward surface normal
n	number of spatial dimensions
n_avg	average number concentration of counterions in the spherical shell
$n(R_\mathrm{cell})$	number concentration of counterions at the outer shell boundary
NOE	nuclear Overhauser effect
NOESY	nuclear Overhauser enhancement spectroscopy
O_D	obstruction factor

OGSE	oscillating gradient spin-echo (sequence)
p	coherence order (*see also* multiple quantum coherences)
P_b	bound (fractional) population (of a ligand)
P_e	Péclet number
P_f	free (fractional) population (of a ligand)
$P_n(x)$	Legendre polynomial of order n
$\bar{P}(\mathbf{R}, t)$	average or mean propagator, the probability that a particle will move a distance $\mathbf{R}$ in time t
$P(a)$	distribution sphere radii
$P(x, y)$	joint probability (e.g., $P(D_{ii}, D_{jj})$, $P(\omega, D)$)
$P(v)$	distribution of velocities
$P(\mathbf{r}_0, \mathbf{r}_1, t)$, P	diffusion propagator, the probability of moving from $\mathbf{r}_0$ to $\mathbf{r}_1$ in time t
P_ϕ, $P_\phi(\phi, t)$	probability of phase distribution = phase distribution function
$[P]_{Total}$	total macromolecule concentration
PFG	pulsed field gradient
PGSE	pulsed gradient spin-echo
PGSTE	pulsed gradient stimulated (spin) echo
$\mathbf{q}$	dephasing strength of gradients when used for measuring displacement (*see* $\mathbf{q}$, q-space)
$\Delta\mathbf{q}$	gradient mismatch
q_{max}	maximum value of q used in an experiment
$\mathbf{r}$	position of a particle (or voxel)
$\Delta\mathbf{r}$	sample movement
$\mathbf{r}_0$	initial position
$\mathbf{r}_1$	finishing position
r_S	Stokes radius (effective hydrodynamic radius)
$\mathbf{R}$	dynamic displacement ($= \mathbf{r}_1 - \mathbf{r}_0$)
R_a	Rayleigh number
R_{cell}	radius of an electroneutral spherical shell
R_S	radius of a charged hard sphere
RMS	root mean square
RMSD	root mean square displacement
RTOP	return to origin probability
$RTOP^0$	RTOP in free isotropic solution
$RTOP^e$	dimensionless RTOP enhancement
rf	radio frequency
$S(\mathbf{q})$	Fourier transform of $\rho(\mathbf{r}_1)$, signal in MRI

$\|S(\mathbf{q})^2\|$	power spectrum of $\rho(\mathbf{r_0})$, elastic incoherent structure factor or form factor
$\|S_0(\mathbf{q})^2\|$	average pore structure factor
SE	spin-echo (sequence) (or Hahn echo (sequence))
SGP	short gradient pulse approximation
SGSE	steady gradient spin-echo
S_P	surface area of a pore
S_P/V_P	surface (area)-to-volume ratio of a pore
STE	stimulated echo (sequence)
STRAFI	stray-field imaging
t_e	delay for eddy current dissipation
t_{echo}	time at which echo formation occurs
T	temperature
T	total time for image acquisition
T_1	spin–lattice relaxation time
T_2	spin–spin relaxation time
T_2^{ZQC}	transverse relaxation time of a zero-quantum coherence
t_r	reptation time
T_{RD}	radiation damping time constant
T_S	singlet relaxation time
TOCSY	total correlation spectroscopy
$\mathbf{u}$	fluctuating part of velocity
v	velocity
v	kinematic viscosity
v^+	velocity of cationic species
v^-	velocity of anionic species
v_{max}	maximum measurable velocity (*see* NMR Imaging – max measurable flow rate)
v_0	constant velocity
V	total volume (of a system)
V_h	partial specific volume of a solvent
V_P	volume of a pore
V_s	partial specific volume of a solute
$\mathbf{v}$	(local) spin velocity (*see also* Lagrangian velocity field)
$\mathbf{v}_i(t)$	velocity of a particle in the barycentric reference frame
$\bar{\mathbf{V}}$	average velocity
$\langle v\rangle$	average velocity
$\langle v^2\rangle$	mean square velocity
$\langle v(0)v(t)\rangle$	velocity auto-correlation function

WATERGATE	a gradient-based water suppression technique
x, y, z	Cartesian coordinates (primes can be used to denote the laboratory frame, if required)
(r, θ, ϕ)	spherical polar coordinates
z_0	initial position for a one-dimensional motion
z_1	finishing position for a one-dimensional motion
Z	one dimensional dynamic displacement ($= z_1 - z_0$)
$\langle Z \rangle$	average displacement
$\langle Z_u^2 \rangle$	positional variance
$Z_{\mathrm{u}}(t)$	displacement fluctuation
$\langle Z^n(\Delta) \rangle$	moments of the average propagator ($\bar{P}(Z, \Delta)$)

Greek

α	thermal expansion
α	time-independent scaling constant
$\Gamma(\Delta)$	time-dependent function characteristic of a geometry
γ	(chemical) activity coefficient
γ	gyromagnetic ratio
γ_{eff}	effective gyromagnetic ratio
δ	duration of the gradient pulse
δ_{h}	hydration/solvation (grams of solvent/grams of solute)
δ_{nm}	Kronecker delta
Δ	difference
Δ	timescale of the diffusion measurement
Δ_{CPMG}	effective diffusion time (i.e., effective Δ) in a CPMG measurement
Δ^{eff}	effective diffusion time (i.e., effective Δ) in an OGSE experiment
η	magic (gradient) ratio
η	viscosity
η_{F}	filling factor
Θ	Spectral function
θ	occupation probability
θ_{a}	exponent of anomalous diffusion
κ	$= 2/d_w$
κ	thermal diffusivity
κ	exponent characterising the time dependence of the mean square displacement
Λ	conductivity
Λ_q	pitch of a magnetisation helix due to a pulse of area '$2\pi q$'

$\rho(\mathbf{r}_0)$, $\rho(\mathbf{r}_0,0)$	equilibrium spin density
λ	separation between adjacent sites
μ	drift velocity
μ_0	permittivity constant (magnetic permeability)
σ	length of oscillating gradient pulse
σ	standard error
τ	a delay in a pulse sequence or discretisation time
τ_b	lifetime of a ligand in the bound state
τ_c	reorientational correlation time
τ_e	extracellular (external to the pore) lifetime
τ_f	lifetime of a ligand in the free state
τ_i	intracellular (internal to the pore) lifetime
τ_J	time between jumps
τ_m	mixing time
τ_v	velocity correlation time
ν	frequency (with respect to spectrometer reference frequency)
ν_l	step rate
$\Delta\nu$	linewidth
$\Delta\nu_{1/2}$	linewidth at half-height
ξ	dimensionless variable (= $D\Delta/a^2$) for characterising restricted diffusion
Φ_{Flow}	change in phase of a spin-echo signal due to flow
ϕ	phase angle (phase of spins)
ϕ	porosity
ϕ	volume fraction
$\langle\phi^2\rangle$	mean square phase distribution
ω	frequency (rad s^{-1})

1

Diffusion and its measurement

1.1 Introduction

This chapter introduces the concept of diffusion and other associated forms of translational motion such as flow, together with their physical significance. Measurements of translational motion and their interpretation are necessarily tied to a mathematical framework. Consequently, a detailed coverage of the mathematics, including the partial differential equation known as the diffusion equation, is presented. Finally, the common techniques for measuring diffusion are discussed.

1.2 Types of translational motion – physical interpretation and significance

'Diffusion' is used in the scientific literature with imprecision and ambiguity as there are a number of types of diffusion. With respect to molecular motion, diffusion is used to denote self-diffusion, mutual diffusion and 'distinct' (not in the sense of individual to a species) diffusion coefficients.[1–4] Confusion arises since, although related and having the same units (i.e., m^2s^{-1}), these phenomena are physically distinct.[5] The confusion is exacerbated in the NMR literature with the term 'spin-diffusion' which is a distinct NMR cross relaxation – based phenomenon involving the random migration of magnetisation via mutual spin flips in neighbouring nuclei,[6,7] even though it can be measured using techniques related to those outlined in this book.[8,9] In this book 'diffusion' signifies self-diffusion, which will also be referred to as translational diffusion, although some consideration will be given to mutual diffusion since many of the alternative methods for measuring diffusion, especially those based on scattering, provide information on mutual diffusion[10] which is often compared with the results of NMR measurements of translational diffusion.

1.2.1 Self-diffusion

Translational diffusion, as depicted in Figure 1.1A, is the (stochastic) random thermal motion of molecules and thus forms the most fundamental form of transport.[2,11–21] It can be thought of as Brownian motion without an applied force and thus on the average no net displacement is observed, yet molecules that start together in the same vicinity will be separated (i.e., dispersed). It is one of the principal factors, other than activation energy and orientation, responsible for almost all chemical reactions since the reacting species must collide before they can react. Indeed, the kinetics of some fast reactions where the activation energy is low are thought to be under diffusion control and a great deal of theory has been developed for such cases.[22,23] For example, before protein association can occur, the protein molecules must align properly via diffusion (and perhaps electrostatic guidance) to form a reaction complex. Diffusion is also behind the formation of complex structures in processes involving diffusion-limited aggregation[24,25] and is involved in the formation of cellular aqueous compartments.[26] More examples illustrating the importance of diffusion are given in Chapter 11. Thus, being able to characterise translational diffusion is of enormous importance in a great number of applications and theoretical considerations.

The stochastic motion of molecules in a pure liquid at thermal equilibrium is termed *self-diffusion* and is characterised by a self-diffusion coefficient D (m^2s^{-1}). For a binary or multicomponent system, the term *tracer diffusion* arises from using mixtures of species with different isotopic labelling (e.g., $H_2{}^{17}O$ in H_2O) to study ('trace') diffusion. This is also known as *intradiffusion*[27] since it relates to the mutual diffusion of two (approximately) chemically equivalent species. The term tracer diffusion is also applied where the labelled species is not isotopic with any one of the components (e.g., ^{23}Na in H_2O). In this book the terms intradiffusion,

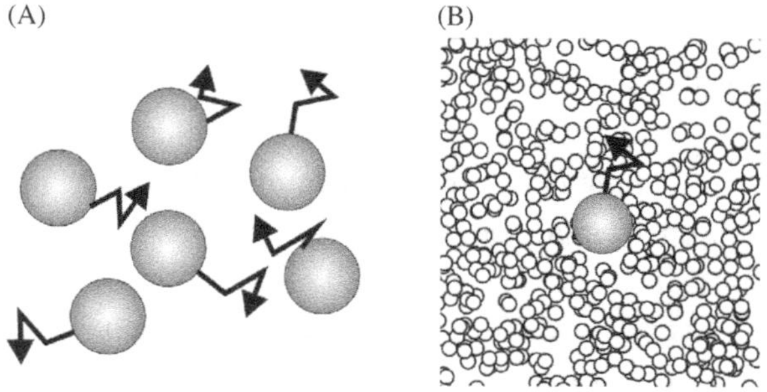

Figure 1.1 Translational diffusion. (A) Diffusion is the random thermal motion of the molecules. It is characterised by a diffusion coefficient D (m^2s^{-1}). (B) Brownian motion of a particle in a liquid. In this example, the molecules constituting the liquid are so small that they effectively form a continuum with respect to the diffusing solute particle.

tracer diffusion and self-diffusion are used synonymously. A key point is that in a multicomponent system each component will have its own individual diffusion coefficient, D_i (m^2s^{-1}), which may not be equal to that of any of the other species. Self-diffusion coefficients typically range from $\sim 10^{-20}\,m^2s^{-1}$ for solids up to $\sim 1\,m^2s^{-1}$ for dilute gases[28,29] and the self-diffusion coefficients of some representative compounds in solution are given in Table 1.1.

The examples in Table 1.1 reflect isotropic (i.e., equal in all directions) free diffusion; however, in general, the physical nature (e.g., internal structures in porous systems) of a system in which a species is diffusing not only affects the rate of its propagation, but also the pattern of its time-dependence.[29] For example, a species diffusing within a biological cell or within brain tissue or water diffusing along protein filaments will be restricted in its motion and similarly a species diffusing within a crystal or liquid crystal will not diffuse isotropically[47] and in more complicated systems such as fractals it may diffuse anomalously.[48–50] Thus, the study of diffusion in anisotropic and restricted systems is a source of information about the underlying microscopic architecture. It is for this reason that MRI-based studies of diffusional anisotropy is of particular current interest since brain white matter looks homogeneous in conventional MRI but the diffusional anisotropy provides a means of studying axonal trajectory (i.e., fibre tracts) and thus brain neuroanatomy. These aspects are covered below.

1.2.2 Mutual and distinct diffusion

Mutual diffusion, also known as *interdiffusion, concentration diffusion* or *transport diffusion*, is characterised by a mutual diffusion coefficient, D_M (m^2s^{-1}). Specifically, a concentration inhomogeneity of any component in a multicomponent system will result in mass fluxes (i.e., cooperative diffusion) to average out the inhomogeneity to achieve thermodynamic equilibrium. Hence, the force behind mutual diffusion is the gradient of the chemical potential (recall that the units of chemical potential are $J\,mol^{-1}$).[51] In the case of a salt concentration gradient in a binary electrolyte system, the requirement for maintaining electrical neutrality results in the positive and negative ions moving from regions of higher to lower concentration at the same speed. Consequently, in a volume-fixed (or mean volume) reference frame there is only one mutual diffusion coefficient – whereas the self-diffusion coefficients of the ions and the solvent are not necessarily the same as there is no charge neutrality requirement. The different reference frames used in studying diffusion have been considered in detail by Brady.[52] Most natural systems contain more than one species and such multicomponent mutual diffusion is described in terms of irreversible thermodynamics[1,2] and in general an n-component system will have $(n-1)^2$ different mutual diffusion coefficients.[2]

Table 1.1 *Selected self-diffusion coefficients of species of biological and chemical significance measured using NMR*

Species	Concentration	pH or p^{2}H; solvent	MW	T (K)	D (m^2s^{-1})	Reference
Xenon (gas)	1 atm	-	131	Ambient	5.7×10^{-6}	30
^2H	-	H_2O	2	296	4.0×10^{-9}	31
^1H_2O	pure	-	18	298	2.3×10^{-9}	32
$^2H_2O^a$	pure	-	20	298	1.87×10^{-9}	33
$H^{13}CO_3^-$	125 mM	7.5; $^2H_2O/H_2O$ 1:3 v:v	62	310	1.3×10^{-9}	34
$H_2PO_2^-$	40 mM	7.4; $^2H_2O/H_2O$ 3:7 v:v	65	310	1.6×10^{-9}	34
[2, ^{13}C]glycine	100 mM	H_2O	76	310	1.2×10^{-9}	35
Uridine	15 mM	DMSO	244	299	5.2×10^{-10}	36
Monensin (free acid form)	infinite dilution^b	C^2HCl_3	688	293	7.8×10^{-10}	37
Monensin (sodium salt)	infinite dilution^b	C^2HCl_3	692	293	7.4×10^{-10}	37
C_{60}	1 mM	$C_6{}^2H_6$	720	298	8.3×10^{-10}	38
Cyclosporin A	21 mM	$C^2H_3O^2H/CO_2$ 7:1 v:v	1,203	293	1.4×10^{-9}	39
BPTI	infinite dilution^b	4.5; ^2H_2O	6,500	274	7.7×10^{-10}	40
Ubiquitin	10 mg/ml	4.0 95% H_2O	8,500	298	1.5×10^{-10}	41
Insulin dimer	0.72 mM	9.3–9.4; ^2H_2O	11,600	298	1.4×10^{-10}	42
Lysozyme	infinite dilution^b	4.6; 90% H_2O	14,400	298	1.1×10^{-10}	43
Ovalbumin	infinite dilution^b	5.5; H_2O	45,000	293	7.9×10^{-11}	44
Haemoglobin	infinite dilution^b	6.9; H_2O	68,000	298	6.8×10^{-11}	45
Polystyrene ($M_w/M_n{}^c < 1.10$)	2%	CCl_4	233,000	301	7.4×10^{-12}	46

^{a} Measured using the diaphragm cell technique.
^{b} Extrapolated to infinite dilution.
^{c} M_w = weight-averaged molar mass, M_n = number-averaged molar mass

Due to the importance of mutual diffusion in many applications there is much interest in describing mutual diffusion in terms of self-diffusion.[1,53] However, the linkage is fraught with difficulties and the mutual and self-diffusion coefficients can only be related in the limit of infinite dilution.[1] This aspect is further considered in Section 1.3.3.

'Distinct' diffusion coefficients D_i ($\mathrm{m^2 s^{-1}}$) – not to be confused with individual diffusion coefficients – are a measure of intermolecular interactions and have definitions related to mutual diffusion coefficients.[53–56] For example, the molecular interactions in a binary system are described by three distinct diffusion coefficients which can be calculated from the self-diffusion coefficients, mutual diffusion coefficient, mole fractions and thermodynamic activity coefficients. Consequently, distinct diffusion coefficients are unable to provide insight into molecular interactions in a multicomponent system under thermodynamic equilibrium. However, as will be seen in the Section 1.3.3, hydrodynamic analysis of self-diffusion coefficients can provide information on molecular interactions.

1.2.3 Flow and higher motion

Flow is the net movement of a species. Diffusion will always be present but when flow occurs in porous media additional mechanisms to diffusion will be involved in increasing the rate of dispersion (i.e., the separation of initially adjacent molecules) of molecules and this flow-induced dispersion will be significantly greater than that resulting from diffusion alone. Plug flow is an ideal case where every part of the system has the same constant velocity and is a reasonable model for the flow of a liquid inside a pipe but close to the entrance at low Reynolds number (the Reynolds number is the ratio of inertial to viscous forces and thus depends on the system under consideration).[20,57] Further into the pipe 'Poiseuille flow' develops in which the fluid velocity varies with radial position, with the velocity being exactly zero at the walls. Laminar flow is where fluid flows in parallel layers but without disruption between the layers. Laminar flows occur when viscous forces dominate and have low Reynolds numbers (< 2100). Couette flow refers to the special case of laminar flow between two surfaces moving with respect to each other (e.g., between planes or between rotating cylinders).[58] Turbulent flow is far more complex to describe and occurs at high Reynolds numbers (> 3000) when inertial forces dominate and produce local flow velocities that fluctuate randomly generating effects such as vortices and eddies. It is also noted that thermal gradients in a sample can induce convective flow ('convection') in a liquid.[59,60]

A Lagrangian (i.e., from the viewpoint of following a particle along its trajectory; in contrast to the Eulerian view of describing the changes at a fixed point) velocity field, $\mathbf{v}$, can be decomposed into a superposition of an average velocity, $\bar{\mathbf{V}}$, and a fluctuating part, $\mathbf{u}$, such that the local instantaneous velocity is given by[61–63]

$$\mathbf{v}(t) = \bar{\mathbf{V}} + \mathbf{u}(t), \tag{1.1}$$

where

$$\bar{\mathbf{V}} = \lim_{t\to\infty} \langle \mathbf{v} \rangle, \tag{1.2}$$

and the angled brackets denote an ensemble average over the distribution of velocity fields. $\mathbf{u}(t)$ reflects the dispersive process in the flow. Equation (1.2) states that the ensemble is stationary. The reasons for the appearance of fluctuating components will become clearer when flow in porous media is examined (see Section 1.9).

1.2.4 Ionic conductivity and translational diffusion

Electrolyte theory provides interrelationships between the mobilities of charged species in electric fields and their translational diffusion coefficients.[64,65] For instance, the molar ionic conductivity of a solution containing a *z*:*z* valent electrolyte is given by the Nernst–Einstein equation[64,65]

$$\Lambda = \frac{zF^2}{RT}(D_{\text{cation}} + D_{\text{anion}}), \tag{1.3}$$

where F is the Faraday constant and D_{cation} and D_{anion} are the diffusion coefficients of the cation and anion, respectively. Indeed, it has been commented that ions undergoing Brownian motion have randomly oriented instantaneous velocities of the order of 10^2 m s^{-1} but with extremely short mean free paths, and in a field of 10^3 volts m^{-1} the velocity of the ions in the direction of the field would be of the order of 10^{-4} – 10^{-6} m s^{-1}.[66] Thus the motion in the electric field represents a very small directional biasing of the Brownian motion. As such it can be termed a directed diffusion process with obvious similarities to mutual diffusion. A key point is that whilst diffusion is an incoherent motion, the directed migration of ions is a coherent process.

1.3 Mathematical modelling of self-diffusion

1.3.1 Introduction

The modelling of self-diffusion can be performed from two perspectives: (i) at the molecular level where the velocities, positions and interparticle collisions of the individual molecules are considered, or (ii) at a more macroscopic level where the particles are taken as a continuum. The continuum approach, which is based on Fick's second law (i.e., the diffusion equation) and appropriate initial and boundary conditions, leads to the concept of diffusion propagators and is considered in Section 1.4. Although it should be noted that a more general approach is to use the Smoluchowski equation with a random force.[67–69]

The molecular level approach is described by an ensemble of random walkers and most numerical techniques (e.g., Brownian dynamics) work at this level whereas most experimental techniques are sensitive to macroscopic behaviour because of the experimental timescale. Molecular level motion is analysed using either the particle position or velocity as a function of time. The particle position approach leads to the propagator formalism. In the velocity approach, the evolution of a particle's velocity is described in terms of a velocity auto-correlation function and the diffusion spectrum. The molecular level details of motion can be related to the macroscopic behaviour. The coverage is begun by considering the molecular level perspective.

1.3.2 Statistical mechanics

Diffusion, as can be comprehended from Figure 1.1, is a many-body problem; however, at sufficiently large times this reduces to a one-body stochastic problem where all of the many-body effects can be characterised by a single number, the self-diffusion coefficient D,[1,53] namely

$$D = \lim_{t\to\infty} \frac{1}{6t} \left\langle [\mathbf{r}_i(t) - \mathbf{r}_i(0)]^2 \right\rangle \tag{1.4}$$

where $\mathbf{r}_i(t)$ is the location of particle i at time t and the angled brackets denote the ensemble average. $\langle[\mathbf{r}_i(t) - \mathbf{r}_i(0)]^2\rangle$ is known as the mean square displacement (MSD) which will be discussed in detail in Section 1.6. This reduction to 'normal' or 'Fickian' diffusion, as implied by the limit in Eq. (1.4), depends on the timescale of the underlying molecular-level motion. Generally, this occurs on timescales far shorter than the timescale accessible with NMR measurements of translational diffusion (typically ms). In the case of water, the language of diffusion can be used for times longer than 10 ps.[70] In highly viscous systems 'anomalous' diffusion may be observed since the underlying motions may occur on a similar timescale to that of the measurement.

Statistical mechanics in the regime of linear response due to a sufficiently weak perturbation shows that any transport coefficient, A, can be written as an integral of the form (Green–Kubo relations[71])

$$A = \int_0^\infty \langle \mathbf{J}_A(0) \cdot \mathbf{J}_A(t) \rangle dt, \tag{1.5}$$

where $\mathbf{J}_A(t)$ is the flux associated with A at time t. Using this formalism, self-diffusion coefficients are defined in terms of the *velocity auto-correlation* function as[21,53]

$$D = \frac{1}{3}\int_0^\infty \langle \mathbf{v}_i(0) \cdot \mathbf{v}_i(t) \rangle dt, \tag{1.6}$$

where $\mathbf{v}_i(t)$ is the velocity of particle i in the barycentric (i.e., centre of mass) reference frame – but as self-diffusion does not result from macroscopic gradients, as does mutual diffusion, the laboratory and volume-fixed frames coincide with the barycentric frame. Equation (1.6) is equivalent to Eq. (1.4), and both equations provide a means for determining self-diffusion coefficients from molecular dynamic simulations (Eq. (1.6) for short times and Eq. (1.4) for long times). In contrast to Eq. (1.6), mutual diffusion would be defined in terms of *velocity cross-correlations*.

The diffusion coefficient spectrum, $\mathbf{D}_\omega(\omega)$, is defined as the Fourier spectrum (or spectral density function) of the velocity auto-correlation function,[72,73]

$$\mathbf{D}_\omega(\omega) = \int_{-\infty}^{\infty} \langle \mathbf{v}(0)\mathbf{v}(t)\rangle \exp(i\omega t)dt. \tag{1.7}$$

Note Eq. (1.6) is the zero frequency component of $\mathbf{D}_\omega(\omega)$. Note, starting from Eq. (1.1), the velocity auto-correlation function is defined as

$$\begin{aligned}\langle v(0)v(t)\rangle &= (\langle \bar{V} + u(0)\rangle\langle \bar{V} + u(t)\rangle) \\ &= \langle u(0)u(t)\rangle + \langle \bar{V}u(0)\rangle + \langle \bar{V}u(t)\rangle + \langle \bar{V}\bar{V}\rangle\end{aligned} \tag{1.8}$$

but as $\bar{V}$ and u are uncorrelated, the two middle terms are zero and, since $\bar{V}$ is a constant, the last term $(\langle \bar{V}\bar{V}\rangle)$ is also a constant. Thus, from the perspective of stochastic motion and without loss of generality $\langle \bar{V}\bar{V}\rangle$ can be set to 0,[74] consequently Eqs. (1.6) and (1.7) could have been expressed in terms of $\mathbf{u}_i(t)$ instead of $\mathbf{v}_i(t)$.

In a bulk liquid $\langle v(0)v(t)\rangle$ is often assumed to decay exponentially (i.e., an Ornstein–Uhlenbeck process[75]), namely[19]

$$\langle v(0)v(t)\rangle = \langle v^2\rangle \exp\left(\frac{-t}{\tau_\mathrm{v}}\right), \tag{1.9}$$

where $\langle v^2\rangle$ is the mean square velocity and τ_v is the velocity correlation time – that is the time taken for a particle to 'forget' its previous velocity and corresponds to the average collision time. Thus, inserting Eq. (1.9) into Eq. (1.7) and taking the real part gives

$$D_\omega(\omega) = \frac{\langle v^2\rangle \tau_\mathrm{v}}{1 + \omega^2\tau_\mathrm{v}^2}. \tag{1.10}$$

If $\omega = 0$, Eq. (1.10) reduces to

$$D_\omega(0) = \langle v^2\rangle \tau_\mathrm{v} = D^0 \tag{1.11}$$

where D^0 is the bulk diffusion coefficient. The diffusion spectrum for a bulk liquid is plotted in Figure 1.2.

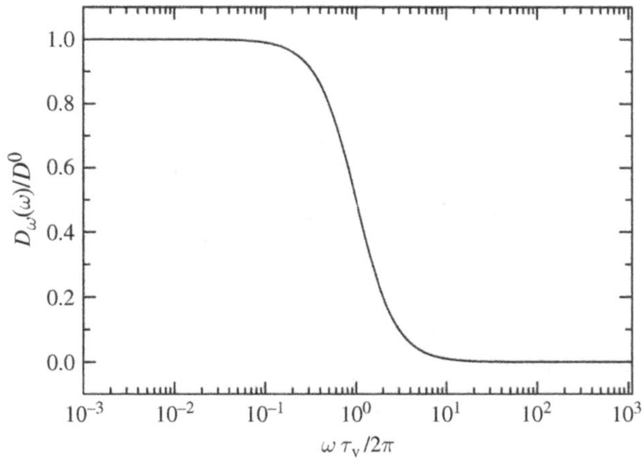

Figure 1.2 The diffusion spectrum of a pure liquid with velocity correlation time τ_v, calculated with Eq. (1.10).

τ_v^{-1} is typically very high for small molecules; however, motions in complex fluids such as tube disengagement in entangled polymers may give rise to features in the diffusion spectrum below τ_v^{-1} related to 'organisational' frequencies.[76] However, the diffusion spectrum contains no features at low frequencies for systems whose local motion is Brownian and have boundaries (e.g., in porous media; see below) that impose constraints on a timescale longer than τ_v, and in such cases the propagator approach is more appropriate for describing the translational dynamics (see Section 1.4.2).

1.3.3 Diffusion and hydrodynamics

1.3.3.1 Diffusion, viscosity and solvation

Viscosity is the manifestation of the internal friction during motion in a liquid (we are concerned with the shear viscosity and not the kinematic viscosity which is the ratio of the viscosity to the density). It gives rise to the irreversible conversion of a part of the energy of motion to heat. Viscosity, and indeed diffusion, are not purely thermodynamic properties and there is at least one order of magnitude difference between various models used to relate them.[77] The Einstein–Sutherland equation will be examined in detail as it is the most commonly used means of relating the two quantities in the literature.

Consider a two-component system (e.g., a macromolecular solute and solvent); since there are no external forces applied in the case of free diffusion, the diffusion force and the frictional force (i.e., the viscosity and hydrodynamics) must be equal, giving the Einstein–Sutherland equation (also known as the Einstein, Stokes–Einstein or Einstein–Smoluchowski equation),[2,3,14,19,20,78–84]

$$D_{\mathrm{M}} = \frac{kT}{f}\left(1 + \frac{\partial(\ln\gamma)}{\partial(\ln c)}\right)(1-\phi), \tag{1.12}$$

where k ($1.380622 \times 10^{-23}\,\mathrm{J\,K^{-1}}$) is the Boltzmann constant, T (K) is temperature, f ($\mathrm{kg\,s^{-1}}$) is the friction coefficient and γ and c are the activity coefficient and concentration (M) of the solute molecule, respectively, and ϕ is the volume fraction (i.e., the volume of a constituent of a mixture divided by the sum of volumes of all constituents prior to mixing). The numerator kT (J) corresponds to the available thermal energy. Equation (1.12) describes mutual diffusion – although it does not describe it accurately in concentrated solutions.[2] At infinite dilution the second term in the first pair of parentheses approaches zero and the equation reduces to describing self-diffusion,

$$D^0 = \frac{kT}{f}. \tag{1.13}$$

The superscript 0 has been added to signify infinite dilution. In theory, the friction factors for self-diffusion and mutual diffusion are different as the friction factor can be concentration-dependent.[85] For the simple case of a spherical particle with an effective hydrodynamic radius (i.e., Stokes radius) r_{S} (m) in a solvent of viscosity η (Pa s $= 10$ Poise; the viscosity of water at 298 K is 8.9×10^{-4} Pa s), the friction coefficient is given by

$$f = f_{\mathrm{sphere}} = b\pi\eta r_{\mathrm{S}} \tag{1.14}$$

where the dimensionless parameter b reflects the boundary conditions for the velocity of the solvent particles at the surface of the solute and is defined by[1,86,87]

$$b = \frac{6(1 + 2\eta/\beta r)}{(1 + 3\eta/\beta r)}. \tag{1.15}$$

The parameter β characterises the friction between the solute and the solvent. In the limit where there is no interaction between the particle and the solvent (the 'slip' boundary condition), β is very small and thus the lower limit of b is 4. The other extreme, where the particle interacts strongly with the solvent molecules such that the solvent layer closest to the surface moves at the same velocity as the particle, is termed the 'stick' boundary condition. In this case β goes to infinity giving an upper limit of $b=6$, and in this case Eq. (1.14) is termed Stokes law. However, as noted below, the underlying assumptions in these equations are not met and experimentally b is often found to be outside the physical range of 4–6. When Eq. (1.14) in the form of the Stokes law is used as the denominator, Eq. (1.13) is most commonly referred to as the Stokes–Einstein equation.

To a first approximation, since for a sphere the volume of a molecule is proportional to its molecular weight, it is clear from Eq. (1.13) that

$$D \propto \frac{1}{\sqrt[3]{\mathrm{MW}}}, \tag{1.16}$$

and thus the diffusion coefficients of two molecules with similar shapes are inversely proportional to the cube roots of their molecular weights:

$$\frac{D_1}{D_2} = \sqrt[3]{\frac{\mathrm{MW}_2}{\mathrm{MW}_1}}. \tag{1.17}$$

Importantly, the Stokes radius represents the size of the diffusing unit and not that of a hard sphere.[82,88] Hence, the friction coefficient f is determined by the overall dimensions of the diffusing species, its hydration (solvation) and the rugosity of the surface exposed to the solvent.[89] In the case of proteins the thickness of hydration layers has been estimated as being in the range of 0.9–1.2 Å thick.[90–93] Including the effects of solvation (hydration), the Stokes radius of a particle is given by[14]

$$r_{\mathrm{solvated}} = \left[\frac{3}{4\pi}\frac{\mathrm{MW}}{N_A}(V_{\mathrm{s}} + \delta_{\mathrm{h}} V_{\mathrm{h}})\right]^{1/3}, \tag{1.18}$$

where N_A is Avogadro's number, δ_{h} is the solvation of the solute (grams of bound solvent per gram of solute) and V_{h} and V_{s} are the partial specific volumes of the solvent and solute, respectively. For example, for proteins in water typical values of V_{h} are 0.70–0.75 $\mathrm{cm^3\,g^{-1}}$ and $\delta_{\mathrm{h}} = 0.3$–$0.4$ g H_2O per g protein;[93] which increases f by 10–20%. Equation (1.18) is very simplistic since it implies a uniform hydration layer, yet the estimated thickness of the protein hydration layer is less than the diameter of a water molecule – indicating that the hydration must be non-uniform. Thus, the real physical form of the hydration is rather more complex than just a simple shell.[94]

Thus, very importantly to many physical studies, diffusion provides information on the interactions and shape of the diffusing molecule and the difference between the hydrodynamic radius obtained from a diffusion measurement and the value calculated from the number density or van der Waals radius can be considerable, especially when there is significant solvation such as in the case of biomolecules.[95] Cantor and Schimmel have considered the limitations of the analysis of diffusion data to determine the shape and hydration of molecules.[14] Differences can also arise out of the neglect of the free volume of the liquid and this is particularly noticeable in hydrogen-bonded liquids and electrolyte friction in the case of charged molecules.[96,97]

Due to the assumptions on which it is based, rigorous data interpretations are difficult with the Einstein–Sutherland equation. In particular, it is a macroscopic relation in that it is derived upon the assumptions that the diffusing species is truly macromolecular and sees the solvent as a continuum (Figure 1.1B) and that the diffusing particle is at infinite dilution (no concentration gradient and (equivalently) no interaction between solute particles). This explains why η refers to the viscosity of the solvent and not that of the solution. The first condition is only met when the solute is much larger than the solvent molecules. Although a recent study has indicated that the equation is valid until the solvent diameter reaches half that of the solute.[98] When the solute is smaller than the solvent (e.g., hydrogen diffusion in water) the Einstein–Sutherland equation tends to underestimate the diffusion coefficient. A number of later empirical derivations and modifications exist (see refs. 2,82,88,99–101 and references therein). Another problem of the Einstein–Sutherland theory is that microscopic attractive and kinetic interactions between the solute and solvent are ignored.[87] For example, in measurements of water diffusion coefficients in aqueous albumin solutions Lamanna *et al.* observed a protein concentration dependence of the water Stokes radius.[102] However, friction coefficients have also been considered for ions and ion-induced inhomogeneity.[103] Further, complementary data on the shape or volume of a molecule greatly assist data analysis.

1.3.3.2 Realistic geometries

Spheres are poor approximations for the shapes of most molecular geometries. In general, frictional coefficients are calculated with the help of the hydrodynamic equation, particularly the Stokes equation for low Reynolds numbers.[104] Apart from the sphere, exact solutions for f are only available for some other simple geometries including cylinders and ellipsoids (see Table 1.2). Ellipsoids are in general more realistic approximations to the shapes of real molecules (e.g., proteins). The surface area of an ellipsoid is greater than that of a sphere of the same volume. Consequently, the friction coefficients are greater for both the prolate and oblate ellipsoids than for the equivalent spheres. The ratio of the friction coefficient of a geometry compared to that of the sphere of equivalent volume, namely

$$F = \frac{f}{f_{\text{sphere}}} \tag{1.19}$$

is known as the Perrin (or shape) factor. The Perrin factor for cylinders and ellipsoids are given in Table 1.2 and plotted in Figure 1.3.

Analytic solutions for frictional coefficients are generally intractable and estimates of the friction coefficients have to be numerically determined, most commonly using bead and shell models based on Kirkwood–Riseman theory[89–91,95,110–121] and more recently boundary elements methods.[93] As an example, for a polymer composed of N identical subunits of radius a with friction coefficient f_{m} where the subunits are widely spaced by frictional linkers, the frictional coefficient would be

$$f = N f_{\text{m}}. \tag{1.20}$$

Table 1.2 *Friction coefficients for three simple geometries in the stick boundary condition*

Shape		Parameters	f
Sphere		a = radius	$6\pi\eta a\ (=f_{\text{sphere}})$
Cylinder[a]		d = diameter l = length $p = l/d$ $a = l\left(\frac{3}{16p^2}\right)^{1/3}$	$f_{\text{sphere}}\frac{\left(\frac{2}{3p^2}\right)^{1/3}}{\ln(p)+\vartheta}$ $\vartheta = 0.312 + 0.565/p - 0.100/p^2$
Ellipsoid	Oblate	b = semi-major axes c = semi-minor axes	$f_{\text{sphere}}\frac{\sqrt{1-p^2}}{p^{1/3}\tan^{-1}\left(\sqrt{1-p^2}/p\right)}$
	Prolate	$p = b/c < 1$ oblate > 1 prolate $a = (b^2c)^{1/3}$ oblate $a = (bc^2)^{1/3}$ prolate	$f_{\text{sphere}}\frac{\sqrt{p^2-1}}{p^{1/3}\tanh^{-1}\left[\left(\sqrt{p^2-1}\right)/p\right]}$

Note: a denotes the radius of the sphere, or in the case of the non-spherical geometries, the radius of the sphere of equivalent volume. The value of a given for each geometry is to be used for calculating the respective value of f_{sphere}.[20,105–109] Unfortunately, there is a great deal of confusion in the literature regarding the friction coefficients for the ellipsoids and also for the coefficient for the cylinder.

[a] This formula is only valid for p in the range of 2–20. An approximate expression with a larger range of axial ratio applicability is given elsewhere.[110] Whether the cylinders are open, closed or capped with hemispheres makes only a small difference even at $p = 2$.[111]

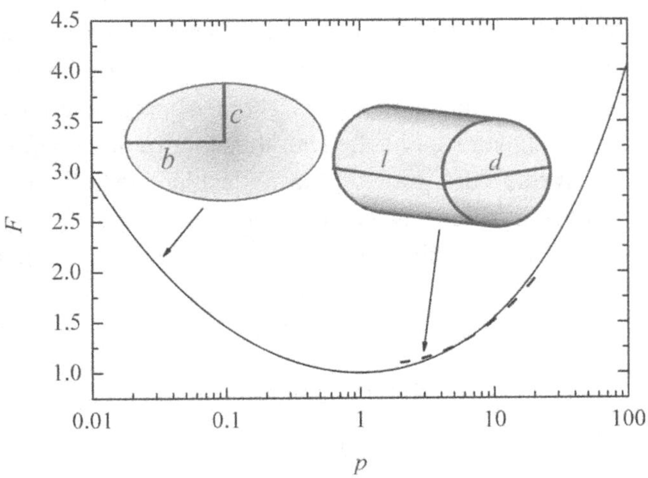

Figure 1.3 Perrin factors for two simple geometries: cylinder (- - -) and ellipsoid (—), using the friction coefficients given in Table 1.2 and Eq. (1.19).

Equation (1.20) corresponds to the frictional coefficient in the free draining limit, that is there is no hydrodynamic interactions between the subunits. This would in general be unrealistic and a more probable situation would be the non-draining limit in which there is maximal interaction between the subunits. In this case, and defining the intersubunit distance, r_{ij}, Kirkwood–Riseman theory gives

$$f = f_{\mathrm{m}} N \left[1 + \frac{1}{N} \sum_{i} \sum_{j \neq i} \alpha_{ij}^{-1} \right]^{-1}, \tag{1.21}$$

where $\alpha_{ij} = r_{ij}/a$. As expected, calculations with Eq. (1.21) show that the most compact structures have the smallest frictional coefficients whereas linear conformations have the largest. For expanded structures such as random coil proteins, the true value would lie somewhere between the two limits. There is also evidence that the diffusion coefficient in solution is larger than that predicted from the molecular structures that have been derived from crystallography and this indicates rearrangement or swelling of the subunits upon hydration.[93]

Polydispersity is commonly encountered in real systems, especially in synthetic polymers, and typically, the exact populations of the oligomers are not known; instead they can only be expressed in terms of a distribution (see Section 2.5). As a first approximation, the subunit-subunit contact in oligomers can be regarded as hard-sphere contact,[89,111] and thus the ratio of the diffusion coefficient of an i-mer, D_i^0, to that of the monomer, D_1^0, can be modelled (at infinite dilution) by

$$D_i^0 = D_1^0 \frac{f_1}{f_i} = D_1^0 F_i \, i^{-1/3}, \tag{1.22}$$

and the values of the geometric factor F_i for various geometries are given by Teller *et al.* (e.g., dimer: 0.9449; trimer [equilateral triangle]: 0.9555; tetramer [tetrahedron]: 0.9772).[89] Whilst there is only one possible geometry for dimer formation, many possibilities exist for higher oligomers. A crude approximation takes all of the oligomers to be hydrodynamically spherical, thus the friction coefficient increases according to the inverse cube root of the molecular weight (cf. Eq. (1.17)).[43] However, the friction coefficients calculated from the Einstein–Sutherland equation for reasonable geometrical possibilities for the oligomeric shapes are all quite close to that obtained for a sphere of equivalent volume.

1.4 Propagation

1.4.1 The diffusion equation

Although we are ultimately interested in propagation due to self-diffusion, we begin by considering the diffusion equation describing mutual diffusion because it will lead to the mathematical formalism needed to describe propagation. In terms of the number of particles per unit volume, $c(\mathbf{r}, t)$ located at $\mathbf{r}$ and the mutual diffusion coefficient, D_M, the flux of a particle is given by Fick's first law of diffusion to be[13,21,122]

$$\mathbf{J}(\mathbf{r}, t) = -D_M \nabla c(\mathbf{r}, t). \tag{1.23}$$

The minus sign indicates that (in isotropic media) the direction of flow is from higher to lower concentration. Because of the conservation of mass, the continuity theorem applies and thus,

$$\frac{\partial c(\mathbf{r}, t)}{\partial t} = -\nabla \cdot \mathbf{J}(\mathbf{r}, t). \tag{1.24}$$

Equation (1.24) states that $\partial c(\mathbf{r}, t)/\partial t$ is the difference between the influx and efflux from the point located at $\mathbf{r}$. Combining Eqs. (1.23) and (1.24) we arrive at Fick's second law of diffusion[13,21,122] (a parabolic partial differential equation)

$$\frac{\partial c(\mathbf{r}, t)}{\partial t} = D_M \nabla^2 c(\mathbf{r}, t), \tag{1.25}$$

where ∇^2 is the Laplace operator. Thus, $c(\mathbf{r}, t)$ changes with time only when the second derivatives of $c(\mathbf{r}, t)$ with respect to position are not all zero.

In general, the diffusion process is anisotropic and the isotropic diffusion coefficient D_M (i.e., a scalar) in Eq. (1.25) is replaced by a rank two diffusion tensor.[122] Note, diffusion tensors will be discussed in more detail below. Thus, for Cartesian coordinates $\mathbf{D}_M$ is given by D_{Mij} where i and j take each of the Cartesian directions. Thus accounting for anisotropic diffusion Eq. (1.23) becomes

$$\mathbf{J}(\mathbf{r}, t) = -\mathbf{D}_M \nabla c(\mathbf{r}, t), \tag{1.26}$$

which is shorthand for

$$\begin{bmatrix} J(x,t) \\ J(y,t) \\ J(z,t) \end{bmatrix} = -\begin{bmatrix} D_{Mxx} & D_{Mxy} & D_{Mxz} \\ D_{Myx} & D_{Myy} & D_{Myz} \\ D_{Mzx} & D_{Mzy} & D_{Mzz} \end{bmatrix} \begin{bmatrix} \frac{\partial c(x,t)}{\partial x} \\ \frac{\partial c(y,t)}{\partial y} \\ \frac{\partial c(z,t)}{\partial z} \end{bmatrix}. \tag{1.27}$$

The diagonal elements of $\mathbf{D}_M$ scale concentration gradients and fluxes in the same direction, the off-diagonal elements couple fluxes and concentration gradients in orthogonal directions. Similarly, Eq. (1.25) becomes

$$\frac{\partial c(\mathbf{r}, t)}{\partial t} = \nabla \cdot \mathbf{D}_M \cdot \nabla c(\mathbf{r}, t). \tag{1.28}$$

If in addition to the diffusion there is a net flow with a velocity $\mathbf{v}$, Eq. (1.28) becomes

$$\frac{\partial c(\mathbf{r},t)}{\partial t} = \nabla \cdot \mathbf{D}_{\mathrm{M}} \cdot \nabla c(\mathbf{r},t) - \nabla \cdot \mathbf{v}\, c(\mathbf{r},t). \tag{1.29}$$

Normally, and as intimated in the above equations, we are concerned with diffusion occurring within Euclidean space. However, the theory and even just the definition of the diffusion equation become more complicated when the space in which diffusion occurs is not Euclidean in nature and contains derivatives of fractional order.[123–125]

1.4.2 Diffusion propagators, average propagators and the equilibrium spin density

As will be seen later, the concept of a diffusion propagator,[126,127] $P(\mathbf{r}_0, \mathbf{r}_1, t)$ (which will sometimes be written in abbreviated form as P), also commonly referred to as the Green function,[128–131] is important for analysing NMR diffusion measurements and its derivation is discussed in detail below. $P(\mathbf{r}_0, \mathbf{r}_1, t)$ is the conditional probability of finding a particle initially at a position $\mathbf{r}_0$, at a position $\mathbf{r}_1$ after a time t and, as will be shown below, is given by the solution of the diffusion equation. Of particular interest is that apart from inelastic neutron scattering, NMR diffusion measurements are the only other experimental means of directly obtaining access to this conditional probability density.[76,132] Since $P(\mathbf{r}_0, \mathbf{r}_1, t)$ is a probability function it obeys the normalisation condition

$$\int P(\mathbf{r}_0, \mathbf{r}_1, t) d\mathbf{r}_1 = 1. \tag{1.30}$$

As will be seen below from the solution to the diffusion equation for the case of free isotropic diffusion (Eq. (1.60)), the propagator is independent of the initial position $\mathbf{r}_0$ and depends only on the displacement $\mathbf{R} = \mathbf{r}_1 - \mathbf{r}_0$ (often referred to as the dynamic displacement). However, in a heterogeneous system the displacement and thus the propagator depends on the starting position.

Another closely related probability is the equilibrium particle (or 'a priori probability') density defined by

$$\rho(\mathbf{r}_0) = \int \lim_{t \to \infty} P(\mathbf{r}_0, \mathbf{r}_1, t)\, d\mathbf{r}_1, \tag{1.31}$$

and thus is independent of $\mathbf{r}_0$, because after infinite time the finishing position of a particle in the system will be independent of the starting position but is representative of the static structure of the diffusion space. $P(\mathbf{r}_0, \mathbf{r}_1, t)$ for free diffusion (see Eq. (1.60) below) approaches 0 as $t \to \infty$, but the 'effective volume' (i.e., the integral over $\mathbf{r}_1$) becomes proportionally larger, consequently $\rho(\mathbf{r}_0)$ stays constant. In the case of diffusion in an enclosed pore (i.e., a restricted geometry) $\rho(\mathbf{r}_0)$ is given by the inverse of the volume. Also, by definition, we must have[133]

$$\int \rho(\mathbf{r}_0) d\mathbf{r}_0 = 1. \tag{1.32}$$

In self-diffusion, in contrast to mutual diffusion, there is no concentration gradient and instead we are concerned with the total probability, $P(\mathbf{r}_1, t)$ of finding a particle at position $\mathbf{r}_1$ at time t. This is given by the product of the initial particle density $\rho(\mathbf{r}_0)$ and $P(\mathbf{r}_0, \mathbf{r}_1, t)$ integrated over all possible starting positions, namely

$$P(\mathbf{r}_1, t) = \int \rho(\mathbf{r}_0) P(\mathbf{r}_0, \mathbf{r}_1, t) d\mathbf{r}_0. \tag{1.33}$$

Clearly the integrand

$$\rho(\mathbf{r}_0) P(\mathbf{r}_0, \mathbf{r}_1, t) \tag{1.34}$$

is the probability of starting from $\mathbf{r}_0$ and moving to $\mathbf{r}_1$ in time t. The ensemble averaged (over all of the starting positions, $\mathbf{r}_0$) probability that an arbitrarily selected particle will displace by $\mathbf{R}$ during the period t is defined as[127,134]

$$\bar{P}(\mathbf{R}, t) = \int \rho(\mathbf{r}_0) P(\mathbf{r}_0, \mathbf{r}_0 + \mathbf{R}, t) d\mathbf{r}_0 . \tag{1.35}$$

$\bar{P}(\mathbf{R}, t)$ has also been termed the average or mean propagator. Consequently, we also have the relation

$$\int \bar{P}(\mathbf{R}, t) d\mathbf{R} = 1. \tag{1.36}$$

In a heterogeneous sample, the probability function of molecular displacement is a function of a position and in such cases $\bar{P}(\mathbf{R}, t)$ reflects a mean value over the sample.

$P(\mathbf{r}_1, t)$ is an ensemble-averaged probability concentration for a single particle and it is thus reasonable to assume that it obeys the diffusion equation.[132] Because the spatial derivatives in Fick's laws refer to $\mathbf{r}_1$, Fick's laws can be rewritten in terms of $P(\mathbf{r}_0, \mathbf{r}_1, t)$ with the initial condition

$$P(\mathbf{r}_0, \mathbf{r}_1, 0) = \delta(\mathbf{r}_1 - \mathbf{r}_0), \tag{1.37}$$

where δ is the delta function[131,135,136] (the significance of this will be explained below). Thus, if in Eq. (1.23) $P(\mathbf{r}_0, \mathbf{r}_1, t)$ is substituted for $c(\mathbf{r}, t)$, $\mathbf{J}$ becomes the conditional probability flux. Similarly, in terms of $P(\mathbf{r}_0, \mathbf{r}_1, t)$, Eq. (1.25) becomes

$$\frac{\partial P(\mathbf{r}_0, \mathbf{r}_1, t)}{\partial t} = D \nabla^2 P(\mathbf{r}_0, \mathbf{r}_1, t), \tag{1.38}$$

where D is now the self-diffusion coefficient.

So far the diffusion propagator has only been considered in terms of general coordinates and to proceed, the equations must be transformed into the most appropriate coordinate system[137] to the problem at hand. For example, for problems involving spherical symmetry, it is convenient to recast Eq. (1.38) into spherical polar coordinates (see Figure 1.4) as

$$\frac{\partial P}{\partial t} = D\left[\frac{1}{r^2}\frac{\partial}{\partial r}\left(r^2\frac{\partial P}{\partial r}\right) + \frac{1}{r^2 \sin\theta}\frac{\partial}{\partial \theta}\left(\sin\theta\frac{\partial P}{\partial \theta}\right) + \frac{1}{r^2 \sin^2\theta}\frac{\partial^2 P}{\partial \phi^2}\right], \tag{1.39}$$

where r is the radius from a point to the origin, θ is the polar angle and ϕ is the azimuthal angle.

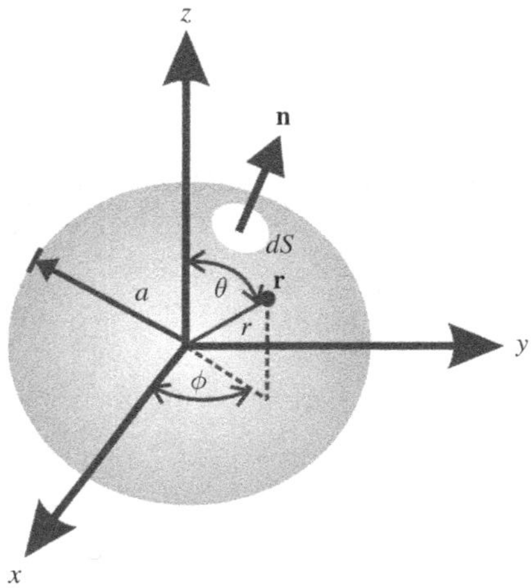

Figure 1.4 The position of a point **r** on a sphere of radius a in spherical polar coordinates. The position of **r** is denoted by (r, θ, ϕ), where r is the radius, θ is the polar angle and ϕ is the azimuth $(0 \leq r \leq a, 0 \leq \theta \leq \pi, 0 \leq \phi \leq 2\pi)$. **n** denotes the outward-pointing normal unit vector to a differential area dS of interest on the sphere surface.

1.4.3 Boundary and initial conditions

When solving the diffusion equation we are typically concerned with linear boundary conditions with respect to the surface of the volume of interest (denoted by S) and they can be divided into the following five classes.[129,136,138]

(i) The first kind (Dirichlet problem)

$$P|_S = f(\mathbf{r}_S, t), \tag{1.40}$$

where $\mathbf{r}_S \in S$ and $f(\mathbf{r}_S, t)$ denote a function over the boundary surface. $f(\mathbf{r}_S, t)$ is usually considered to be continuous on S.

(ii) The second kind (Neumann conditions)

$$\left.\frac{\partial P}{\partial n}\right|_S = f(\mathbf{r}_S, t), \tag{1.41}$$

where $\frac{\partial}{\partial n}$ denotes differentiation along the outward directed normal (see Figure 1.4); the special case of (homogeneous boundary condition of the second kind)

$$\left.\frac{\partial P}{\partial n}\right|_S = 0 \tag{1.42}$$

is very commonly encountered in the literature and represents reflection on the boundary surface.

(iii) The third kind (Robin problem)

$$\left.\left(L\frac{\partial P}{\partial n} + MP\right)\right|_S = f(\mathbf{r}_S, t), \tag{1.43}$$

for some constants L and M. This boundary condition is a linear combination of (i) and (ii) and defines a conservation law for flux between the bulk and through the boundary. Setting $f(\mathbf{r}_S, t) = 0$, Eq. (1.43) represents the radiative or relaxation boundary condition,[139–141] and M specifies the rate of relaxivity or permeability depending upon the context. $M = 0$ corresponds to zero relaxivity (or no transport) and thus Eq. (1.43) reduces to Eq. (1.42).

(iv) The fourth kind

$$P|_{S^-} = P|_{S^+} \tag{1.44}$$

and

$$D^- \left.\frac{\partial P}{\partial n}\right|_{S^-} = -D^+ \left.\frac{\partial P}{\partial n}\right|_{S^+}, \tag{1.45}$$

where the superscripts '–' and '+' denote on the inside and outside of the boundary S, respectively. Boundary conditions of the fourth kind pertain to diffusion in systems containing permeable boundaries[142] and Eq. (1.44) states that the diffusion propagator, P, is continuous on the boundary between the internal domain and external medium and Eq. (1.45) states that the diffusive flux is continuous on this boundary. These boundary conditions are relevant to some of the models considered in Chapter 4.

(v) Mixed boundary conditions

In this case, different parts of the surface are subject to different boundary conditions, for example a Dirichlet condition over part of the boundary

$$P|_{S_1} = f_1(\mathbf{r}_S, t) \quad \mathbf{r}_S \in S_1, \tag{1.46}$$

and a Neumann problem over a different part of the boundary

$$\left.\frac{\partial P}{\partial n}\right|_{S_2} = f_2(\mathbf{r}_S, t) \quad \mathbf{r}_S \in S_2, \tag{1.47}$$

where S_1 and S_2 are different parts of the surface such that $S_1 \cup S_2 = S$. Such boundary conditions are commonly encountered for example in modelling diffusion-controlled reactions between chemically anisotropic reactants.[23]

The initial condition that will normally be encountered below (i.e., for studying tracer diffusion) is given in terms of a delta function[135,136]

$$P(\mathbf{r}_0, \mathbf{r}_1, t = 0) = \delta(\mathbf{r}_0 - \mathbf{r}_1). \tag{1.48}$$

The imposition of a delta function initial condition means that the solution at time t will depend only on the particle that was at starting position $\mathbf{r}_0$ at time $t = 0$.

1.5 Solving the diffusion equation

Apart from numerical solutions (e.g., finite element analysis),[143] many analytical methods for solving the diffusion equation exist,[13,122,128–131,136,144–146] and it should be noted that the mathematics of heat conduction, after making the appropriate changes of notation, is identical to that for describing diffusion.[13,144] In general, analytical solutions exist for only the simplest geometries due to the complexity of handling the boundary conditions at the pore or obstacle interface. However, semi-analytical solutions exist for particles diffusing in periodic three-dimensional porous media.[147–149]

Consideration of how to solve the diffusion equation is begun by considering the one-dimensional diffusion equation for the case of free diffusion. Then some consideration is given to eigenfunction expansion solutions to solving diffusion in bounded systems. Eigenfunction expansions form a critical part of the mathematical toolbox needed to understand diffusion in restricted geometries such as spheres, which are used to model real systems such as biological cells, and to calculate the relevant diffusion propagators. Finally, propagators for some relevant simple geometries are given.

1.5.1 Free isotropic diffusion

We wish to solve the one-dimensional diffusion equation for free isotropic diffusion

$$\frac{\partial P}{\partial t} = D\frac{\partial^2 P}{\partial z^2}, -\infty < z < \infty, t > 0 \tag{1.49}$$

(i.e., $P = P(z, t)$) subject to the initial condition

$$P(z, 0) = P_0(z) = f(z). \tag{1.50}$$

The Fourier transform[136,150] of $P(z, t)$ with respect to z is defined by

$$\mathrm{FT}\{P(z,t)\} = \tilde{P}(k,t) = \frac{1}{\sqrt{2\pi}}\int_{-\infty}^{\infty} P(z,t)e^{-ikz}dz, \tag{1.51}$$

where the transform variable k is real. We remark that there is no general agreement on the definition of the Fourier transform. Some authors take the kernel to be e^{-ikz} so that the kernel of the inverse transform (see below) becomes e^{ikz}. Similarly, there is no consensus on the choice of the constant before the integral. Thus, applying Eq. (1.51) to Eq. (1.49) to obtain

$$\frac{d\tilde{P}(k,t)}{dt} = -k^2 D\tilde{P}(k,t). \tag{1.52}$$

This is an ordinary differential equation with solution

$$\tilde{P}(k,t) = \tilde{P}_0(k)e^{-k^2 Dt} \tag{1.53}$$

where $\tilde{P}_0(k)$ is the Fourier transform of the initial condition (Eq. (1.50))

$$\tilde{P}_0(k) = \frac{1}{\sqrt{2\pi}} \int_{-\infty}^{\infty} f(z)e^{-ikz}dz. \tag{1.54}$$

Taking the inverse Fourier transform of Eq. (1.53) we have

$$\begin{aligned} P(z,t) &= \frac{1}{\sqrt{2\pi}} \int_{-\infty}^{\infty} \tilde{P}(k,t)e^{ikz}dk \\ &= \frac{1}{2\pi} \int_{-\infty}^{\infty} f(z_0) \int_{-\infty}^{\infty} e^{ik(z-z_0)}e^{-k^2 Dt}dk\, dz_0 . \end{aligned} \tag{1.55}$$

The integral over k is

$$G(z,t;z_0) = \frac{1}{\sqrt{4\pi Dt}}\exp\left(\frac{-(z-z_0)^2}{4Dt}\right). \tag{1.56}$$

We remark that $G(z, t; z_0)$ is called the Green function for the above Cauchy problem or fundamental solution to Eq. (1.49). And so, Eq. (1.55) becomes

$$P(z,t) = \int_{-\infty}^{\infty} f(z_0)\frac{1}{\sqrt{4\pi Dt}}\exp\left(\frac{-(z-z_0)^2}{4Dt}\right)dz_0. \tag{1.57}$$

In our case we are interested in a delta function initial condition given by

$$P(z,0) = P_0(z) = f(z) = \delta(z-z_0). \tag{1.58}$$

This represents a (geometrical) plane source located at z_0 (see Figure 1.5), and thus

$$P(z_0,z,t) = \frac{1}{\sqrt{4\pi Dt}}\exp\left(\frac{-(z-z_0)^2}{4Dt}\right). \tag{1.59}$$

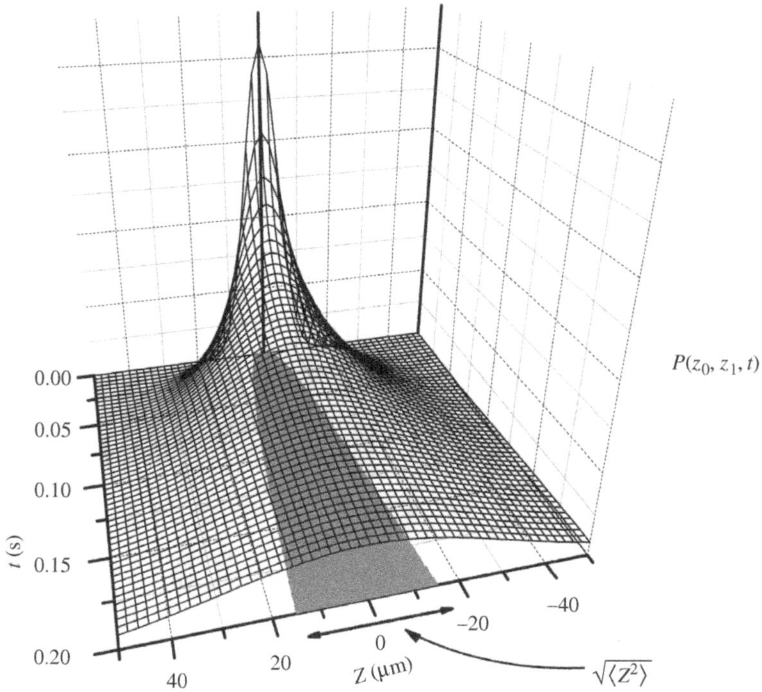

Figure 1.5 A plot of the displacement distribution function for $Z\ (=z_1-z_0)$ versus time for one-dimensional diffusion calculated using Eq. (1.59). The initial condition (i.e., a delta function at $t=0$) corresponds to the bold line at $Z=0$. The evolving Gaussian nature of the distribution is clearly evident. The corresponding MSD (calculated using Eq. (1.103)) is also depicted in dark grey on the base of the plot. The diffusion coefficient used was that of water at 298 K ($2.3\times10^{-9}\ \mathrm{m^2 s^{-1}}$). Note $\int_{-\infty}^{\infty} P(Z,t)dZ = 1$.

For the case of (three-dimensional) diffusion in an isotropic and homogeneous medium (i.e., boundary condition $P(\mathbf{r}_0, \mathbf{r}_1, t) \to 0$ as $\mathbf{r}_1 \to \infty$), $P(\mathbf{r}_0, \mathbf{r}_1, t)$ can be determined from Eq. (1.38) using Fourier transforms or simply by multiplying three orthogonal one-dimensional solutions of the form of Eq. (1.59), namely

$$\begin{aligned} P(\mathbf{r}_0, \mathbf{r}_1, t) &= P(x_0, x_1, t)P(y_0, y_1, t)P(z_0, z_1, t) \\ &= \frac{1}{(4\pi Dt)^{3/2}} \exp\left(-\frac{(\mathbf{r}_1 - \mathbf{r}_0)^2}{4Dt}\right). \end{aligned} \tag{1.60}$$

Equations (1.59) and (1.60) state that the radial distribution function of the species in an infinitely large system with regard to an arbitrary reference time is Gaussian. From Eq. (1.60) (and (1.59)) it is evident that $P(\mathbf{r}_0, \mathbf{r}_1, t)$ does not depend on the initial position, $\mathbf{r}_0$, but depends only on the net displacement, $\mathbf{r}_1 - \mathbf{r}_0$ (= $\mathbf{R}$). This reflects the Markovian nature of Brownian motion.[75] And for the case of free diffusion the average propagator (Eq. (1.35))

$$\begin{aligned}\bar{P}(\mathbf{R},t) &= \int \rho(\mathbf{r}_0)P(\mathbf{r}_0,\mathbf{r}_0+\mathbf{R},t)d\mathbf{r}_0 \\ &= P(\mathbf{r}_0,\mathbf{r}_0+\mathbf{R},t)\int \rho(\mathbf{r}_0)d\mathbf{r}_0 \\ &= \frac{1}{(4\pi Dt)^{3/2}}\exp\left(-\frac{\mathbf{R}^2}{4Dt}\right)\end{aligned} \tag{1.61}$$

which is identical to Eq. (1.60). In this derivation $\rho(\mathbf{r}_0) = 1$ (see above) and the reason that the propagator ($P(\mathbf{r}_0, \mathbf{r}_1, t)$) does not appear in the integrand in the second step of the derivation is because, as noted above, the propagator does not depend on $\mathbf{r}_0$ but only the difference $\mathbf{R} = \mathbf{r}_1 - \mathbf{r}_0$. For other geometries, as will be seen below, the situation is not so simple. For completeness, had we started the analysis including the effects of flow (see Eq. (1.29)), an average propagator

$$\bar{P}(\mathbf{R},t) = \frac{1}{(4\pi Dt)^{3/2}}\exp\left(-\frac{(\mathbf{R}-\mathbf{v}t)^2}{4Dt}\right) \tag{1.62}$$

would have been obtained. Some plots of the one-dimensional propagator are given in Figure 1.5.

1.5.2 Anisotropic Gaussian diffusion

In the case of anisotropic diffusion, instead of starting with Eq. (1.38) we would start with (cf. Eq. (1.28))

$$\frac{\partial P(\mathbf{r}_0,\mathbf{r}_1,t)}{\partial t} = \nabla\cdot\mathbf{D}\cdot\nabla P(\mathbf{r}_0,\mathbf{r}_1,t), \tag{1.63}$$

where $\mathbf{D}$ is the self-diffusion tensor

$$\mathbf{D} = \begin{pmatrix} D_{xx} & D_{xy} & D_{xz} \\ D_{yx} & D_{yy} & D_{yz} \\ D_{zx} & D_{zy} & D_{zz} \end{pmatrix}. \tag{1.64}$$

$\mathbf{D}$ is positive definite and symmetric (i.e., $D_{ij} = D_{ji}$) and thus, there are only six independent elements. We note, however, that $\mathbf{D}$ would not be symmetric in the case of charged species placed in a magnetic field but this has an insignificant effect for our purposes.[151–153]

Since $\mathbf{D}$ is a tensor, the values of the elements change depending upon the choice of reference frame. The Cartesian directions could correspond to the laboratory frame (i.e., those used to define the direction of diffusion measurement in the first instance), but depending upon the system, the form of **D** may be simpler by choosing some three-dimensional rotation of the reference frame. For example, in the case of liquid crystals, it makes sense to align the x, y or z axis along the director axis (i.e., the direction of maximum alignment); by so doing **D** would become a diagonal matrix containing only the principal diffusivities (i.e., D_{xx}, D_{yy} and D_{zz}) with the off-diagonal elements all being 0. As an aside we note that in liquid crystals with axial symmetry, two of the three principal components will be equal and the symbols $D_{\parallel}$ and $D_{\perp}$ denote diffusion parallel and perpendicular to the director, respectively. Such a situation commonly occurs in solid crystals too.[154] Similarly, to a first approximation, biological cells can be considered to have a cylindrical or ellipsoidal shape and thus the principal directions of **D** correspond to the orthotopic directions of the cell (or tissue).

To prevent later confusion, where necessary, we will use primed coordinates to denote laboratory frame (x', y', z') and unprimed coordinates to denote the principal axes frame (x, y, z) in which the diffusion tensor is diagonalised. Thus, formalising our comments in the previous paragraph, we can state that a diffusion tensor as measured in the laboratory frame, $\mathbf{D}^{\mathrm{lab}}$, can be decomposed as

$$\mathbf{D}^{\mathrm{lab}} = \mathbf{U}\mathbf{D}^{\mathrm{pr}}\mathbf{U}^{-1}, \tag{1.65}$$

where $\mathbf{U} = (\varepsilon_1 \mid \varepsilon_2 \mid \varepsilon_3)$ is a matrix of the column of (orthogonal) eigenvectors (principal coordinate directions) which performs the transformation from $(x', y', z') \rightarrow (x, y, z)$ and $\mathbf{D}^{\mathrm{pr}}$ is the (diagonalised) diffusion tensor in the principal eigenvalue matrix given by

$$\mathbf{D}^{\mathrm{pr}} = \begin{pmatrix} \lambda_x & & \\ & \lambda_y & \\ & & \lambda_z \end{pmatrix}, \tag{1.66}$$

where λ_x, λ_y and λ_z are the (real) eigenvalues and correspond to the principal diffusivities (i.e., D_{xx}, D_{yy} and D_{zz}, respectively). Thus, if the coordinate system is chosen such that it is oriented along the principal axes of the diffusion ellipsoid (see Section 1.6), also known as the principal reference frame, **D** becomes a diagonal matrix since the translational displacements appear to be uncorrelated.

We note that in general for real systems, neither the principal diffusivities nor the orientation of the principal reference frame (it is not normally coincident with the laboratory frame of reference) are known a priori. Accordingly, many more experimental measurements are required to characterise anisotropic systems. As a further complication, which will be further explored below, many real samples are not uniformly aligned (e.g., a suspension of red blood cells). The measured $\mathbf{D}^{\mathrm{lab}}$ is then an average of the different orientations. Measuring diffusion in three orthogonal directions so as to determine the trace ('Tr') of the diffusion tensor

$$\frac{1}{3}\mathrm{Tr}(\mathbf{D}^{\mathrm{lab}}) = \frac{1}{3}\left(D^{\mathrm{lab}}_{x'x'} + D^{\mathrm{lab}}_{y'y'} + D^{\mathrm{lab}}_{z'z'}\right) = \frac{1}{3}\mathrm{Tr}(\mathbf{D}^{\mathrm{pr}}) = \frac{1}{3}\left(D^{\mathrm{pr}}_{xx} + D^{\mathrm{pr}}_{yy} + D^{\mathrm{pr}}_{zz}\right) = D_{av} \tag{1.67}$$

is a means of overcoming the anisotropy problems.[155,156] Importantly, since the trace is invariant under rotations, the orientational dependence is removed and consequently the trace of the diffusion tensor is the same in all reference frames. However, where the diffusion tensor is observation time-dependent, as in restricted diffusion (see below), the measured trace can differ from the true trace of the effective diffusion tensor and the measured quantity is not completely rotationally invariant.[157]

Solving Eq. (1.63) gives the solution[134,155,158,159]

$$P(\mathbf{r}_0, \mathbf{r}_1, t) = \frac{1}{(4\pi|\mathbf{D}|t)^{3/2}} \exp\left(-\frac{(\mathbf{r}_1 - \mathbf{r}_0)^{\mathrm{T}}\mathbf{D}^{-1}(\mathbf{r}_1 - \mathbf{r}_0)}{4t}\right), \tag{1.68}$$

where the superscript 'T' denotes transpose and $|\mathbf{D}|$ denotes the determinant of **D**. Thus, if the coordinate system is chosen such that it is oriented along the principal axes of the diffusion ellipsoid, Eq. (1.68) becomes

$$P(\mathbf{r}_0, \mathbf{r}_1, t) = \frac{1}{\left(64\pi^3 D_{xx} D_{yy} D_{zz} t^3\right)^{1/2}} \exp\left(-\frac{(x_1 - x_0)^2}{4D_{xx}t} - \frac{(y_1 - y_0)^2}{4D_{yy}t} - \frac{(z_1 - z_0)^2}{4D_{zz}t}\right). \tag{1.69}$$

Of course, in the case of free isotropic diffusion **D** is diagonal with all three elements equalling D and thus Eq. (1.69) reverts to Eq. (1.60). We will consider further the analysis of anisotropic diffusion in Section 1.6 where we consider the case that **D** might be time-dependent (i.e., $\mathbf{D}(t)$).

1.5.3 Eigenfunction expansions

Let us begin by assuming that we can solve Eq. (1.38) in a bounded domain V in the form of an eigenfunction (aka eigenmode) expansion (or spectral decomposition) in terms of eigenmodes and eigenfrequencies as obtained, for example, using the separation of variables technique. Then we will obtain a solution in the form of the sum (with respect to the eigenvalues, $\lambda_m \geq 0$) of a product of the spatial eigenfunctions $\psi_m(\mathbf{r})$ relevant to (the Helmholtz equation for) the geometry,[137,160] e.g., Bessel, Legendre and trigonometric in the case of diffusion in a sphere, and an appropriate function of time $T_m(t)$ which will simply be an exponential function, namely

$$P(\mathbf{r}, t, f(\mathbf{r})) = \sum_{m=0}^{\infty} \underbrace{c_m \psi_m(\mathbf{r})}_{\text{Spatial}} \underbrace{T_m(t)}_{\text{Time}}$$
$$= \sum_{m=0}^{\infty} c_m \psi_m(\mathbf{r})\, e^{-D\lambda_m t}. \tag{1.70}$$

The most general solution is a sum or superposition (i.e., a series solution) over all possible eigenvalues. If Eq. (1.70) converges uniformly the constants c_m will be determined from the initial conditions

$$f(\mathbf{r}) = \sum_{m=0}^{\infty} c_m \psi_m(\mathbf{r}). \tag{1.71}$$

A critically important point is that the eigenfunctions are orthogonal over the domain V and the inner product of the eigenfunctions is defined as

$$(\psi_m, \psi_n) = \int_V \psi_m(\mathbf{r})\psi_n^*(\mathbf{r})d\mathbf{r} = \delta_{mn}\|\psi_m\|^2, \tag{1.72}$$

where we assume that the weight function is unity and the asterisk denotes complex conjugate (in our case the eigenfunctions are always real valued and so $\psi_m(\mathbf{r}) = \psi_m^*(\mathbf{r})$) and δ_{mn} is the *Kronecker* delta (i.e., $\delta_{mn} = 1$ if $m = n$, otherwise $\delta_{mn} = 0$). The corresponding norm of $\psi_n(\mathbf{r})$ is

$$\|\psi_n\| = (\psi_n, \psi_n)^{1/2}. \tag{1.73}$$

The orthogonality property of the eigenfunctions (Eq. (1.72)) allows the values of c_m to be determined as follows: starting from Eq. (1.71), and multiplying by $\psi_m^*(\mathbf{r})$ and integrating over the volume, thus

$$c_m = \|\psi_m\|^{-2}\int_V f(\mathbf{r})\psi_m(\mathbf{r})d\mathbf{r}. \tag{1.74}$$

It is expedient to define normalised spatial eigenfunctions (i.e., such that their norms are unity)

$$K_m(\mathbf{r}) = \frac{\psi_m(\mathbf{r})}{\|\psi_m\|}. \tag{1.75}$$

Thus, written in terms of normalised eigenfunctions and substituting Eq. (1.74) containing the delta function initial condition (i.e., Eq. (1.48)) into Eq. (1.70) gives

$$P(\mathbf{r}, t, f(\mathbf{r})) = \sum_{m=0}^{\infty} K_m(\mathbf{r}) \int_V \delta(\mathbf{r} - \mathbf{r}_0)K_m(\mathbf{r})d\mathbf{r}\, e^{-D\lambda_m t}. \tag{1.76}$$

Now, noting the property of the delta function

$$\int_V K(\mathbf{r})\delta(\mathbf{r} - \mathbf{r}_0)d\mathbf{r} = K(\mathbf{r}_0), \tag{1.77}$$

Eq. (1.76) becomes

$$P(\mathbf{r}, t, f(\mathbf{r})) = G(\mathbf{r}_0, \mathbf{r}, t) = \sum_{m=0}^{\infty} K_m(\mathbf{r}_0) K_m(\mathbf{r})\, e^{-D\lambda_m t}. \tag{1.78}$$

Equation (1.78) has special significance in mathematics and is commonly termed the Green function for the posed Cauchy problem or fundamental solution $G(\mathbf{r}_0, t_0; \mathbf{r}, t)$.[128,130,131] If $t=0$, Eq. (1.78) becomes

$$P(\mathbf{r}_0, \mathbf{r}, 0) = G(\mathbf{r}_0, \mathbf{r}, 0) = \delta(\mathbf{r} - \mathbf{r}_0) = \sum_{m=0}^{\infty} K_m(\mathbf{r}) K_m(\mathbf{r}_0). \tag{1.79}$$

Thus, the above Green function is the solution to the diffusion equation given a delta function initial condition. We note in passing that instead of using Fourier transforms, the free diffusion equation (Section 1.5.1) could have been solved using eigenfunction expansions.[161]

By taking the expression for the Green function (Eq. (1.78)) and setting $\mathbf{r} = \mathbf{r}_0$ (i.e., the same initial and final position) and noting the normalisation condition of the eigenfunctions it is possible to define another useful expression termed the Spectral function[162–164]

$$\Theta(t) = \int_V G(\mathbf{r}_0, \mathbf{r}_0, t) d\mathbf{r}_0 = \sum_{m=0}^{\infty} e^{-D\lambda_m t}, \tag{1.80}$$

which is related to the return to origin probability (RTOP; see Section 1.8.3).

1.5.4 Diffusion in isolated pores

Here we list the diffusion propagators for the three most commonly used 'closed' geometries for studying restricted diffusion in isolated pores (i.e., single-site models) including the case of relaxation boundary conditions. The propagators for many other geometries can be found in the literature.[144] Propagators for multi-site systems with porous boundaries are discussed in Section 1.8.

1.5.4.1 Parallel planar boundaries

For diffusion between infinitely long reflecting planes (i.e., homogeneous boundary condition of the second kind (Eq. (1.42))) separated by a distance $2a$, i.e.,

$$\left.\frac{\partial P}{\partial z}\right|_{z=0,\, 2a} = 0, \tag{1.81}$$

with a delta function initial condition (Eq. (1.37)), solving the one-dimensional diffusion equation (Eq. (1.49)), the normalised eigenfunctions are given by

$$K_n(z) = \begin{cases} \sqrt{\frac{1}{2a}} & n = 0 \\ \sqrt{\frac{1}{a}}\cos\left(\frac{n\pi z}{2a}\right) & n = 1, 2, 3 \ldots \end{cases} \tag{1.82}$$

with corresponding eigenvalues

$$\lambda_n = \frac{n^2\pi^2}{(2a)^2}, \tag{1.83}$$

and the diffusion propagator is given by[144]

$$P(z_0, z_1, t) = \frac{1}{2a} + \frac{1}{a}\sum_{n=1}^{\infty}\cos\left(\frac{n\pi z_0}{2a}\right)\cos\left(\frac{n\pi z_1}{2a}\right)\exp\left(-\frac{n^2\pi^2 Dt}{(2a)^2}\right), \tag{1.84}$$

where the first (and time-independent) term results from the zero eigenvalue. Parallel planes are an example of a closed pore. An obvious, but nevertheless important point is that propagators such as Eq. (1.84) are only valid within their domain of definition and are equal to zero for values outside this. Plots of the diffusion propagator and the respective average propagator for diffusion between reflecting planes are given in Figure 1.6.

For diffusion between planes subject to the radiative or relaxation boundary condition (Eq. (1.43))

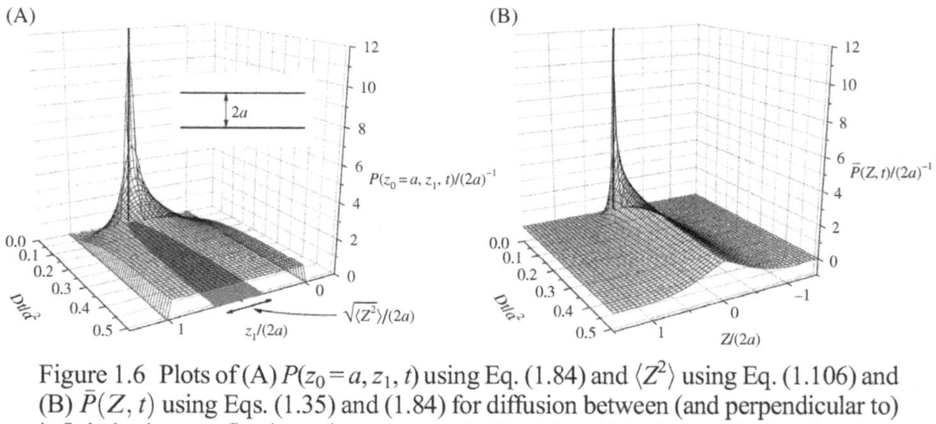

Figure 1.6 Plots of (A) $P(z_0 = a, z_1, t)$ using Eq. (1.84) and $\langle Z^2 \rangle$ using Eq. (1.106) and (B) $\bar{P}(Z, t)$ using Eqs. (1.35) and (1.84) for diffusion between (and perpendicular to) infinitely long reflecting planes at $z = 0$, $2a$. Note the time axis is plotted in dimensionless units (i.e., Dt/a^2). At very short timescales the diffusion appears isotropic; however, even at rather short timescales the effects of the boundaries (i.e., at $z_0/a = 0$, 2) come into effect and the long-time limit is essentially reached before $Dt/a^2 \sim 0.4$ and thus it can be understood that at long times the propagator reflects the pore shape and this results in an upper limit on the MSD of $(2a)^2/6$ (see Eq. (1.106)) and is in contrast to free diffusion where there is no upper limit (Figure 1.5).

$$(D\mathbf{n} \cdot \nabla P + MP)|_{z=a} = 0 \tag{1.85}$$

where $\mathbf{n}$ is the outward surface normal, $z = |\mathbf{z}_1|$ and the positive constant M specifies the relaxivity, the diffusion propagator is given by[140]

$$P(z_0, z_1, t) = a^{-1}\left[\sum_{n=0}^{\infty}\left(1+\frac{\sin(2\alpha_n)}{2\alpha_n}\right)^{-1}\cos\left(\frac{\alpha_n z_0}{a}\right)\cos\left(\frac{\alpha_n z_1}{a}\right)\exp\left(-\frac{\alpha_n^2 Dt}{a^2}\right)\right.$$
$$\left.+\sum_{m=0}^{\infty}\left(1-\frac{\sin(2\beta_m)}{2\beta_m}\right)^{-1}\sin\left(\frac{\beta_m z_0}{a}\right)\sin\left(\frac{\beta_m z_1}{a}\right)\exp\left(-\frac{\beta_m^2 Dt}{a^2}\right)\right], \tag{1.86}$$

and the eigenvalues α_n and β_m are determined by the equations

$$\alpha_n \tan(\alpha_n) = \frac{Ma}{D} \tag{1.87}$$

and

$$\beta_m \cot(\beta_m) = -\frac{Ma}{D}, \tag{1.88}$$

respectively. Plots of the diffusion propagator and the respective average propagator for diffusion between partially absorbing planes are given in Figure 1.7.

1.5.4.2 Spherical pores

For diffusion in a reflecting sphere of radius a,

$$\left.\frac{\partial P}{\partial r}\right|_{r=a} = 0 \tag{1.89}$$

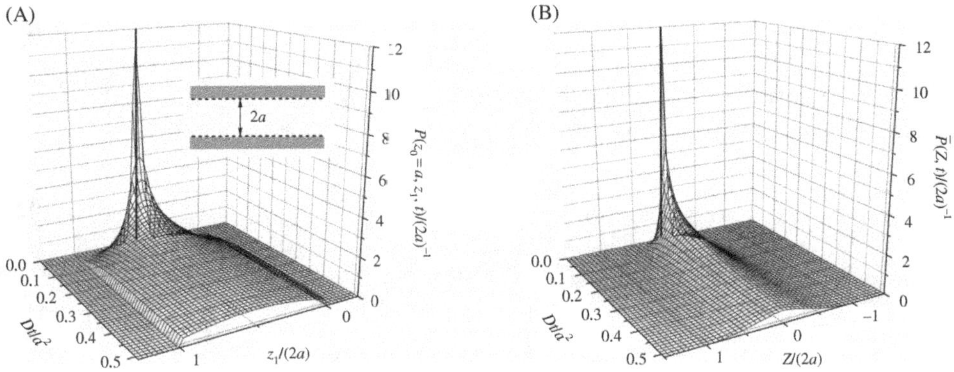

Figure 1.7 Plots of (A) $P(z_0 = a, z_1, t)$ using Eq. (1.86) and (B) $P(Z, t)$ using Eqs. (1.35) and (1.86) for diffusion between partially absorbing planes (i.e., $Ma/D = 2$) at $z_1 = 0, 2a$. The effects of the absorbing boundaries are evident in comparison with Figure 1.6. Note that although Eq. (1.86) pertains to the limits $z = a$, the position and displacement axes of these two plots have been shifted to facilitate comparison with Figure 1.6.

where $r = |\mathbf{r}_1|$. Solving the diffusion equation in spherical coordinates ($\mathbf{r} = (r, \theta, \phi)$ as in Fig. 1.4) gives the propagator[144,165]

$$P(\mathbf{r}_0, \mathbf{r}_1, t) = \frac{3}{4\pi a^3} + \sum_{n>0}^{\infty}\sum_{m=0}^{\infty}\left(\frac{(2n+1)\alpha_{nm}^3(\alpha_{nm}r_0)^{-\frac{1}{2}}J_{n+\frac{1}{2}}(\alpha_{nm}r_0)P_n(\mu_0)(\alpha_{nm}r_1)^{-\frac{1}{2}}J_{n+\frac{1}{2}}(\alpha_{nm}r_1)P_n(\mu_1)}{2\pi\left(\alpha_{nm}^2a^2-n^2-n\right)\left[J_{n+\frac{1}{2}}(\alpha_{nm}a)\right]^2} \times \exp\left(-D\alpha_{nm}^2 t\right)\right) \tag{1.90}$$

where $P_n(x)$ is a Legendre polynomial, J_n are Bessel functions of the first kind and $\mu = \cos(\theta)$, where θ is the angle between $\mathbf{r}$ and the direction of the measurement (see Figure 1.4; the direction of measurement is taken as being along the z axis). Note there is no azimuthal dependence and α_{nm} is the root of the Bessel function equation

$$\frac{n}{\alpha_{nm}a}J_{n+\frac{1}{2}}(\alpha_{nm}a) = J_{n+\frac{3}{2}}(\alpha_{nm}a). \tag{1.91}$$

For diffusion in a sphere subject to the radiative boundary condition (i.e., Eq. (1.43))

$$(D\mathbf{n}\cdot\nabla P + MP)|_{r=a} = 0, \tag{1.92}$$

where $r = |\mathbf{r}_1|$ and the diffusion propagator is given by[140]

$$P(\mathbf{r}_0, \mathbf{r}_1, t) = \sum_{n=0}^{\infty}\sum_{m=0}^{\infty}\frac{j_n(\alpha_{nm}r_0/a)P_n(\mu_0)j_n(\alpha_{nm}r_1/a)P_n(\mu_1)}{j_n^2(\alpha_{nm}) - j_{n-1}(\alpha_{nm})j_{n+1}(\alpha_{nm})}\frac{2n+1}{2\pi a^3}\exp\left(-\frac{\alpha_{nm}^2 Dt}{a^2}\right), \tag{1.93}$$

where the j_n are spherical Bessel functions $\left(\text{note } j_n(x) = \sqrt{\frac{\pi}{2x}}\,J_{n+\frac{1}{2}}(x)\right)$.[160] The eigenvalues are determined by

$$\alpha_{nm}\frac{j_n'(\alpha_{nm})}{j_n(\alpha_{nm})} = -\frac{Ma}{D} \tag{1.94}$$

where $j_n'(\alpha_{nm}) = \frac{n}{\alpha_{nm}}j_n(\alpha_{nm}) - j_{n+1}(\alpha_{nm})$.[160]

1.5.4.3 Cylindrical pores

For diffusion in a reflecting cylinder of radius a,

$$\left.\frac{\partial P}{\partial r}\right|_{r=a} = 0 \tag{1.95}$$

where $r = |\mathbf{r}_1|$. Solving the diffusion equation gives the propagator[144,165]

$$P(\mathbf{r}_0, \mathbf{r}_1, t) = \frac{1}{\pi a^2} + \sum_{n>0}^{\infty}\sum_{m=0}^{\infty}\left(\frac{2J_n(\alpha_{nm}r_0)J_n(\alpha_{nm}r_1)\cos(n\theta_1 - n\theta_0)}{\pi a^2\left(1-\frac{n^2}{\alpha_{nm}^2 a^2}\right)J_n(\alpha_{nm}a)^2}\exp(-D\alpha_{nm}^2 t)\right), \tag{1.96}$$

where n is an integer, θ is the angle between $\mathbf{r}$ and the direction of the measurement, and α_{nm} is the root of the Bessel function equation,

$$J_n'(\alpha_{nm}a) = 0. \tag{1.97}$$

Similarly, for diffusion in a cylindrical pore subject to the radiative boundary condition the diffusion propagator is given by[140]

$$\begin{aligned} P(\mathbf{r}_0, \mathbf{r}_1, t) = &\frac{1}{\pi a^2}\sum_{k=0}^{\infty}\frac{\alpha_{0k}^2/J_0^2(\alpha_{0k})}{\alpha_{0k}^2 + (Ma/D)^2}J_0\left(\frac{\alpha_{0k}r_0}{a}\right)J_0\left(\frac{\alpha_{0k}r_1}{a}\right)\exp\left(-\frac{\alpha_{0k}^2 Dt}{a^2}\right) \\ &+\frac{2}{\pi a^2}\sum_{n\neq 0}^{\infty}\sum_{k=0}^{\infty}\frac{\alpha_{nk}^2/J_n^2(\alpha_{nk})}{\alpha_{nk}^2 - n^2 + (Ma/D)^2}J_n\left(\frac{\alpha_{nk}r_0}{a}\right)J_n\left(\frac{\alpha_{nk}r_1}{a}\right) \\ &\times \cos(n\theta_0)\cos(n\theta_1)\exp\left(-\frac{\alpha_{nk}^2 Dt}{a^2}\right), \end{aligned} \tag{1.98}$$

where the eigenvalues are determined by

$$\alpha_{nk}J_n'(\alpha_{nk})/J_n(\alpha_{nk}) = -\frac{Ma}{D}. \tag{1.99}$$

1.6 Mean square displacement, time-dependent diffusion coefficients and the diffusion ellipsoid

Self-diffusion is a random process and thus the mean displacement, $\langle \mathbf{r}_1 - \mathbf{r}_0\rangle$ or $\langle \mathbf{R}\rangle$, is equal to zero. A more useful measure of the width of the distribution is given by the MSD[19,166]

$$\left\langle(\mathbf{r}_1 - \mathbf{r}_0)^2\right\rangle = \int_{-\infty}^{\infty}(\mathbf{r}_1 - \mathbf{r}_0)^2\rho(\mathbf{r}_0)P(\mathbf{r}_0, \mathbf{r}_1, t)d\mathbf{r}_0 d\mathbf{r}_1, \tag{1.100}$$

or equivalently (as the second moment of the average propagator)

$$\left\langle \mathbf{R}(t)^2\right\rangle = \int \bar{P}(\mathbf{R}, t)\,\mathbf{R}^2 d\mathbf{R}. \tag{1.101}$$

Equation (1.100) (or (1.101)) presents a relationship between the molecular displacement due to diffusion and the diffusion equation.

By substituting the propagator for 1D free isotropic diffusion (Eq. (1.59)) into Eq. (1.100) in terms of motion along the z axis only, noting $\rho(z_0) = 1$, and using the standard integral (e.g., 3.462.8 in ref. 167)

$$\int_{-\infty}^{\infty} x^2 e^{-\mu x^2 + 2\nu x} dx = \frac{1}{2\mu}\sqrt{\frac{\pi}{\mu}}\left(1 + 2\frac{\nu^2}{\mu}\right) e^{\frac{\nu^2}{\mu}} \quad [|\arg \nu| < \pi, \ \mathrm{Re}\ \mu > 0] \tag{1.102}$$

to obtain

$$\left\langle (z_1 - z_0)^2 \right\rangle = \left\langle Z^2 \right\rangle = 2Dt. \tag{1.103}$$

Equation (1.100) can be calculated for two and three dimensions by writing in polar and spherical polar coordinates, respectively, and by using appropriate standard integrals. The MSD for free diffusion is shown graphically in Figure 1.5. Ultimately, the MSD for free diffusion in one, two and three dimensions can be written in the compact form

$$\left\langle (r_1 - r_0)^2 \right\rangle = \left\langle R^2 \right\rangle = nDt, \tag{1.104}$$

where $n = 2, 4$ or 6 for one, two or three dimensions, respectively. Equation (1.104) provides an alternative definition of the diffusion coefficient and is equivalent to Fick's first and second laws; it states that for free isotropic diffusion the MSD changes linearly with time. Diffusive processes for which the MSD scales linearly with time are said to be 'Fickian'.

From Eq. (1.104), it can be seen that if the MSD is known it is possible to define a time-dependent diffusion coefficient (cf. Eq. (1.4))[168]

$$D(t) = \frac{\left\langle R^2 \right\rangle}{nt}. \tag{1.105}$$

This definition emphasises that according to how the MSD alters with time the value of the diffusion coefficient determined may depend on the value of t. As can be understood from the discussion above, the diffusion coefficient determined by this method is a constant for free diffusion but is not time invariant in the case of restricted diffusion. In the literature a diffusion coefficient determined from Eq. (1.105) is sometimes referred to as an 'apparent' diffusion coefficient; however, we prefer to refer to this as a 'time-dependent' diffusion coefficient. This aspect of nomenclature is further discussed at the beginning of Chapter 2. As an example, the MSD for diffusion between planes is calculated using the propagator (Eq. (1.84)) and Eq. (1.100) (after rewriting in Cartesian coordinates) to be

$$\langle Z^2(t)\rangle = \frac{(2a)^2}{6} - \frac{16(2a)^2}{\pi^4}\sum_{n=0}^{\infty}\frac{\exp\left(-\frac{(2n+1)^2\pi^2 Dt}{(2a)^2}\right)}{(2n+1)^4} \tag{1.106}$$

(note in the derivation the summation index has changed from n to $2n+1$ to account for the effects of an alternating series and thus the lower limit of the summation has also changed). Due to the effects of the planar boundaries, the long-time limit of $\langle Z^2(t)\rangle$ (i.e., $\langle Z^2(\infty)\rangle$) is bounded and is given by the first term which arises from $\rho(z_0)$ – since at long times all of the particles have lost memory of their initial conditions and are now uniformly distributed. Substituting $\langle Z^2(t)\rangle$ into Eq. (1.105), the time-dependent diffusion coefficient is then given by

$$D(t) = \frac{(2a)^2}{12t} - \frac{8(2a)^2}{\pi^4 t}\sum_{n=0}^{\infty}\frac{\exp\left(-\frac{(2n+1)^2\pi^2 Dt}{(2a)^2}\right)}{(2n+1)^4}. \tag{1.107}$$

$\langle Z^2(t)\rangle$ and $D(t)$ are plotted in Figure 1.8. In contrast to free diffusion where $D(\infty) = D^0$, in a completely restricted geometry $D(t) \rightarrow 0$ as $1/t \rightarrow 0$.

One-dimensional diffusion is often modelled as a random walk with a step rate of ν_l and a step length of l. Hence, the MSD after N steps is then[11,21]

$$\langle Z^2\rangle = Nl^2 = N\frac{2D}{\nu_l}, \tag{1.108}$$

and thus

$$D = \frac{l^2\nu_l}{2}. \tag{1.109}$$

The MSD can also be determined from the velocity auto-correlation function[21,169]

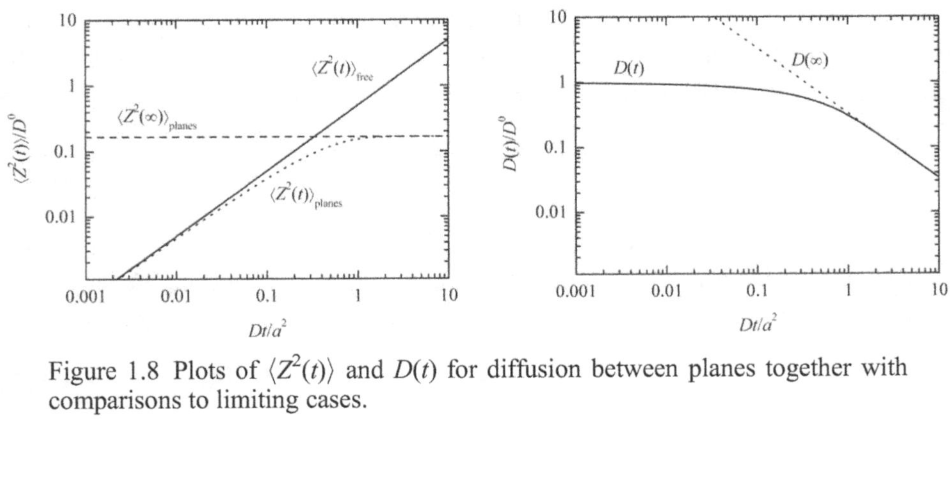

Figure 1.8 Plots of $\langle Z^2(t)\rangle$ and $D(t)$ for diffusion between planes together with comparisons to limiting cases.

$$\langle Z^2(t)\rangle = 2\int_0^t\int_0^{t'}\langle v(0)v(t'')\rangle dt''dt', \tag{1.110}$$

or from the diffusion spectrum[73]

$$\langle Z^2(t)\rangle = \frac{2t^2}{\pi}\int_0^\infty D_\omega(\omega)\mathrm{sinc}^2\left(\frac{\omega t}{2}\right)d\omega. \tag{1.111}$$

Further, combining Eq. (1.111) with Eq. (1.105) provides the connection between D and $D(\omega)$

$$D(t) = \frac{t}{\pi}\int_0^\infty D_\omega(\omega)\mathrm{sinc}^2\left(\frac{\omega t}{2}\right)d\omega. \tag{1.112}$$

In an anisotropic system, it is possible to define an MSD tensor,[170,171] namely

$$\mathbf{H} = \begin{pmatrix} \langle R_x^2\rangle & \langle R_xR_y\rangle & \langle R_xR_z\rangle \\ \langle R_yR_x\rangle & \langle R_y^2\rangle & \langle R_yR_z\rangle \\ \langle R_zR_x\rangle & \langle R_zR_y\rangle & \langle R_z^2\rangle \end{pmatrix}, \tag{1.113}$$

where following Eq. (1.101) the elements of **H** are defined as

$$\langle R_iR_j\rangle = \int R_iR_j\bar{P}(\mathbf{R},t)d\mathbf{R} \qquad i,j = x,y,z. \tag{1.114}$$

Further, in analogy to Eq. (1.105), a time-dependent (apparent or effective) diffusion tensor can be defined[170]

$$\mathbf{D}(t) = \mathbf{H}/2t = \begin{pmatrix} D_{xx}(t) & D_{xy}(t) & D_{xz}(t) \\ D_{yx}(t) & D_{yy}(t) & D_{yz}(t) \\ D_{zx}(t) & D_{zy}(t) & D_{zz}(t) \end{pmatrix}. \tag{1.115}$$

Sometimes it is useful to graphically represent $\mathbf{D}(t)$ by a 'diffusion ellipsoid' in the principal reference frame[13,155,172] (see Figure 1.9) which can be constructed by setting the quadratic form in Eq. (1.68) to $-1/2$ such that

$$\frac{(\mathbf{r}_1-\mathbf{r}_0)^{\mathrm{T}}\mathbf{D}(t)^{-1}(\mathbf{r}_1-\mathbf{r}_0)}{2t} = 1. \tag{1.116}$$

This defines a surface of constant mean translational motion of diffusing species at time t. Transforming from the reference frame of measurement (i.e., the laboratory frame, denoted by $\mathbf{r}'$) to the principal frame of reference (denoted by $\mathbf{r}$) using

$$\mathbf{r} = \mathbf{U}^{\mathrm{T}}(\mathbf{r}'_1-\mathbf{r}'_0), \tag{1.117}$$

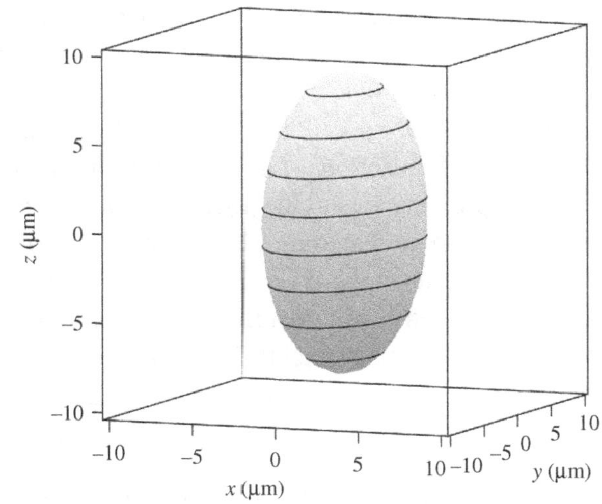

Figure 1.9 An example of an effective diffusion ellipsoid in the principal axis frame calculated using Eq. (1.119). The parameters used in the simulation were $t = 20$ ms, $D_{xx} = 0.4 \times 10^{-9}\,\mathrm{m^2s^{-1}}$, $D_{yy} = 1.4 \times 10^{-9}\,\mathrm{m^2s^{-1}}$, $D_{zz} = 1.8 \times 10^{-9}\,\mathrm{m^2s^{-1}}$. The major axes of the ellipsoid correspond to the RMSDs in each of the principal axes, respectively. The extremely anisotropic diffusion parameters were chosen to allow easy visualisation of the ellipsoidal shape.

to obtain

$$\frac{\mathbf{r}'^{\mathrm{T}}\mathbf{D}^{\mathrm{pr}-1}\mathbf{r}'}{2t} = 1, \tag{1.118}$$

which when expanded becomes (cf. Eq. (1.69))

$$\left(\frac{x'}{\sqrt{2\lambda_1 t}}\right)^2 + \left(\frac{y'}{\sqrt{2\lambda_2 t}}\right)^2 + \left(\frac{z'}{\sqrt{2\lambda_3 t}}\right)^2 = 1. \tag{1.119}$$

We note from Eq. (1.104) that the major axes of the ellipsoids in Eq. (1.119) are the root mean square displacement (RMSD) in the three principal directions (e.g., $\sqrt{2\lambda_1 t} = \sqrt{2D_{xx}t} = \sqrt{\langle x'^2 \rangle}$, etc.). An example of a diffusion ellipsoid is given in Figure 1.9. Determining the principal diffusivities is insufficient for characterising anisotropic Gaussian diffusion, since the off-diagonal elements of $\mathbf{D}(t)$ are necessary too if the size, shape and orientation of the diffusion ellipsoid are to be specified.

In ordered structures such as muscle fibres and white matter the diffusional anisotropy is primarily due to restricted diffusion (e.g., consider diffusion within a cylindrical geometry). Consequently, at very short times the diffusion appears isotropic (cf. Figure 1.6) and thus the ellipsoid would be spherical, but as the effects of the boundaries become significant on the translational motion the ellipsoid would become increasingly non-spherical and more evidently prolate. Thus, the principal frame must coincide with the orthotropic directions of the tissue with the largest diffusivity corresponding to the tissue's fibre tract axis.[155] Given that, with the exception of liquid crystals, anisotropic diffusion arises from the presence of anisotropically arranged barriers such as cell membranes which render the diffusion non-Gaussian, the concept of anisotropic Gaussian diffusion and hence diffusion ellipsoids are thus only approximations to the actual diffusion propagator.[173]

Finally, we mention that some materials can be locally anisotropic but macroscopically isotropic and are termed polycrystalline. Thus, below some length scale, termed the domain size, anisotropy is manifested in locally ordered crystalline domains. Examples of such systems include elastomers, polydomain lyotropic liquid crystals, polymer melts and semi-dilute solutions and porous media.[174]

1.7 Measurement timescales and restricted diffusion

Translational diffusion measurements have associated with them a characteristic timescale (an 'observation' or 'diffusion' time), Δ, and is generally synonymous with t in the theory above. In enclosed pores, the duration of Δ determines (and limits) the type of information obtainable on the dynamics of the system being studied. To illustrate this, consider a particle undergoing: (i) isotropic free diffusion and (ii) restricted diffusion in a sphere (i.e., closed spherical pore) of radius a as shown in Figure 1.10. To characterise the effects of restriction on diffusion measurements the dimensionless variable, ξ, is defined from Eq. (1.104) (i.e., at $n=1$, $t=\Delta$), namely

$$\xi = \frac{D\Delta}{a^2}. \tag{1.120}$$

Diffusion measurements in restricted geometries can be split into three regimes according to the size of the MSD compared to the pore size: (i) the 'short time' or 'free diffusion' limit ($\xi \ll 1$ or short values of Δ); in this limit the particle does not diffuse far enough during Δ to feel the effects of restriction and measurements performed within this timescale lead to the true diffusion coefficient (i.e., D^0). (ii) the 'crossover regime' from the short- to long-time diffusion $\xi \approx 1$; in this regime some of the particles feel the effects of restriction and the diffusion coefficient measured within this timescale will be a function of Δ (i.e., $D(\Delta)$). The fraction of particles that feel the effects of the boundary will be dependent on the surface-to-volume ratio of the pore, S_P/V_P. (iii) the 'long-time' limit ($\xi \gg 1$ or large values of Δ); in this regime all of the diffusing particles feel the effects of restriction and now the displacement of the particle is independent of Δ and depends only on a. It is also instructive to view these comments with respect to Figure 1.6. Given the timescales of NMR diffusion measurements (10 –100 ms) and for small rapidly diffusing molecules like water, this typically means that restricted diffusion will become apparent for structures with length scales 100 μm or smaller.

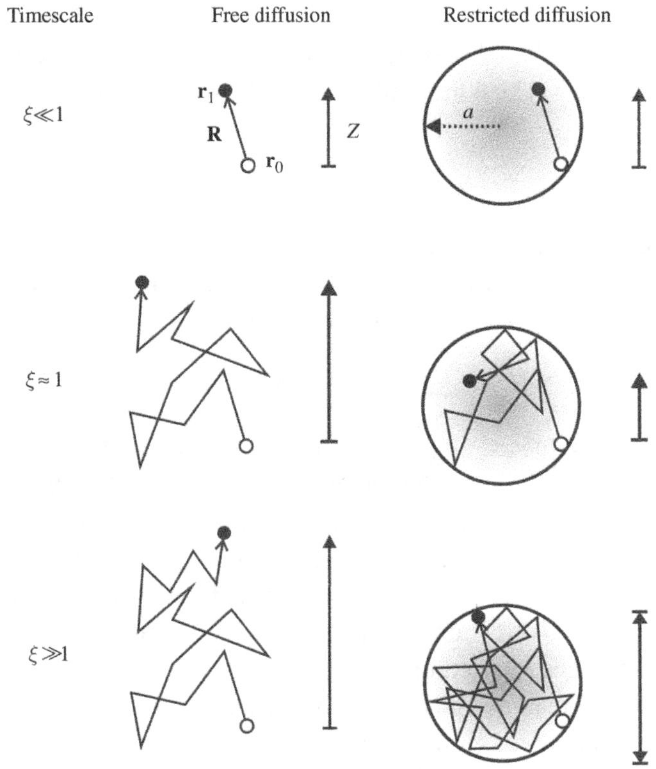

Figure 1.10 A schematic diagram comparing the effect of measurement timescale when measuring free diffusion with diffusion in a restricted geometry. Imagine an experiment in which the displacement of a species in the z direction (vertical) is measured by observing its starting position $\mathbf{r}_0$ (○) at $t=0$ (open circle) and then at position $\mathbf{r}_1$ (●) at $t=\Delta$ (closed circle). **R** denotes the displacement between the initial and final positions and the vertical arrows represent displacement, Z, in the vertical direction. At long times (i.e., $\xi \gg 1$) the maximum displacement in the restricted geometry is limited by the boundaries. Importantly the trajectory of the motion between these two observation points is unknown. Modified from Price.[175]

As depicted in Figure 1.10, the MSD scales linearly with time (see Eq. (1.104)) in the case of free diffusion and as such reflects the true diffusion coefficient for all values of Δ. However, for a confined particle the behaviour is very different and the behaviour in the three time domains is considered in the following three subsections.

We note that Sen has reviewed the use of the time-dependent diffusion coefficient for probing restricted geometries.[176]

1.7.1 Short-time limit

In the short-time limit the measured diffusion coefficient will be the same as that observed for the freely diffusing species (i.e., D^0) and the diffusion propagator describing diffusion in the restricted system (e.g., Eqs. (1.84), (1.86), (1.90), (1.93), (1.96) and (1.98)), although perhaps not easily seen, will reduce to that for free diffusion (Eq. (1.60)). Thus, the MSD is given by Eq. (1.104).

1.7.2 Intermediate times

As Δ increases the values measured for the free and restricted cases increasingly diverge as more of the confined particles feel the effects of the boundary and the MSD will not scale linearly with Δ, for example[177–179]

$$\langle R^2 \rangle = \alpha \Delta^{\kappa}, \tag{1.121}$$

where α is a time-independent scaling constant and the exponent κ is sometimes written as $2/d_{\mathrm{w}}$ where d_{w} is termed the random walk (or path or trail) dimension or $\kappa = 2/(2+\theta_{\mathrm{a}}) = 2H$, where θ_{a} is termed the exponent of anomalous diffusion and H is the Hurst exponent. κ generally lies between 0 and 1. The special case of $\kappa = 1$ (i.e., $d_{\mathrm{w}} = 2$) and $\alpha = 6D$ corresponds to unrestricted free diffusion (cf. Eq. (1.104)). For $\kappa \neq 1$ the *measured* diffusion coefficient (i.e., $D(\Delta)$) will be Δ-dependent and the diffusion is sometimes referred to as anomalous.

For an isotropic medium analysis of the change in D as a function of Δ for (non-fractal) boundaries with a finite surface relaxivity (M) leads to the relationship in terms of powers of the diffusion-length, $\sqrt{D^0 t}$ (i.e., a scale parameter),[176,180–185] (note ref. 176 has a particularly lucid derivation of this equation)

$$\begin{aligned} D(\Delta) = D^0 \Bigg\{ 1 &- \frac{4}{3d\sqrt{\pi}} \frac{S_P}{V_P} \sqrt{D^0 \Delta} - \frac{S_P}{2dV_P} \left[\frac{1}{2} \left\langle \frac{1}{R_1} + \frac{1}{R_2} \right\rangle - \frac{M S_P}{V_P} \right] (D^0 \Delta) \\ &+ \frac{S_P}{V_P} o\left(\left(D^0 \Delta \right)^{3/2} \right) \Bigg\}, \end{aligned} \tag{1.122}$$

where d is the number of spatial dimensions, R_1 and R_2 are the principal radii of the curvature of the interface – chosen so that they are positive for the outside of a sphere, and $o(\ldots)$ represents higher-order terms. Only particles within the (diffusion) length $\sqrt{D^0 \Delta}$ from the surface have had their diffusive path affected by the boundaries. Consequently, the higher-order terms in $\sqrt{D^0 \Delta}$ become increasingly important at larger Δ since an increasingly large proportion of the spins are so affected. Thus, $D(\Delta)$ deviates from D^0 approximately linearly with $\Delta^{1/2}$. As noted by Candela and Wong,[186] these equations require modification when applied to fractal surfaces.

Anomalous diffusion in the context of general porous media is further considered in Section 1.8.3.

1.7.3 Long-time limit

In the long-time limit in a perfectly reflecting geometry (i.e., at $M=0$) the MSD is limited by the boundaries and eventually becomes independent of the dynamics of the system (and thus Δ and D^0) and instead reflects only the shape, dimensions and orientation of the restricting geometry itself. Thus, as t (equiv. Δ) $\rightarrow \infty$, the (time-dependent) second terms in the three propagators for the reflecting boundary conditions (i.e., Eqs. (1.84), (1.90) and (1.96)) disappear (due to the presence of the negative exponentials), leaving only the first terms which result from the zero-eigenvalue. Indeed, the diffusing particles trapped within a fully enclosing pore (as in Figure 1.10) lose memory of their starting position (i.e., their final position becomes independent of their starting position) which is physically reasonable since the diffusing species cannot escape their confinement and at long times they completely sample their 'universe'. Thus at long times the probability of moving from any point to any other within the geometry is 1/(volume) ($\equiv \rho(\mathbf{r}_0)$; the equilibrium particle density) as shown in Eq. (1.84), (1.90) and (1.96), and since in these cases $P(\mathbf{r}_0, \mathbf{r}_1, t)$ is conservative

$$P(\mathbf{r}_0, \mathbf{r}_1, \infty) = \rho(\mathbf{r}_0) = \rho(\mathbf{r}_1). \tag{1.123}$$

In such cases the average propagator in the long-time limit becomes

$$\bar{P}(\mathbf{R}, \infty) = \int \rho(\mathbf{r}_0)\, \rho(\mathbf{r}_0 + \mathbf{R})\, d\mathbf{r}_0. \tag{1.124}$$

Thus, $\bar{P}(\mathbf{R}, \infty)$ is the auto-correlation function of the molecular density $\rho(\mathbf{r}_0)$.

Although similar in behaviour at short times, the three propagators for the relaxation boundary condition (Eqs. (1.86), (1.93) and (1.98)) go to zero at long times since all spins are ultimately quenched by contact with the boundaries. Of course, Eqs. (1.84), (1.90) and (1.96) are identically equal to the $M=0$ cases of Eqs. (1.86), (1.93) and (1.98), respectively.

1.8 Diffusion in heterogeneous, porous and polymer systems

1.8.1 Introduction

Diffusive processes in real materials and solutions are in general complex. Real materials are often porous in that they are in essence a 'solid' with holes and the boundary conditions are more complex (Figure 1.11). A porous material is composed of at least two phases: the porous matrix and the pore space which is filled with a liquid or gas. The porous matrix is not necessarily solid in a rigid sense. Examples include biological tissues (e.g., brain[18]), cellulose fibres, ceramics, concentrated emulsions, gels, porous rocks, soil and solid catalysts, and zeolites. Consequently, apart from free solutions, almost all samples of interest come under the porous material category including concentrated macromolecular solutions that

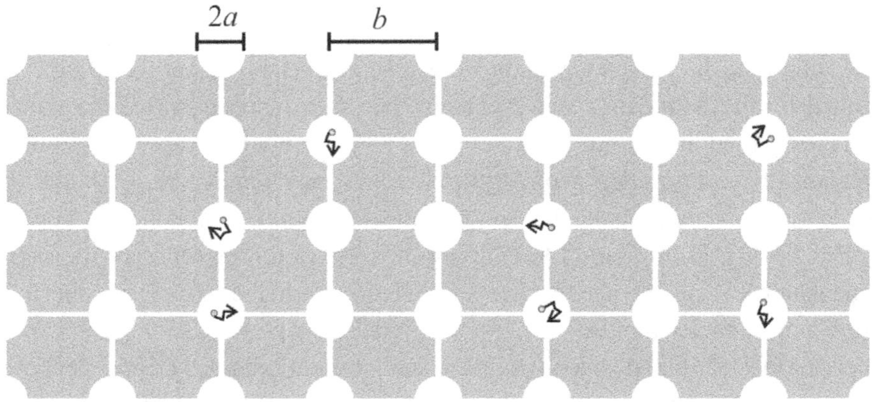

Figure 1.11 A two-dimensional representation of a (in reality three-dimensional) porous medium with a regular structure. This could also be termed an open geometry since the pores are not isolated and transport between the pores is possible. Real porous structures, except zeolites and other regular structures, are of course disordered and irregular. In this diagram roughly circular pores of radius a with interpore spacings of b ($> 2a$) are joined by narrower throats. If three diffusing molecules were to be in the same throat it would be a case of single file diffusion since the diffusion of the molecule in the middle would depend largely on the movement of its neighbours.

will exhibit properties such as obstruction (see below). Most porous substances have a disordered morphology and it is difficult to specify the geometry of the interface between the pore space and the matrix. The transport and thermodynamic properties of fluids in porous media are closely related to the size and connectivity of the pores.[187] A myriad of different physical phenomena occur in porous media ranging from permeation catalysis to biological perfusion and permeation which relate to a diverse range of areas including chemistry, physiology and oil recovery.

The length scale being studied is typically very important in porous media since below some length scale the local structure will be relatively heterogeneous, whereas over larger length scales it will appear homogeneous. The pore size at a specific point in a material can be characterised as the radius, a, of the largest sphere that fits into the pore space. In general there will be a distribution of such pore sizes in a material. The International Union of Pure and Applied Chemistry (IUPAC) classifies porous materials into three categories: micropores of less than 2 nm in diameter, mesopores between 2 and 50 nm and macropores of greater than 50 nm.[188] In the literature, nanoporous materials are defined as those porous materials with pore diameters less than 100 nm. Porosity is a measure of the volume fraction of the void space. Thus, the porosity of a porous material is defined as

$$\phi = V_{\mathrm{P}}/V, \qquad (1.125)$$

where V is the total volume and V_P is the volume of the pore space. Whilst ϕ for a rigid porous material like rock is a constant, for a soft porous material it can be a function of the amount of liquid in the pores. In many cases (e.g., catalysts) the surface area of a porous material is important and it is characterised by the surface-to-volume ratio S_P/V_P.

In a pure liquid (e.g., water) the measured diffusion coefficient corresponds to the bulk diffusion coefficient (i.e., D^0). Following from the above discussion it is clear that in isolated pores the measured diffusion coefficient in a porous medium will be less than D^0 and also time-dependent. As conceptualised in Figure 1.11, most porous media consist of interconnected pores (i.e., an 'open' geometry) and the transport between pores is typically of enormous importance. The porosity and the connectivity in porous media is characterised by the tortuosity, $\Im$, which, as will be detailed in the following section, relates the short-time diffusion coefficient (i.e., D^0) to the long-time value (i.e., D^∞; sometimes referred to in the literature as an effective diffusion coefficient, D^{eff}).

A related, but nevertheless distinct phenomenon to tortuosity that also leads to a decrease in the measured diffusion coefficient is that of obstruction. At infinite dilution, as was the case for the Einstein–Sutherland equation, interactions between large particles were neglected and the solvent particles were treated as a continuum and dealt with as a randomly fluctuating force; thus the problem was reduced to a one-body problem. However, the situation is more complex in a macromolecular solution (e.g., cell cytoplasm, polymer solution, protein solution, emulsions) where smaller molecules (e.g., water) have to skirt around the larger and generally irregularly shaped 'obstructing' molecules thereby lengthening the diffusion path of the smaller molecule (Figure 1.12). Indeed, in biological cells proteins are typically present at volume fractions, ϕ, ranging from 0.2 to 0.3. As well as its direct effects on diffusion, such macromolecular 'crowding' can have enormous thermodynamic consequences.[189,190] Thus, obstruction is a complicated many-body problem which is in essence restriction by a time-dependent geometry (the obstructing molecules are moving too) in which the interactions between the particles need to be considered. Obstruction results in the observed diffusion coefficient, $D(t)$, being reduced by some factor, O_D $(=D(t)/D^0)$ which will depend on the concentration (i.e., ϕ) of the particles as well as perhaps factors such as electrolyte friction[96] and solvation. The effects leading to obstruction generally operate on very short time and length scales and are consequently typically well averaged on experimentally available timescales. Thus, from the perspective of diffusion, obstruction and viscosity are similar in their effects and closely related and this helps to explain why the measured viscosity can depend on the size of the probe molecule.[35]

In many porous media (e.g., chromatographic media) both obstruction and tortuosity will be active and lead to reductions in the measured diffusion coefficients. Similarly, diffusion in polymer systems has elements of diffusion in porous media

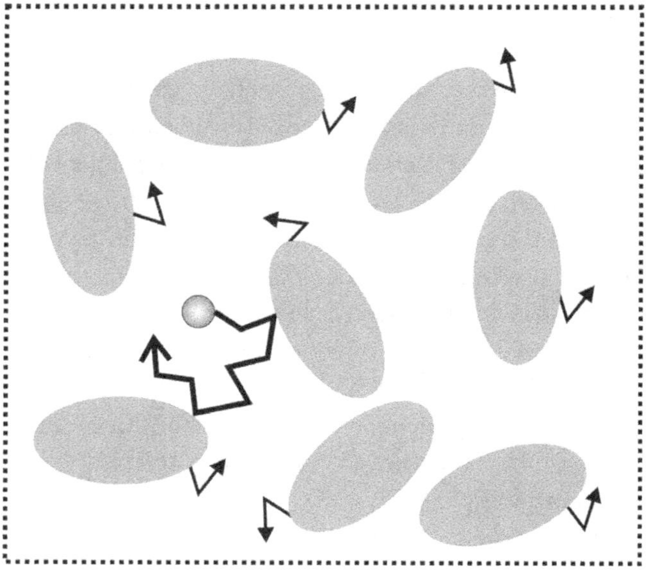

Figure 1.12 A schematic model of obstruction. In this case the diffusive path of the observed species (small circle) is affected by the presence of the larger species (solid ellipses) that it has to skirt around. In reality the system is three-dimensional and the volume fraction of the obstructing species is represented by ϕ.

and obstruction. We also note that models for time-dependent diffusion in a dilute suspension of spheres with partially absorbing boundary conditions have been developed.[191] Depending on the timescale of the measurement the effects of obstruction can result in anomalous diffusion.

1.8.2 Porous media, measurement timescales and tortuosity

The behaviour of diffusion propagators in permeable porous media is an extension of Section 1.7. In materials containing permeable pores the crossover from short to long times is often described in terms of anomalous diffusion and this is discussed in the Section 1.8.3. At very short timescales the diffusion process is insensitive to the macroscopic morphology of the system and thus a diffusion measurement would not distinguish between open and closed geometries.[192]

As the observation time increases, the time-dependence of $D(t)$ increasingly diverges for impermeable and permeable pores. In contrast to impermeable pores, in the long-time limit in materials with interconnected media as in Figure 1.11, $D(t)$ decreases to a plateau value, D^{∞}, reflecting the long range connectivity of the medium and is related to the tortuosity of the porous medium. Tortuosity describes the average hindrance of a complex medium relative to an obstacle-free medium and is thus interpreted as a path-length multiplication factor for molecules that have to find their way around obstacles and is defined by (the relative diffusion coefficient)[69,184,193–197]

$$\Im = \frac{D^0}{D^\infty}. \tag{1.126}$$

Unfortunately, there is more than one definition for tortuosity and in neurological literature the definition is identical to Eq. (1.126) except that the square root is taken on the right-hand side.[18]

As noted by Valiullin and Skirda,[69] $\Im$ is a measure of the part of the specific surface that is explored by the diffusing spins before they lose their memory about the orientation of the surface at their initial position and thus reflects both the connectivity and curvature of the surfaces in the porous network.[195–197]. $\Im$ increases with increasing curvature and constrictions in the pore space. Although ϕ and S_P/V_P are scalar properties, $\Im$ is in general a symmetric second rank tensor. In many cases $\Im$ has cylindrical symmetry (e.g., in wood). Being able to characterise $\Im$ has proved very important in understanding many processes in the brain, ranging from ischaemia and osmotic stress to delivery of nutrients and drugs.[18,198] Indeed, the permeability of a porous medium, which is a transport coefficient relating flux to a pressure gradient, is clearly related to the tortuosity. Tortuosity can be calculated for uniform colloidal objects[199,200] and periodic porous materials,[197,200] but is in general difficult to calculate especially in biologically relevant geometries.[198] In disordered geometries $\Im$ has to be determined experimentally.

1.8.3 Return to origin probability

Another commonly used means of interpreting diffusion in restricted geometries is via the Return To Origin Probability (RTOP), which is related to the spectral function (Eq. (1.80)) by normalising with respect to the value for unrestricted diffusion[164]

$$\mathrm{RTOP}(\Delta) = \frac{(4\pi D^0\Delta)^{3/2}}{V_P}\Theta(\Delta). \tag{1.127}$$

In the case of partially absorbing boundary conditions it is given by[164,201,202]

$$\mathrm{RTOP}(\Delta) = 1 + \frac{\sqrt{\pi}}{2}\frac{S_P}{V_P}\sqrt{D^0\Delta} - \left[\frac{1}{3}\left\langle\frac{1}{R_1}+\frac{1}{R_2}\right\rangle + \frac{M}{D^0}\right]\frac{S_P}{V_P}\left(D^0\Delta\right) + o\left[\left(D^0\Delta\right)^{3/2}\right]. \tag{1.128}$$

Like the expression for $D(\Delta)$ (Eq. (1.122)), the expression of RTOP depends only on the surface-to-volume ratio. However, the signs of the corrections in the two equations differ, for example in Eq. (1.128) the sign of the first term including

S_P/V_P is positive since the effect of the confinement by the pore is to increase RTOP. In contrast, the sign of the corresponding term in Eq. (1.122) is negative since diffusive motion in the pore space is restricted resulting in a reduction in the MSD travelled by the particles.[164] It has been noted that in the long-time limit RTOP is related to electrical conductivity.[164]

Whilst at short times both $D(\Delta)$ and RTOP provide information on the local properties of the pore surface, at longer times they provide information on pore connectivity. For realistic geometries it is difficult to derive an expression linking the short-time behaviour to the long-time asymptotic behaviour. However, a Padé approximant can be found that has the correct limiting behaviour and captures the crossover accurately.[184,192,203]

It has been suggested that the asymptotic form of the normalised RTOP is given for the case of $M=0$ in periodic three-dimensional geometries by[201]

$$\mathrm{RTOP}(\Delta \to 0) = \frac{1}{\phi}\left(\frac{D^0}{D^\infty}\right)^{3/2}. \tag{1.129}$$

1.8.4 Anomalous diffusion

Anomalous diffusion and transport properties are in fact quite common[125,204] with examples including the diffusion of protein hydration water,[205] diffusion in percolation clusters,[206] water diffusion in human tissue,[207] polymers[208] and surface diffusion.[209] The importance of being able to study anomalous diffusion cannot be underestimated since it is thought that, for example, the conventional equations for biochemical pathways fail to describe the reactions in vivo since the reactions follow fractal-like kinetics on account of anomalous diffusion and mixing of the biochemical species.[210]

The time-dependence of the MSD of anomalous diffusion often obeys a power law relation,[29,50,211] as shown in Eq. (1.121). When $\kappa=3$ the motion is termed 'turbulent', $\kappa=2$ is 'ballistic', $\kappa>1$ is 'superdiffusive', $\kappa=1$ is 'normal', $0<\kappa<1$, as is generally found in fractal networks and random percolation clusters and is termed 'subdiffusive', and finally $\kappa=0$ is termed 'localised'. Thus it is evident why anomalous diffusion is sometimes referred to as fractional Brownian motion. Anomalous diffusion is not invariant against time translation and the MSD does not change linearly with the observation time. Consequently, the origin of the timescale cannot be set arbitrarily. This is in contrast to ordinary diffusion, which is a Gaussian stochastic process, whose increments are uncorrelated. Valiullin and Skirda[69] have commented that the term 'anomalous diffusion' should be reserved for cases where the underlying molecular processes truly exhibit non-Gaussian statistics, thereby excluding cases commonly termed as anomalous in which Gaussian diffusion occurs in a regular array of obstructions.

Anomalous diffusion can be subdivided into two categories based on the origin:[177] (i) Gaussian propagation but where obstruction leads to subdiffusive behaviour as in 'single file' diffusion (see below) and (ii) anomalous diffusion due to geometrical restriction, since as the effects of boundaries become significant, the propagation is no longer Gaussian. If the diffusion occurs in a space with a non-Euclidean fractal-like nature the resulting diffusion is often termed anomalous.[50] However, the diffusion may only appear anomalous at some observation time or length scales.[212] It is emphasised that non-integer scaling (see Eq. (1.121)) is not synonymous with fractality and, for example, diffusion between planes leads to a non-integral exponent.[176]

Where molecules are diffusing in a fractal volume, the fractal diffusion propagator has been suggested to take the form[213,214]

$$P(R,\Delta) = \frac{d_{\mathrm{w}}}{d_{\mathrm{s}}\Gamma(d_{\mathrm{s}}/d_{\mathrm{w}})}\left(\frac{1}{D_{\kappa}d_{\mathrm{w}}^{2}\Delta}\right)^{-d_{\mathrm{f}}/d_{\mathrm{w}}}\exp\left(-\frac{R^{d_{\mathrm{w}}}}{D_{\kappa}d_{\mathrm{w}}^{2}\Delta}\right), \tag{1.130}$$

where Γ is the gamma function,[167] D_{κ} plays the role of the diffusion coefficient and d_{f} is the fractal dimension which is related to the walk and spectral (fracton) dimension, d_{s}, by[50,207]

$$d_{\mathrm{f}} = \frac{d_{\mathrm{w}}d_{\mathrm{s}}}{2}. \tag{1.131}$$

For normal diffusion $d_{\mathrm{s}} = d$, where d (integer) is the embedding dimension. Starting from the largely heuristically derived fractional diffusion equation of Metzler *et al.*,[123] Damion and Packer[215] derived the fractal propagator for diffusion in one dimension to be

$$\begin{aligned} P(Z,\Delta) = {} & \frac{d_{\mathrm{w}}\left(2D_{\kappa}^{1/2}\Delta^{\kappa/2}\right)^{2-d_{\mathrm{f}}}}{\Gamma\left(1-\frac{d_{\mathrm{f}}}{d_{\mathrm{w}}}+\frac{d_{\mathrm{f}}}{2}\right)\Gamma\left(\frac{d_{\mathrm{f}}}{2}-1\right)} Z^{d_{\mathrm{f}}-3} \\ & \times \sum_{n=0}^{\infty}\frac{(-1)^{n}}{n!}\left[\frac{\Gamma\left(\frac{d_{\mathrm{f}}}{d_{\mathrm{w}}}-2-n\right)}{\Gamma\left(1-\frac{d_{\mathrm{f}}}{d_{\mathrm{w}}}+\frac{2}{d_{\mathrm{w}}}\left(1-\frac{d_{\mathrm{f}}}{d_{\mathrm{w}}}+n\right)\right)}\left(\frac{Z}{2D_{\kappa}^{1/2}\Delta^{\kappa/2}}\right)^{2\left(2-\frac{d_{\mathrm{f}}}{d_{\mathrm{w}}}+n\right)}\right. \\ & \left. + \frac{\Gamma\left(2-\frac{d_{\mathrm{f}}}{d_{\mathrm{w}}}-n\right)}{\Gamma\left(1-\frac{d_{\mathrm{f}}}{d_{\mathrm{w}}}+\frac{2}{d_{\mathrm{w}}}(n-1)\right)}\left(\frac{Z}{2D_{\kappa}^{1/2}\Delta^{\kappa/2}}\right)^{2n}\right]. \end{aligned} \tag{1.132}$$

For completeness we note that RTOP is given by[50,207]

$$\mathrm{RTOP}(\Delta) = P(R=0,\Delta) \propto \Delta^{-d_{\mathrm{s}}/2}. \tag{1.133}$$

A particularly interesting case of anomalous diffusion is the translational diffusion in one-dimensional channels in materials such as zeolites which is known as single file diffusion. Essentially, if mutual passages of molecules in a channel are excluded, the confinement of a species by the other diffusing species (i.e., self-obstruction) results in anomalous diffusion of the species even though the pore network itself may be regular. This is so because a species can only displace in one direction if the species in front of it also displaces in the same direction. Thus, in the event of a displacement, there will effectively be a higher concentration in front of the particle than behind it – and this difference will lead to a higher probability of the subsequent displacement being in the opposite direction. Taking the displacement to be along the z-direction, one obtains[29]

$$\left\langle Z^2(\Delta)\right\rangle = 2K\Delta^{1/2}, \tag{1.134}$$

where the mobility factor K is given by

$$K = \lambda^2 \frac{1-\theta}{\theta}\frac{1}{\sqrt{2\pi\tau_J}}, \tag{1.135}$$

where λ is the separation between adjacent sites with a mean time τ_J between subsequent jump attempts and θ is the occupation probability. A jump is only successful if the target site is vacant.

1.8.5 Propagators in permeable porous systems and the pore-hopping model

As mentioned previously, in the case of single pores contained within a permeable boundary, a first approximation in developing a diffusion propagator is to institute a relaxation boundary condition on the boundary and to relate the relaxation properties to transport. However, in biological tissues such an approach is in general too simplistic as the reality is diffusion in compartmentalised spaces with interchange between the compartments. From the above discussion it is clear that in a permeable porous system the propagator is Gaussian in both the short- and long-time limits with the diffusion coefficients D^0 and D^∞, respectively. Whilst the propagators of isolated pores have relatively sharp cutoffs, this does not occur with a permeable porous material.

1.8.5.1 Permeable porous systems

An early model of a porous system was that of Tanner[216] in which diffusion occurred between a series of equally spaced planes of arbitrary permeability (see Figure 1.13), for which the propagator is given by[216]

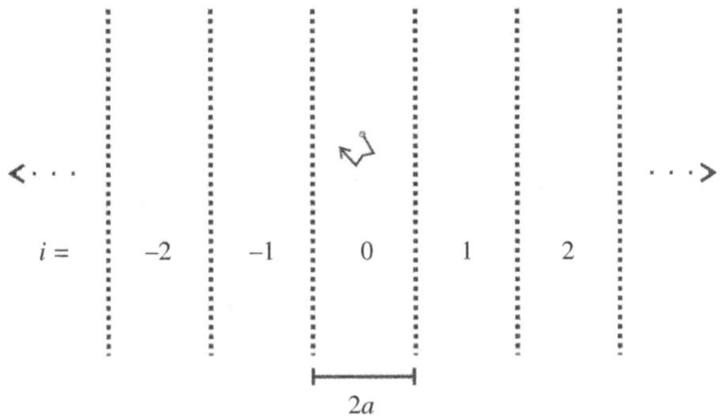

Figure 1.13 A system of equally spaced paralllel planes as a model for diffusion in a biological system. The permeability of the planes is given by M.

$$P(x_0, x_1, \Delta) = \sum_{n=1}^{\infty} \frac{X_{0n}(x_0) X_{in}(x_1)}{\sum\limits_{\text{all } i} \int\limits_{i\text{th}} X_{in}^2(x_1)dx} \exp(-\beta_n^2 D^0 t), \tag{1.136}$$

where the layers are numbered by i and the X_{in} are the infinite set of eigenfunctions and β_n are eigenvalues of $\left(\frac{\partial^2}{\partial x^2} + \beta_n^2\right) X_{in} = 0$, which is discontinuous at the boundaries, and of the form $X_{in} = c_{in} \cos \beta_n x_1 + d_{in} \sin \beta_n x_1$. Tanner used this to give an approximate solution for $D(\Delta)$ in a system with 19 layers. $D(\Delta)$, in the direction normal to the planes, decreases monotonically from the free solution value to its asymptotic value[217,218]

$$D^{\infty} = \frac{D^0 M 2a}{D^0 + M2a}, \tag{1.137}$$

where $2a$ is the interplanar spacing and M is the permeability. In a later work[219] exact solutions for the Laplace transforms of the MSD and $D(\Delta)$ were derived for particles diffusing in the system shown in Figure 1.13.

1.8.5.2 Interconnected pores

For more general porous media as depicted in Figure 1.11, the propagator approach outlined in Section 1.4.2 needs to be extended. Taking $\mathbf{r}_{0i}$ to be the position of the ith pore, a structure of N pores can be represented by a spin density function given by a superposition of normalised local density functions $\rho_{0i}(\mathbf{r}_1 - \mathbf{r}_{0i})$ of the form[132,220,221]

$$\rho(\mathbf{r}_1) = \frac{1}{N} \sum_{i=1}^{N} \rho_{0i}(\mathbf{r}_1 - \mathbf{r}_{0i}), \tag{1.138}$$

where $\mathbf{r}_{0i}$ is the position of the ith pore centre. The connectivity of the pores can be described by the local conditional probability[132,220–222]

$$P_i(\mathbf{r}_0, \mathbf{r}_1, \Delta) = \sum_{j=1}^{N} C_{ij}(\Delta)\rho_{0j}(\mathbf{r}_1 - \mathbf{r}_{0j}), \tag{1.139}$$

where the $C_{ij}(\Delta)$ are connection matrix elements describing the probability that a particle will move from pore i to j during Δ. For isolated pores $C_{ij}(\Delta)$ would be a diagonal matrix as the off-diagonal elements would be zero. Since $2a$ is taken to be less than b, it is presumed that the particles will have equilibrated in each pore in a much shorter time than that required for migration between pores.[220] Assuming a regular lattice as in Figure 1.11 and sufficiently large distances and long times $C_{ij}(\Delta)$ has the (Gaussian) form[132,220,222]

$$C_{ij}(\Delta) = \frac{b}{\sqrt{4\pi D^{\infty}\Delta}} \exp\left(\frac{-(i-j)^2 b^2}{4D^{\infty}\Delta}\right). \tag{1.140}$$

Note that this is the one-dimensional form of the matrix since the measurements we will ultimately be concerned with measure diffusion in only one direction (at a time). In the case of a disordered lattice in which there is no correlation between lattice spacing and pore size, and defining the one-dimensional displacement in the direction of the measurement given by $Z = (z_1 - z_0)$, it is possible to define an average pore structure factor, $C(Z, \Delta)$, equivalent to Eq. (1.140). The relative positions of the pores can be quantified with a 'microstructural or lattice correlation function', $L(Z)$, which represents the probability that a lattice site will be found at a displacement Z from a starting point. Using this assumption the propagator can be written as[132,220,222]

$$P(Z, \Delta) = \bar{P}(Z, \infty) \otimes [C(Z, \Delta)L(Z)], \tag{1.141}$$

where $\bar{P}(Z, \infty)$ is the auto-correlation function of a single pore (i.e., Eq. (1.124)) and $\otimes$ denotes convolution. Simplistically $C(Z, \Delta)$ can be assumed to be Gaussian; however, a more sophisticated and accurate approach would be to use the pore-hopping formalism[220] that is consistent with the distances between the pores. Indeed, non-Gaussian features result in systems in which there is a well-defined interpore distance b. For example, in a 1D porous system in which $L(Z)$ is a set of delta functions, the propagator has local maxima at multiples of b.

Although porous media are likely to appear macroscopically isotropic, the pore structure might nevertheless be microscopically anisotropic. Consequently, it is possible to define higher propagators such as $P(x, y, \Delta)$.[223–225]

1.8.6 Obstruction

1.8.6.1 Hard spheres

In general the mathematical problem of calculating the obstruction factor involves the solution to the steady-state diffusion equation (i.e., Laplace's equation; Eq. (1.38) but with the LHS set to 0) under the appropriate boundary conditions.[200] Except for trivial cases this represents a formidable mathematical problem and a common means of approaching this problem is that of the multipole expansion.[200,226] Approximate solutions are known only for some simple cases[5,85,97,199,227–233] and these are often only valid at low volume fractions (ϕ). In general, analysis is performed using numerical simulations (e.g., off-lattice random flight simulations[200]) – such simulations can also be used to check the validity of the various approximate solutions for the obstruction factor.

Here we consider some commonly encountered approximations. Neglecting binding effects, using a cell model Jönsson *et al.* derived the obstruction effect for a system containing evenly spaced spherical mono-disperse obstructing particles is found to be[199]

$$O_D = \frac{D(t)}{D^0} = \frac{1}{1 + \phi/2}. \tag{1.142}$$

Thus, in this case the limiting value of D at $\phi = 1$ is $2D^0/3$. Modelled using the memory effect or a thermodynamic relaxation approach, Lekkerkerker and Dhont arrived at the relation[227]

$$O_D = 1 - 2\phi. \tag{1.143}$$

Using scaled particle theory Han and Herzfeld derived the relationship[230]

$$O_D = \exp\left(-\frac{l}{R}\left(3\frac{\phi}{1-\varphi} + \frac{9}{2}\frac{\phi^2}{(1-\phi)^2} + \frac{9}{4}\frac{\phi^3}{(1-\phi)^3}\right)\right), \tag{1.144}$$

where l is the step size and, from the Smoluchowski equation, $l/a = 2/3$ where a is the radius of the diffusing particle. Starting from the Navier–Stokes equation Tokuyama and Oppenheim calculated the relationship[234]

$$O_D = \frac{1 - \dfrac{9\phi}{32}}{1 + H(\phi) + \dfrac{\phi/\phi_0}{(1 - \phi/\phi_0)^2}}, \tag{1.145}$$

where

$$H(\phi) = \frac{2b(\phi)^2}{(1 - b(\phi))} - \frac{c(\phi)}{(1 + 2c(\phi))} - \frac{b(\phi)c(\phi)(2 + c(\phi))}{(1 + c(\phi))(1 - b(\phi) + c(\phi))},$$

$$b(\phi) = \sqrt{9\,\phi/8},$$

$$c(\phi) = 11\,\phi/16,$$

and

$$\phi_0 = \frac{\left(\frac{4}{3}\right)^3}{7\ln 3 - 8\ln 2 + 2}.$$

Free volume theories have been widely used especially in studying the diffusion of small molecules in polymer solution.[97,232,233] Within this framework the diffusion of a small molecule depends on the probability of encountering a suitable-sized hole through which it can diffuse through resulting in an obstruction factor of the form[97,232,233]

$$O_D = \exp\left[\frac{B_g}{f_s}\left(1 - \frac{1}{1-\phi}\right)\right] \tag{1.146}$$

where B_g is a measure of the size of the diffusing molecule and f_s is the free volume contributed by the solvent and ϕ .

Plots of four of these obstruction models as a function of ϕ are given in Figure 1.14. As can be seen from both the plots and the forms of the equations there is considerable difference between the degrees of obstruction predicted. Importantly, none of these models account for the presence of an aggregation process and thus will likely progressively overestimate the reduction in diffusion as the concentration (and degree of aggregation) increases. Nevertheless, these equations provide one means of estimating D^0 from diffusion experiments at a finite concentration. Further, since the degree of obstruction is dependent on the shapes of the obstructing

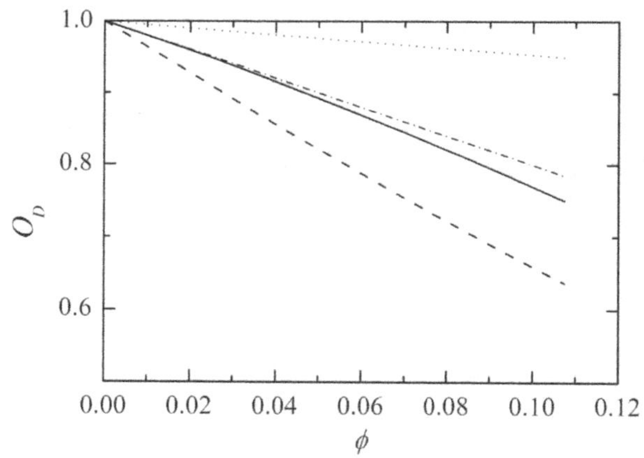

Figure 1.14 Simulations of the obstruction factor, O_D, as a function of the volume fraction, ϕ, using the models of Jönsson *et al.* (······; Eq. (1.142)), Lekkerkerker and Dhont (–·–·–·; Eq. (1.143)), Han and Herzfeld (——; Eq. (1.144)), Tokuyama and Oppenheim (– – –; Eq. (1.145)). Although there is considerable discrepancy between the models, the simulations show that as ϕ increases the diffusion coefficient is significantly reduced by the effects of obstruction.

particles, better models coupled with very accurate diffusion measurements provide an additional source of solution structural information.

1.8.6.2 Charged species

In reality many molecules cannot be treated as hard spheres and often the effects of charge should be taken into account such as in studies involving colloids, which include protein studies.[43] One approach for doing this is based on the Poisson–Boltzmann–Smoluchowski model for an electroneutral spherical shell of radius R_{cell} in which a charged hard sphere of radius R_{S} is surrounded by a charged sphere of monovalent (oppositely charged) counterions distributed according to a Poisson–Boltzmann distribution.[235,236] For monovalent counterions the reduction in diffusion is given by

$$O_D = \chi(R_{\text{cell}})\frac{n(R_{\text{cell}})}{n_{\text{avg}}}, \tag{1.147}$$

where $n(R_{\text{cell}})$ is the number of counterions at the outer cell boundary, n_{avg} is the average number of counterions in the spherical shell and $\chi(R_{\text{cell}})$ is determined from the first-order differential equation

$$r\chi' + \chi\left(1 - \chi - r\frac{\partial \phi_{\text{e}}}{\partial r}\right) - 2 = 0, \tag{1.148}$$

with the boundary condition $\chi(R_{\text{S}}) = 0$, where ϕ_{e} is the electrostatic potential and r denotes the distance from the centre of the spherical cell.

More recently, Darwish *et al.*[237] presented a phenomenological modification of the geometric obstruction model of Amsden[238] to model the diffusion of ions in polyelectrolyte gels. They incorporated electrostatic obstruction effects by adopting an effective chain diameter from Dobrynin scaling theory for polyelectrolytes and by adding an exclusion layer around the charged chains to account for the electrostatic interaction between the ions and the chains.

1.8.6.3 Solvated species

If solvent molecules are diffusing in the presence of macromolecules then, in addition to obstruction, should they bind to (or be structured around[102]) the macromolecules their diffusion will be even further decreased. For macromolecules represented as ellipsoids with semi-major axis a and semi-minor axis b, combining both effects, the reduction in solvent diffusion is given by[239,240]

$$O_D = \frac{(1 - \bar{a}\phi_{\text{h}})(1 - \delta_{\text{h}})}{1 - \phi_{\text{h}}}, \tag{1.149}$$

where

$$\phi_{\mathrm{h}} = \frac{V_{\mathrm{S}} + \dfrac{\delta_{\mathrm{h}}}{d^0}}{V_{\mathrm{S}} + \dfrac{1}{d^0}\left(\dfrac{1-w}{w}\right)}, \tag{1.150}$$

$$f = 1 - \frac{w}{1-w}\delta_{\mathrm{h}}, \tag{1.151}$$

where d^0 is the density of the solvent and w is the weight fraction of anhydrous macromolecules in solution. The shape factor $\bar{\alpha}$ is defined by

$$\bar{\alpha} = \frac{1}{3}(\alpha_a + \alpha_b + \alpha_c). \tag{1.152}$$

For the case of a sphere $\alpha_a = \alpha_b = \alpha_c = 1.5$. In the case of oblate ellipsoids,

$$\alpha_a = \left\{1 - \frac{p}{1-p^2}\left[(1-p^2)^{-\frac{1}{2}}\left(\tan^{-1}\left(\frac{p}{\sqrt{1-p^2}}\right) - \frac{\pi}{2}\right) + \frac{1}{p}\right]\right\}^{-1} \tag{1.153}$$

and

$$\alpha_b = \alpha_c = 2\left\{\frac{2-p^2}{1-p^2} - \frac{p}{\sqrt[3]{1-p^2}}\left[\left(\frac{\pi}{2} - \tan^{-1}\left(\frac{p}{\sqrt{1-p^2}}\right)\right) + \frac{1}{p}\right]\right\}^{-1}. \tag{1.154}$$

For prolate ellipsoids

$$\alpha_a = \left\{\frac{p^2}{p^2-1} - \frac{p}{2(p^2-1)^{3/2}}\ln\left(\frac{p+\sqrt{p^2-1}}{p-\sqrt{p^2-1}}\right)\right\}^{-1} \tag{1.155}$$

and

$$\alpha_b = \alpha_c = 2\left\{\frac{p^2-2}{p^2-1} - \frac{p}{2(p^2-1)^{3/2}}\ln\left(\frac{p+\sqrt{p^2-1}}{p-\sqrt{p^2-1}}\right)\right\}^{-1}, \tag{1.156}$$

where $p = a/b$.

Using a cell model including the effects of obstruction and binding (i.e., hydration) Jönsson *et al.*[199] determined the observed self-diffusion coefficient which is given by

$$O_D = \frac{1}{1 - \left(1 - \dfrac{C_1}{C_2}\right)\phi}\,\frac{1-\beta\phi}{1+\dfrac{\beta\phi}{2}}, \tag{1.157}$$

with

$$\beta = \frac{D_2C_2 - D_1C_1}{D_2C_2 + 0.5D_1C_1}, \tag{1.158}$$

where C_1 and D_1 are the water concentration and self-diffusion coefficient inside the spherical particles and C_2 and D_2 are the same properties but surrounding the particle. Equation (1.157) reduces to Eq. (1.142) in the limit $C_1 \to 0$.

1.8.7 Diffusion in polymers

Diffusion in polymers embraces a huge area of work and we give only a token coverage. Estimates of friction coefficients for polymers can be obtained using Kirkwood–Riseman theory as outlined in Section 1.3.3. Discussion of polymer diffusion is typically categorised into dilute, semi-dilute, and concentrated (or melt) regimes. Further specialisation can be made for diluents and penetrants in polymers, large or flexible molecules dissolved in polymers and polymer gels. It is emphasised that the diffusion of polymers, especially in concentrated solution and melts, is complicated by entanglement of the polymer chains. Detailed discussion on diffusion in polymer systems can be found elsewhere.[2,21,104,177,233,241]

Various phenomenological models, free volume models and molar mass (M) scaling laws have been presented.[233] The well-known Rouse model states that the diffusion of high molecular weight polymers in a dilute solution is expected to scale as[242]

$$D \sim M^{-1}. \tag{1.159}$$

Reptation theory was introduced by de Gennes[104,243] to model diffusion when the polymer chains are entangled. In this model the polymer chain is considered to be constrained by the surrounding polymer molecules. Since the polymer molecule is surrounded by other polymer molecules, the central part of the polymer remains restricted for a longer time than the extremities of the chain; the polymer is said to be constrained within a tube and performs wormlike displacements. Consequently, lateral motion is ignored and only tubular motion is considered. Whereas for high molecular weight polymers in either concentrated solutions or melts, the resultant tube diffusion is expected to scale as[243]

$$D \sim M^{-2}. \tag{1.160}$$

For short observation times (i.e., small Δ), the observed displacement may be smaller than the end-to-end distance of the polymer chain and thus a polymer system can be considered as being heterogeneous at short times. The diffusion of the polymer molecules is restricted by instantaneous tubes formed by the bulk phase of the surrounding molecules. The reptation time, t_r, is defined as the time required for a polymer molecule to cover a curvilinear diffusion path of the order of its contour length. For $\Delta > t_r$, the confining tube will be completely uncorrelated with the previous tube and the observed diffusion coefficient will appear to be constant (i.e., it will not scale with time). Conversely, for $\Delta < t_r$, the motion of the polymer segments is subject to a correlated confinement. Accordingly, the apparent diffusion coefficient will decrease with increasing Δ.

In some polymer gel systems the diffusion is described by the particles diffusing in an isotropic medium while being harmonically bound to an attractive centre (see Section 4.6).[158,159,244]

To better describe the temperature dependence of diffusion, diffusion models which can be considered as thermodynamic models are also used. For example, Arrhenius equations of the form

$$D = A \exp\left(-\frac{E_{\mathrm{A}}}{\mathrm{R}T}\right) \tag{1.161}$$

where A is a pre-exponential factor, E_{A} is the activation energy and R is the gas constant are often used to estimate the temperature dependence of the diffusion coefficient of polymer systems. It is noted that the usage of Eq. (1.161) is not just restricted to polymer systems.

1.9 Flow in porous media

We can define the velocity correlation time τ_{v} for the duration of flow around the characteristic length scale in a porous medium (Figure 1.15), d_{p}, by[245,246]

$$\tau_{\mathrm{v}} = \frac{d_{\mathrm{p}}}{\langle V \rangle}. \tag{1.162}$$

The ratio of the time taken to diffuse across a pore to the time taken to flow across a pore is given by the Péclet number[20]

$$P_{\mathrm{e}} = \frac{l\langle V \rangle}{D^0}, \tag{1.163}$$

where, in the case of packed beads, l is given by the effective pore diameter

$$l = \phi d_{\mathrm{p}}/(1 - \phi), \tag{1.164}$$

where ϕ is the porosity. At large Péclet numbers (i.e., $P_{\mathrm{e}} \gg 1$) the dominating source of the dispersion of the particle trajectories is the tortuosity determined by the pore space geometry. In this case incoherent displacements due to tortuous flow dominate Brownian displacements and particle transport tends to be subdiffusive (see Section 1.8.4).

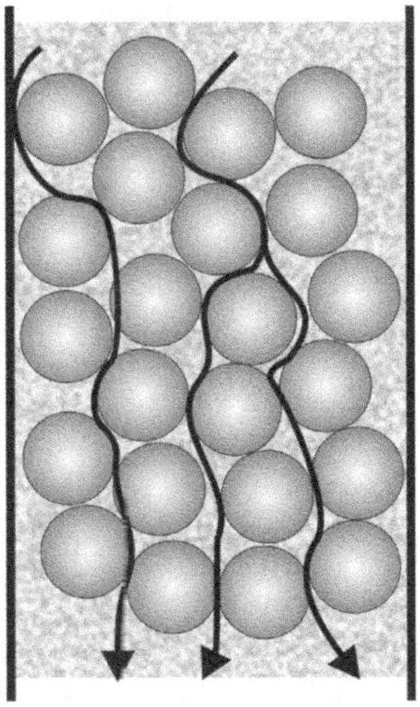

Figure 1.15 A schematic representation of flow through a porous medium. The flow is associated with an average velocity (see Eq. (1.1)). The dispersive nature of the flow can be visualised from the diagram and includes mechanical dispersion (e.g., bifurcation of streamlines and turbulence) and Taylor dispersion (diffusion across velocity shear).[247]

1.10 Techniques for measuring diffusion

1.10.1 Traditional techniques

Numerous methods exist for measuring diffusion[2] including NMR, light scattering,[10,21] neutron scattering,[248] capacity intermittent titration techniques (CITT),[249] capillary methods, fluorescence,[250,251] attenuated total reflection infrared spectroscopy (ATRIR)[252] and centrifuge studies.[253] The non-NMR methods in particular are generally of limited application (e.g., concentration range), extremely tedious, fraught with experimental difficulty or are invasive in nature.[2,14] Neutron scattering, however, is confined to making measurements on the scale of only 0.01–10 nm and thus is usually confined to single scattering events as it measured on the timescale of 10^{-9}–10^{-12} s. Dynamic light scattering (DLS) depends on concentration fluctuations and measures mutual diffusion and is non-invasive and non-perturbative.[2,10]

1.10.2 NMR-based techniques

Due to its non-invasive nature and generally not requiring labelled probe molecules, NMR spectroscopy is a unique tool for studying molecular dynamics in chemical and biological systems. Importantly, it allows measurements to be performed under physical conditions, such as high pressure[254] and temperature, where other methods may be precluded. In contrast to virtually all of the traditional methods, NMR diffusion measurements are applicable at any concentration that provides a sufficient signal-to-noise ratio (S/N) and being of non-invasive nature, they do not perturb the thermodynamics of the system.

There are two main ways in which NMR may be used to study self-diffusion coefficients: (i) through analysis of nuclear spin relaxation data and (ii) through magnetic field gradient NMR. Relaxation-based measurements of diffusion are microscopic in nature since relaxation is sensitive to motions on the timescale of the reorientational correlation time of the species (τ_c). In the solution state this corresponds to motions occurring on the ps to ns timescale. In this approach, relaxation data is analysed to determine τ_c of a probe species.[255] τ_c can then be related to the solution viscosity and ultimately to the translational diffusion coefficient by using the Debye equation[256–258]

$$\tau_c = \frac{4\pi\eta r_S^3}{3kT}, \tag{1.165}$$

and the Einstein–Sutherland equation (i.e., Eq. (1.13)). However, a number of assumptions, which depending upon the system being studied may or may not be justified, need to be made in performing this analysis. First, the relaxation mechanism of the probe species needs to be known and it is required that the intermolecular contributions to the relaxation can be separated from the intra-molecular contributions.[259,260] Secondly, only if the molecule is spherical can its reorientational dynamics be properly characterised by a single correlation time. Thirdly, depending on the size of the probe molecules compared to the molecules of the bulk solution, the probe molecules may not see the solution as being continuous, and as a consequence one of the basic requirements for the validity of the Debye equation is violated.[261,262] Thus, serious assumptions are involved in applying this method to, for example, biological milieux when a small probe species is used since the solution normally has a large macromolecular component (e.g., a large part of the cytoplasm of red blood cells is composed of haemoglobin). The final problem with this method is that the Stokes radius of the probe molecule needs to be known and this, as has been noted above, is in general far from straightforward as will be seen below.

PGSE NMR provides estimates of the translational self-diffusion coefficient as against methods such as scattering techniques which actually provide estimates of the mutual diffusion coefficients. Although we will see in Chapter 9 that NMR imaging can be used to measure mutual diffusion. The remainder of this book is predominantly concerned with the use of magnetic field gradients to measure diffusion and other aspects of translational motion.

We remark that the two NMR methods are complementary and study motion on very different timescales. Specifically, gradient-based NMR measurements probe spin displacement distributions over Δ (i.e. the MSD, the average propagator, the time-dependent diffusion coefficient and the diffusion spectrum) and diffusion is just a special case of what can be measured. Gradient-based methods can also be used to probe $D_\omega(\omega)$.

References

1. H. J. V. Tyrrell and K. R. Harris, *Diffusion in Liquids: A Theoretical and Experimental Study*. (London: Butterworths, 1984).
2. E. L. Cussler, *Diffusion Mass Transfer in Fluid Systems*, 2nd edn. (Cambridge: Cambridge University Press, 1997).
3. B. E. Poling, J. M. Prausnitz, and J. P. O'Connell, *The Properties of Gases and Liquids*, 5th edn. (New York: McGraw Hill, 2001).
4. R. B. Bird, E. N. Lightfoot, and W. E. Stewart, *Transport Phenomena*, 2nd edn. Revised (New York: Wiley, 2007).
5. S. Hanna, W. Hess, and R. Klein, Self-Diffusion of Spherical Brownian Particles with Hard-Core Interaction. *Physica A* **111** (1982), 181–99.
6. A. Kalk and H. J. C. Berendsen, Proton Magnetic Relaxation and Spin Diffusion in Proteins. *J. Magn. Reson.* **24** (1976), 343–66.
7. N. Fatkullin, Theory of Stimulated Spin Echo in Polymer System. *Sov. Phys. JETP* **72** (1991), 563–69.
8. W. Zhang and D. G. Cory, First Direct Measurement of the Spin Diffusion Rate in a Homogeneous Solid. *Phys. Rev. Lett.* **80** (1998), 1324–7.
9. M. E. Komlosh and P. T. Callaghan, Spin Diffusion in Semidilute Random Coil Polymers Studied by Pulsed Gradient Spin-Echo NMR. *Macromolecules* **33** (2000), 6824–7.
10. C. S. Johnson, Jr. and D. A. Gabriel, *Laser Light Scattering*. (New York: Dover, 1994).
11. S. Chandrasekhar, Stochastic Problems in Physics and Astronomy. *Rev. Mod. Phys.* **15** (1943), 1–89.
12. M. Kac, Random Walk and the Theory of Brownian Motion. *Am. Math. Mon.* **54** (1947), 369–91.
13. J. Crank, *The Mathematics of Diffusion*, 2nd edn. (Oxford: Oxford University Press, 1975).
14. C. R. Cantor and P. R. Schimmel, *Biophysical Chemistry, Part II: Techniques for the Study of Biological Structure and Function*. (New York: W. H. Freeman, 1980).
15. M. A. Lauffer, *Motion in Biological Systems*. (New York: Alan R. Liss, Inc., 1989).
16. H. C. Berg, *Random Walks in Biology*. (New Jersey: Princeton University Press, 1993).
17. R. Glaser, *Biophysics*, Revised (Berlin: Springer-Verlag, 2000).
18. C. Nicholson, Diffusion and Related Transport Mechanisms in Brain Tissue. *Rep. Prog. Phys.* **64** (2001), 815–84.
19. K. A. Dill and S. Bromberg, *Molecular Driving Forces*. (New York: Garland Science, 2003).
20. G. A. Truskey, F. Yuan, and D. F. Katz, *Transport Phenomena in Biological Systems*. (New York: Prentice Hall, 2003).

21. P. F. Green, *Kinetics, Transport, and Structure in Hard and Soft Materials*. (Baton Rouge: CRC Press, 2005).
22. A. V. Barzykin, K. Seki, and M. Tachiya, Kinetics of Diffusion-Assisted Reactions in Microheterogeneous Systems. *Adv. Colloid Interface Sci.* **89–90** (2001), 47–140.
23. S. D. Traytak and W. S. Price, Exact Solution for Anisotropic Diffusion-Controlled Reactions with Partially Reflecting Conditions. *J. Chem. Phys.* **127** (2007), 184508-1–184508-8.
24. T. C. Halsey, Diffusion-Limited Aggregation: A Model for Pattern Formation. *Phys. Today* **53** (2000), 36–41.
25. L. M. Sander, Diffusion-Limited Aggregation: A Kinetic Critical Phenomenon? *Contemp. Phys.* **41** (2000), 203–18.
26. A. S. Verkman, Solute and Macromolecular Diffusion in Cellular Aqueous Compartments. *Trends Biochem. Sci.* **27** (2002), 27–33.
27. J. G. Albright and R. Mills, A Study of Diffusion in the Ternary System, Labeled Urea-Urea-Water, at 25° by Measurements of the Intradiffusion Coefficients of Urea. *J. Phys. Chem.* **69** (1965), 3120–6.
28. M. E. Glicksman, *Diffusion in Solids: Field Theory, Solid-State Principles, and Applications*. (New York: Wiley, 2000).
29. J. Kärger and F. Stallmach, PFG NMR Studies of Anomalous Diffusion. In *Diffusion in Condensed Matter*, ed. P. Heitjans and J. Kärger. (Berlin: Springer, 2006), pp. 417–59.
30. R. W. Mair, D. G. Cory, S. Peled, C.-H. Tseng, S. Patz, and R. L. Walsworth, Pulsed-Field-Gradient Measurements of Time-Dependent Gas Diffusion. *J. Magn. Reson.* **135** (1998), 478–86.
31. P. W. Kuchel, B. E. Chapman, and A. J. Lennon, Diffusion of Hydrogen in Aqueous Solutions Containing Protein. Pulsed Field Gradient NMR Measurements. *J. Magn. Reson. A* **103** (1993), 329–31.
32. H. Weingärtner, Self-Diffusion in Liquid Water. A Reassessment. *Z. Phys. Chem.* **132** (1982), 129–49.
33. R. Mills, Self-Diffusion in Normal and Heavy Water in the Range 1–45°. *J. Phys. Chem.* **77** (1973), 685–8.
34. W. S. Price and P. W. Kuchel, Restricted Diffusion of Bicarbonate and Hypophosphite Ions Modulated by Transport in Suspensions of Red Blood Cells. *J. Magn. Reson.* **90** (1990), 100–10.
35. W. S. Price, B. E. Chapman, B. A. Cornell, and P. W. Kuchel, Translational, Diffusion of Glycine in Erythrocytes Measured at High Resolution with Pulsed Field Gradients. *J. Magn. Reson.* **83** (1989), 160–6.
36. W. H. Gmeiner, C. J. Hudalla, A. M. Soto, and L. Marky, Binding of Ethidium to DNA Measured Using a 2D Diffusion-Modulated Gradient COSY NMR Experiment. *FEBS Lett.* **465** (2000), 148–52.
37. K. Adachi, M. Natsuisaka, and A. Tanioka, Measurements of Self-Diffusion Coefficients of Monensin in Chloroform Solution by PFG-NMR. *J. Chem. Soc., Faraday Trans.* **93** (1997), 3347–50.
38. T. Kato, K. Kikuchi, and Y. Achiba, Measurement of the Self-Diffusion Coefficient of C_{60} in Benzene-d_6 Using ^{13}C Pulsed-Gradient Spin-Echo. *J. Phys. Chem.* **97** (1993), 10251–3.
39. S. Gaemers, C. J. Elsevier, and A. Bax, NMR of Biomolecules in Low Viscosity, Liquid CO_2. *Chem. Phys. Lett.* **301** (1999), 138–44.
40. H. Pan, G. Barany, and C. Woodward, Reduced BPTI is Collapsed. A Pulsed Field Gradient NMR Study of Unfolded and Partially Folded Bovine Pancreatic Trypsin Inhibitor. *Protein Sci.* **6** (1997), 1985–92.

41. J. P. Mackay, G. L. Shaw, and G. F. King, Backbone Dynamics of the c-Jun Leucine Zipper: ^{15}N NMR Relaxation Studies. *Biochemistry* **35** (1996), 4867–77.
42. M. Lin and C. K. Larive, Detection of Insulin Aggregates with Pulsed-Field Gradient Nuclear Magnetic Resonance Spectroscopy. *Anal. Biochem.* **229** (1995), 214–20.
43. W. S. Price, F. Tsuchiya, and Y. Arata, Lysozyme Aggregation and Solution Properties Studied Using PGSE NMR Diffusion Measurements. *J. Am. Chem. Soc.* **121** (1999), 11503–12.
44. S. J. Gibbs, A. S. Chu, E. N. Lightfoot, and T. W. Root, Ovalbumin Diffusion at Low Ionic Strength. *J. Phys. Chem.* **95** (1991), 467–71.
45. C. H. Everhart and C. S. Johnson, Jr., The Determination of Tracer Diffusion Coefficients for Proteins by Means of Pulsed Field Gradient NMR with Applications to Hemoglobin. *J. Magn. Reson.* **48** (1982), 466–74.
46. P. T. Callaghan and D. N. Pinder, Influence of Polydispersity on Polymer Self-Diffusion Measurements by Pulsed Field Gradient Nuclear Magnetic Resonance. *Macromolecules* **18** (1985), 373–9.
47. I. Furó and S. V. Dvinskikh, NMR Methods Applied to Anisotropic Diffusion. *Magn. Reson. Chem.* **40** (2002), S3–14.
48. D. Le Bihan, Molecular Diffusion, Tissue Microdynamics and Microstructure. *NMR Biomed.* **8** (1995), 375–86.
49. E. W. Hsu, N. R. Aiken, and S. J. Blackband, A Study of Diffusion Isotropy in Single Neurons by Using NMR Microscopy. *Magn. Reson. Med.* **37** (1997), 624–27.
50. D. Ben-Avraham and S. Havlin, *Diffusion and Reactions in Fractals and Disordered Systems*. (Cambridge: Cambridge University Press, 2000).
51. G. Job and F. Herrmann, Chemical Potential – A Quantity in Search of Recognition. *Eur. J. Phys.* **27** (2006), 353–71.
52. J. B. Brady, Reference Frames and Diffusion Coefficients. *Am. J. Sci.* **275** (1975), 954–83.
53. H. Weingärtner, The Molecular Description of Mutual Diffusion Processes in Liquid Mixtures. In *Diffusion in Condensed Matter*, ed. P. Heitjans and J. Kärger (Berlin: Springer, 2005), pp. 555–78.
54. H. L. Friedman and R. Mills, Velocity Cross Correlations in Binary Mixtures of Simple Fluids. *J. Solution Chem.* **10** (1981), 395–409.
55. H. L. Friedman, F. O. Raineri, and M. D. Wood, Distinct Diffusion Coefficients, Probes of Ion-Ion Interactions. *Chem. Scr.* **29A** (1989), 49–59.
56. E. Hawlicka, Self-Diffusion in Multicomponent Liquid Systems. *Chem. Soc. Rev.* **24** (1995), 367–77.
57. B. Newling, Gas Flow Measurements by NMR. *Prog. NMR Spectrosc.* **52** (2008), 31–48.
58. J. Abshagen, A. Schulz, and G. Pfister, The Couette-Taylor Flow: A Paradigmatic System for Instabilities, Pattern Formation and Routes to Chaos. In *Nonlinear Physics of Complex Systems*. ed. J. Parisi, S. C. Müller, and W. Zimmermann, Lect. Notes Phys. vol. 476 (Berlin: Springer, 1997), pp. 63–72.
59. H. Bénard, Les tourbillons celluaries dans une nappe liquide transportant de la chaleur par convection en regime permanent. *Ann. Chim. Phys.* **23** (1901), 62–144.
60. Lord Rayleigh, On Convection Currents on a Horizontal Layer of Fluid When the Higher Temperature is on the Under Side. *Philos. Mag.* **32** (1916), 529–46.
61. D. L. Koch and J. F. Brady, A Non-Local Description of Advection-Diffusion with Application to Dispersion in Porous Media. *J. Fluid. Mech.* **180** (1987), 387–403.
62. P. T. Callaghan, S. L. Codd, and J. D. Seymour, Spatial Coherence Phenomena Arising from Translational Spin Motion in Gradient Spin Echo Experiments. *Concepts Magn. Reson.* **11** (1999), 181–202.

63. P. T. Callaghan, Some Perspectives on Dispersion and the Use of Ensemble-Averaged PGSE NMR. *Magn. Reson. Imaging* **23** (2005), 133–7.
64. P. Turq, J. Barthel, and M. Chemla, *Transport, Relaxation and Kinetic Processes in Electrolyte Solutions*. (Berlin: Springer-Verlag, 1992).
65. J. O. Bockris and A. K. N. Reddy, *Modern Electrochemistry*, 2nd edn. (New York: Plenum Press, 1998).
66. M. Holz, Field-Assisted Diffusion Studied by Electrophoretic NMR. In *Diffusion in Condensed Matter*, ed. J. Kärger and P. Heitjans. (Berlin: Springer, 2005), pp. 717–42.
67. N. Fatkullin, Theory of Diffusive Damping of Spin-Echo Signal in a Medium with Random Obstacles. *Sov. Phys. JETP* **71** (1990), 1141–4.
68. A. Fulinski, On Marian Smoluchowski's Life and Contribution to Physics. *Acta Physiol. Pol. B* **29** (1998), 1523–37.
69. R. Valiullin and V. Skirda, Time Dependent Self-Diffusion Coefficient of Molecules in Porous Media. *J. Chem. Phys.* **114** (2001), 452–58.
70. A. M. Berezhkovskii and G. Sutmann, Time and Length Scales for Diffusion in Liquids. *Phys. Rev. E* **65** (2002), 060201-1–060201-4.
71. R. Kubo, The Fluctuation-Dissipation Theorem. *Rep. Prog. Phys.* **29** (1966), 255–84.
72. J. Stepišnik, Analysis of NMR Self-Diffusion Measurements by a Density Matrix-Calculation. *Physica B & C* **104** (1981), 350–64.
73. J. Stepišnik, Time-Dependent Self-Diffusion by NMR Spin-Echo. *Physica B* **183** (1993), 343–50.
74. S. D. Traytak and A. G. Kudryavtsev, On the Coagulation Equation for a Passive Admixture in a Stochastic Medium. *Physica A* **260** (1998), 381–90.
75. C. W. Gardiner, *Handbook of Stochastic Methods: For Physics, Chemistry and the Natural Sciences*, 2nd edn. (Berlin: Springer-Verlag, 1996).
76. P. T. Callaghan and J. Stepišnik, Generalized Analysis of Motion Using Magnetic Field Gradients. *Adv. Magn. Opt. Reson.* **19** (1996), 325–88.
77. I. Avramov, Viscosity in Disordered Media. *J. Non-Cryst. Solids* **351** (2005), 3163–73.
78. W. Sutherland, A Dynamical Theory of Diffusion for Nonelectrolytes and the Molecular Mass of Albumin. *Philos. Mag.* **S.6, 9** (1905), 781–85.
79. A. Einstein, *Investigations on the Theory of Brownian Movement*. (New York: Dover, 1956).
80. R. J. Bearman, Statistical Mechanical Theory of the Diffusion Coefficients in Binary Liquid Solutions. *J. Chem. Phys.* **32** (1960), 1308–13.
81. C. Tanford, *Physical Chemistry of Macromolecules*. (New York: Wiley, 1961).
82. J. T. Edward, Molecular Volume and the Stokes–Einstein Equation. *J. Chem. Educ.* **47** (1970), 261–70.
83. P. Kruus, *Liquids and Solutions Structure and Dynamics*. (New York: Marcel Dekker, 1977).
84. J. Renn, Einstein's Invention of Brownian Motion. *Ann. Physik.* **14** (2005), 23–7.
85. J. A. Marqusee and J. M. Deutch, Concentration Dependence of the Self-Diffusion Coefficient. *J. Chem. Phys.* **73** (1980), 5396–7.
86. L. G. MacDowell, B. Garzón, S. Calero, and S. Lago, Dynamical Properties and Transport Coefficients of Kihara Linear Fluids. *J. Chem. Phys.* **106** (1997), 4753–67.
87. C. Branca, S. Magazù, G. Maisano, P. Migliardo, and E. Tettamanti, Anomalous Translational Diffusive Processes in Hydrogen-Bonded Systems Investigated by Ultrasonic Technique, Raman Scattering and NMR. *Physica B* **291** (2000), 180–9.
88. U. Balucani, R. Vallauri, and T. Gaskell, Generalized Stokes–Einstein Relation. *Ber. Bunsenges. Phys. Chem.* **94** (1990), 261–4.
89. D. C. Teller, E. Swanson, and C. De Haën, The Translational Friction Coefficient of Proteins. *Methods Enzymol.* **61** (1979), 104–24.

90. H.-X. Zhou, Calculation of Translational Friction and Intrinsic Viscosity. II. Application to Globular Proteins. *Biophys. J.* **69** (1995), 2298–303.
91. J. García de la Torre, M. L. Huertas, and B. Carrasco, Calculation of Hydrodynamic Properties of Globular Proteins from their Atomic-Level Structure. *Biophys. J.* **78** (2000), 719–30.
92. J. García de la Torre, Hydration from Hydrodynamics. General Considerations and Applications of Bead Modelling to Globular Proteins. *Biophys. Chem.* **93** (2001), 159–70.
93. S. Aragon and D. K. Hahn, Precise Boundary Element Computation of Protein Transport Properties: Diffusion Tensors, Specific Volume, and Hydration. *Biophys. J.* **91** (2006), 1591–603.
94. B. Halle and M. Davidovic, Biomolecular Hydration: From Water Dynamics to Hydrodynamics. *Proc. Natl. Acad. Sci. U.S.A.* **100** (2003), 12135–40.
95. O. Byron, Construction of Hydrodynamic Bead Models from High-Resolution X-Ray Crystallographic or Nuclear Magnetic Resonance Data. *Biophys. J.* **72** (1997), 408–15.
96. S. Allison, C. Chen, and D. Stigter, The Length Dependence of Translational Diffusion, Free Solution Electrophoretic Mobility, and Electrophoretic Tether Force of Rigid Rod-Like Model Duplex DNA. *Biophys. J.* **81** (2001), 2558–68.
97. J. A. Davies and P. C. Griffiths, A Phenomenological Approach to Separating the Effects of Obstruction and Binding for the Diffusion of Small Molecules in Polymer Solutions. *Macromolecules* **36** (2003), 950–2.
98. M. Sharma and S. Yashonath, Breakdown of the Stokes–Einstein Relationship: Role of Interactions in the Size Dependence of Self-Diffusivity. *J. Phys. Chem. B* **110** (2006), 17207–11.
99. V. A. Gierer and K. Wirtz, Molekulare Theorie der Mikroreibung. *Z. Naturforsch.* **8a** (1953), 532–8.
100. H.-C. Chen and S.-H. Chen, Diffusion of Crown Ethers in Alcohols. *J. Phys. Chem.* **88** (1984), 5118–21.
101. U. Balucani, G. Nowotny, and G. Kahl, The Generalized Stokes–Einstein Relation for Liquid Sodium. *J. Phys. Condens. Matter* **9** (1997), 3371–6.
102. R. Lamanna, M. Delmelle, and S. Cannistraro, Solvent Stokes–Einstein Violation in Aqueous Protein Solutions. *Phys. Rev. E* **49** (1994), 5878–80.
103. N. V. Brilliantov and P. L. Krapivsky, Stokes Laws for Ions in Solutions with Ion-Induced Inhomogeneity. *J. Phys. Chem.* **95** (1991), 6055–7.
104. M. Doi and S. F. Edwards, *The Theory of Polymer Dynamics*. (Oxford: Clarendon Press, 1986).
105. S. H. Koenig, Brownian Motion of an Ellipsoid. A Correction to Perrin's Results. *Biopolymers* **14** (1975), 2421–3.
106. J. Happel and H. Brenner, *Low Reynolds Number Hydrodynamics – With Special Applications to Particulate Matter*. (Dordrecht: Kluwer, 1983).
107. R. F. Probstein, *Physicochemical Hydrodynamics*, 2nd edn. (New York: Wiley, 2003).
108. M. M. Tirado and J. García de la Torre, Translational Friction Coefficients of Rigid, Symmetric Top Macromolecules. Application to Circular Cylinders. *J. Chem. Phys.* **71** (1979), 2581–7.
109. M. M. Tirado, C. López Martínez, and J. García de la Torre, Comparison of Theories for the Translational and Rotational Diffusion Coefficients of Rod-Like Macromolecules. Application to Short DNA Fragments. *J. Chem. Phys.* **81** (1984), 2047–52.
110. S. Hansen, Translational Friction Coefficients for Cylinders of Arbitrary Axial Ratios Estimated by Monte Carlo Simulation. *J. Chem. Phys.* **121** (2004), 9111–15.

111. J. García de la Torre and V. A. Bloomfield, Hydrodynamic Properties of Complex, Rigid, Biomolecules: Theory and Applications. *Q. Rev. Biophys.* **14** (1981), 81–139.
112. V. Bloomfield, W. O. Dalton, and K. E. Van Holde, Frictional Coefficients of Multisubunit Structures. I. Theory. *Biopolymers* **5** (1967), 135–48.
113. J. García de la Torre, Hydrodynamic Properties of Macromolecular Assemblies. In *Dynamic Properties of Biomolecular Assemblies*, ed. S. E. Harding and A. J. Rowe. (Cambridge: Royal Society of Chemistry, 1989), pp. 3–31.
114. S. E. Harding, On the Hydrodynamic Analysis of Macromolecular Conformation. *Biophys. Chem.* **55** (1995), 69–93.
115. C. H. Robert, Estimating Friction Coefficients of Mixed Globular/Chain Molecules, such as Protein/DNA Complexes. *Biophys. J.* **69** (1995), 840–8.
116. H.-X. Zhou, Calculation of Translational Friction and Intrinsic Viscosity. I. General Formulation for Arbitrarily Shaped Particles. *Biophys. J.* **69** (1995), 2286–97.
117. S. E. Harding, The Intrinsic Viscosity of Biological Macromolecules. Progress in Measurement, Interpretation and Application to Structure in Dilute Solution. *Prog. Biophys. Mol. Biol.* **68** (1997), 207–62.
118. J. García de la Torre and B. Carrasco, Intrinsic Viscosity and Rotational Diffusion of Bead Models for Rigid Macromolecules and Bioparticles. *Eur. Biophys. J.* **27** (1998), 549–57.
119. B. Carrasco and J. García de la Torre, Hydrodynamic Properties of Rigid Particles: Comparison of Different Modeling and Computational Procedures. *Biophys. J.* **76** (1999), 3044–57.
120. J. García de la Torre, S. E. Harding, and B. Carrasco, Calculation of NMR Relaxation, Covolume, Scattering-Related Properties of Bead Molecules Using the SOLPRO Computer Program. *Eur. Biophys. J.* **28** (1999), 119–32.
121. J. García de la Torre, A. Ortega, H. E. Perez Sanchez, and J. G. Hernandez Cifre, MULTIHYDRO and MONTEHYDRO: Conformational Search and Monte Carlo Calculation of Solution Properties of Rigid or Flexible Bead Models. *Biophys. Chem.* **116** (2005), 121–8.
122. G. Arfken and H. J. Weber, *Mathematical Methods for Physicists*, 4th edn. (New York: Academic Press, 1995).
123. R. Metzler, W. G. Glöcke, and T. F. Nonnenmacher, Fractional Model Equation for Anomalous Diffusion. *Physica A* **211** (1994), 13–24.
124. V. E. Arkhincheev, Anomalous Diffusion and Charge Relaxation on Comb Model: Exact Solutions. *Physica A* **280** (2000), 304–14.
125. R. Metzler, The Restaurant at the End of the Random Walk: Recent Developments in the Description of Anomalous Transport by Fractional Dynamics. *J. Phys. A Math. Gen.* **37** (2004), R161–208.
126. P. A. Egelstaff, *An Introduction to the Liquid State*, 2nd edn. (Oxford: Oxford Science Publications, 1994).
127. J. Kärger and W. Heink, The Propagator Representation of Molecular Transport in Microporous Crystallites. *J. Magn. Reson.* **51** (1983), 1–7.
128. G. F. Roach, *Green's Functions*, 2nd edn. (Cambridge: Cambridge, 1982).
129. M. N. Özisik, *Boundary Value Problems of Heat Conduction*. (New York: Dover, 1989).
130. G. Barton, *Elements of Green's Functions and Propagation Potentials, Diffusion and Waves*. (Oxford: Oxford University Press, 1995).
131. D. G. Duffy, *Green's Functions With Applications*. (Boca Raton: CRC, 2001).
132. P. T. Callaghan, *Principles of Nuclear Magnetic Resonance Microscopy*. (Oxford: Clarendon Press, 1991).

133. N. G. Van Kampen, *Stochastic Processes in Physics and Chemistry*, 3rd edn. (Amsterdam: North Holland, 2001).
134. J. Kärger, H. Pfeifer, and W. Heink, Principles and Applications of Self-Diffusion Measurements by Nuclear Magnetic Resonance. *Adv. Magn. Reson.* **12** (1988), 1–89.
135. R. F. Hoskins, *Delta Functions: An Introduction to Generalised Functions*. (New York: Albion/Horwood Publishing House, 1999).
136. D. G. Duffy, *Advanced Engineering Mathematics With MATLAB*, 2nd edn. (Boca Raton: CRC, 2003).
137. P. Moon and D. E. Spencer, *Field Theory Handbook*, 2nd edn. (Berlin: Springer-Verlag, 1971).
138. I. N. Sneddon, *Mixed Boundary Value Problems in Potential Theory*. (Amsterdam: North-Holland Publishing Company, 1966).
139. K. R. Brownstein and C. E. Tarr, Importance of Classical Diffusion in NMR Studies of Water in Biological Cells. *Phys. Rev. A* **19** (1979), 2446–53.
140. P. T. Callaghan, Pulsed Gradient Spin Echo NMR for Planar, Cylindrical and Spherical Pores under Conditions of Wall Relaxation. *J. Magn. Reson. A* **113** (1995), 53–59.
141. A. V. Barzykin, W. S. Price, K. Hayamizu, and M. Tachiya, Pulsed Field Gradient NMR of Diffusive Transport Through a Spherical Interface into an External Medium Containing a Relaxation Agent. *J. Magn. Reson. A* **114** (1995), 39–46.
142. W. S. Price, A. V. Barzykin, K. Hayamizu, and M. Tachiya, A Model for Diffusive Transport Through a Spherical Interface Probed by Pulsed-Field Gradient NMR. *Biophys. J.* **74** (1998), 2259–71.
143. G. R. Buchanan, *Finite Element Analysis*. (New York: McGraw-Hill, 1995).
144. H. S. Carslaw and J. C. Jaeger, *Conduction of Heat in Solids*, 2nd edn. (Oxford: Oxford University Press, 1959).
145. D. Zwillinger, *Handbook of Differential Equations*, 3rd edn. (Boston: Academic Press, 1998).
146. D. G. Duffy, *Transform Methods for Solving Partial Differential Equations*, 2nd edn. (Boca Raton: CRC Press, 2004).
147. D. J. Bergman and K.-J. Dunn, Self-Diffusion in a Periodic Porous Medium With Interface Absorption. *Phys. Rev. E* **51** (1995), 3401–16.
148. D. J. Bergman, K.-J. Dunn, L. M. Schwartz, and P. P. Mitra, Self-Diffusion in a Periodic Porous Medium: A Comparison of Different Approaches. *Phys. Rev. E* **51** (1995), 3393–400.
149. K.-J. Dunn and D. J. Bergman, Self Diffusion of Nuclear Spins in a Porous Medium with a Periodic Microstructure. *J. Chem. Phys.* **102** (1995), 3041–54.
150. A. I. Zayed, *Handbook of Generalized Function Transformations*. (Boca Raton: CRC, 1996).
151. L. Onsager, Reciprocal Relations in Irreversible Processes. I. *Phys. Rev.* **37** (1931), 405–26.
152. L. Onsager, Reciprocal Relations in Irreversible Processes. II. *Phys. Rev.* **38** (1931), 2265–79.
153. S. R. de Groot and P. Mazur, *Non-Equilibrium Thermodynamics*. (New York: Dover, 1984).
154. S. D. Traytak, A. V. Barzykin, and M. Tachiya, Effect of Anisotropic Diffusion and External Electric Field on the Rate of Diffusion-Controlled Reactions. *J. Chem. Phys.* **120** (2004), 10111–17.
155. P. J. Basser, J. Mattiello, and D. Le Bihan, MR Diffusion Tensor Spectroscopy and Imaging. *Biophys. J.* **66** (1994), 259–67.

156. P. van Gelderen, D. DesPres, P. C. M. Van Zijl, and C. T. W. Moonen, Evaluation of Restricted Diffusion in Cylinders. Phosphocreatine in Rabbit Leg Muscle. *J. Magn. Reson. B* **103** (1994), 255–60.
157. T. M. De Swiet and P. P. Mitra, Possible Systematic Errors in Single-Shot Measurements of the Trace of the Diffusion Tensor. *J. Magn. Reson. B* **111** (1996), 15–22.
158. M. C. Wang and G. E. Uhlenbeck, On the Theory of the Brownian Motion II. *Rev. Mod. Phys.* **17** (1945), 323–42.
159. E. O. Stejskal, Use of Spin Echoes in a Pulsed Magnetic-Field Gradient to Study Anisotropic Restricted Diffusion and Flow. *J. Chem. Phys.* **43** (1965), 3597–603.
160. M. Abramowitz, and I. A. Stegun, *Handbook of Mathematical Functions.* (New York: Dover, 1970).
161. W. S. Veeman, Diffusion in a Closed Sphere. In *Annual Reports on NMR Spectroscopy*, ed. G. A. Webb. vol. 50 (London: Elsevier, 2003), pp. 201–16.
162. M. Kac, Can One Hear the Shape of a Drum? *Am. Math. Mon.* **73** (1966), 1–23.
163. M. D. Hürlimann, L. M. Schwartz, and P. N. Sen, Probability of Return to the Origin at Short Times: A Probe of Microstructure in Porous Media. *Phys. Rev. B* **51** (1995), 14936–40.
164. L. M. Schwartz, M. D. Hürlimann, K.-J. Dunn, P. P. Mitra, and D. J. Bergman, Restricted Diffusion and the Return to the Origin Probability at Intermediate and Long Times. *Phys. Rev. E* **55** (1997), 4225–34.
165. C. H. Neuman, Spin Echo of Spins Diffusing in a Bounded Medium. *J. Chem. Phys.* **60** (1974), 4508–11.
166. R. Lenk, *Fluctuations, Diffusion and Spin Relaxation*. (Amsterdam: Elsevier, 1986).
167. I. S. Gradshteyn and I. M. Ryzhik, *Table of Integrals, Series, and Products*, 7th edn. (New York: Academic Press, 2007).
168. D. E. Woessner, N. M. R. Spin-Echo Self-Diffusion Measurements on Fluids Undergoing Restricted Diffusion. *J. Phys. Chem.* **67** (1963), 1365–67.
169. D. Chandler, *Introduction to Modern Statistical Mechanics*, (Oxford: Oxford University Press, 1987).
170. P. J. Basser, Relationships Between Diffusion Tensor and *q*-Space MRI. *Magn. Reson. Med.* **47** (2002), 392–7.
171. D. Topgaard, Probing Biological Tissue Microstructure with Magnetic Resonance Diffusion Techniques. *Curr. Opin. Colloid Interface Sci.* **11** (2006), 7–12.
172. P. J. Basser, Inferring Microstructural Features and the Physiological State of Tissues from Diffusion-Weighted Images. *NMR Biomed.* **8** (1995), 333–44.
173. L. Minati and W. P. Węglarz, Physical Foundations, Models, and Methods of Diffusion Magnetic Resonance Imaging of the Brain: A Review. *Concepts Magn. Reson.* **30A** (2007), 278–307.
174. P. T. Callaghan and M. E. Komlosh, Locally Anisotropic Motion in a Macroscopically Isotropic System: Displacement Correlations Measured Using Double Pulsed Gradient Spin-Echo NMR. *Magn. Reson. Chem.* **40** (2002), S15–19.
175. W. S. Price, Pulsed Field Gradient NMR as a Tool for Studying Translational Diffusion, Part I. Basic Theory. *Concepts Magn. Reson.* **9** (1997), 299–336.
176. P. N. Sen, Time-Dependent Diffusion Coefficient as a Probe of Geometry. *Concepts Magn. Reson.* **23A** (2004), 1–21.
177. R. Kimmich, Strange Kinetics, Porous Media, and NMR. *Chem. Phys.* **284** (2002), 253–85.
178. I. Ardelean and R. Kimmich, Principles and Unconventional Aspects of NMR Diffusometry. In *Annual Reports on NMR Spectroscopy*, ed. G. A. Webb. vol. 49. (London: Elsevier, 2003), pp. 43–115.

179. G. Vojta and U. Renner, Diffusion in Fractals. In *Diffusion in Condensed Matter*, ed. J. Kärger and P. Heitjans. (Berlin: Springer-Verlag, 2005), pp. 793–811.
180. P. P. Mitra and P. N. Sen, Effects of Surface Relaxation on NMR Pulsed Field Gradient Experiments in Porous Media. *Physica A* **186** (1992), 109–14.
181. P. P. Mitra, P. N. Sen, L. M. Schwartz, and P. Le Doussal, Diffusion Propagator as a Probe of the Structure of Porous Media. *Phys. Rev. Lett.* **68** (1992), 3555–8.
182. P. P. Mitra and P. N. Sen, Effects of Microgeometry and Surface Relaxation on NMR Pulsed-Field-Gradient Experiments: Simple Pore Geometries. *Phys. Rev. B* **45** (1992), 143–56.
183. P. P. Mitra, P. N. Sen, and L. M. Schwartz, Short-Time Behaviour of the Diffusion Coefficient as a Geometrical Probe of Porous Media. *Phys. Rev. B* **47** (1993), 8565–74.
184. L. L. Latour, P. P. Mitra, R. L. Kleinberg, and C. H. Sotak, Time-Dependent Diffusion Coefficient of Fluids in Porous Media as a Probe of Surface-to-Volume Ratio. *J. Magn. Reson. A* **101** (1993), 342–6.
185. P. N. Sen, Time-Dependent Diffusion Coefficient as a Probe of Permeability of the Pore Wall. *J. Chem. Phys.* **119** (2003), 9871–6.
186. D. Candela and P.-Z. Wong, Using NMR to Measure Fractal Dimensions. *Phys. Rev. Lett.* **90** (2003), 039601-1.
187. J. R. Banavar and L. M. Schwartz, Probing Porous Media with Nuclear Magnetic Resonance. In *Molecular Dynamics in Restricted Geometries*, ed. J. Klafter and J. M. Drake. (New York: Wiley, 1989), pp. 273–309.
188. K. S. W. Sing, D. H. Everett, R. A. W. Haul, L. Moscou, R. A. Pierotti, J. Rocquérol, and T. Siemieniewska, Reporting Physisorption Data for Gas/Solid Systems with Special Reference to the Determination of Surface Area and Porosity (Recommendations 1984) Commission on Colloid and Surface Chemistry Including Catalysis. *Pure Appl. Chem.* **57** (1985), 603–19.
189. S. B. Zimmerman and A. P. Minton, Macromolecular Crowding: Biochemical, Biophysical, and Physiological Consequences. *Annu. Rev. Biophys. Biomol. Struct.* **22** (1993), 27–65.
190. P. Bernadó, J. García de la Torre, and M. Pons, Macromolecular Crowding in Biological Systems: Hydrodynamics and NMR Methods. *J. Mol. Recognit.* **17** (2004), 397–407.
191. J. Qian and P. N. Sen, Time Dependent Diffusion in a Disordered Medium with Partially Absorbing Walls: A Perturbative Approach. *J. Chem. Phys.* **125** (2006), 194508-1–194508-6.
192. L. J. Zielinski and P. N. Sen, Effects of Finite Width Pulses in the Pulsed Field Gradient Measurement of the Diffusion Coefficient in Connected Porous Media. *J. Magn. Reson.* **165** (2003), 153–61.
193. J. Bear, *Dynamics of Fluids in Porous Media*. (New York: Dover, 1988).
194. F. A. L. Dullien, *Porous Media: Fluid Transport and Pore Structure*, 2nd edn. (New York: Academic Press, 1992).
195. L. L. Latour, R. L. Kleinberg, P. P. Mitra, and C. H. Sotak, Pore-Size Distributions and Tortuosity in Heterogeneous Porous Media. *J. Magn. Reson. A* **112** (1995), 83–91.
196. S. R. Heil and M. Holz, Electrical Transport in a Disordered Medium: NMR Measurement of Diffusivity and Electrical Mobility of Ionic Charge Carriers. *J. Magn. Reson.* **135** (1998), 17–22.
197. U. Hizi and D. J. Bergman, Molecular Diffusion in Periodic Porous Media. *J. Appl. Phys.* **87** (2000), 1704–11.
198. J. Hrabe, S. Hrabtová, and K. Segeth, A Model of Effective Diffusion and Tortuosity in the Extracellular Space of the Brain. *Biophys. J.* **87** (2004), 1606–17.

199. B. Jönsson, H. Wennerström, P. G. Nilsson, and P. Linse, Self-Diffusion of Small Molecules in Colloidal Systems. *Colloid Polym. Sci.* **264** (1986), 77–88.
200. H. Jóhannesson and B. Halle, Solvent Diffusion in Ordered Macrofluids: A Stochastic Simulation Study of the Obstruction Effect. *J. Chem. Phys.* **104** (1996), 6807–17.
201. P. P. Mitra, L. L. Latour, R. L. Kleinberg, and C. H. Sotak, Pulsed-Field-Gradient NMR Measurements of Restricted Diffusion and the Return-to-the Origin Probability. *J. Magn. Reson. A* **114** (1995), 47–58.
202. P. P. Mitra, Diffusion in Porous Materials as Probed by Pulsed Gradient NMR Measurements. *Physica A* **241** (1997), 122–7.
203. M. D. Hürlimann, K. G. Helmer, L. L. Latour, and C. H. Sotak, Restricted Diffusion in Sedimentary Rocks. Determination of Surface-Area-to-Volume Ratio and Surface Relaxivity. *J. Magn. Reson. A* **111** (1994), 169–78.
204. M. F. Schlesinger, G. M. Zaslavsky, and J. Klafter, Strange Kinetics. *Nature* **363** (1993), 31–7.
205. A. R. Bizzari and S. Cannistraro, Molecular Dynamics Simulation Evidence of Anomalous Diffusion of Protein Hydration Water. *Phys. Rev. E* **53** (1996), R3040–3.
206. A. Klemm, R. Metzler, and R. Kimmich, Diffusion on Random-Site Percolation Clusters: Theory and NMR Microscopy Experiments with Model objects. *Phys. Rev. E* **65** (2002), 021112-1–021112-11.
207. E. Özarslan, P. J. Basser, T. M. Shepherd, P. E. Thelwall, B. C. Vemuri, and S. J. Blackband, Observation of Anomalous Diffusion in Excised Tissue by Characterizing the Diffusion-time Dependence of the MR Signal. *J. Magn. Reson.* **183** (2006), 315–23.
208. W. Paul, Anomalous Diffusion in Polymer Melts. *Chem. Phys.* **284** (2003), 59–66.
209. O. V. Bychuk and B. O'Shaugnessy, Anomalous Surface Diffusion: A Numerical Study. *J. Chem. Phys.* **101** (1994), 772–80.
210. S. Schnell and T. E. Turner, Reaction Kinetics in Intracellular Environments with Macromolecular Crowding: Simulations and Rate Laws. *Prog. Biophys. Molec. Biol.* **85** (2004), 235–60.
211. R. Kimmich, *NMR: Tomography, Diffusometry, Relaxometry.* (Berlin: Springer-Verlag, 1997).
212. J. R. Banavar, M. Lipsicas, and J. F. Willemsen, Determination of the Random-Walk Dimension of Fractals by Means of NMR. *Phys. Rev. B* **32** (1985), 6066.
213. H. G. E. Hentschel and I. Procaccia, Relative Diffusion in Turbulent Media: The Fractal Dimension of Clouds. *Phys. Rev. A* **29** (1984), 1461–70.
214. B. O'Shaugnessy and I. Procaccia, Analytical Solutions for Diffusion on Fractal Objects. *Phys. Rev. Lett.* **54** (1985), 455–8.
215. R. A. Damion and K. J. Packer, Predictions for Pulsed-Field-Gradient NMR Experiments of Diffusion in Fractal Spaces. *Proc. R. Soc. London A* **453** (1997), 205–11.
216. J. E. Tanner, Transient Diffusion in a System Partitioned by Permeable Barriers. Application to NMR Measurements with a Pulsed Field Gradient. *J. Chem. Phys.* **69** (1978), 1748–54.
217. F. Crick, Diffusion in Embryogenesis. *Nature* **225** (1970), 420–2.
218. R. L. Cooper, D. B. Chang, A. C. Young, C. J. Martin, and B. Ancker-Johnson, Restricted Diffusion in Biophysical Systems: Experiment. *Biophys. J.* **14** (1974), 161–77.
219. O. K. Dudko, A. M. Berezhkovskii, and G. H. Weiss, Diffusion in the Presence of Periodically Spaced Permeable Membranes. *J. Chem. Phys.* **121** (2004), 11283–8.
220. P. T. Callaghan, A. Coy, T. P. J. Halpin, D. MacGowan, K. J. Packer, and F. O. Zelaya, Diffusion in Porous Systems and the Influence of Pore Morphology in Pulsed Field

Gradient Spin-Echo Nuclear Magnetic Resonance Studies. *J. Chem. Phys.* **97** (1992), 651–62.

221. P. T. Callaghan and A. Coy, PGSE NMR and Molecular Translational Motion in Porous Media. In *NMR Probes and Molecular Dynamics*, ed. R. Tycko. (Dordrecht: Kluwer, 1994), pp. 489–523.
222. P. T. Callaghan, D. MacGowan, K. J. Packer, and F. O. Zelaya, High-Resolution *q*-Space Imaging in Porous Structures. *J. Magn. Reson.* **90** (1990), 177–82.
223. K. J. Packer, S. Stapf, J. J. Tessier, and R. A. Damion, The Characterisation of Fluid Transport in Porous Solids by Means of Pulsed Magnetic Field Gradient NMR. *Magn. Reson. Imaging* **16** (1998), 463–9.
224. S. Stapf, K. J. Packer, R. G. Graham, J.-F. Thovert, and P. M. Adler, Spatial Correlations and Dispersion for Fluid Transport Through Packed Glass Beads Studied by Pulsed Field-Gradient NMR. *Phys. Rev. E* **58** (1998), 6206–21.
225. R. G. Graham, W. M. Holmes, C. De Panfilis, and K. J. Packer, Characterisation of Locally Anisotropic Structures Within Isotropic Porous Solids Using 2-D Pulsed Field Gradient NMR. *Chem. Phys. Lett.* **332** (2000), 319–23.
226. J. W. Rayleigh, On the Influence of Obstacles Arranged in Rectangular Order Upon the Properties of a Medium. *Philos. Mag.* **34** (1892), 481–502.
227. H. N. W. Lekkerkerker and J. K. G. Dhont, On the Calculation of the Self-Diffusion Coefficient of Interacting Brownian Particles. *J. Chem. Phys.* **80** (1984), 5790–92.
228. N. Muramatsu and A. P. Minton, Tracer Diffusion of Globular Proteins in Concentrated Protein Solutions. *Proc. Natl. Acad. Sci. U.S.A.* **85** (1988), 2984–8.
229. M. H. Blees and J. C. Leyte, The Effective Translational Self-Diffusion Coefficient of Small Molecules in Colloidal Crystals of Spherical Particles. *J. Colloid Interface Sci.* **166** (1994), 118–27.
230. J. Han and J. Herzfeld, Macromolecular Diffusion in Crowded Solutions. *Biophys. J.* **65** (1993), 1155–61.
231. O. Z. Bezrukov, A. E. Lukyanov, V. K. Pozdyshev, and A. V. Struts, Analytical Calculations of Self-Diffusion Coefficients in a Medium with Obstacles. *Physica Scr.* **64** (2001), 382–5.
232. H. Fujita, Diffusion in Polymer-Diluent Systems. *Adv. Polym. Sci.* **3** (1961), 1–47.
233. L. Masaro and X. X. Zhu, Physical Models of Diffusion for Polymer Solutions, Gels and Solids. *Progress in Polymer Science* **24** (1999), 731–75.
234. M. Tokuyama and I. Oppenheim, Dynamics of Hard-Sphere Suspensions. *Phys. Rev. E* **50** (1994), R16–19.
235. G. M. Bell, Self-Diffusion of Ions in the Electric Fields of Spherical Particles. *Trans. Faraday Soc.* **60** (1964), 1752–9.
236. F. J. M. Schipper and J. C. Leyte, Mass Transport in Polyelectrolyte Solutions. *J. Phys. Condens. Matter* **11** (1999), 1409–21.
237. M. I. M. Darwish, J. R. C. van der Maarel, and P. L. J. Zitha, Ionic Transport in Polyelectrolyte Gels: Model and NMR Investigations. *Macromolecules* **37** (2005), 2307–12.
238. B. Amsden, An Obstruction-Scaling Model for Diffusion in Homogeneous Hydrogels. *Macromolecules* **32** (1999), 874–9.
239. J. H. Wang, Theory of the Self-Diffusion of Water in Protein Solutions. A New Method for Studying the Hydration and Shape of Protein Molecules. *J. Am. Chem. Soc.* **76** (1954), 4755–63.
240. M. E. Clark, E. E. Burnell, N. R. Chapman, and J. A. M. Hinke, Water in Barnacle Muscle. *Biophys. J.* **39** (1982), 289–99.

241. I. Teraoka, *Polymer Solutions: An Introduction to Physical Properties*. (New York: Wiley, 2002).
242. P. E. Rouse, Jr., A Theory of the Linear Viscoelastic Properties of Dilute Solutions of Coiling Polymers. *J. Chem. Phys.* **21** (1953), 1272–80.
243. P. G. De Gennes, Reptation of a Polymer Chain in the Presence of Fixed Obstacles. *J. Chem. Phys.* **55** (1971), 572–9.
244. E. M. Terentjev, P. T. Callaghan, and M. Warner, Pulsed Gradient Spin–Echo Nuclear Magnetic Resonance of Confined Brownian Particles. *J. Chem. Phys.* **102** (1995), 4619–24.
245. J. D. Seymour and P. T. Callaghan, Generalized Approach to NMR Analysis of Flow and Dispersion in Porous Media. *AIChE J.* **43** (1997), 2096–111.
246. P. T. Callaghan and A. A. Khrapitchev, Time-Dependent Velocities in Porous Media Dispersive Flow. *Magn. Reson. Imaging* **19** (2001), 301–5.
247. C. Van den Broeck, Taylor Dispersion Revisited. *Physica A* **168** (1990), 677–96.
248. B. Crist, Polymer Self-Diffusion Measurements by Small-Angle Neutron Scattering. *J. Non-Cryst. Solids* **131–133** (1991), 709–14.
249. X.-C. Tang, X.-W. Song, P.-Z. Shen, and D. Z. Jia, Capacity Intermittent Titration Technique (CITT): A Novel Technique for Determination of Li^+ Solid Diffusion Coefficient of $LiMn_2O_4$. *Electrochim. Acta.* **50** (2007), 5581–7.
250. A. J. W. G. Visser and M. A. Hink, New Perspectives of Fluorescence Correlation Spectroscopy. *J. Fluoresc.* **9** (1999), 81–7.
251. K. Braeckmans, L. Peeters, N. N. Sanders, S. C. De Smedt, and J. DeMeester, Three-Dimensional Fluorescence Recovery after Photobleaching with the Confocal Scanning Laser Microscope. *Biophys. J.* **85** (2003), 2240–52.
252. S. B. Lyon and L. T. E. Phillipe, Direct Measurements of Ionic Diffusion in Protective Organic Coatings. *Trans. Inst. Met. Finish.* **84** (2006), 23–7.
253. A. P. Minton, Analytical Centrifugation with Preparative Ultracentrifuges. *Anal. Biochem.* **176** (1989), 209–16.
254. E. W. Lang and H.-D. Lüdemann, Density Dependence of Rotational and Translational Molecular Dynamics in Liquids Studies by High Pressure NMR. *Prog. NMR Spectrosc.* **25** (1993), 507–633.
255. J. McConnell, *Nuclear Magnetic Relaxation in Liquids*. (Cambridge: Cambridge University Press, 1987).
256. P. Debye, *Polar Molecules*. (New York: Dover Publications, 1945).
257. N. Bloembergen, E. M. Purcell, and R. V. Pound, Relaxation Effects in Nuclear Magnetic Resonance Absorption. *Phys. Rev.* **73** (1948), 679–712.
258. J. W. Hennel and J. Klinowski, *Fundamentals of Nuclear Magnetic Resonance*. (Essex: Longman Scientific & Technical, 1993).
259. W. S. Price, P. W. Kuchel, and B. A. Cornell, Microviscosity of Human Erythrocytes Studied with Hypophosphite and ^{31}P- NMR. *Biophys. Chem.* **33** (1989), 205–15.
260. W. S. Price, B.-C. Perng, C.-L. Tsai, and L.-P. Hwang, Microviscosity of Human Erythrocytes Studied using Hypophosphite Two-Spin Order Relaxation. *Biophys. J.* **61** (1992), 621–30.
261. D. E. Woessner, Brownian Motion and Correlation Times. In *Encyclopedia of Nuclear Magnetic Resonance*, ed. D. M. Grant and R. K. Harris. vol. 2. (New York: Wiley, 1996), pp. 1068–84.
262. D. E. Woessner, Relaxation Effects of Chemical Exchange. In *Encyclopedia of Nuclear Magnetic Resonance*, ed. D. M. Grant and R. K. Harris. vol. 6. (New York: Wiley, 1996), pp. 4018–28.

2

Theory of NMR diffusion and flow measurements

2.1 Introduction

As soon as the spin-echo was discovered by Hahn in 1950 it was realised that it could form the basis of self-diffusion measurements.[1] Indeed, certainly within the next decade the concept of spin-echo-based diffusion measurements using static magnetic gradients (i.e., Steady Gradient Spin-Echo or SGSE NMR) had become widespread and used in quite sophisticated measurements such as on water[2] and ^{3}He.[3] Many of the experimental limitations of static gradient measurements were removed with the suggestion in 1963 by McCall, Douglass and Anderson[4] and experimental introduction in 1965 by Stejskal and Tanner[5] of applying the magnetic gradients as pulses in the spin-echo sequence (i.e., Pulsed Gradient Spin-Echo NMR or PGSE NMR). Carr and Purcell[6] were the first to discuss NMR flow measurements and in 1960 NMR flow measurements were considered for the purpose of measuring sea-water motion.[7]

Virtually all contemporary NMR diffusion (and flow experiments) are based on some form of spin-echo.[1,6,8–10] Indeed, for all but the simplest cases the dependence of the observed echo amplitudes on diffusion rapidly becomes very complicated and this can be exacerbated in pulse sequences where the magnetisation is kept in a steady state.[9,11,12] However, in the following discussions we will assume, unless otherwise noted, that all pulse sequences start with the spin system being in thermal equilibrium (i.e., M_0). As the diffusing species necessarily contains a nuclear spin, the terms spin and particle will henceforth become synonymous. This chapter explains why spin-echoes in NMR are sensitive to translational motion and how, with the imposition of appropriately placed magnetic field gradient pulses, they form the basis of NMR diffusion and flow measurements.

The key theory relating spin-echo attenuation to diffusion and the experimental parameters (e.g., magnetic gradient pulses) is encapsulated in the Bloch–Torrey equations – however, full solutions starting from these equations become mathematically intractable for anything but the simplest systems. Thus, various approximations are used of which the two most commonly encountered are the Short Gradient Pulse (SGP) and the Gaussian Phase Distribution (GPD) approximations. The solutions are derived for the case of free isotropic diffusion using each of these three methods. All three approaches involve solving the diffusion equation subject to the correct boundary conditions. However, some of the earliest studies evaluated the effects of diffusion on echo sequences by the conceptually appealing but cumbersome approach of approximating diffusion as a random walk of discrete steps.[6,9,13] The relationship between the experimental variables and the workings of the PGSE NMR experiment is most easily and conceptually expressed using the SGP approximation – consequently, this is the first method for relating the spin-echo attenuation to the experimental parameters that we examine.

Consideration is given to the most commonly encountered sequences used in PGSE NMR, namely the Hahn spin-echo sequence (SE) and the stimulated echo sequence (STE). Coverage is also given of Carr–Purcell–Meiboom–Gill type sequences (CPMG), and the related oscillating gradient experiments (OGSE). OGSE sequences, although in essence a subset of the family of spin-echo experiments, probe the shorter timescale of the diffusion spectrum and consequently can be analysed by a different means; thus their form and analysis is treated separately.

Diffusion measurements of multicomponent systems, although no more difficult to perform than single component systems, present special difficulties in their analysis. This is covered in considerable detail. The last section of this chapter gives a brief overview of the distant dipolar field and multiple echoes.

Finally, we note that although measurements can be performed with either B_0 (magnetic) or B_1 (rf) gradients, B_0 gradients are more commonly used and thus they form the focus of this chapter. Generally, the theory for B_0 and B_1 gradients is analogous; some salient points of B_1 gradients and B_1 gradient methods are covered in Chapter 10.

2.2 Nuclear spins, magnetic gradients and motion

2.2.1 NMR coherence orders and signal detection

The analysis of NMR diffusion pulse sequences involves following nuclear spin coherence pathways through the sequence using density matrix or product operator formalisms,[14–23] and accounting for the effects of diffusion and magnetic gradients on the bulk magnetisation. Starting from thermal equilibrium, the (initial) longitudinal magnetisation has coherence order $p = 0$ with the coherence order signifying the multiple of the Larmor frequency describing the free precession of the respective density matrix element. Only rf pulses can change p and a $\pi/2$ rf pulse will transform longitudinal into transverse magnetisation, which corresponds to $p=+1$ and -1 in equal amounts. The evolution of these two transverse coherence orders is followed through their respective coherence pathways as specified in the pulse sequence and ultimately only those signals corresponding to $p=-1$ give rise to a detectable signal. Importantly, π pulses can change the sign of p. Some examples of pertinent coherence pathways are given in the depictions of the pulse sequences below (coherence pathways not contributing to the detected signal are omitted). In a system of coupled spins or in which the spin quantum number is greater than ½, although the initial $\pi/2$ pulse only generates coherences with $p=\pm 1$, subsequent $\pi/2$ pulses can generate multiple quantum coherences (i.e., $|p|>1$).

The NMR signal detected at the end of a pulse sequence is the projection of the precessing spin magnetisation (from the $p=-1$ coherences) onto the x-y plane. Quadrature detection[24,25] detects a complex signal (i.e., the detected signal has a real and a complex component) which importantly allows the sense of the nuclear spin precession to be determined. Recalling Euler's formula

$$e^{i\phi} = \cos\phi + i\sin\phi, \tag{2.1}$$

it can be seen that the amplitude of the signal from a spin will be proportional to the real component, i.e., cos ϕ, where ϕ is the phase angle of the spin in the transverse plane. In an NMR experiment there is an ensemble of spins and at time t after the initial excitation of the spins, the signal acquired, $S(t)$, will be the volume integral of the product of the bulk magnetisation at a point $\mathbf{r}$, $M(\mathbf{r})$, times the instantaneous phase of the magnetisation $\phi(\mathbf{r},t)$

$$S(t) = \int M(\mathbf{r})e^{i\phi(\mathbf{r},t)}dV. \tag{2.2}$$

$M(\mathbf{r})$ is proportional to $B_0^{7/4}$,[13,26] but there must be sufficient phase coherence of the magnetisation to detect a signal. And similarly, the volume of integration must be large enough to obtain a measurable signal – this can become a particular problem in MRI-based experiments when seeking spatially localised information.

In a structurally homogeneous system, the attenuation of the NMR signal due to the distribution of ϕ is given by[6]

$$E(t) = \int_{-\infty}^{\infty} P_\phi(\phi, t)e^{i\phi}d\phi, \tag{2.3}$$

where $P_\phi(\phi, t)$ is the (relative) phase-distribution function; $P_\phi(\phi, t)$ is normalised and thus,

$$\int_{-\infty}^{\infty} P_\phi(\phi, t)d\phi = 1. \tag{2.4}$$

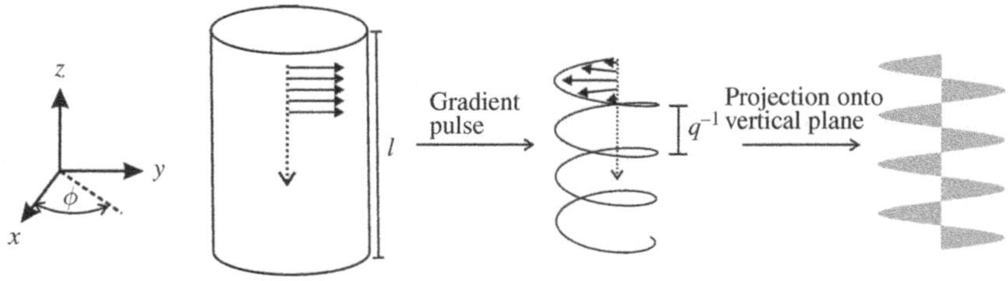

Figure 2.1 Initially coherent transverse magnetisation (spin isochromats, represented as the five arrows all pointing in the same direction = maximum signal) in a cylindrical sample of length l is transformed ('spatially encoded') into a helix (i.e., a magnetisation grating)[30,32] with pitch q^{-1} (m) by the application of gradient pulse of dephasing strength q (see Eq. (2.14)). The solid line marking the helix is a guide for the eye. The projection of the helix on to a plane containing the z axis is depicted on the right. The acquired signal corresponds to the vector sum of the magnetisation and no signal will be obtained when $q^{-1} = l/m$ where m is an integer.

Thus, when all of the spins are coherently oriented along $\phi = 0$ (see Figure 2.1), which corresponds to $P_\phi(\phi, t) = \delta(\phi)$ where δ is the delta function, there will be zero attenuation (i.e., maximum signal), namely

$$E(t) = \int_{-\infty}^{\infty} \delta(\phi) e^{i\phi} d\phi = 1. \tag{2.5}$$

As will be seen below, a likely situation in NMR diffusion measurements is where the phase distribution is Gaussian, namely

$$P_\phi(\phi, t) = \left(2\pi\langle\phi^2\rangle\right)^{-1/2} \exp\left(\frac{-\phi^2}{2\langle\phi^2\rangle}\right), \tag{2.6}$$

where $\langle\phi^2\rangle$ is the mean-square phase distribution. Using the Gaussian distribution, and noting the standard integral (3.323 2 in ref. 27)

$$\int_{-\infty}^{\infty} e^{-a^2x^2 \pm bx} dx = e^{\frac{b^2}{4a^2}} \frac{\sqrt{\pi}}{a} \quad [\mathrm{Re}\, a^2 > 0], \tag{2.7}$$

the signal attenuation (Eq. (2.3)) becomes

$$E(t) = \exp\left(-\langle\phi^2\rangle/2\right). \tag{2.8}$$

2.2.2 Magnetic gradients, nuclear spins and translational motion

In the following it is assumed that $\mathbf{B}_0$ is oriented in the z-direction as it is in most (non-whole body) superconducting magnets. A spatially homogeneous magnetic field will, according to the Larmor equation, result in all spins of the same type having the same Larmor (or resonance) frequency, ω_0, irrespective of their position, $\mathbf{r}$ (m), throughout the sample

$$\omega_0 = p\gamma B_0, \tag{2.9}$$

where γ is the gyromagnetic ratio. Formally, Eq. (2.9) should be negative to give the correct sense of precession about $\mathbf{B}_0$ but for simplicity it is neglected as it is unimportant so long as consistency is maintained in subsequent calculations.

In almost all that follows we will be concerned with the imposition of a spatially homogeneous (or constant) gradient, $\mathbf{g}$ ($\mathrm{T\,m^{-1}}$) throughout the sample. The gradient strength $g(=|\mathbf{g}|)$ is often expressed in $\mathrm{G\,cm^{-1}}$ ($1\,\mathrm{G\,cm^{-1}} = 0.01\,\mathrm{T\,m^{-1}}$). It will generally be the case that $|\mathbf{g}\cdot\mathbf{r}| \ll |\mathbf{B}_0|$, and thus the magnetic field components perpendicular to $\mathbf{B}_0$ can be neglected (i.e., $\mathbf{B} = (0, 0, B_z)$). Thus, $\mathbf{g}$ is defined by the **grad** of the magnetic field component parallel to $\mathbf{B}_0$

$$\mathbf{g} = \nabla B_z = \frac{\partial B_z}{\partial x}\mathbf{i} + \frac{\partial B_z}{\partial y}\mathbf{j} + \frac{\partial B_z}{\partial z}\mathbf{k}, \tag{2.10}$$

where $\mathbf{i}$, $\mathbf{j}$ and $\mathbf{k}$ are unit vectors of the laboratory frame of reference. Hence in the presence of such a gradient, the magnitude of the magnetic field at any point is

$$B(\mathbf{r}) = B_0 + \mathbf{g}\cdot\mathbf{r}. \tag{2.11}$$

Thus, a constant gradient leads to a linear spatial change in the magnetic field in the direction of the gradient resulting from the scalar (or dot) product ($\mathbf{g}\cdot\mathbf{r}$). In the literature $\mathbf{g}$ is often incorrectly referred to as ‘linear’, but a linear gradient would lead to a curved (i.e., quadratically varying with position) magnetic field.

By modifying the Larmor equation to include the presence of $\mathbf{g}$, ω becomes a function of the position $\mathbf{r}$,

$$\omega(\mathbf{r}) = p(\omega_0 + \gamma(\mathbf{g}\cdot\mathbf{r})). \tag{2.12}$$

Thus, if a homogeneous gradient of known magnitude is imposed through the sample, the Larmor frequency becomes a spatial label with respect to the direction of the gradient. This is the foundation of all gradient-based methods and it also shows that whilst zero quantum transitions are unaffected by the presence of the gradient, successively higher quantum transitions are more sensitive to the effects of the gradient. For heteronuclear multiple quantum transitions, Eq. (2.12) must be modified to account for the coherent spins (see Section 8.3).

For a spin located at $\mathbf{r}$ after the application of a ‘rectangular’ magnetic gradient pulse of duration δ, the phase ϕ will be given by

$$\phi(\mathbf{r}) = p\gamma\delta\mathbf{g}\cdot\mathbf{r}. \tag{2.13}$$

$\gamma\delta g$ can be looked on as the 'area' or the 'dephasing strength' of the gradient pulse with γ acting as a proportionality constant for the nucleus in question. Due to the dot product only the component of the motion parallel to the direction of the gradient will cause a change in the phase of the spin. The inclusion of the coherence order, p, also leads to the concept of an 'effective gradient', $g^{\text{eff}} = pg$, since the phase change resulting from a gradient pulse acting on a coherence of order $p = -1$ is the opposite of that induced by the same gradient pulse on a coherence of order $p = 1$. Consequently, it is important to account for the effects of π rf pulses in pulse sequences when determining the effects of gradient pulses.

The notation

$$\mathbf{q} = \frac{1}{2\pi}\gamma\delta\mathbf{g}\ (\text{m}^{-1}) \tag{2.14}$$

is often used to characterise gradient pulses involved in measuring a dynamic property and leads to the concept of q-space. $\mathbf{q}$ and $\mathbf{g}$ are often used synonymously. We note that the definition of $\mathbf{q}$ is identical to the definition of $\mathbf{k}$ (and k-space) as used in conventional (i.e., MRI) imaging literature.[28] However, in contrast to $\mathbf{q}$ and q-space, $\mathbf{k}$ and k-space relate to the gradient pulse being used to provide information on spatial location. Thus, as will be shown below, q-space involves the spatial spectrum of nuclear spin displacements and k-space involves the spatial spectrum of nuclear spin positions.

So far we have only thought of gradient pulses in terms of rectangular pulses; however, for technical and other reasons other gradient shapes ranging from trapezoidal to oscillating sinusoidal gradients are sometimes used. Generalising Eq. (2.14) to include the possibility of time-dependent gradients and coherence orders gives

$$\begin{aligned}\mathbf{q}(t) &= \frac{1}{2\pi}\gamma\int_0^t p(t')\mathbf{g}(t')dt' \\ &= \frac{1}{2\pi}\gamma\int_0^t \mathbf{g}^{\text{eff}}(t')dt'.\end{aligned} \tag{2.15}$$

Although a magnetic gradient can in theory be in any direction, unless otherwise noted, it will be assumed that any applied gradient is parallel with $\mathbf{B}_0$ (i.e., $\mathbf{g} = g_z\mathbf{k}$).

Although often written as such, a gradient pulse does not 'randomise' magnetisation but instead precisely spatially encodes it. For example, as depicted in Figure 2.1 a gradient pulse applied along the long axis of a cylindrical sample of length l immediately after a $\pi/2$ excitation pulse twists the magnetisation into a helix[29] (i.e., circularly polarised transverse magnetisation) with a pitch of

$$\Lambda_q = \frac{2\pi}{\gamma\delta g} = q^{-1}\,(\mathrm{m}). \tag{2.16}$$

During an NMR pulse sequence consisting of rf pulses and magnetic field gradient pulses the magnetisation vector field has a complicated temporal and spatial dependence and is sometimes referred to as a magnetisation grating or multiple magnetisation modulation due to its periodic structure.[10,30–34] For example, the helix in Figure 2.1 when projected onto a vertical plane has the form $A\cos(qz+\phi_0)$ where A is a constant and ϕ_0 is the phase offset. Indeed, analogies exist between gradient-based NMR diffusion measurements and the diffraction of light from concentration gratings[35] and transient refractive index gratings in optical holographic relaxation experiments.[30,36]

If signal acquisition was to be performed for a cylindrical sample with the gradient along the long axis, which corresponds to integrating the phase of the spins over the sample volume, starting from Eq. (2.13) we find that the signal attenuation due to the phase twist is given by

$$E_{\mathrm{Phase}}(g) = \frac{\int_{z_0}^{z_1} e^{i\gamma\delta g z}\,dz}{z_1 - z_0} = \frac{\int_{-l/2}^{l/2} e^{i\gamma\delta g z}\,dz}{l} = \mathrm{sinc}\left(\frac{\gamma g \delta l}{2}\right), \tag{2.17}$$

where $\mathrm{sinc}(x) = \sin(x)/x$. Thus, when the spread in phase across l is 360° the vector sum of the magnetisation is zero and thus no signal is observed when

$$g = \frac{2m\pi}{\gamma\delta l}, \tag{2.18}$$

or, equivalently, using Eq. (2.14)

$$ql = m, \tag{2.19}$$

where m is an integer. Thus the signal attenuation due to a gradient pulse depends on the gradient amplitude and direction.

The concept of a magnetisation helix formation is central to the use of gradients for coherence selection, phase cycling and solvent suppression,[37–39] but perhaps most importantly, the transformation from coherent transverse magnetisation to a spatially well-defined helix allows the detection of translational motion. The net component of any translational motion along the direction of the helix (i.e., the same direction as the gradient) will alter the helix, whereas motion orthogonal to the direction of the helix will leave the helix unchanged. Specifically, we will see that diffusive motion will reduce the helix diameter while flow results in a net movement of the helix. The sensitivity of these geometrical changes of the helix by motion will be increased with more tightly wound helices (i.e., higher q). Indeed, due to the finite length of a gradient pulse (δ), the transition from coherent transverse magnetisation to magnetisation helix does not happen instantaneously and the maximum q and maximum sensitivity to diffusive motion occur at the end of the pulse. Motional restriction (diffusion inside a restricting geometry such as a biological cell) will impart a characteristic distortion signature onto the helix. In essence this book is about how to generate such a magnetisation helix, how translational motion affects the helix and how such a geometrically altered or distorted helix can be interrogated to reveal information on the translational motion. Singer has presented an intuitive description of diffusion and flow measurements termed 'spin phase graphing'.[40]

Helix formation is affected if motion occurs during the gradient pulse (e.g., diffusion would result in an attenuation of the helix diameter but the helix will have the same pitch as if no diffusion had occurred); consequently, the spin dynamics are more conceptually easy to visualise in the short gradient pulse (SGP) limit (rigorously, a gradient pulse has the form of a delta function, that is $\delta \rightarrow 0$ and $|\mathbf{g}| \rightarrow \infty$ while their product remains finite). Further, the SGP limit also means that any motional corruption during the gradient pulses can be ignored.

More generally the phase shift at time t of a nuclear spin following a path $\mathbf{r}(t)$ in a gradient $\mathbf{g}(t)$ is given by

$$\begin{aligned}\phi(t) &= \gamma \int_0^t p(t')\mathbf{g}(t') \cdot \mathbf{r}(t') dt' \\ &= \gamma \int_0^t \mathbf{g}^{\text{eff}}(t') \cdot \mathbf{r}(t') dt'. \end{aligned} \tag{2.20}$$

Only the component of the spin's motion in the direction of the gradient, $z(t)$, is relevant and this can be expanded in a Taylor series[41]

$$z(t) = z_0 + \left(\frac{\partial z}{\partial t}\right)_{t=0} t + \frac{1}{2}\left(\frac{\partial^2 z}{\partial t^2}\right)_{t=0} t^2 + \frac{1}{6}\left(\frac{\partial^3 z}{\partial t^3}\right)_{t=0} t^3 \cdots \tag{2.21}$$

The terms on the right-hand side of Eq. (2.21) correspond to the position z_0, velocity $v_0 = \left(\frac{\partial z}{\partial t}\right)_{t=0}$, acceleration $a_0 = \left(\frac{\partial^2 z}{\partial t^2}\right)_{t=0}$, jerk $j_0 = \left(\frac{\partial^3 z}{\partial t^3}\right)_{t=0}$ and higher terms and thus Eq. (2.20) can be rewritten as

$$\phi(t) = \gamma z_0 M_0 + \gamma v_0 M_1 + \frac{1}{2}\gamma a_0 M_2 + \frac{1}{6}\gamma j_0 M_3 + \cdots \tag{2.22}$$

where

$$M_n = \int_0^t p(t') g_z(t')(t')^n dt'. \tag{2.23}$$

M_n is termed the nth moment of $g_z(t)$ with respect to t. An important point that will become evident in measurements of flow, or conversely when particular translational motions are to be excluded from measurements, is that the moments can be individually manipulated by appropriate design in NMR pulse sequences. Indeed Eq. (2.23) reveals that a sequence that has an even number of positive and negative (or effectively positive and negative due to change in coherence order) gradient pulses will cause the zeroth moment to become zero.

In the case of uniform flow alone, which would correspond to all moments apart from M_1 being zero in Eq. (2.22), the phase relation is given by

$$\phi(t) = \gamma v_0 M_1 \tag{2.24}$$

or

$$v_0 = \frac{\phi(t)}{\gamma M_1} \tag{2.25}$$

indicating that the velocity is proportional to the measured phase.

Often in PGSE measurements we must also consider the effects of 'background' gradients ($\mathbf{g}_0$) that arise: (i) out of intentional application, (ii) as a result of an imperfect $\mathbf{B}_0$ field (or in MRI-based diffusion measurements from imaging gradients[42]) or (iii) due to the sample being magnetically heterogeneous resulting in internal magnetic field gradients ($\mathbf{g}_{\text{int}}$ or IMFG). Typically, such background gradients are viewed as a source of artefacts to be accounted; however we will see in Section 4.8 that the internal gradients can be used to study diffusion in porous media.

2.2.3 *Effects of diffusion and flow in gradient echoes in the short gradient pulse (SGP) approximation*

The conceptually simplest measurement of translational motion is a gradient spin-echo in the SGP limit as depicted in Figure 2.2. For the moment we neglect the effects of spin relaxation. In this NMR sequence, a $\pi/2$ rf pulse is used to prepare coherent transverse magnetisation which is then twisted into a helix by the first gradient pulse. After a delay Δ, a second identical gradient pulse but with negative amplitude to the first is used to spatially unencode the helix into transverse magnetisation. Importantly, the echo occurs when (i.e., $t = t_{\text{echo}}$) the integral (i.e., Eq. (2.20)) of the effects of the gradient (pulses and or static) is equal to zero, namely

$$\mathbf{q}(t_{\text{echo}}) = 0. \tag{2.26}$$

Providing the nuclear spins have not moved during Δ (or have returned to their position at the time of the first gradient pulse) the effects of the first gradient pulse

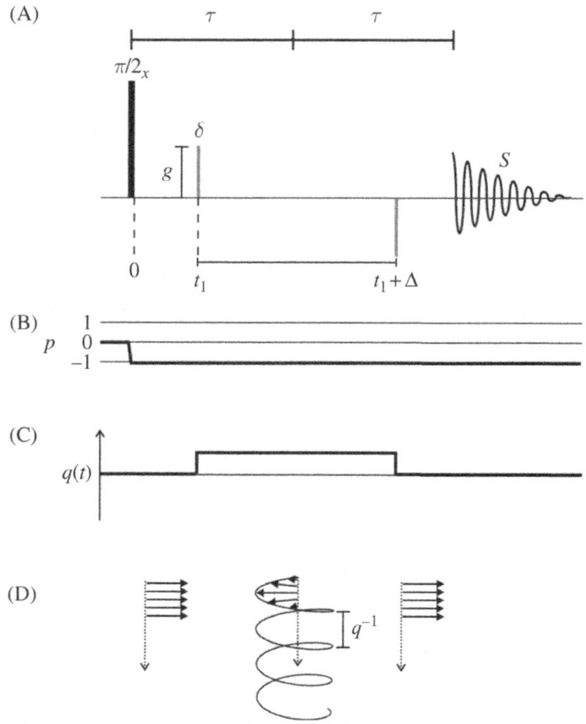

Figure 2.2 Schematic diagram of (A) the gradient spin-echo pulse sequence experiment, (B) the relevant coherence levels, (C) the change in $q(t)$ in the short gradient pulse limit and (D) the effects of this pulse sequence on a sample of nuclear spins (arrows). The $\pi/2$ rf pulse prepares coherent transverse magnetisation. Each rf pulse is typically very short $\sim 10\,\mu s$. $t=0$ is set as immediately after the $\pi/2$ pulse. The first gradient pulse is applied a time t_1 later and winds the magnetisation into a helix. After a delay Δ, which defines the timescale of the measurement, an identical gradient pulse but of opposite magnitude refocusses the transverse magnetisation which is then detected (i.e., the FID) at $t_{echo}=2\tau$. The amplitude of the NMR signal is proportional to the vector sum of the magnetisation (i.e., the arrows), which would be a maximum in this case as they are all pointing in the same direction at the beginning of signal acquisition. With foresight to consistency with later discussion we define the start of echo signal acquisition as being at time 2τ.

will be completely counteracted by the second gradient pulse and all the magnetisation is returned to a coherent transverse state (i.e., an echo) and is detected giving a maximum echo signal.

Here we consider the effects of diffusion and flow on the gradient spin-echo signal. We will see that both phenomena have significant – but distinctly different – effects on the refocussing mechanism. First consider the case of plug flow (Figure 2.3) in which *all* spins will move a distance $v_0\Delta$ between the two gradient

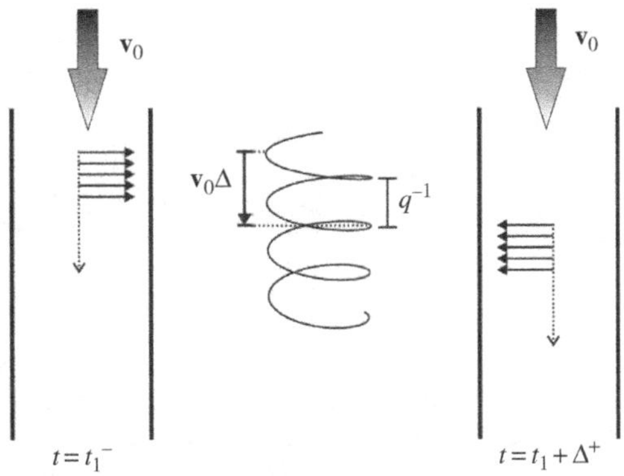

Figure 2.3 A conceptual diagram of the effects of flow at velocity $\mathbf{v}_0$ in the direction of the gradient on a liquid containing nuclear spins in the gradient spin-echo shown in Figure 2.2 showing the orientation of the spins before the first gradient pulse (t_1^-) and after the second ($t_1 + \Delta^+$). The spins will move a distance of $\mathbf{v}_0\Delta$ m in the interval Δ between the gradient pulses which, in the present example, is approximately equal to $1.5\,q^{-1}$ which corresponds to a phase shift of $\Delta\phi \approx 3\pi$. Importantly, only the orientation with respect to the x–y plane changes while the vector sum of the spins remains the same.

pulses resulting in a net phase shift ($\Delta\phi$; here Δ means difference) of (from Eq. (2.24))

$$\Delta\phi(\Delta v_0) = \gamma\delta\mathbf{g} \cdot \mathbf{v}_0\Delta = 2\pi\mathbf{q} \cdot \mathbf{v}_0\Delta. \tag{2.27}$$

Consequently, the effect of flow on signal amplitude is given by noting Eq. (2.2)

$$\begin{aligned} \Phi_{\text{Flow}}(\Delta v_0) &= \exp(i\gamma\delta\mathbf{g} \cdot \mathbf{v}_0\Delta) \\ &= \exp(i2\pi\mathbf{q} \cdot \mathbf{v}_0\Delta). \end{aligned} \tag{2.28}$$

As this is a complex-valued exponential, it is evident that uniform flow (i.e., coherent motion) does not cause signal attenuation, but instead a phase change since plug flow merely translates the helix along its long axis. If the distance flowed along the long axis is equal to an integer multiple of the helix pitch the magnetisation will be completely refocussed by the second gradient pulse. However, non-integral displacements will result in perfect refocussing of coherent transverse magnetisation but in a different direction. Thus, experimentally the NMR signal will be observed to have a complex phase modulation. The sequence becomes more sensitive to flow at higher q values since the pitch of the helix is tighter.

Diffusion, which is an incoherent motion with each spin moving independently, decreases the amplitude (or diameter) of the helix so that after the second gradient pulse, although the magnetisation is transverse it is less coherent resulting in a reduced signal being detected (see Figure 2.4). Clearly the diameter of a tighter helix

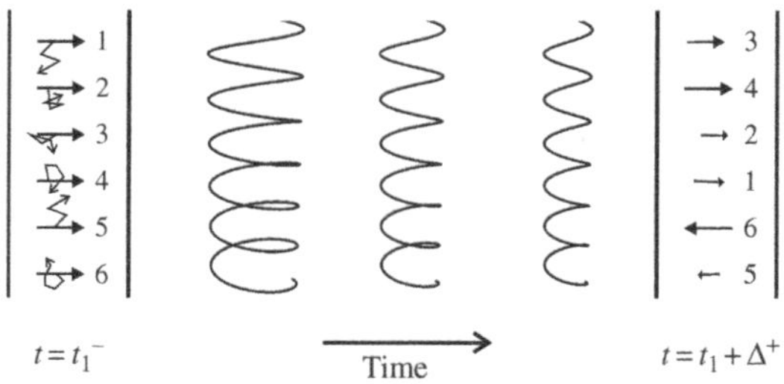

Figure 2.4 A conceptual diagram of the effects of diffusion on a sample containing nuclear spins, represented as arrows, in the gradient spin-echo shown in Figure 2.2, showing the coherent orientation of the spins before the first gradient pulse (t_1^-) and decoherent orientation after the second ($t_1 + \Delta^+$). The diffusive motion between the gradient pulses attenuates the magnetisation helix in an irreversible way (note that there is really an ensemble of spins generating the helix and not just a small number as depicted at $t = t_1^-$ and $t_1 + \Delta^+$). Specifically, the diameter of the helix is attenuated but the pitch of the helix is left unchanged. Only the component of the diffusive motion in the direction of the magnetic gradient (i.e., along the long axis of the helix) is of consequence. After the second gradient pulse the residual coherent magnetisation is refocussed. Consequently, the vector sum at $t_1 + \Delta^+$ is less than that at t_1^- since the spins are now pointing in different directions and this is reflected as an attenuated echo signal.

will decay more rapidly due to diffusion since less displacement is required for spins to come into regions containing spins with substantially different phases and thereby reducing the local magnetisation coherence (i.e., vector sum). Thus, in contrast to the effects of flow which are microscopically reversible, the effects of diffusion has changed what was a reversible process into an irreversible one. One can imagine that for any one small segment of the helix the coherence of the magnetisation will be lost in an incoherent Gaussian-like decay due to the diffusion of the spins in analogy with the change in displacement probability as depicted in Figure 1.5 and this ultimately leads to a reduction in the diameter of the helix. Due to the large number of spins involved and the Gaussian nature of diffusion there is no net phase change and only signal attenuation, E_{Diff}, is observed. The relationship between E_{Diff}, the experimental variables and diffusion will be considered in more detail below. In general, we shall omit the subscript 'Diff'. Since flow and diffusion have unique signatures, both effects can be detected simultaneously.

As a prelude of what is to come in our discussion of diffusion in restricted geometries in later chapters, it can be conceptualised that were this helix winding and unwinding to occur inside a pore, the geometry of the pore itself will be reflected in the distortion of the helix and that it will be a more complicated process than a Gaussian decay of the helical diameter.

We now consider what constitutes the NMR signal in more detail. If the initial magnetisation is taken to be M_0, the signal detected at time 2τ will be given by

$$S(q, 2\tau) = M_0 E_{\mathrm{Relax}}(2\tau) E_{\mathrm{Diff}}(q, 2\tau) \Phi_{\mathrm{Flow}}(q, 2\tau), \tag{2.29}$$

where $E_{\mathrm{Relax}}(2\tau)$ is the attenuation of the signal due to relaxation (that we have previously ignored), which in the present case is simply that due to spin–spin relaxation (time constant T_2), namely

$$E_{\mathrm{Relax}}(2\tau) = \exp\left(-\frac{2\tau}{T_2}\right). \tag{2.30}$$

Importantly, providing the length of the sequence is constant, the attenuation due to relaxation will also be constant. Although $E_{\mathrm{Diff}}(q, 2\tau)$ and $\Phi_{\mathrm{Flow}}(q, 2\tau)$ are defined in Eq. (2.29) as a function of 2τ, providing that only the two purposely applied gradient pulses as shown in Figure 2.2 need to be considered, both are really only functions of Δ since only motion during Δ is of consequence.

The derivation of E_{Diff} proceeds as follows: starting from Eq. (2.13) the phase change of a spin which was at position $\mathbf{r}_0$ during the first gradient pulse and at position $\mathbf{r}_1$ during the second is given by

$$\Delta\phi(\mathbf{r}_1 - \mathbf{r}_0) = 2\pi\mathbf{q} \cdot (\mathbf{r}_1 - \mathbf{r}_0), \tag{2.31}$$

as depicted in Figure 2.5.

The probability density of a spin starting from $\mathbf{r}_0$ and moving to $\mathbf{r}_1$ in time Δ is given by (recall Eq. (1.34))

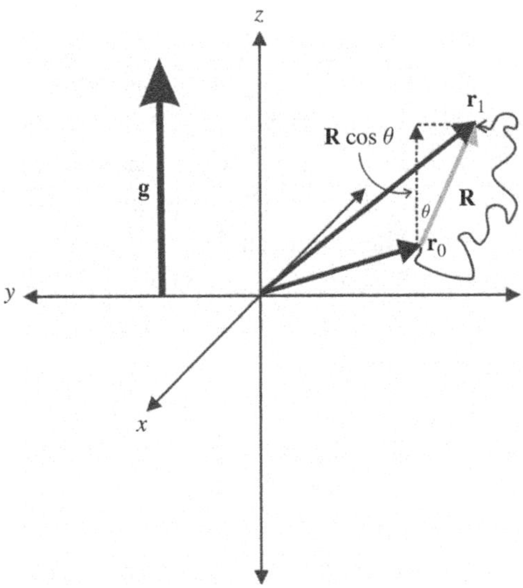

Figure 2.5 Schematic representation of a spin diffusing from $\mathbf{r}_0$ to $\mathbf{r}_1$ and the resulting dynamic displacement vector, $\mathbf{R}$ $(= \mathbf{r}_1 - \mathbf{r}_0)$. θ is the angle between $\mathbf{R}$ and the applied magnetic field gradient direction. Here the gradient direction is taken to be along z. It is the magnitude of the projection of $\mathbf{R}$ onto the direction of the gradient (i.e., $\mathbf{R}$ cos θ) that is used in determining the phase change, $\Delta\phi$, experienced by the spin.

$$\rho(\mathbf{r}_0)P(\mathbf{r}_0, \mathbf{r}_1, \Delta). \tag{2.32}$$

In the context of a PGSE sequence, $\rho(\mathbf{r}_0)$ is really the starting spin density $\rho(\mathbf{r}_0, 0)$ which is the quantity mapped in conventional (i.e., k-space) imaging, and strictly the assumption that $\rho(\mathbf{r}_0, 0) = \rho(\mathbf{r}_0)$ requires that the first gradient pulse is applied sufficiently quickly after the initial rf excitation (i.e., the delay t_1 is small in Figure 2.2) so that the effects of spin relaxation can be ignored.[43] The delta function initial condition also means that the starting spin density is spatially uniform (i.e., $M(\mathbf{r}_0, t = 0^+) = 1/V_P$).[44]

Noting Eq. (2.1) the NMR signal is proportional to the vector sum of the transverse components of the magnetisation and so the signal from one spin is given by

$$\rho(\mathbf{r}_0)P(\mathbf{r}_0, \mathbf{r}_1, \Delta)\, e^{i2\pi\mathbf{q}\cdot(\mathbf{r}_1-\mathbf{r}_0)}. \tag{2.33}$$

But in NMR the signal results from an ensemble of spins and thus integrating over all possible starting and finishing positions to obtain the SGP master equation[45,46]

$$E_{\text{Diff}}(\mathbf{q}, \Delta) = \int\int \rho(\mathbf{r}_0)P(\mathbf{r}_0, \mathbf{r}_1, \Delta)\, e^{i2\pi\mathbf{q}\cdot(\mathbf{r}_1-\mathbf{r}_0)} d\mathbf{r}_0\, d\mathbf{r}_1, \tag{2.34}$$

thus the *total* signal is a superposition of signals (transverse magnetisations), in which each phase term is weighted by the probability for a spin to begin at $\mathbf{r}_0$ and move to $\mathbf{r}_1$ during Δ. Note a negative exponent is commonly encountered in the literature depending on the definition of the Larmor equation as noted before. In subsequent discussion we shall assume, unless otherwise noted, that E and E_{Diff} are synonymous. Equation (2.34) defines a Fourier relationship between $E(\mathbf{q}, \Delta)$ and $P(\mathbf{r}_0, \mathbf{r}_1, \Delta)$ (appropriately weighted over the nuclear ensemble).[47] Another important property of Eq. (2.34) is that the particle motion, which is contained in $P(\mathbf{r}_0, \mathbf{r}_1, \Delta)$ is separated from the phase evolution behaviour. Finally, we note that if the medium being studied is isotropic it is permissible to write $E(q, \Delta)$ for $E(\mathbf{q}, \Delta)$. Assuming that the sample is homogeneous such that any variations in the structure of $\rho(\mathbf{r}_0)$ across the sample must occur on a length scale much shorter than q^{-1}, it is possible to rewrite the SGP equation (Eq. (2.34)) in terms of the average propagator (Eq. (1.35))[48,49]

$$E(\mathbf{q}, \Delta) = \int \bar{P}(\mathbf{R}, \Delta)\, e^{i2\pi\mathbf{q}\cdot\mathbf{R}} d\mathbf{R} = \text{FT}\{\bar{P}(\mathbf{R}, \Delta)\}. \tag{2.35}$$

Equation (2.35) is one of the key equations in NMR diffusion theory and so-called q-space imaging and much of the following discussion stems from noting that $E(\mathbf{q}, \Delta)$ and $\bar{P}(\mathbf{R}, \Delta)$ are Fourier conjugates.

In conventional (k-space) imaging, k-space is sampled over a finite raster giving[28]

$$S(\mathbf{k}) = \int \rho(\mathbf{r}) e^{i2\pi \mathbf{k}\cdot\mathbf{r}} d\mathbf{r}, \tag{2.36}$$

showing that in conventional imaging $S(\mathbf{k})$ and $\rho(\mathbf{r})$ are Fourier conjugates. Importantly, as the maximum value of $\mathbf{k}$ is increased to provide higher resolution and thus a finer raster, $S(\mathbf{k})$ rapidly decreases since the volume elements ('voxels') are correspondingly smaller. Thus, being able to obtain sufficient S/N per voxel so that it is observable is in general the limiting factor in obtaining high resolution in conventional imaging. In stark contrast, we could have written Eq. (2.34) (or (2.35)) as[50]

$$E(\mathbf{q}, \Delta) = \langle \exp(i2\pi \mathbf{q}\cdot\mathbf{R}) \rangle, \tag{2.37}$$

emphasising that the signal in q-space imaging is an (ensemble) average over the *whole* sample and not a voxel. The consequences and significance of the Fourier relationship and the ensemble average implied by Eq. (2.37) will be further discussed in Chapter 3.

Inserting the diffusion propagator for isotropic free diffusion (Eq. (1.60)) into Eq. (2.34)

$$E(q, \Delta) = \frac{1}{(4\pi D\Delta)^{3/2}} \int\int \rho(\mathbf{r}_0) e^{-\frac{(\mathbf{r}_1-\mathbf{r}_0)^2}{4D\Delta}} e^{i2\pi \mathbf{q}\cdot(\mathbf{r}_1-\mathbf{r}_0)} d\mathbf{r}_1 d\mathbf{r}_0. \tag{2.38}$$

In the present case $\rho(\mathbf{r}_0) = 1$ and $\mathbf{q} = q_z$ (we will drop the subscript z, however). Using spherical polar coordinates (see Figure 1.4) and noting that $d\mathbf{R} = R^2 \sin\theta\, dR\, d\theta\, d\phi$; also since θ is the angle between $\mathbf{R}$ and $\mathbf{q}$ we have

$$E(q, \Delta) = \frac{1}{(4\pi D\Delta)^{3/2}} \int_0^{2\pi} d\phi \int_0^{\infty} e^{-\frac{R^2}{4D\Delta}} R^2 \int_0^{\pi} e^{i2\pi q R\cos\theta} \sin\theta\, d\theta\, dR. \tag{2.39}$$

As there is no ϕ dependence it can be integrated out

$$\frac{2\pi}{(4\pi D\Delta)^{3/2}} \int_0^{\infty} e^{-\frac{R^2}{4D\Delta}} R^2 \int_0^{\pi} e^{i2\pi q R\cos\theta} \sin\theta\, d\theta\, dR, \tag{2.40}$$

then by noting that $d\cos\theta = -\sin\theta\, d\theta$ we get

$$\frac{2\pi}{(4\pi D\Delta)^{3/2}} \int_0^{\infty} e^{-\frac{R^2}{4D\Delta}} R^2 \int_{-1}^{1} e^{i2\pi q R\theta} d\theta\, dR. \tag{2.41}$$

The integral over θ is then performed and evaluated using Euler's formula Eq. (2.1) resulting in

$$\frac{4\pi}{\gamma\delta g(4\pi D\Delta)^{3/2}} \int_0^{\infty} R e^{-\frac{R^2}{4D\Delta}} \sin(2\pi q R)\, dR. \tag{2.42}$$

We note from a table of standard integrals (e.g., 3.952 1 in ref. 27) that

$$\int_0^\infty xe^{-c^2x^2}\sin(ax)\,dx = \frac{a\sqrt{\pi}}{4c^3}e^{-\frac{a^2}{4c^2}}. \tag{2.43}$$

In our case $x = R$, $c = (4D\Delta)^{-1/2}$ and $a = 2\pi q$ and thus we obtain the final result

$$E(q,\Delta) = \exp\left(-\gamma^2 g^2 D\delta^2\Delta\right) = \exp\left(-(2\pi q)^2 D\Delta\right). \tag{2.44}$$

This is the solution for the PGSE signal attenuation in the SGP limit for spins undergoing free diffusion and in contrast to the expression for flow this function is a real exponential. Thus, inserting Eqs. (2.30), (2.28) (assuming that flow is collinear with g) and (2.44) into Eq. (2.29) the signal amplitude at $t = 2\tau$ is

$$S(q,2\tau) = M_0\exp\left(-\frac{2\tau}{T_2}\right)\exp\left(-(2\pi q)^2 D\Delta\right)\exp(i2\pi q v_0\Delta). \tag{2.45}$$

Normally experimental PGSE data are normalised to the echo intensity obtained when $g = 0$ thereby removing the effects of relaxation (and therefore the dependence on τ)

$$E(q,\Delta) = \frac{S(q,2\tau)}{S(0,2\tau)} = \exp\left(-(2\pi q)^2 D\Delta\right)\exp(i2\pi q v_0\Delta). \tag{2.46}$$

Using Eq. (1.103) this can be rewritten in terms of the MSD,

$$E(q,\Delta) = \exp\left(-(2\pi q)^2\frac{\langle Z^2(\Delta)\rangle}{2}\right)\exp(i2\pi q v_0\Delta). \tag{2.47}$$

The MSD has been denoted by $\langle Z^2(\Delta)\rangle$ to make it explicit that it is the MSD on the timescale of Δ.

So far only an ideal case has been considered and many aspects have been neglected including: (i) the tacit assumption that all of the flowing spins stay within the receiver coil during the measurement, (ii) it is impossible to practically meet the requirements for the SGP approximation, (iii) chemical shift effects have been ignored, (iv) the presence of background gradients have been ignored and (v) we have not considered the effects of restricted diffusion. These aspects are considered below.

2.2.4 Pulsed and steady gradient spin-echoes

2.2.4.1 Hahn and stimulated spin-echoes

More generally practical than a gradient echo for measuring diffusion and indeed the most commonly used approach is to use either Hahn spin-echo (SE or

'Stejskal and Tanner' sequence) or stimulated-echo (or three-pulse Hahn echo) sequence (STE) modified to include magnetic field gradient pulses as depicted in Figure 2.6.[1] Detailed descriptions of the basis of spin-echo phenomena are given elsewhere.[1,51,52] The experiment in which a streak of dye is placed in a viscous liquid between concentric cylinders is made to refocus by reverse rotation of one of the cylinders provides an intuitive feeling for echo phenomena.[53] However, this simplistic model is incapable of showing the essential non-linearity imposed on the oscillating magnetic moments.[51] As a note of clarification in his original 1950 paper,[1] Hahn used $\pi/2_x - \tau - \pi/2_x - \tau -$ Acq. (eight-ball echo) and $\pi/2_x - \tau_1 - \pi/2_x - \tau_2 - \pi/2_x - \tau_1 -$ Acq. for the spin-echo and stimulated echo, respectively. It was Carr and Purcell[6] who introduced the sequence $\pi/2_x - \tau - \pi_x - \tau -$ Acq., which is commonly referred to as the Hahn spin-echo. Importantly, compared to the gradient echo, both the Hahn and stimulated echo sequences refocus static effects such as constant background gradients and chemical shift effects. The Hahn spin-echo can be viewed as being the STE sequence in the limit of $\tau_2 \rightarrow 0$. However, whilst the Hahn spin-echo sequence only generates a single echo signal, the STE sequence produces as many as five echoes (four spin echoes as well as the stimulated echo)[1,52,54] and extensive phase cycling is required to remove the effects of the other echoes, although the phase cycle can be greatly reduced by inserting a homospoil (or 'purge') pulse into the τ_2 delay.[54–57]

Whereas the Hahn-echo based sequence was seen to generate a magnetisation helix, the first two $\pi/2$ pulses in the stimulated echo sequence generate a z-magnetisation grid.[30,32] Although, the Hahn and STE sequences are by far the most commonly used, in principle any type of spin-echo is suitable for measuring diffusion including multiple quantum coherence transfer echoes (see Section 8.3). Kimmich[57] has presented a particularly lucid overview of spin-echoes and their origin.

In contrast to the gradient echo, in the Hahn echo PGSE sequence a π rf pulse is inserted at $t = \tau$ (i.e., the middle of the sequence) and both gradient pulses now have the same sign. The signal starting from 2τ (i.e., the second half of the echo) is used as the FID as shown in Figure 2.6. The π rf pulse acts as a reflection transformation and reverses the sign of the phase angle acquired during the first τ period (this is in contrast to a gradient echo which is a reversal phenomenon). Consequently, in addition to effectively negating the sign of the first gradient pulse – because the coherence order p is changed from +1 to −1, the π pulse also negates the phase angle acquired due to chemical shift evolution and frequency dispersion due to residual B_0 inhomogeneity and magnetic susceptibility effects in heterogeneous samples. In contrast, in a gradient-echo, only the phase dispersion resulting from the gradient pulses is refocussed.

The amplitude of the Hahn spin-echo signal including the effects of the finite length of the gradient pulses is given by[5]

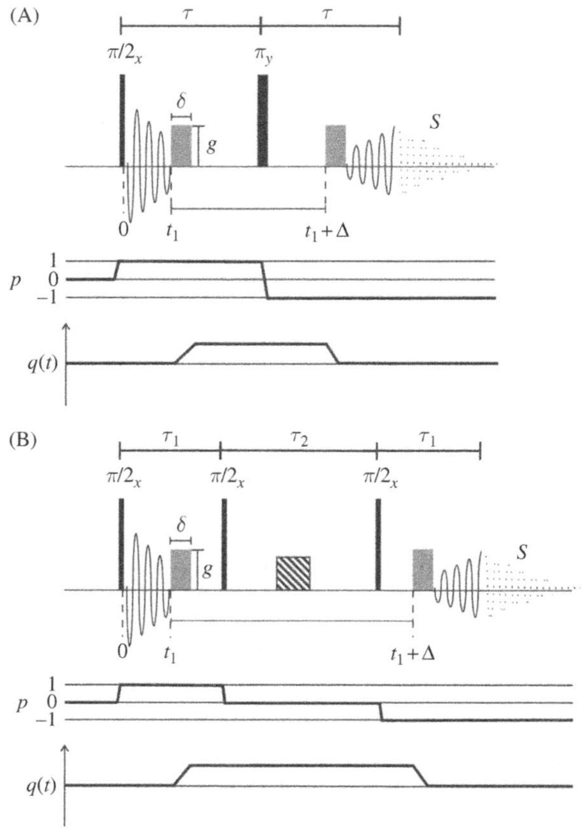

Figure 2.6 (A) The Hahn spin-echo (SE) and (B) stimulated (spin-) echo (STE)-based PGSE pulse sequences.[54,55,58] The relevant coherence levels (p) and the change in $q(t)$ are also shown. A notable feature of the STE sequence is that for most of Δ (specifically during the period τ_2), the magnetisation is aligned along z-axis (STE). In both cases the gradient pulses are represented as having finite length. The second half of the echo (i.e., starting at $t=2\tau$) is digitised (indicated by dots) and used as the FID. Hereinafter all signals will be illustrated using solid lines and it will be assumed from the context that it is clear which part of the signal is being used as the FID. From the perspective of diffusion measurements, the π pulse in the Hahn spin-echo and the two $\pi/2$ pulses in the stimulated echo is to negate the amplitude of the first gradient pulse in each sequence. An equivalent way of describing this is that due to the change in coherence order from $p=1$ for the period in which the first gradient pulse is applied to $p=-1$ for the second gradient pulse is applied, it is as if the two gradient pulses have opposite polarity with the second gradient pulse effectively being negative. The striped gradient 'purge' pulse in the stimulated echo removes any transverse relaxation and allows a significant shortening of the phase cycle.

$$S(g, 2\tau) = M_0 \exp\left(-\frac{2\tau}{T_2}\right) \exp\left(-\gamma^2 g^2 D\delta^2(\Delta - \delta/3)\right). \tag{2.48}$$

Or by normalising with respect to the echo intensity obtained when $g = 0$, to obtain (the 'Stejskal and Tanner' equation)

$$\begin{aligned} E(g, \Delta) = E(g, 2\tau) &= \frac{S(g, 2\tau)}{S(g = 0, 2\tau)} = \exp\left(-\gamma^2 g^2 D\delta^2(\Delta - \delta/3)\right) \\ &= \exp(-bD). \end{aligned} \tag{2.49}$$

Although expressed as $E(g, \Delta)$, on account of Eq. (2.14) (or equiv. Eq. (2.15)), it could equally well be written as $E(q, \Delta)$. Often the arguments will be dropped when the dependence is clear. In the second line of Eq. (2.49) all of the experimental parameters have been incorporated into the term b, known as the 'gradient' or 'diffusion weighting factor' (sometimes also written as k) – a variable commonly encountered in the clinically oriented literature.

In theory, diffusion can be probed by varying any of the three experimental variables (i.e., δ, Δ or g) and observing the effects on signal attenuation. An example set of PGSE attenuation data for CCl_4 is presented in Figure 2.7 together with its analysis giving the diffusion coefficient and its standard error (i.e., $D \pm \sigma$). Diffusion coefficients down to $7.5 \times 10^{-16}\,m^2\,s^{-1}$ have been measured; however, under favourable circumstances, it is likely that the technique has the potential to measure diffusion down to less than $10^{-17}\,m^2\,s^{-1}$.[59]

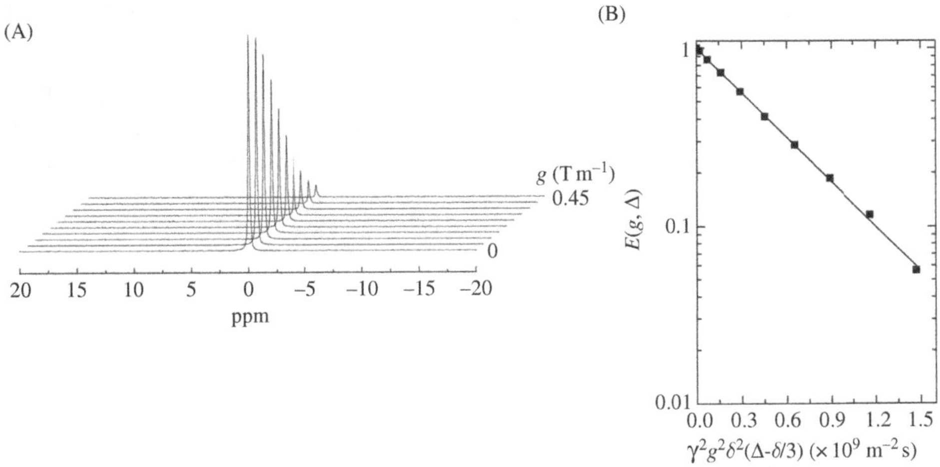

Figure 2.7. (A) ^{13}C PGSE NMR spectra of $^{13}CCl_4$ at 303 K obtained using the Hahn spin-echo pulse sequence (Figure 2.6). The spectra were acquired with $\Delta = 100$ ms, $\delta = 4$ ms and g ranging from 0 to 0.45 T m^{-1} in 0.05 T m^{-1} increments. The spectra are presented in phase-sensitive mode with a line broadening of 5 Hz. As the intensity of the gradient increases the echo intensity decreases due to the effects of diffusion. From Price.[60] (B) Regression of Eq. (2.49) onto this data gives a diffusion coefficient of $1.92 \pm 0.01 \times 10^{-9}\,m^2\,s^{-1}$. The slope of the line is $-D$.

The methods for relating signal attenuation to the pulse sequence and translational dynamics including the complete derivation of Eq. (2.48) are considered in Section 2.2.4.2. Compared to the SGP expression (Eq. (2.45)), Eq. (2.48) contains a correction factor of $-\delta/3$ to account for the finite length of the gradient pulses. If the width of the gradient pulse is expanded to fill the τ delay (i.e., $\delta = \Delta = \tau$) thereby becoming the steady gradient spin-echo sequence (see Figure 2.8), and Eq. (2.48) reduces to the well-known expression for the intensity of the Hahn spin-echo sequence in the presence of a steady (i.e., time-independent) gradient[1,6,9,61]

$$S(g, 2\tau) = M_0 \exp\left(-\frac{2\tau}{T_2}\right) \exp\left(-\gamma^2 g^2 D \frac{2\tau^3}{3}\right). \tag{2.50}$$

Simple analysis of the sequence[62] using Eq. (2.24) allows inclusion of the effects of flow

$$S(g, 2\tau) = M_0 \exp\left(-\frac{2\tau}{T_2}\right) \exp\left(-\gamma^2 g^2 D \frac{2\tau^3}{3}\right) \exp\left(i\gamma v_0 g \tau^2\right). \tag{2.51}$$

Applying the magnetic field gradient in pulses instead of continuously as in the steady gradient experiment circumvents a number of experimental limitations:[5] (i) Chemical shift information is retained since the applied gradient is off during acquisition and thus the method is suitable for measuring the diffusion coefficient of more than one species simultaneously. (ii) The rf power does not have to be increased to cope with a gradient broadened spectrum. (iii) Smaller diffusion coefficients can be measured since it is possible to use larger gradients. (iv) As the gradient pulse parameters are independent of τ it is possible to separate the effects of diffusion from spin–spin relaxation in the case of a single diffusing species (i.e., see Eq. (2.46)). (v) The time over which diffusion is measured is better defined because the gradient is applied in pulses; this is of particular importance to

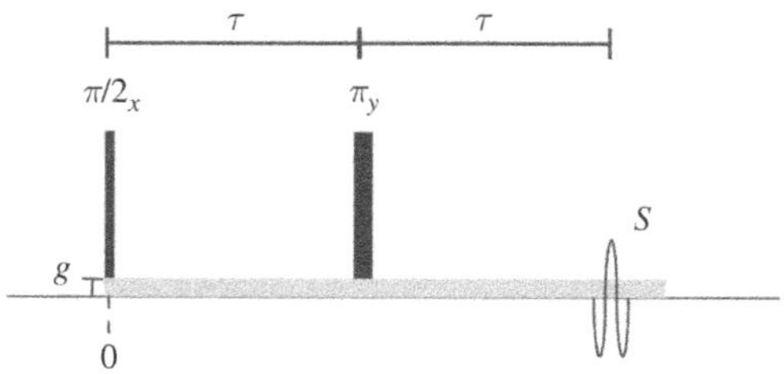

Figure 2.8 The steady gradient spin-echo sequence. The applied gradient is on for the entire sequence including the rf pulses and acquisition. Consequently, much stronger rf pulses are needed to adequately excite what is now a gradient broadened spectrum and the echo decay becomes increasingly short with increasing gradient strength with concomitant loss of chemical shift information. Further, as the diffusion time is now equal to 2τ, any extension of the diffusion time necessarily entails a greater attenuation of the echo signal due to relaxation.

studies of restricted diffusion. The reason is, as noted above, that in the steady gradient experiment – which can be viewed as very long gradient pulses in each τ period – it takes time for the helix to both wind up and unwind, consequently the sequence is most sensitive to diffusive motion when it occurs near the π rf pulse.

The STE sequence has similar benefits to the Hahn spin-echo, but the relaxation attenuation component of the signal additionally includes spin-lattice relaxation (time constant T_1), namely[55,58]

$$S(g, 2\tau) = \frac{M_0}{2}\exp\left(-\frac{2\tau_1}{T_2} - \frac{\tau_2}{T_1}\right)\exp(-\gamma^2 g^2 D\delta^2(\Delta - \delta/3)), \qquad (2.52)$$

where the delays are defined in Figure 2.6. The additional loss of a factor of 2 in the STE sequence compared to the Hahn spin-echo arises because the second $\pi/2$ pulse only stores half of the magnetisation;[1,13] however, as discussed in Section 6.2, the superior relaxation properties of the STE sequence are often sufficient to outweigh the initial 50% loss in initial magnetisation. The steady gradient solution for the STE pulse sequence analogous to that shown for the Hahn echo in Figure 2.8, is[9]

$$S(g, 2\tau) = \frac{M_0}{2}\exp\left(-\frac{2\tau_1}{T_2} - \frac{\tau_2}{T_1} - \gamma^2 g^2 D\tau_1^2\left(\tau_2 + \frac{2\tau_1}{3}\right)\right). \qquad (2.53)$$

2.2.4.2 *CPMG type sequences*

If the single π pulse in the steady gradient spin-echo sequence is replaced by a train of N π pulses to give the well-known CPMG sequence[6,63] (see Figure 2.9A), the signal intensity is then given by[6]

$$S(g, t_{\text{echo}} = 2N\tau) = M_0 \exp\left(-\frac{t_{\text{echo}}}{T_2}\right)\exp\left(-\gamma^2 g^2 D\frac{(t_{\text{echo}})^3}{12N^2}\right). \qquad (2.54)$$

Clearly, increasing N and proportionally decreasing τ whilst keeping t_{echo} constant decreases the diffusive attenuation of the CPMG echo signal. This can be conceptually reasoned by noting that when τ is short only a magnetisation helix with a large pitch can form which is less sensitive to diffusion.[30]

Interestingly, analysis of the sequence[6,62] using Eq. (2.24) reveals that the even echoes are unaffected by the effects of flow (or convection) since the first moment of the gradient (see Section 2.2.2) is nulled at these positions.

The CPMG sequence can be transformed into a PGSE sequence by inserting a gradient pulse between the π pulses at times $m\tau$ and $(m+2)\tau$, where m is an odd integer and a second gradient pulse between the π pulses at times $(m+4n+2)\tau$ and $(m+4n+4)\tau$, where n is an integer. Thus, setting $\Delta_{\text{CPMG}} = (4n+2)\tau$ and the timing ensures that the second gradient pulse occurs after an odd number of π pulses,[64] the spin-echo attenuation is given by

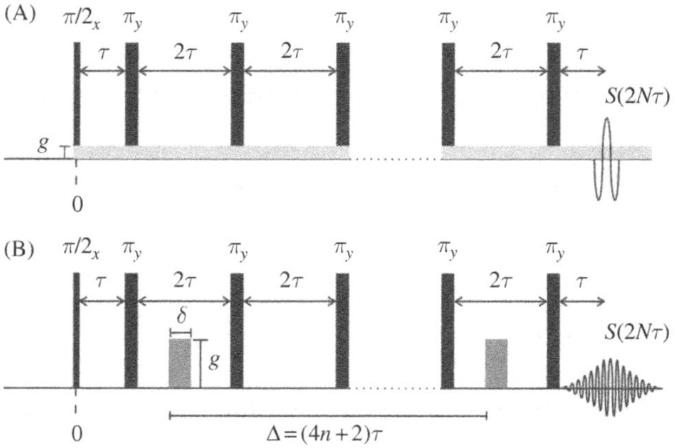

Figure 2.9 The steady gradient (A) and pulsed gradient (B) versions of the CPMG pulse sequence. The application of π pulses at odd multiples of τ results in echoes (not shown) at even multiples of τ – recall that the coherence order alternates between +1 and −1 with each π pulse. The echo centred at $2N\tau$, where N is the total number of π pulses in the sequence, is used as the FID. The CPMG pulse sequence is a variation of the Hahn echo in which a series of π pulses, each separated by 2τ are appended to the sequence. When used as a pulsed gradient spin-echo sequence it is a requirement that there is an odd number of π pulses in the interval between the two gradient pulses.

$$E(g, t_{\text{echo}} = 2N\tau) = M_0 \exp\left(-\frac{t_{\text{echo}}}{T_2}\right) \exp\left(-\gamma^2 g D \delta^2 \left(\Delta_{\text{CPMG}} - \frac{\delta}{3}\right)\right). \quad (2.55)$$

2.3 Correlating the PGSE signal attenuation with diffusion

2.3.1 Introduction

The SGP approximation, as outlined in Section 2.2.3, is an ideal case and does not account for the necessarily finite nature of the gradient pulses. Although quantum mechanical (i.e., density matrix) formulations exist,[13,65,66] a complete analysis normally involves the solution of the (macroscopic) Bloch equations including terms for diffusion (the Bloch–Torrey equations).[67,68] However, analytically solving the equations becomes intractable for all but the simplest cases. Generally it is necessary to resort to various approximations or even numerical methods. Solutions for free diffusion using the Bloch equation and GPD approximation are given below. Solutions to more complicated geometries and further approximations and numerical solutions are given in Chapter 3. How the approximations involved in the SGP and GPD approaches result in deviation from exact solutions are considered in Section 3.9.

2.3.2 Macroscopic description and solution for free diffusion

Typically in liquid state NMR the spatial inhomogeneities in the spin state of molecules is very small over the distance between nearest neighbours and also with respect to the MSD of the Brownian motion of the molecules. Consequently, a hydrodynamical description is appropriate which, as noted by Jeener,[69] states that: (i) each molecule carries its spin state in Brownian motion and bulk flow and (ii) with respect to the distance and timescales the probability distribution for a future position of any molecule depends only on its present position (and normally very weakly on its present spin state). By combining the Bloch equations for the macroscopic nuclear magnetisation with Fick's second law (Eq. (1.28)), we obtain[45,67–70]

$$\frac{\partial \mathbf{M}(\mathbf{r},t)}{\partial t} = \gamma \mathbf{M} \times \mathbf{B}(\mathbf{r},t) - \frac{M_x \mathbf{i} + M_y \mathbf{j}}{T_2} - \frac{(M_z - M_0)\mathbf{k}}{T_1} + \nabla \cdot \mathbf{D} \cdot \nabla \mathbf{M} - \nabla \cdot \mathbf{v}\mathbf{M}, \tag{2.56}$$

where the magnetisation, $\mathbf{M}(\mathbf{r},t)$, is considered to be both a time- and space-dependent function. M_x, M_y, and M_z denote the three orthogonal components of the magnetisation and M_0 is the equilibrium magnetisation due to a static magnetic field B_0 oriented in the z-direction (i.e., $\mathbf{B}_0 = (0, 0, B_0)$). This is the key equation for NMR measurements of translational motion.

In the case of isotropic diffusion, the tensor $\mathbf{D}$ is replaced by the isotropic (i.e., scalar) diffusion coefficient D and the second last term simplifies to $D\nabla^2\mathbf{M}$ which accounts for the contribution of diffusion to the rate of change of the magnetisation represented as a macroscopic fluid. The final term accounts for flow where $\mathbf{v}$ is the velocity of the spins due to the flow of the medium in which they are contained. The solution to Eq. (2.56) must be integrated over the entire sample to obtain the time-dependent magnetisation, $\mathbf{M}(t)$. We note that Jeener also extended Torrey's treatment of molecular diffusion in NMR to polyatomic molecules by introducing a position-dependent average density operator.[69] Jeener emphasised that intermolecular nuclear Overhauser effects (NOE) and the collective effects of radiation damping and the long-range dipolar field (see Section 7.3) can couple together the spin evolution of molecules of different species.

As an example, the case of a free isotropic solution studied using the Hahn spin-echo-based PGSE sequence in the presence of a constant background gradient (Figure 2.10) will be analysed using this macroscopic approach.

If $\mathbf{B}_0$ is superposed by a gradient $\mathbf{g}$ vanishing at the origin which is parallel to $\mathbf{B}_0$ and assuming that the inhomogeneities caused by $\mathbf{g}$ are much smaller than $\mathbf{B}_0$ we obtain

$$B_x = 0,\ B_y = 0,\ B_z = B_0 + \mathbf{g} \cdot \mathbf{r} = B_0 + g_z z. \tag{2.57}$$

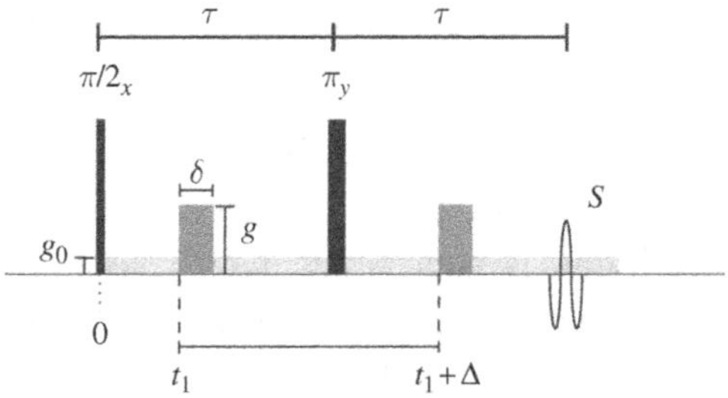

Figure 2.10 The Hahn spin-echo based PGSE sequence including a background gradient $\mathbf{g}_0$. Simplistically it is assumed that the background gradient is uniform in magnitude and direction throughout the entire sample during the sequence. The coherence levels are given in Figure 2.6A.

Noting

$$\mathbf{M} \times \mathbf{B} = (M_y B_z - M_z B_y)_x\,\mathbf{i} + (M_z B_x - M_x B_z)_y\,\mathbf{j} + (M_x B_y - M_y B_x)_z\,\mathbf{k}, \tag{2.58}$$

and defining the transverse magnetisation as

$$m = M_x + iM_y, \tag{2.59}$$

Eq. (2.56), excluding the flow term, simplifies to

$$\frac{\partial m}{\partial t} = -i\omega_0 m - i\gamma \mathbf{g} \cdot \mathbf{r}\, m - m/T_2 + D\nabla^2 m. \tag{2.60}$$

In the absence of diffusion m relaxes exponentially with a time constant T_2 and thus,

$$m = \psi e^{-i\omega_0 t - t/T_2}, \tag{2.61}$$

where ψ represents the amplitude of the precessing magnetisation unaffected by the effects of relaxation. Substituting Eq. (2.61) into Eq. (2.60) we obtain

$$\frac{\partial \psi}{\partial t} = -i\gamma \mathbf{g} \cdot \mathbf{r}\, \psi + D\nabla^2 \psi, \tag{2.62}$$

or, equivalently,

$$\left(\frac{\partial}{\partial t} + i\gamma \mathbf{g} \cdot \mathbf{r} - D\nabla^2\right)\psi = 0. \tag{2.63}$$

As noted by Grebenkov,[71] it is the inclusion of the gradient term (i.e., $i\gamma\mathbf{g}\cdot\mathbf{r}$) that complicates the analysis, since it makes the operator on the left non-Hermitian (Hermitian operators have real eigenvalues and orthogonal eigenfunctions). Without this term, Eq. (2.63) merely reverts to the diffusion equation.

In the absence of diffusion the solution for the period from the $\pi/2$ pulse to the π pulse is

$$\psi(r, t) = S \exp(-i2\pi \mathbf{q}(t) \cdot \mathbf{r}) \tag{2.64}$$

where the constant S corresponds to the value of ψ immediately after the $\pi/2$ pulse and $\mathbf{q}(t)$ is given by Eq. (2.15) (i.e., including the coherence order, p, which is equal to +1).

However, if S is a function of t in Eq. (2.64) and this is substituted into Eq. (2.62) directly and solving for S to obtain

$$S(t) = S(0)\exp\left(-(2\pi)^2\int_0^t \mathbf{q}^2(t')\,dt' D\right). \tag{2.65}$$

Sometimes, especially in medically oriented literature, Eq. (2.65) is written as

$$S(t) = S(0)\exp(-bD), \tag{2.66}$$

where b is defined by

$$b = (2\pi)^2\int_0^t \mathbf{q}^2(t')\,dt'. \tag{2.67}$$

In evaluating Eq. (2.65) the change in coherence order from $p = +1$ to -1 induced by application of the π pulse must be accounted for. The form of $\mathbf{g}(t)$ and $p(t)$ (and thus, using Eq. (2.15), $\mathbf{q}(t)$) for the pulse sequence in Figure 2.10 is specified in Table 2.1.

In evaluating Eq. (2.65) it should also be noted that the lower limit of integration refers to the start of the sequence. For example, using the definition of $\mathbf{g}(t)$ in Table 2.1, $q(t)$ for $t_1 + \Delta < t \leq t_1 + \Delta + \delta$ is calculated using Eq. (2.15) as

$$\begin{aligned} q(t) &= \frac{\gamma}{2\pi}\left[\int_0^{t_1} g_0\,dt + \int_{t_1}^{t_1+\delta} g_0 + g\,dt + \int_{t_1+\delta}^{\tau} g_0\,dt + \int_{\tau}^{t_1+\Delta} -g_0\,dt + \int_{t_1+\Delta}^{t} -g_0 - g\,dt\right] \\ &= \frac{\gamma}{2\pi}[g(\Delta + \delta + t_1 - t) + g_0(2\tau - t)]. \end{aligned} \tag{2.68}$$

The use of Eq. (2.65) to calculate the echo attenuation resulting from the effects of diffusion and applied gradients is straightforward but tedious. An example of the use of the symbolic algebra package MAPLE[72] to evaluate Eq. (2.65) for the sequence in Figure 2.10 is given in the Appendix and from this we obtain the result[5]

Table 2.1 *Time-dependence of the applied and background gradients and coherence levels in the Hahn spin-echo-based PGSE sequence (see Figures 2.6A and 2.10)*

Subinterval of pulse sequence	$\mathbf{g}(t)$	$p(t)$
$0 < t \leq t_1$	g_0	1
$t_1 < t \leq t_1 + \delta$	$g_0 + g$	1
$t_1 + \delta < t \leq \tau$	g_0	1
$\tau < t \leq t_1 + \Delta$	g_0	−1
$t_1 + \Delta < t \leq t_1 + \Delta + \delta$	$g_0 + g$	−1
$t_1 + \Delta + \delta < t \leq 2\tau$	g_0	−1

$$S(2\tau) = M_0 \exp\left(\begin{array}{l} \underbrace{-\gamma^2 g^2 D\delta^2(\Delta - \delta/3)}_{g\ \text{term}} \\ \underbrace{+\gamma^2 \mathbf{g}\cdot\mathbf{g}_0 D\delta\left[t_1^2 + t_2^2 + \delta(t_1+t_2) + \frac{2}{3}\delta^2 - 2\tau^2\right]}_{\mathbf{g}\cdot\mathbf{g}_0\ \text{cross term}} \underbrace{-\gamma^2 g_0^2 D\frac{2}{3}\tau^3}_{g_0\ \text{term}} \end{array} \right), \tag{2.69}$$

where $t_2 = 2\tau - (t_1 + \Delta + \delta)$, $S(0) = M_0$ and the effects of relaxation are neglected. If, instead, the signal was normalised to the signal acquired at $S(2\tau)$ in the absence of applied gradients, the attenuation of the signal would be given by

$$\begin{aligned} E(2\tau) &= \frac{S(2\tau)}{S(2\tau)_{g=0}} \\ &= \exp\left(-\gamma^2 g^2 D\delta^2(\Delta - \delta/3) + \gamma^2 \mathbf{g}\cdot\mathbf{g}_0 D\delta\left[t_1^2 + t_2^2 + \delta(t_1+t_2) + \frac{2}{3}\delta^2 - 2\tau^2\right]\right). \end{aligned} \tag{2.70}$$

In the literature there is occasional confusion resulting from whether the normalisation is performed using (i) the echo signal in the absence of applied gradients (i.e., $g = 0$) as in Eq. (2.70) – the normal procedure, (ii) the signal immediately after the initial excitation (i.e., $S(0)$), or (iii) the echo signal but excluding the effects of diffusion.[73] To understand the difference that these three approaches make, let us rewrite Eq. (2.69) in 'code' and include the effects of relaxation, we have

$$S(2\tau) = M_0 E_{\text{Relax}} E_g E_{\mathbf{g}\cdot\mathbf{g}_0} E_{g_0}, \tag{2.71}$$

where E_{Relax} is attenuation due to spin relaxation and the final three terms represent the three terms in the exponent in Eq. (2.69). The three different normalisation approaches retain different terms, namely

$$E(2\tau) = \frac{S(2\tau)}{S(2\tau)_{g=0}} = E_g E_{\mathbf{g}\cdot\mathbf{g}_0} \tag{2.72}$$

$$E_0(2\tau) = \frac{S(2\tau)}{S(0)} = E_{\text{Relax}} E_g E_{\mathbf{g}\cdot\mathbf{g}_0} E_{g_0} \tag{2.73}$$

$$E_{\text{Diff}}(2\tau) = \frac{S(2\tau)}{S(2\tau)_{D=0}} = E_g E_{\mathbf{g}\cdot\mathbf{g}_0} E_{g_0}. \tag{2.74}$$

Note, if relaxation is ignored $S(0) = S(2\tau)_{D=0}$. Generally the condition $\mathbf{g} \gg \mathbf{g}_0$ holds and thus the $\mathbf{g}\cdot\mathbf{g}_0$ cross terms can be neglected in Eq. (2.70). Thus, in the absence of background gradients ($\mathbf{g}_0 = 0$) the attenuation only depends on the motion during the applied gradient pulses and the separation between them

$$\begin{aligned} E(2\tau) &= E(\Delta) = \exp(-\gamma^2 g^2 D\delta^2(\Delta - \delta/3)) \\ &= \exp(-bD), \end{aligned} \tag{2.75}$$

where in this particular case and with reference to Eq. (2.66), $b = \gamma^2 g^2 \delta^2 (\Delta - \delta/3)$.

So far only rectangular gradient pulses have been considered. Despite being harder to handle mathematically, non-rectangular shapes (e.g., half sine) are technically easier to generate and produce less side effects (i.e., eddy currents – see Section 7.7). Using equation (2.65) the effects of arbitrarily shaped gradient pulses can be considered[74] and the computations can be conveniently performed by simple modification of the MAPLE worksheet given in the Appendix.

Solutions to the Bloch-Torrey equations including flow are considered in Section 4.7.

2.3.3 The Gaussian phase distribution (GPD) approximation

A commonly used method for determining the spin-echo attenuation is by the Gaussian Phase Distribution (GPD) method which is also sometimes referred to as the Gaussian phase approximation or truncated cumulant expansion (see Appendix).[1,75–78] As an illustration the method will be applied to the free diffusion case using the Hahn spin-echo sequence in the following section.

First, noting that $\mathbf{g} = g_z$, the total phase shift of spin i at the end of the echo sequence is given by

$$\begin{aligned} \phi_i(2\tau) &= \underbrace{\left\{\gamma \mathbf{B}_0\tau + \gamma g \int_{t_1}^{t_1+\delta} z_i(t)dt\right\}}_{1st\ \tau \text{ period}} - \underbrace{\left\{\gamma \mathbf{B}_0\tau + \gamma g \int_{t_1+\Delta}^{t_1+\Delta+\delta} z_i(t')dt'\right\}}_{2nd\ \tau \text{ period}} \\ &= \gamma g \left\{\int_{t_1}^{t_1+\delta} z_i(t)dt - \int_{t_1+\Delta}^{t_1+\Delta+\delta} z_i(t')dt'\right\}. \end{aligned} \tag{2.76}$$

The phase distribution resulting from the effects of the echo sequence is derived by noting that $z_i(t)$ is described by the one-dimensional diffusion equation which is a Gaussian for the case of unbounded diffusion (see Eq. (1.59)). Since, as a consequence of the central limit theorem, the probability density for the integral of a variable (in the present case $z_i(t)$), which itself has a Gaussian probability density, is Gaussian,[6,79–81] we have (Eq. (2.6))

$$P_\phi(\phi, t) = \left(2\pi\langle\phi^2\rangle\right)^{-1/2} \exp\left(\frac{-\phi^2}{2\langle\phi^2\rangle}\right), \tag{2.77}$$

where $\langle\phi^2\rangle$ denotes the mean-squared phase change at $t = 2\tau$, which is given by

$$\langle \phi^2 \rangle = \gamma^2 g^2 \left\langle \left\{ \int_{t_1}^{t_1+\delta} z_i(t)dt - \int_{t_1+\Delta}^{t_1+\Delta+\delta} z_i(t)dt \right\}^2 \right\rangle. \tag{2.78}$$

Introducing dummy variables of integration, t_a and t_b,

$$\begin{aligned}\langle \phi^2 \rangle = &\gamma^2 g^2 \left\{ \int_{t_1}^{t_1+\delta} \int_{t_1}^{t_1+\delta} dt_a dt_b - 2 \int_{t_1}^{t_1+\delta} \int_{t_1+\Delta}^{t_1+\Delta+\delta} dt_a dt_b + \int_{t_1+\Delta}^{t_1+\Delta+\delta} \int_{t_1+\Delta}^{t_1+\Delta+\delta} dt_a dt_b \right\} \\ &\langle z(t_a) z(t_b) \rangle .\end{aligned} \tag{2.79}$$

Hence, computation of $\langle \phi^2 \rangle$ is separated into a spatial part given by the MSD in the direction of the gradient, $\langle z(t_a)\ z(t_b) \rangle$, and a temporal part. $\langle z(t_a)\ z(t_b) \rangle$ which is expressed as the products of the probability of each motion times the corresponding displacement in the direction of the gradient, giving[77]

$$\begin{aligned}\langle z(t_a) z(t_b) \rangle = &\iiint (\mathbf{r}_1 - \mathbf{r}_0)_z (\mathbf{r}_2 - \mathbf{r}_0)_z\, \rho(\mathbf{r}_0) P(\mathbf{r}_0, \mathbf{r}_1, t_a) \\ &\times P(\mathbf{r}_1, \mathbf{r}_2, t_b - t_a) d\mathbf{r}_0 d\mathbf{r}_1 d\mathbf{r}_2 \quad t_b > t_a.\end{aligned} \tag{2.80}$$

Equation (2.80) is then calculated after changing to a coordinate system appropriate for the problem at hand. Equation (2.79) is then substituted into Eq. (2.8) to give the echo attenuation.

Or, combining altogether in a more succinct form, the GPD approach can be written as[82]

$$E(\Delta) = \exp\left[-\frac{g^2}{2} \int_0^{\Delta} \int_0^{\Delta} f(t') f(t'') K(t' - t'') dt' dt'' \right], \tag{2.81}$$

where the correlator is defined as

$$K(t) = \iint \rho(\mathbf{r}_0) z_0 z_1 P(\mathbf{r}_0, \mathbf{r}_1, t) d\mathbf{r}_0 d\mathbf{r}_1. \tag{2.82}$$

2.3.4 GPD solution for free diffusion

Using the Green function solution for one-dimensional free diffusion (Eq. (1.59)), Eq. (2.80) becomes

$$\begin{aligned}\langle z(t_a) z(t_b) \rangle = &\int_{-\infty}^{\infty} \int_{-\infty}^{\infty} \int_{-\infty}^{\infty} \rho(z_0)(z_1 - z_0)(z_2 - z_0)(4\pi D t_a)^{-1/2} \exp\left(-\frac{(z_1 - z_0)^2}{4 D t_a} \right) \\ &\times (4\pi D (t_b - t_a))^{-1/2} \exp\left(-\frac{(z_2 - z_1)^2}{4 D (t_b - t_a)} \right) dz_0 dz_1 dz_2.\end{aligned} \tag{2.83}$$

Setting $Z_1 = z_1 - z_0$ and $Z_2 = z_2 - z_0$ and thus

$$\int_{-\infty}^{\infty} \rho(z_0) dz_0 \int_{-\infty}^{\infty} Z_1 (4\pi D t_a)^{-1/2} \exp\left(-\frac{Z_1^2}{4Dt_a}\right) dZ_1 \int_{-\infty}^{\infty} Z_2 (4\pi D(t_b - t_a))^{-1/2} \times \exp\left(-\frac{(Z_2 - Z_1)^2}{4D(t_b - t_a)}\right) dZ_2.$$

Noting Eq. (1.32) the integral over z_0 is removed and making the substitution $Z_2' = Z_2 - Z_1$ we obtain

$$\int_{-\infty}^{\infty} Z_1 (4\pi D t_a)^{-1/2} \exp\left(-\frac{Z_1^2}{4Dt_a}\right) dZ_1 \int_{-\infty}^{\infty} (Z_2' + Z_1)(4\pi D(t_b - t_a))^{-1/2} \times \exp\left(-\frac{Z_2'^2}{4D(t_b - t_a)}\right) dZ_2'. \tag{2.84}$$

The integral over Z_2' in Eq. (2.84) can be rewritten as

$$(4\pi D(t_b - t_a))^{-1/2} \left\{ Z_1 \int_{-\infty}^{\infty} \exp\left(-\frac{Z_2'^2}{4D(t_b - t_a)}\right) dZ_2' + \int_{-\infty}^{\infty} Z_2' \exp\left(-\frac{Z_2'^2}{4D(t_b - t_a)}\right) dZ_2' \right\}. \tag{2.85}$$

The first integral in Eq. (2.85) can be evaluated with the standard integral Eq. (2.7). The second integral in Eq. (2.85) can be evaluated using the standard integral (3.462 6 in ref. 27)

$$\int_{-\infty}^{\infty} x e^{-cx^2 + 2dx} dx = \frac{d}{c} \sqrt{\frac{\pi}{c}}\, e^{d^2/c^2} \quad [\mathrm{Re}\, c > 0]. \tag{2.86}$$

In our case $x = Z_2'$, $c = (4D(t_b - t_a))^{-1}$ and $d = 0$ and so this integral equals 0. Hence Eq. (2.85) reduces to Z_1 and Eq. (2.84) becomes

$$\langle z(t_a) z(t_b) \rangle = \int_{-\infty}^{\infty} Z_1^2 (4\pi D t_a)^{-1/2} \exp\left(-\frac{Z_1^2}{4Dt_a}\right) dZ_1 \tag{2.87}$$

Finally, noting Eq. (2.86), we obtain

$$\langle z(t_a) z(t_b) \rangle = 2Dt_a, \tag{2.88}$$

which is equal to the MSD for the one-dimensional diffusion equation (see Eq. (1.104)).

We now come to a subtle point in evaluating Eqs. (2.79) and (2.80): we have to interchange t_a for t_b in Eq. (2.80) depending on whether $t_a > t_b$ or $t_a < t_b$. This can be understood by noting that the exponentials in Eq. (2.80) must have negative exponents (recall the validity of Eq. (2.80)), hence we get[4]

$$\begin{aligned}\langle z(t_a)z(t_b)\rangle &= \langle z^2(t_a)\rangle = 2Dt_a \text{ if } t_a < t_b \\ \langle z(t_a)z(t_b)\rangle &= \langle z^2(t_b)\rangle = 2Dt_b \text{ if } t_a > t_b.\end{aligned} \tag{2.89}$$

Hence, Eq. (2.78) becomes

$$\begin{aligned}\langle \phi^2 \rangle &= \gamma^2 g^2 \left\{ \begin{array}{l} \int_{t_1}^{t_1+\delta} \left[\int_{t_1}^{t_a} 2Dt_b\, dt_b + \int_{t_a}^{t_1+\delta} 2Dt_a\, dt_b \right] dt_a - 2 \int_{t_1}^{t_1+\delta} \int_{t_1+\Delta}^{t_1+\Delta+\delta} 2Dt_a\, dt_b\, dt_a + \\ \int_{t_1+\Delta}^{t_1+\Delta+\delta} \left[\int_{t_1+\Delta}^{t_a} 2Dt_b\, dt_b + \int_{t_a}^{t_1+\Delta+\delta} 2Dt_a\, dt_b \right] dt_a \end{array} \right\} \\ &= \gamma^2 g^2 2D\, \delta^2 (\Delta - \delta/3).\end{aligned} \tag{2.90}$$

Finally, substituting Eq. (2.90) into Eq. (2.8) gives

$$\ln(E) = -\gamma^2 g^2 D \delta^2 (\Delta - \delta/3), \tag{2.91}$$

which is identical to Eq. (2.75) that was derived starting from the Bloch–Torrey equations. Thus, despite its limitations, which are further discussed in Section 3.8, a particular virtue of the GPD approach is that, in contrast to the SGP approach, it accounts for the finite nature of δ. However, apart from free diffusion, the predictions for the signal attenuation given by the three approaches diverge and this is further considered in Section 3.8.

2.4 Oscillating gradients, higher-order gradient pulse trains and probing $D_\omega(\omega)$

So far only constant, two-pulse gradient (i.e., SE and STE) and CPMG pulsed gradient modulations have been considered, which are sensitive to displacements during Δ, but other types of sequence are possible and these are sensitive to $D_\omega(\omega)$ and dispersive processes due to flow. Here we consider some more complicated sequences.

Instead of applying gradients as discrete pulses it is also possible to apply them in a contiguously oscillating sequence (OGSE; oscillating gradient spin-echo, also known as modulated gradient spin-echo spectroscopy or MGSE) as shown in Figure 2.11.[20,66,83–89] The use of such rapidly time-modulated gradients periodically modulates the spin phase and consequently the echo amplitude becomes sensitive to motion occurring at a certain frequency thereby allowing frequency-dependent diffusion coefficient information (i.e., $D(\omega)$), that is unobtainable with normal PGSE experiments (normal two pulse PGSE experiments mainly probe the zero-frequency lobe with a frequency width of $1/\Delta$). In contrast to traditional PGSE sequences with discrete gradient pulses, the gradient pulses in OGSE sequences do not contribute to the selection of the coherence transfer pathways in a sequence and consequently appropriate phase cycling such as EXORCYCLE[90] must be used.

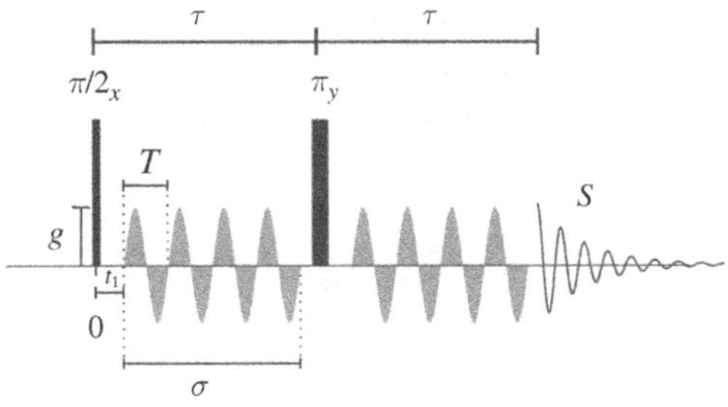

Figure 2.11 A simple oscillating gradient pulse sequence. A sinusoidal oscillating pulse of duration σ and period T (frequency $\omega = 2\pi/T$) is contained in each τ period. This is just one example and other modulated schemes have been examined.[20,84]

For an OGSE sequence consisting of n periods of an oscillating sinusoidal gradient with each 'pulse' having a total length of σ, the echo attenuation is given by[83,87]

$$S(g, 2\tau) = M_0 \exp\left(-\frac{2\tau}{T_2}\right) \exp\left(-\gamma^2 g^2 D \frac{3\sigma^3}{4\pi^2 n^2}\right). \tag{2.92}$$

Importantly, the attenuation due to diffusion is measured on the shorter timescale of σ not Δ since, as inspection of Figure 2.11 reveals that unlike the pulse sequences above, the integral of the gradient 'pulse' in each τ period (i.e., Eq. (2.15)) is zero. Indeed, by comparison of the 'b-value' (Eq. (2.67)) for one gradient oscillation lobe to give an effective diffusion time, $\Delta^{\text{eff}} = 3T/8$ where T is the period of the sine-wave.[83,88] Due to the instrumental limitations on gradient switching times, OGSE can probe the diffusion spectrum at frequencies of less than or on the order of 10^5 Hz.[20] The π pulse is necessary in such modulated gradient schemes in which the magnetisation is encoded for motion rather than position in order to refocus any phase shift due to absolute spin position.[20] Were there to be a steady gradient g_0 also along the z-direction in addition to the oscillating gradient (i.e., a combination of Figures 2.8 and 2.11), the echo attenuation would be given by[83]

$$E(g) = \exp\left(-\gamma^2 D \left[g^2 \frac{3\sigma^3}{4\pi^2 n^2} + g g_0 \frac{n\sigma^2}{\pi n}\left(\frac{n}{\gamma} + 2t_1 - \tau\right)\right]\right) \tag{2.93}$$

and we note that its inclusion introduces a τ-dependence into the attenuation function.

Oscillating gradients are useful for probing the short-time limit of $D^{\text{app}}(t)$ to timescales below 1 ms in porous materials and the spectrum of the velocity auto-correlation function in the presence of slow motions (e.g., tube disengagement in entangled polymers) or wall collisions in restricted geometries.[20,85–88,91] The velocity auto-correlation function has been derived for diffusion and flow in restricted geometries.[92,93]

In contrast to the above analysis it is possible to analyse OGSE sequences from an entirely different perspective using frequency domain analysis.[13,20,66,84,85] We begin by noting that the local spin velocity is given by

$$\mathbf{v} = \frac{\partial \mathbf{r}}{\partial t}, \tag{2.94}$$

and using this, Eq. (2.20) can be integrated by parts to give

$$\begin{aligned}\phi(t) &= \gamma \int_0^t \mathbf{g}^{\text{eff}}(t') \cdot \mathbf{r}(t')\, dt' \\ &= -\int_0^t \mathbf{F}(t') \cdot \mathbf{v}(t')\, dt',\end{aligned} \tag{2.95}$$

where $\mathbf{F}(t) = 2\pi\mathbf{q}(t)$ (see Eq. (2.15)) and recall the condition for echo formation (Eq. (2.26)). The echo attenuation will be given by the ensemble average over the sample

$$E(t) = \exp\left\langle i \int_0^t \mathbf{F}(t') \cdot \mathbf{v}(t')\, dt' \right\rangle. \tag{2.96}$$

Under the assumption of stochastic spin motion without sudden local phase changes, it can be shown that the (ensemble) phase distribution from such motion is Gaussian[20] and thus the exponential in Eq. (2.96) can be written in terms of a cumulant expansion truncated to second order, giving[20,50,66,84,85]

$$E(t) = \exp(i\alpha(t) - \beta(t)), \tag{2.97}$$

where $\alpha(t)$ is the phase shift arising from the mean flow or drift and $\beta(t)$ is an attenuation factor arising from random particle migration (i.e., the velocity auto-correlation function). These two terms are defined as

$$\alpha(t) = -\int_0^t \mathbf{F}(t') \cdot \langle \mathbf{v}(t') \rangle\, dt' \tag{2.98}$$

and

$$\begin{aligned}\beta(t) &= \frac{1}{2}\int_0^t \int_0^t \mathbf{F}(t') \cdot \langle \mathbf{v}(t')\mathbf{v}(t'') \rangle \cdot \mathbf{F}(t'')\, dt' dt'' \\ &= \frac{1}{2}\int_0^t \int_0^t \mathbf{F}(t') \cdot \langle \mathbf{u}(t')\mathbf{u}(t'') \rangle \cdot \mathbf{F}(t'')\, dt' dt''.\end{aligned} \tag{2.99}$$

Noting Eq. (1.7), Eq. (2.99) becomes

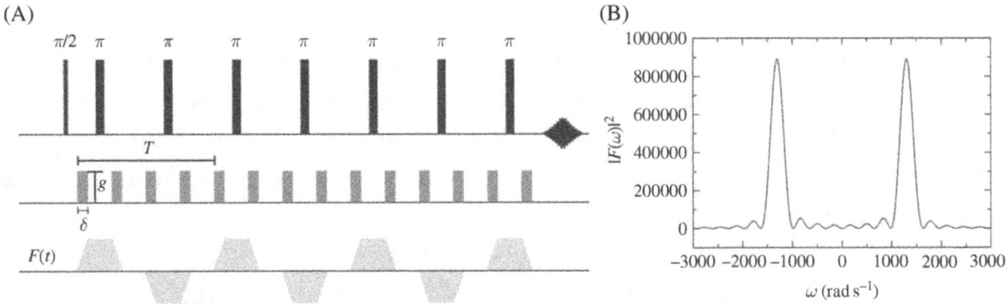

Figure 2.12 (A) A CPMG-based 'single lobe' alternating rectangular modulation OGSE experiment. (B) A plot of how $|F(\omega)|^2$ samples the diffusion spectrum via Eq. (2.101). Note this modulation scheme samples the diffusion spectrum at a single frequency. In contrast a normal PGSE sequence results in a lobe at zero frequency.

$$\beta(t) = \frac{1}{\pi}\int_0^{\infty} \mathbf{F}(\omega)\mathbf{D}(\omega)\mathbf{F}(-\omega)\, d\omega, \tag{2.100}$$

which for isotropic diffusion reduces to

$$\beta(t) = \frac{1}{\pi}\int_0^{\infty} D(\omega)|\mathbf{F}(\omega)|^2 d\omega, \tag{2.101}$$

where $\mathbf{F}(\omega)$ is the gradient modulation spectrum defined by

$$\mathbf{F}(\omega) = \int_0^{t} \mathbf{F}(t')\exp(i\omega t')\, dt'. \tag{2.102}$$

The echo-attenuation factor for the modulation scheme shown in Figure 2.12A is given by

$$\beta \approx \frac{1}{2} NT\gamma^2 g^2 \delta^2 D(2\pi/T) \tag{2.103}$$

and the associated frequency sampling spectrum is given in Figure 2.12B.

We will see later that frequency domain-modulated gradients have particular applications in the study of frequency-dependent dispersion coefficients for flow in porous media in the frequency range from a few Hz to several kHz.[94]

2.5 PGSE of freely diffusing multicomponent systems

So far only measurements of a sample containing a single species have been considered. Yet, very often we are interested in studying mixtures of species of various molecular weights. Further, macromolecular systems such as polymers both synthetic and natural (e.g., proteins) tend to occur as, or form with increasing concentration, polydisperse systems and this size-polydispersity presents particular problems to PGSE measurements.[95–99] In this section we consider the form of the PGSE data obtained from samples which are mixtures of freely diffusing species. For the present, we assume that spins or spin-bearing entities are not exchanging among species with different diffusion coefficients.

Since the diffusion coefficient of a molecule is related to its molecular weight through its effective hydrodynamic radius (at least at high dilution where intermolecular effects can be ignored) via the Einstein–Sutherland equation (Eq. (1.13)), diffusion measurements open up the possibility of working with mixtures of different species and polydisperse systems of the same species (see Figure 2.13). Thus, the differences in attenuation between the components can be used to resolve[100,101] (or edit) the spectra of complex systems and has been termed 'NMR Chromatography'. Note that although no 2D displays were presented in the 1981 paper of Stilbs,[100] this spectral separation based on component-specific PGSE spectral attenuation is actually the basis of what has later become known as DOSY (*D*iffusion *O*rdered *S*pectroscop*Y*) which is described below. Although this

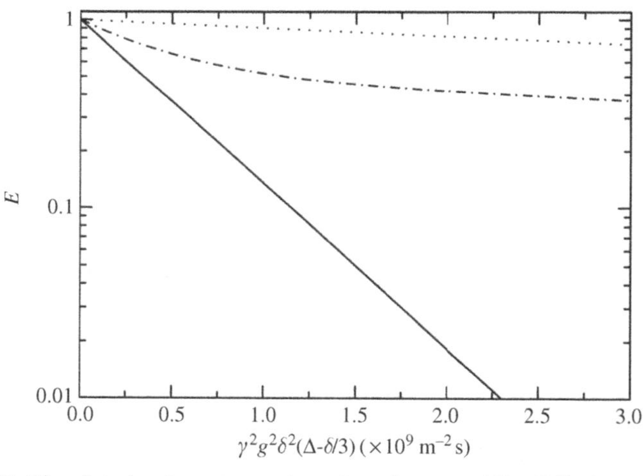

Figure 2.13 Simulated echo attenuation plots for a rapidly diffusing species (—) with a diffusion coefficient similar to water (2×10^{-9} m^{2}s^{-1}) and a more slowly diffusing species (· · · ·) with a diffusion coefficient of 1×10^{-10} m^{2}s^{-1}, which is similar to that of a medium-sized protein such as lysozyme. Also shown is the sum of the two attenuation curves (- · · - · -) as would be observed if the resonances of the two species were overlapped and at equal concentrations. As can be seen, the attenuation curve of the sum of the two species is distinctly non-exponential. The simulations were performed using Eq. (2.75) with $\gamma = \gamma(^1\mathrm{H}) = 2.6571 \times 10^8$ rad s^{-1} T^{-1}, $\Delta = 50$ ms and $\delta = 2$ ms and g ranging from 0 to 0.5 T m^{-1}.

approach is not as powerful as HPLC, it is more convenient and in principle it is possible to combine stopped-flow HPLC-NMR with diffusion NMR to provide even greater separation.[102]

The analysis is simplest when the different species appear as isolated resonances in the spectrum in which case the analysis is no more difficult than that shown in Figure 2.7. In general, however, the resonances of different species will be overlapped. For a discrete system of N_D freely diffusing species with individual diffusion coefficient D_i, assuming that each species is characterised by a single relaxation time, T_{2i}, the echo signal amplitude acquired with a Hahn spin-echo-based sequence is described by a decaying multi-exponential function

$$S(q, 2\tau) = \sum_{i}^{N_D} M_{0,i} \exp\left(\frac{-2\tau}{T_{2i}}\right) \exp(-bD_i), \tag{2.104}$$

where $M_{0,i}$ denotes the equilibrium magnetisation for the ith species. In the case of a polydisperse sample composed of a single monomer species (such that the same resonance from each monomer will overlap regardless of aggregation state) $M_{0,i} \propto \mathrm{MW}_i\, n_i$; where MW_i is the molar mass of the ith aggregate species, n_i is the number of such molecules present. The analysis of polymeric samples is especially complicated since in general the spectra of the different oligomers are completely superimposable – although there may be linewidth differences. If a stimulated echo sequence was used the effects of T_1 and T_2 would have to be included (cf. Eq. (2.52)). In some cases, however, it is permissible to neglect the effects of relaxation since the dependence of spin relaxation on molecular weight is very weak as local motions in macromolecules constitute the dominant spin-relaxation mechanism.[103,104]

As indicated by Eq. (2.104) the PGSE attenuation curves of mixtures and polydisperse[105] samples tend to be non-exponential (see also Figure 2.13) – although depending on the sample (i.e., diffusion differences, degree of polydispersity and relative populations) this may not be apparent. The different molecular weights and molecular mobility of polydisperse species lead to different relaxation rates and hence, the observed echo signal is not weighted by the respective concentrations alone. Normalising Eq. (2.104) we obtain

$$E(q, \Delta) = \frac{S(q, 2\tau)}{S(0, 2\tau)} = \frac{\sum_{i}^{N_D} \mathrm{MW}_i\, n_i \exp\left(\frac{-2\tau}{T_{2i}}\right) \exp(-bD_i)}{\sum_{i}^{N_D} \mathrm{MW}_i\, n_i \exp\left(\frac{-2\tau}{T_{2i}}\right)}. \tag{2.105}$$

Intermolecular dynamics such as obstruction and, depending on the species, entanglement become operative at finite concentrations. Interestingly, although non-exponential decays indicative of polydispersity have been observed in synthetic polymer systems, it is not always so[107] and to the author's knowledge, curvature has not been noted in any PGSE measurement of protein systems and only a single 'average' diffusion coefficient can be obtained from the data (e.g., see Figure 2.14) – even for quite concentrated samples of proteins that are known to aggregate such as lysozyme.[108]

Thus, at least in some systems, it appears likely that there is some process resulting in ensemble averaging of the diffusion coefficients of the different oligomeric species on the microscopic scale. Such an averaging process has been noted in polymer systems.[103] Neglecting the effects of spin-relaxation, the averaging process can be approximated by taking the cumulant expansion of Eq. (2.105) to second order,

$$\ln(E(q,\Delta)) = -b\langle D\rangle_W + \frac{b^2}{2}\left(\langle D\rangle_W^2 - \langle D^2\rangle_W\right), \tag{2.106}$$

where $\langle D\rangle_W$ is the mass-averaged diffusion coefficient defined by

$$\langle D\rangle_W = \frac{\sum_{i=1}^{N_D} \mathrm{MW}_i n_i D_i}{\sum_{i=1}^{N_D} \mathrm{MW}_i n_i}. \tag{2.107}$$

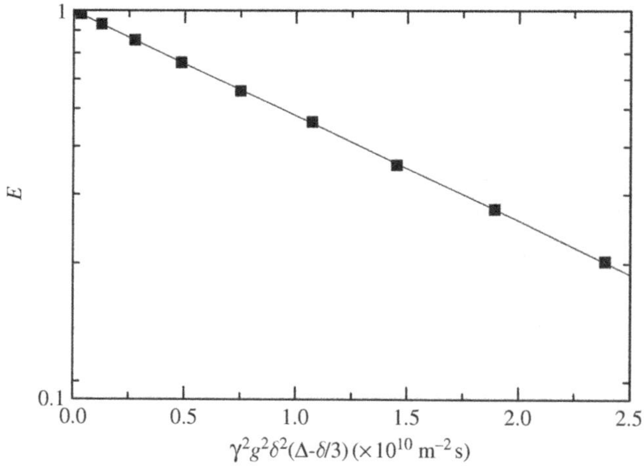

Figure 2.14 A plot of the ^{1}H PGSE attenuation of the integral over the aliphatic region (~0.4–1.7 ppm) of lysozyme (10 mM) in 0.15 M NaCl at pH 6 at 298 K acquired at 500 MHz. The experimental parameters were $\Delta = 34$ ms, $\delta = 5$ ms and g ranging in equal increments from 0 to 0.64 T m^{-1}. The solid line represents regression of Eq. (2.49) onto the data giving a diffusion coefficient of $6.25 \pm 0.02 \times 10^{-11}$ m^{2}s^{-1}. Under these experimental conditions the lysozyme molecules should be distributed amongst various aggregation states and, in the absence of any microscopic ensemble averaging, the decay curve should be multiexponential; however, the experimental attenuation curve is extremely linear when plotted on a logarithmic scale (R = 0.9999). Modified from Price.[106]

The second term (i.e., the 'variance') reflects the degree of polydispersity and may become evident as non-linearity in the attenuation plot (i.e., ln(E) vs. b) especially at large values of b. Neglecting the quadratic terms, Eq. (2.106) shows that a PGSE experiment yields the apparent diffusion coefficient, $\langle D\rangle_W$, or more accurately $\langle D\rangle_W^C$, where the superscript C signifies the inclusion of obstruction effects. Interestingly, although exchange between different large oligomeric units (e.g., proteins) is likely to be slow on the timescale of Δ, the final equation for the apparent diffusion coefficient (i.e., Eq. (2.106) without the quadratic terms) is mathematically equivalent to that for the case of fast exchange

$$E(q,\Delta) = e^{-b\langle D\rangle_W}. \tag{2.108}$$

The effects of exchange on PGSE experiments will be considered in more detail in Chapter 4.

As an example, an isodesmic distribution[109] was used to model the association of lysozyme (see Figure 2.15).[108] In this model aggregates grow by the addition of a monomer unit, L_1, thus

$$L_{i-1} + L_1 \overset{K_e}{\leftrightarrow} L_i \quad i = 2..\infty \tag{2.109}$$

where K_e $(= K_2 = K_3 = \ldots = K_i)$ is the equilibrium constant. The total concentration of lysozyme (in units of monomers) is given by

$$C = \sum_{i=1}^{\infty} c_i\, i, \tag{2.110}$$

where c_i is the concentration of the ith oligomeric state. Hence the mole fraction of the i-mer is given by

$$\alpha_i = i\,(K_eC)^{i-1}\left(\frac{2K_eC + 1 - \sqrt{1+4K_eC}}{2(K_eC)^2}\right)^i. \tag{2.111}$$

Hence, the theoretical apparent diffusion coefficient for a polydisperse oligomeric system is given by (MW$_i$ $n_i \propto c_i\, i$)

$$\langle D\rangle_W^C = \sum_{i=1}^{N_D} \alpha_i\, D_1^0\, i^{-1/3}\, O_D(\phi), \tag{2.112}$$

where D_1^0 is the infinite dilution diffusion coefficient of the monomer and O_D represents a suitable expression for self-obstruction (see Section 1.8.6).

It is not always possible to neglect relaxation and, in conjunction with the effects of obstruction, its neglect can lead to seemingly counterintuitive results as shown in

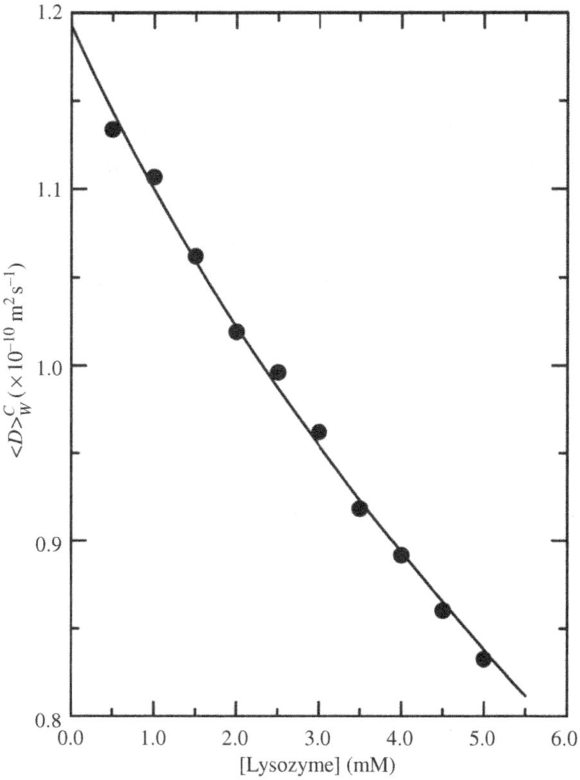

Figure 2.15 Change in lysozyme diffusion coefficients at 298 K and pH 4.6 in the presence of 0.5 M NaCl. The solid line represents the non-linear least squares regression of Eq. (2.112) using the isodesmic model in conjunction with the Tokuyama and Oppenheim correction for obstruction (see Section 1.8.6). The results gave $D_1^0 = 1.19 \pm 0.01 \times 10^{-10}\,\mathrm{m^2 s^{-1}}$ and $K_e = 118 \pm 12\,\mathrm{M^{-1}}$. Modified from Price *et al.*[108]

Figure 2.16 where the measured diffusion coefficient of a saturated lysozyme solution undergoing self-association increases with time. The reason for this effect is that as large aggregates form they become NMR invisible due to relaxation weighting leaving only the smaller oligomers still visible. Also, the effects of obstruction decrease as aggregation proceeds since more and more of the lysozyme is contained in the larger aggregates. Such PGSE data provides a means of characterising the kinetics of crystallisation.[110]

In the case of a truly polydisperse system, in which case there will be a continuum of D values that must be considered, the summation sign is replaced by an integral over all of the species and the observed echo signal would then, neglecting relaxation, be of the form[99,103–105,111]

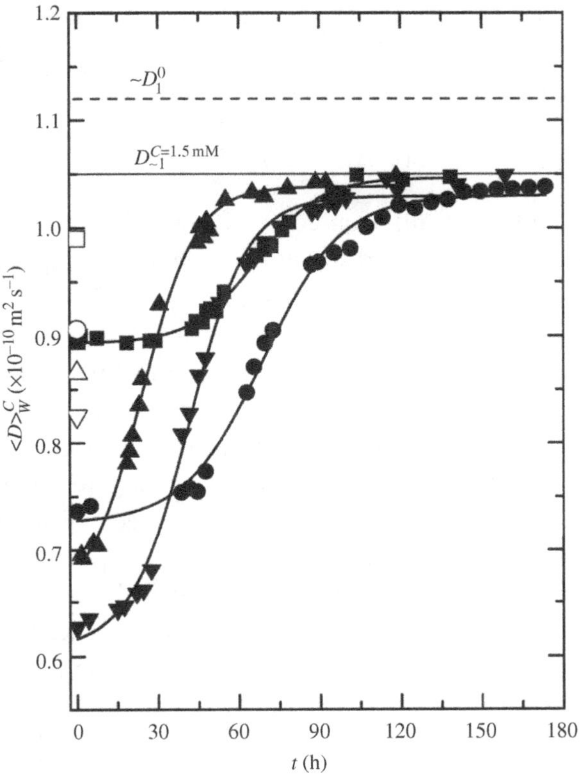

Figure 2.16 The change in the PGSE NMR-determined diffusion coefficients, $\langle D\rangle_W^C$, of lysozyme samples (3 mM, ■; 5 mM, ●; 6 mM, ▲; 7 mM, ▼) with time at pH 6 and 298 K in 0.5 M NaCl. The open symbols at $t=0$ h represent the corresponding D_1^C values for these concentrations. The horizontal solid line indicates $D_{\sim 1}^{C=1.5\,\mathrm{mM}}$ and the horizontal dashed line indicates $\sim D_1^0$. The solid lines represent the regression of a sigmoidal function onto the data. From Price *et al.*[110]

$$E(q,\Delta) = \frac{\int_0^\infty \mathrm{MW}\, n(\mathrm{MW}) \exp(-bD(\mathrm{MW}))\, d\mathrm{MW}}{\int_0^\infty \mathrm{MW}\, n(\mathrm{MW})\, d\mathrm{MW}} \tag{2.113}$$

where the distribution function $n(\mathrm{MW})$ is commonly taken as being log-normal

$$n(\mathrm{MW}) = \frac{1}{\sqrt{\pi\sigma^2}} \exp\left[-\left(\frac{\ln(\mathrm{MW}) - \ln(\mathrm{MW}_0)}{\sigma}\right)^2\right] \tag{2.114}$$

where MW_0 is the most probable molar mass and σ is the standard deviation of the distribution width. The molecular weight-dependence of the diffusion coefficient is typically accounted for using some form of scaling relationship (e.g., $D(\mathrm{MW}) \sim \mathrm{MW}^{\alpha}$ where α is a constant depending on the system; see Section 1.8.7). Poisson distributions have also been used in the case of some polymers.[107] Another approach is to use a stretched exponential (originally introduced by Williams and Watts[112] to fit distributions of dielectric relaxation times) to describe the distribution of diffusion coefficients in which case the attenuation is given by[111,113,114]

$$E(q, \Delta) = \exp(-bD^{\text{app}})^{\beta} \tag{2.115}$$

where D^{app} is the apparent diffusion coefficient and β is the stretching exponent which describes the width of the distribution. It should be kept in mind that when applied to PGSE data, this approach statistically fits $1/D$ distributions rather than those of D-values.

In a PGSE experiment of a sample containing a mixture of species the spectrum (or 'bandshape') of each component, excluding exchangeable resonances, is a constant and only the spectral intensity of each component is a function of q according to the respective diffusion coefficient (see Figure 2.17).

Thus, in a polydisperse sample (consisting of N_D freely diffusing species), apart from perhaps increased S/N, there is no advantage to analysing more than one peak in the spectrum; however, in a sample where not all peaks are equally overlapped it can be advantageous to perform a multivariate analysis instead of just analysing a single channel (i.e., univariate analysis). Thus, considering that each species contributes a conventional 1D NMR spectrum, $S_D(\nu)_i$ where ν is frequency, containing n_i resonances the observed spectrum (an extension of Eq. (2.104)) is

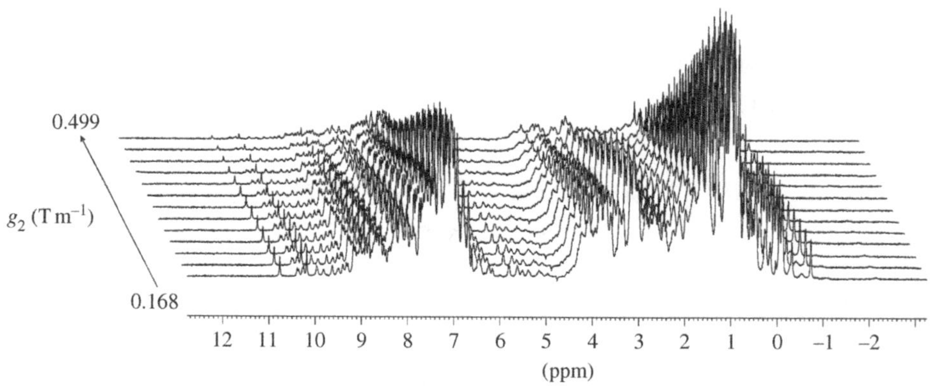

Figure 2.17 A series of 500 MHz ^{1}H PGSTE–WATERGATE spectra (see Figure 6.2) of a sample containing 2 mM lysozyme in water (10:90 $^2H_2O:H_2O$) at 298 K. Acquisition parameters were number of scans = 32, $\Delta = 93.7$ ms, and $\delta = 4$ ms with the strengths of g_1 at 0.1 T m^{-1} and g_2 varying from 0.168 T m^{-1} to 0.499 T m^{-1} in equal increments. The inter-pulse delay in the 'W5' binomial pulses[115] was set to 250 μs. It can be clearly seen that apart from the signal attenuation with increasing gradient strength that each spectrum is identical to the others. From Zheng *et al.*[116]

$$S(\nu, q) = \sum_{i=1}^{N_D} \exp(-bD_i) S_D(\nu)_i$$
$$= \sum_{i=1}^{N_D} \exp(-bD_i) \sum_{j=1}^{n_i} S(\nu)_{ij}, \tag{2.116}$$

where $S(\nu)_{ij}$ represents the lineshape (i.e., including population and relevant relaxation factors) of the jth resonance of the ith species. Thus, Eq. (2.116) for a single resonance is identical to Eq. (2.104). Similar to the above, for a truly polydisperse system with a continuum of D values that must be considered, the summation sign is replaced by an integral and becomes

$$S(\nu, q) = \int_0^\infty S_D(\nu, D) \exp(-bD)\, dD. \tag{2.117}$$

Thus, $S(\nu, q)$ is the Laplace transform of the Laplace spectrum of diffusion coefficients, $S_D(\nu, D)$ (Eq. (2.116) is a discrete representation of the Laplace transform). As an aside we recall the definition of a Laplace transform[41]

$$\mathcal{L}\{F(t)\} = \tilde{F}(k) = \int_0^\infty F(t)e^{-kt}dt, \tag{2.118}$$

where k is the transform variable and the key point is that the kernel is now real (cf. the Fourier transform in Eq. (1.51) which has a complex kernel). Real exponential functions are not orthogonal whereas complex exponential functions (i.e., trigonometric functions) are, and it is the non-orthogonality of the real exponential functions which defeats easy analysis of the inverse Laplace transform. Analysis of the data from a PGSE experiment of a mixture then proceeds by resolution of Eq. (2.116) in the case of discrete components or, in the case of a polydisperse system, the Laplace inversion of Eq. (2.117), to obtain a 'spectrum' of diffusion coefficients which is typically displayed as a 2D 'DOSY' plot with chemical shift on the one axis and the diffusion coefficient on the other (see Figure 2.18).[117–120] In the ideal case inversion would lead to delta functions in diffusion space at the diffusion coefficients of the species. We note that the inverse Laplace transform is a special case of solving a Fredholm integral equation of the first kind.[41,121] And a number of reviews focussing on the resolution of complex mixtures have been published.[102,120,122–128] Cases in which relatively complex mixtures give well-resolved spectra where most peaks result from single species allowing good resolution with respect to diffusion are sometimes referred to as 'high resolution

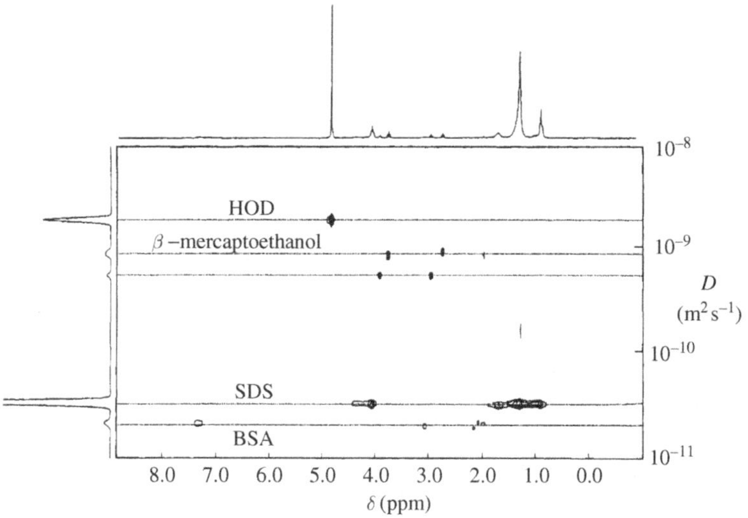

Figure 2.18 DOSY contour plot for a solution containing bovine serum albumin (BSA) (2 g dl^{-1}), sodium dodecyl sulphate (SDS, 2 g dl^{-1}) and β-mercaptoethanol (0.01 M) in phosphate buffer. The unlabelled line represents the reaction product, $HOCH_2CH_2SSCH_2CH_2OH$. Reproduced from Chen *et al.*[129] with permission.

DOSY' and conversely, cases when the resonances are heavily overlapped or superimposed limiting the resolution of the diffusion analysis are sometimes termed 'low resolution DOSY'.[102]

In the simple case of a discrete sample, each individual resonance arises from a single freely diffusing and non-exchangeable species and can be fit to a single exponential as described in the previous section. A 2D plot is then constructed based on the lineshape properties of each individual peak in the chemical shift dimension and a normalised Gaussian shape in the diffusion dimension[117,118]

$$S_D(\nu, D) = \sum_{i=1}^{N_D} \frac{S_D(\nu)_i}{\sqrt{2\pi\sigma_i^2}} \exp\left[\frac{-(D - D_i)^2}{2\sigma_i^2}\right]. \tag{2.119}$$

The Gaussian lineshape is of course artificial. The 2D plot in a DOSY display is distinctly different from that in conventional 2D NMR. Whereas in conventional 2D NMR the positions of the signals in the indirectly detected dimension (F_1) are encoded as phase and amplitude oscillations as a function of an evolution time t_1 which can be Fourier transformed and thus separated since the data is in essence complex, in DOSY the positions must be determined by exponential deconvolution ('curve stripping)' of Eq. (2.116) or via inverse Laplace transform of Eq. (2.117). Although regression of a single exponential function (e.g., Eq. (2.75)) on single exponential PGSE data is straightforward, deconvolution of even high S/N data which is the sum of two exponentials is fraught with difficulty unless the time constants of the two exponentials differ by a factor of three or so.[130] Even more challenging is the inverse Laplace transform of Eq. (2.117) (and indeed in related problems with multiexponential relaxation) which is a non-trivial mathematical/computational exercise as generally the inversion is mathematically ill-conditioned and without additional information leads to multiple solutions[131–133] and thus small changes in the amount of noise in the data can lead to very different results and the approximate bounds of the resolution of Laplace inversion have recently been considered.[134] As a further complication, in general, without prior knowledge of the sample, it will not be clear whether Eq. (2.116) or Eq. (2.117) pertains to the data set. Consequently, the amount of additional information that can be supplied such as: (i) whether the distribution of diffusion coefficients is discrete or continuous (e.g., an associating polymer system), (ii) if it is discrete how many components, or (iii) if continuous what is the functional form of the distribution of the intensity variation, is crucial to the success of the data analysis.[135] It is clearly unfortunate that the numerical inverse Laplace transform is unstable as ideally model-free data analysis is preferable to having to assume a particular diffusion coefficient distribution.

In many cases, it is possible to remove or at least simplify the analysis of overlapping resonances spectroscopically by incorporating additional chemical shift information through homonuclear or heteronuclear correlation experiments, that is by using pulse sequences in which diffusion encoding is added to usual multi-dimensional NMR sequences (e.g., COSY). These sequence aspects are covered in Section 8.6.

2.6 The long-range dipolar field and multiple echoes

The equilibrium magnetisation of a strong sample (e.g., water) at room temperature directly generates a field termed the long-range dipolar field (as will be seen below 'long-range' refers to the length scale beyond which there is effective motional averaging of intermolecular dipolar couplings in an NMR experiment), which is also known as the 'distant dipolar field', 'bulk susceptibility effect' or 'demagnetising field'. This field is parallel to $\mathbf{B}_0$ thereby adding a very weak field to the externally applied field and can shift the solute signals by ~1 Hz depending on whether the bulk magnetisation is oriented parallel or anti-parallel to $\mathbf{B}_0$.[136–138] This effect depends strongly on the shape of the sample and the spatial distribution of the spin magnetisation, and consequently can be local or through the entire sample.

The strength of a dipolar interaction is proportional to

$$3\cos^2\theta - 1 \tag{2.120}$$

where θ is the angle between the interspin vector and $\mathbf{B}_0$. Over short distances, molecular diffusion rapidly averages the interaction between a pair of spins to zero;

the averaging becomes less complete as the distance between the spin pairs increases. Although the strength of the interaction decreases dramatically with distance, the number of contributing spins increases drastically with distance and thus, the net effects of the couplings between (distant) spins can still be significant. Ordinarily, the orientation of the magnetisation vector is uniform throughout the sample and so averaging over all spins at a certain distance makes the net magnetic field due to the couplings vanish. However, if anisotropy is introduced by, for example, the application of a magnetic gradient, the long range couplings might not be completely averaged. The classical demagnetising field is embodied in the collectivity of these couplings. The demagnetising field can cause a number of enigmatic effects.[138,139] Flow is also known to affect the distant dipolar field.[140]

In Eq. (2.12) the Larmor frequency was taken as being linearly related to the (external) magnetic field gradient; however, due to the effect of the demagnetising field, the Larmor frequency can become non-linearly related. Essentially, the internal magnetic flux density of the sample is modified during the pulse sequence. Pertinent to the subject matter of this book, provided that a magnetic field gradient is present, multiple echoes can occur in addition to Hahn-type echoes after the Hahn and STE sequences described in Section 2.2.4.[29,57,141–145] Importantly, these effects become more relevant at higher static magnetic fields. Such effects can have deleterious effects on PGSE measurements (see Section 7.3), but they can also be used to measure diffusion (see Section 8.10).

References

1. E. L. Hahn, Spin Echoes. *Phys. Rev.* **80** (1950), 580–94.
2. J. H. Simpson and H. Y. Carr, Diffusion and Nuclear Spin Relaxation in Water. *Phys. Rev.* **111** (1958), 1201–2.
3. R. L. Garwin and H. A. Reich, Self-Diffusion and Nuclear Relaxation in He^3. *Phys. Rev.* **115** (1959), 1478–92.
4. D. W. McCall, D. C. Douglass, and E. W. Anderson, Self-Diffusion Studies by Means of Nuclear Magnetic Resonance Spin-Echo Techniques. *Ber. Bunsenges. Phys. Chem.* **67** (1963), 336–40.
5. E. O. Stejskal and J. E. Tanner, Spin Diffusion Measurements: Spin Echoes in the Presence of a Time-Dependent Field Gradient. *J. Chem. Phys.* **42** (1965), 288–92.
6. H. Y. Carr and E. M. Purcell, Effects of Diffusion on Free Precession in Nuclear Magnetic Resonance Experiments. *Phys. Rev.* **94** (1954), 630–8.
7. E. L. Hahn, Detection of Sea-Water Motion by Nuclear Precession. *J. Geophys. Res.* **65** (1960), 776–7.
8. T. P. Das and A. K. Saha, Mathematical Analysis of the Hahn Spin-Echo Experiment. *Phys. Rev.* **93** (1954), 749–56.
9. D. E. Woessner, Effects of Diffusion in Nuclear Magnetic Resonance Experiments. *J. Chem. Phys.* **34** (1961), 2057–61.

10. R. J. Nelson, Y. Maguire, D. F. Caputo, G. Leu, Y. Kang, M. Pravia, D. Tuch, Y. S. Weinstein, and D. G. Cory, Counting Echoes: Application of a Complete Reciprocal-Space Description of Spin Dynamics. *Concepts Magn. Reson.* **10** (1998), 331–41.
11. R. Kaiser, E. Bartholdi, and R. R. Ernst, Diffusion and Field-Gradient Effects in NMR Fourier Spectroscopy. *J. Chem. Phys.* **60** (1974), 2966–79.
12. E. X. Wu and R. B. Buxton, Effect of Diffusion on the Steady-State Magnetization with Pulsed Field Gradients. *J. Magn. Reson.* **90** (1990), 243–53.
13. P. T. Callaghan, *Principles of Nuclear Magnetic Resonance Microscopy.* (Oxford: Clarendon Press, 1991).
14. O. W. Sørenson, G. W. Eich, M. H. Levitt, G. Bodenhausen, and R. R. Ernst, Product Operator Formalism for the Description of NMR Pulse Experiments. *Prog. NMR Spectrosc.* **16** (1983), 163–92.
15. H. Kessler, M. Gehrke, and C. Griesinger, Two-Dimensional NMR Spectroscopy: Background and Overview of the Experiments. *Angew. Chem.* (*Int. Ed. Engl*) **27** (1988), 490–536.
16. P. B. Kingsley, Product Operators, Coherence Pathways, and Phase Cycling. Part I: Product Operators, Spin-Spin Coupling, and Coherence Pathways. *Concepts Magn. Reson.* **7** (1995), 29–47.
17. P. B. Kingsley, Product Operators, Coherence Pathways, and Phase Cycling. Part II: Coherence Pathways in Multipulse Sequences: Spin Echoes, Simulated Echoes, and Multiple-Quantum Coherences. *Concepts Magn. Reson.* **7** (1995), 115–36.
18. P. B. Kingsley, Product Operators, Coherence Pathways, and Phase Cycling. Part III: Phase Cycling. *Concepts Magn. Reson.* **7** (1995), 167–92.
19. K. Blum, *Density Matrix Theory and Applications*, 2nd edn. (New York: Plenum Press, 1996).
20. P. T. Callaghan and J. Stepišnik, Generalized Analysis of Motion Using Magnetic Field Gradients. *Adv. Magn. Opt. Reson.* **19** (1996), 325–88.
21. P. Allard, M. Helgstrand, and T. Härd, The Complete Homogeneous Master Equation for a Heteronuclear Two-Spin System in the Basis of Cartesian Product Operators. *J. Magn. Reson.* **134** (1998), 7–16.
22. M. Levitt, *Spin Dynamics – Basic Principles of NMR Spectroscopy*, 2nd edn. (New York: Wiley, 2008).
23. A. D. Bain, Operator Formalisms: An Overview. *Concepts Magn. Reson.* **28A** (2006), 369–83.
24. D. D. Traficante, Phase-Sensitive Detection. Part II: Quadrature Phase Detection. *Concepts Magn. Reson.* **2** (1990), 181–95.
25. A. D. Bain and I. W. Burton, Quadrature Detection in One or More Dimensions. *Concepts Magn. Reson.* **8** (1996), 191–204.
26. E. Fukushima, Nuclear Magnetic Resonance as a Tool to Study Flow. *Annu. Rev. Fluid Mech.* **31** (1999), 95–123.
27. I. S. Gradshteyn and I. M. Ryzhik, *Table of Integrals, Series, and Products*, 7th edn. (New York: Academic Press, 2007).
28. P. T. Callaghan, D. MacGowan, K. J. Packer, and F. O. Zelaya, High-Resolution *q*-Space Imaging in Porous Structures. *J. Magn. Reson.* **90** (1990), 177–82.
29. G. Deville, M. Bernier, and J. M. Delrieux, NMR Multiple Echoes Observed in Solid ^{3}He. *Phys. Rev. B* **19** (1979), 5666–88.
30. T. R. Saarinen and C. S. Johnson, Jr., Imaging of Transient Magnetization Gratings in NMR. Analogies with Laser-Induced Gratings and Applications to Diffusion and Flow. *J. Magn. Reson.* **78** (1988), 257–70.

31. W. E. Maas and D. G. Cory, Discrete Magnetization Gratings in NMR Spectroscopy. *Chem. Phys. Lett.* **254** (1995), 165–9.
32. R. Kimmich and E. Fischer, One- and Two-Dimensional Pulse Sequences for Diffusion Experiments in the Fringe Field of Superconducting Magnets. *J. Magn. Reson. A* **106** (1994), 229–35.
33. A. Sodickson and D. G. Cory, A Generalized *k*-Space Formalism for Treating the Spatial Aspects of a Variety of NMR Experiments. *Prog. NMR Spectrosc.* **33** (1998), 77–108.
34. Y.-Q. Song and X. Tang, A One-Shot Method for Measurement of Diffusion. *J. Magn. Reson.* **170** (2004), 136–48.
35. P. Mansfield and P. K. Grannell, 'Diffraction' and Microscopy in Solids and Liquids by NMR. *Phys. Rev. B* **12** (1975), 3618–34.
36. L. S. Lever, M. S. Bradley, and C. S. Johnson, Jr., Comparison of Pulsed Field Gradient NMR and Holographic Relaxation Spectroscopy in the Study of Diffusion of Photochromic Molecules. *J. Magn. Reson.* **68** (1986), 335–44.
37. J. Keeler, R. T. Clowes, A. L. Davis, and E. D. Laue, Pulsed-Field Gradients: Theory and Practice. *Methods Enzymol.* **239** (1994), 145–207.
38. W. S. Price, Gradient NMR. In *Annual Reports on NMR Spectroscopy*, ed. G. A. Webb. vol. 32. (London: Academic Press, 1996), pp. 51–142.
39. W. S. Price, Water Signal Suppression in NMR Spectroscopy. In *Annual Reports on NMR Spectroscopy*, ed. G. A. Webb. vol. 38 (London: Academic Press, 1999), pp. 289–354.
40. J. R. Singer, NMR Diffusion and Flow Measurements and Introduction to Spin Phase Graphing. *J. Phys. E: Sci. Instrum.* **11** (1978), 281–91.
41. G. Arfken and H. J. Weber, *Mathematical Methods for Physicists*, 4th edn. (New York: Academic Press, 1995).
42. X. Hong and W. T. Dixon, Measuring Diffusion in Inhomogeneous Systems in Imaging Mode Using Antisymmetric Sensitizing Gradients. *J. Magn. Reson.* **99** (1992), 561–70.
43. P. T. Callaghan, Pulsed Gradient Spin Echo NMR for Planar, Cylindrical and Spherical Pores under Conditions of Wall Relaxation. *J. Magn. Reson. A* **113** (1995), 53–9.
44. M. D. Hürlimann, L. M. Schwartz, and P. N. Sen, Probability of Return to the Origin at Short Times: A Probe of Microstructure in Porous Media. *Phys. Rev. B* **51** (1995), 14936–40.
45. E. O. Stejskal, Use of Spin Echoes in a Pulsed Magnetic-Field Gradient to Study Anisotropic Restricted Diffusion and Flow. *J. Chem. Phys.* **43** (1965), 3597–603.
46. J. E. Tanner and E. O. Stejskal, Restricted Self-Diffusion of Protons in Colloidal Systems by the Pulsed-Gradient, Spin-Echo Method. *J. Chem. Phys.* **49** (1968), 1768–77.
47. P. T. Callaghan, Pulsed Field Gradient Nuclear Magnetic Resonance as Probe of Liquid State Molecular Organization. *Aust. J. Phys.* **37** (1984), 359–87.
48. J. Kärger and W. Heink, The Propagator Representation of Molecular Transport in Microporous Crystallites. *J. Magn. Reson.* **51** (1983), 1–7.
49. J. Kärger and F. Stallmach, PFG NMR Studies of Anomalous Diffusion. In *Diffusion in Condensed Matter*, ed. P. Heitjans and J. Kärger. (Berlin: Springer, 2006), pp. 417–59.
50. P. T. Callaghan, How Two Pairs of Gradient Pulses Give Access to New Information about Molecular Dynamics. In *Diffusion Fundamentals*, ed. J. Kärger, F. Grinberg, and P. Heitjans. (Leipzig: University of Leipzig, 2005), pp. 321–38.
51. E. L. Hahn, Echo Phenomena and Non-Linearity. *Concepts Magn. Reson.* **6** (1994), 193–9.
52. P. Shkarin and R. G. S. Spencer, Direct Simulation of Spin Echoes by Summation of Isochromats. *Concepts Magn. Reson.* **8** (1996), 253–68.
53. R. G. Brewer and E. L. Hahn, Atomic Memory. *Sci. Am.* **251** (1984), 42–9.
54. D. Burstein, Stimulated Echoes: Description, Applications, Practical Hints. *Concepts Magn. Reson.* **8** (1996), 269–78.

55. J. E. Tanner, Use of the Stimulated Echo in NMR Diffusion Studies. *J. Chem. Phys.* **52** (1970), 2523–6.
56. D. Wu, A. Chen, and C. S. Johnson, Jr., An Improved Diffusion-Ordered Spectroscopy Experiment Incorporating Bipolar-Gradient Pulses. *J. Magn. Reson. A* **115** (1995), 260–4.
57. R. Kimmich, *NMR: Tomography, Diffusometry, Relaxometry.* (Berlin: Springer Verlag, 1997).
58. J. E. Tanner, Erratum: Use of the Stimulated Echo in NMR Diffusion Studies. *J. Chem. Phys.* **57** (1972), 3586.
59. P. T. Callaghan, M. E. Komlosh, and M. Nydén, High Magnetic Field Gradient PGSE NMR in the Presence of a Large Polarizing Field. *J. Magn. Reson.* **133** (1998), 177–82.
60. W. S. Price, Pulsed Field Gradient NMR as a Tool for Studying Translational Diffusion, Part I. Basic Theory. *Concepts Magn. Reson.* **9** (1997), 299–336.
61. A. G. Avent, Spin Echo Spectroscopy of Liquid Samples. In *Encyclopedia of Nuclear Magnetic Resonance*, ed. D. M. Grant and R. K. Harris. vol. 7. (New York: Wiley, 1996), pp. 4524–30.
62. K. J. Packer, The Study of Slow Coherent Molecular Motion by Pulsed Nuclear Magnetic Resonance. *Mol. Phys.* **17** (1969), 355–68.
63. S. Meiboom and D. Gill, Modified Spin-Echo Method for Measuring Nuclear Relaxation Times. *Rev. Sci. Instrum.* **29** (1958), 688–91.
64. K. J. Packer, C. Rees, and D. J. Tomlinson, A Modification of the Pulsed Magnetic Field-Gradient Spin Echo Method of Studying Diffusion. *Mol. Phys.* **18** (1970), 421–3.
65. J. Kärger, H. Pfeifer, and W. Heink, Principles and Applications of Self-Diffusion Measurements by Nuclear Magnetic Resonance. *Adv. Magn. Reson.* **12** (1988), 1–89.
66. J. Stepišnik, Analysis of NMR Self-Diffusion Measurements by a Density Matrix-Calculation. *Physica B & C* **104** (1981), 350–64.
67. H. C. Torrey, Bloch Equations with Diffusion Terms. *Phys. Rev.* **104** (1956), 563–5.
68. A. Abragam, *The Principles of Nuclear Magnetism.* (Oxford: Clarendon Press, 1961).
69. J. Jeener, Macroscopic Molecular Diffusion in Liquid NMR, Revisited. *Concepts Magn. Reson.* **14** (2002), 79–88.
70. I. J. Lowe, The Measurement of Diffusion Using Pulsed NMR. *Bull. Magn. Reson.* **3** (1981), 163–71.
71. D. S. Grebenkov, NMR Survey of the Reflected Brownian Motion. *Rev. Mod. Phys.* **79** (2007), 1077–136.
72. Maple 12, Waterloo Maple Inc., Waterloo, ON, Canada (2008).
73. R. F. Karlicek, Jr. and I. J. Lowe, A Modified Pulsed Gradient Technique for Measuring Diffusion in the Presence of Large Background Gradients. *J. Magn. Reson.* **37** (1980), 75–91.
74. W. S. Price and P. W. Kuchel, Effect of Nonrectangular Field Gradient Pulses in the Stejskal and Tanner (Diffusion) Pulse Sequence. *J. Magn. Reson.* **94** (1991), 133–9.
75. D. C. Douglass and D. W. McCall, Diffusion in Paraffin Hydrocarbons. *J. Phys. Chem.* **62** (1958), 1102–7.
76. J. S. Murday and R. M. Cotts , Self-Diffusion Coefficient of Liquid Lithium. *J. Chem. Phys.* **48** (1968), 4938–45.
77. C. H. Neuman, Spin Echo of Spins Diffusing in a Bounded Medium. *J. Chem. Phys.* **60** (1974), 4508–11.
78. P. W. Kuchel, A. J. Lennon, and C. J. Durrant, Analytical Solutions and Simulations for Spin-Echo Measurements of Diffusion of Spins in a Sphere with Surface and Bulk Relaxation. *J. Magn. Reson. B* **112** (1996), 1–17.
79. W. Feller, *An Introduction to Probability Theory and Its Applications*, vol. 1, 3rd edn. (New York: Wiley, 1968).

80. C. W. Gardiner, *Handbook of Stochastic Methods: For Physics, Chemistry and the Natural Sciences*, 2nd edn. (Berlin: Springer-Verlag, 1996).
81. N. G. Van Kampen, *Stochastic Processes in Physics and Chemistry*, 3rd edn. (Amsterdam: North Holland, 2001).
82. A. V. Barzykin, Theory of Spin Echo in Restricted Geometries Under a Step-Wise Gradient Pulse Sequence. *J. Magn. Reson.* **139** (1999), 342–53.
83. B. Gross and R. Kosfeld, Anwendung der Spin-Echo-Methode bei der Messung der Selbstdiffusion. *Messtechnik* **7** (1969), 171–7.
84. P. T. Callaghan and J. Stepišnik, Frequency-Domain Analysis of Spin Motion Using Modulated-Gradient NMR. *J. Magn. Reson. A* **117** (1995), 118–22.
85. J. Stepišnik, Measuring and Imaging Flow. *Prog. NMR Spectrosc.* **17** (1985), 187–209.
86. C. S. Johnson, Jr., Diffusion Measurements by Magnetic Field Gradient Methods. In *Encyclopedia of Nuclear Magnetic Resonance*, ed. D. M. Grant and R. K. Harris. vol. 3. (New York: Wiley, 1996), pp. 1626–44.
87. M. Schachter, M. D. Does, A. W. Anderson, and J. C. Gore, Measurements of Restricted Diffusion Using an Oscillating Gradient Spin-Echo Sequence. *J. Magn. Reson.* **147** (2000), 232–7.
88. E. C. Parsons, M. D. Does, and J. C. Gore, Modified Oscillating Gradient Pulses for Direct Sampling of the Diffusion Spectrum Suitable for Imaging Sequences. *Magn. Reson. Imaging* **21** (2003), 279–85.
89. K. I. Momot, P. W. Kuchel, and B. E. Chapman, Acquisition of Pure-Phase Diffusion Spectra Using Oscillating-Gradient Spin Echo. *J. Magn. Reson.* **176** (2005), 151–59.
90. G. Bodenhausen, R. Freeman, and D. L. Turner, Suppression of Artifacts in Two-Dimensional J Spectroscopy. *J. Magn. Reson.* **27** (1977), 511–14.
91. J. Stepišnik and P. T. Callaghan, The Long Time Tail of Molecular Velocity Correlation in a Confined Fluid: Observation by Modulated Gradient Spin-Echo NMR. *Physica B* **292** (2000), 296–301.
92. I. Oppenheim and P. Mazur, Brownian Motion in Systems of Finite Size. *Physica* **30** (1964), 1833–45.
93. R. P. Kennan, J.-H. Gao, J. Zhong, and J. C. Gore, A General Model of Microcirculatory Blood Flow Effects in Gradient Sensitized MRI. *Med. Phys.* **21** (1994), 539–45.
94. P. T. Callaghan and S. L. Codd, Flow Coherence in a Bead Pack Observed Using Frequency Domain Modulated Gradient Nuclear Magnetic Resonance. *Phys. Fluids* **13** (2001), 421–7.
95. E. D. von Meerwall, Interpreting Pulsed-Gradient Spin-Echo Diffusion Experiments in Polydisperse Specimens. *J. Magn. Reson.* **50** (1982), 409–16.
96. E. von Meerwall and K. R. Bruno, Pulsed-Gradient Spin-Echo Diffusion Study of Polydisperse Paraffin Mixtures. *J. Magn. Reson.* **62** (1985), 417–27.
97. E. von Meerwall and P. Palunas, The Effects of Polydispersity on Pulsed-Gradient NMR Diffusion Experiments in Polymer Melts. *J. Polym. Sci.* **25** (1987), 1439–57.
98. P. T. Callaghan and D. N. Pinder, A Pulsed Field Gradient NMR Study of Self-Diffusion in a Polydisperse Polymer System: Dextran in Water. *Macromolecules* **16** (1983), 968–73.
99. G. Fleischer, D. Geschke, J. Kärger, and W. Heink, Peculiarities of Self-Diffusion Studies on Polymer Systems by the Pulsed Field Gradient Technique. *J. Magn. Reson.* **65** (1985), 429–43.
100. P. Stilbs, Molecular Self-Diffusion Coefficients in Fourier Transform Nuclear Magnetic Resonance Spectrometric Analysis of Complex Mixtures. *Anal. Chem.* **53** (1981), 2135–7.

101. P. Stilbs, Fourier Transform Pulsed-Gradient Spin-Echo Studies of Molecular Diffusion. *Prog. NMR Spectrosc.* **19** (1987), 1–45.
102. G. A. Morris, Diffusion-Ordered Spectroscopy (DOSY). In *Encyclopedia of Nuclear Magnetic Resonance*, ed. D. M. Grant and R. K. Harris. vol. 9. (New York: Wiley, 2002), pp. 35–44.
103. P. T. Callaghan and D. N. Pinder, Influence of Polydispersity on Polymer Self-Diffusion Measurements by Pulsed Field Gradient Nuclear Magnetic Resonance. *Macromolecules* **18** (1985), 373–9.
104. P. T. Callaghan and J. Lelievre, The Size and Shape of Amylopectin: A Study Using Pulsed Field Gradient Nuclear Magnetic Resonance. *Biopolymers* **24** (1985), 441–60.
105. G. Fleischer, The Effect of Polydispersity on Measuring Polymer Self-Diffusion with the N.M.R. Pulsed Field Gradient Technique. *Polymer* **26** (1985), 1677–82.
106. W. S. Price, NMR Gradient Methods in the Study of Proteins. In *Annual Reports on the Progress in Chemistry Section C*, ed. G. A. Webb. vol. 96. (Cambridge: Royal Society of Chemistry, 2000), pp. 3–53.
107. A. Vergara, L. Paduano, G. D'Errico, and R. Sartorio, Network Formation in Polyethyleneglycol Solutions. An Intradiffusion Study. *Phys. Chem. Chem. Phys.* **1** (1999), 4875–9.
108. W. S. Price, F. Tsuchiya, and Y. Arata, Lysozyme Aggregation and Solution Properties Studied Using PGSE NMR Diffusion Measurements. *J. Am. Chem. Soc.* **121** (1999), 11503–12.
109. R. B. Martin, Comparisons of Indefinite Self-Association Models. *Chem. Rev.* **96** (1996), 3043–64.
110. W. S. Price, F. Tsuchiya, and Y. Arata, Time-Dependence of Aggregation in Crystallizing Lysozyme Solutions Probed Using NMR Self-Diffusion Measurements. *Biophys. J.* **80** (2001), 1585–90.
111. M. Nydén and O. Söderman, An NMR Self-Diffusion Investigation of Aggregation Phenomena in Solutions of Ethyl(hydroxyethyl)cellulose. *Macromolecules* **31** (1998), 4990–5002.
112. G. Williams and D. C. Watts, Non-Symmetrical Dielectric Relaxation Behaviour Arising from a Simple Empirical Decay Function. *J. Chem. Soc., Faraday Trans.* **66** (1970), 80–5.
113. Bo. Nyström, H. Walderhaug, and F. K. Hansen, Dynamic Crossover Effects Observed in Solutions of a Hydrophobically Associating Water-Soluble Polymer. *J. Phys. Chem.* **97** (1993), 7743–52.
114. H. Walderhaug, F. K. Hansen, S. Abrahmsén, K. Persson, and P. Stilbs, Associative Thickeners: NMR Self-Diffusion and Rheology Studies of Aqueous Solutions of Hydrophobically Modified Poly(oxyethylene) Polymers. *J. Phys. Chem.* **97** (1993), 8336–42.
115. M. Liu, X.-A. Mao, C. Ye, He. Huang, J. K. Nicholson, and J. C. Lindon, Improved WATERGATE Pulse Sequences for Solvent Suppression in NMR Spectroscopy. *J. Magn. Reson.* **132** (1998), 125–9.
116. G. Zheng, T. Stait-Gardner, P. G. Anil Kumar, A. M. Torres, and W. S. Price, PGSTE-WATERGATE: A Stimulated-Echo-Based PGSE NMR Sequence with Excellent Solvent Suppression. *J. Magn. Reson.* **191** (2008), 159–63.
117. K. F. Morris and C. S. Johnson, Jr., Diffusion-Ordered Two-Dimensional Nuclear Magnetic Resonance Spectroscopy. *J. Am. Chem. Soc.* **114** (1992), 3139–41.
118. K. F. Morris and C. S. Johnson, Jr., Resolution of Discrete and Continuous Molecular Size Distributions by Means of Diffusion-Ordered 2D NMR Spectroscopy. *J. Am. Chem. Soc.* **115** (1993), 4291–99.

119. D. P. Hinton and C. S. Johnson, Jr., Diffusion Ordered 2D NMR Spectroscopy of Phospholipid Vesicles: Determination of Vesicle Size Distributions. *J. Phys. Chem.* **97** (1993), 9064–72.
120. C. S. Johnson, Jr., Diffusion Ordered Nuclear Magnetic Resonance Spectroscopy: Principles and Applications. *Prog. NMR Spectrosc.* **34** (1999), 203–56.
121. B. Davies, *Integral Transforms and Their Applications*, 3rd edn. (Berlin: Springer-Verlag, 2002).
122. G. A. Morris and H. Barjat, High Resolution Diffusion Ordered Spectroscopy. In *Methods for Structure Elucidation by High-Resolution NMR*, ed. G. Batta, K. E. Köver, and C. Szántay, Jr. (Amsterdam: Elsevier, 1997), pp. 209–26.
123. M. D. Pelta, H. Barjat, G. A. Morris, A. L. Davis, and S. J. Hammond, Pulse Sequences for High-Resolution Diffusion-Ordered Spectroscopy (HR-DOSY). *Magn. Reson. Chem.* **36** (1998), 706–14.
124. J. S. Gounarides, A. Chen, and M. J. Shapiro, Nuclear Magnetic Resonance Chromatography: Applications of Pulse Field Gradient Diffusion NMR to Mixture Analysis and Ligand – Receptor Interactions. *J. Chromatogr. B* **725** (1999), 79–90.
125. B. Antalek, Using Pulsed Gradient Spin Echo NMR for Chemical Mixture Analysis: How to Obtain Optimum Results. *Concepts Magn. Reson.* **14** (2002), 225–58.
126. B. Antalek, J. M. Hewitt, W. Windig, P. D. Yacobucci, T. Mourey, and K. Le, The Use of PGSE NMR and DECRA for Determining Polymer Composition. Magn. *Reson. Chem.* **40** (2002), S60–71.
127. R. Huo, R. Wehrens, J. van Duynhoven, and L. M. C. Buydens, Assessment of Techniques for DOSY NMR Data Processing. *Anal. Chim. Acta* **490** (2003), 231–51.
128. J. C. Cobas, P. Groves, M. Martin-Pastor, and A. De Capua, New Applications, Processing Methods and Pulse Sequences Using Diffusion NMR. *Current Analytical Chemistry* **2** (2005), 289–306.
129. A. Chen, D.-H. Wu, and C. S. Johnson, Jr., Determination of the Binding Isotherm and Size of the Bovine Serum Albumin–Sodium Dodecyl Sulfate Complex by Diffusion-Ordered 2D NMR. *J. Phys. Chem.* **99** (1995), 828–34.
130. V. V. Apanasovich and E. G. Novikov, The Method of Fluoresence Decays Simultaneous Analysis. *Rev. Sci. Instrum.* **67** (1996), 48–54.
131. S. W. Provencher, An Eigenfunction Expansion Method for the Analysis of Exponential Recovery Decay Curves. *J. Chem. Phys.* **64** (1976), 2772–7.
132. G. C. Borgia, R. J. S. Brown, and P. Fantazzini, Uniform-Penalty Inversion of Multiexponential Decay Data. *J. Magn. Reson.* **132** (1998), 65–77.
133. A. I. Istratov and O. F. Vyvenko, Exponential Analysis in Physical Phenomena. *Rev. Sci. Instrum.* **70** (1999), 1233–57.
134. Y.-Q. Song, L. Venkataramanan, and L. Burcaw, Determining the Resolution of Laplace Inversion Spectrum. *J. Chem. Phys.* **122** (2005), 104104-1–104104-10.
135. P. Stilbs, Component Separation in NMR Imaging and Multidimensional Spectroscopy Through Global Least-Squares Analysis, Based on Prior Knowledge. *J. Magn. Reson.* **135** (1998), 236–41.
136. H. T. Edzes, The Nuclear Magnetization as the Origin of Transient Changes in the Magnetic Field in Pulse NMR Experiments. *J. Magn. Reson.* **86** (1990), 293–303.
137. M. P. Augustine and K. W. Zilm, Observation of Bulk Susceptibility Effects in High-Resolution Nuclear Magnetic Resonance. *J. Magn. Reson. A* **123** (1996), 145–56.
138. M. H. Levitt, Demagnetization Field Effects in Two-Dimensional Solution NMR. *Concepts Magn. Reson.* **8** (1996), 77–103.

139. W. S. Warren, S. Y. Huang, S. Ahn, and Y.-Y. Lin, Understanding Third-Order Dipolar Effects in Solution Nuclear Magnetic Resonance: Hahn Echo Decays and Intermolecular Triple-Quantum Coherences. *J. Chem. Phys*. **116** (2002), 2075–84.
140. P. Loureiro de Sousa, D. Gounot, and D. Grucker, Flow Effects in Long-Range Dipolar Field MRI. *J. Magn. Reson*. **162** (2003), 356–63.
141. R. Bowtell, R. M. Bowley, and P. Glover, Multiple Spin Echoes in Liquids in a High Magnetic Field. *J. Magn. Reson*. **88** (1990), 643–51.
142. A. S. Bedford, R. Bowtell, and R. M. Bowley, Multiple Spin Echoes in Multicomponent Liquids. *J. Magn. Reson*. **93** (1991), 516–32.
143. D. Einzel, G. Eska, Y. Hirayoshi, T. Kopp, and P. Wölfle, Multiple Spin Echoes in a Normal Fermi Liquid. *Phys. Rev. Lett*. **53** (1984), 2312–15.
144. I. Ardelean, R. Kimmich, S. Stapf, and D. E. Demco, Multiple Nonlinear Stimulated Echoes. *J. Magn. Reson*. **127** (1997), 217–24.
145. I. Ardelean and R. Kimmich, Principles and Unconventional Aspects of NMR Diffusometry. In *Annual Reports on NMR spectroscopy*, ed. G. A. Webb. vol. 49. (London: Elsevier, 2003), pp. 43–115.

3

PGSE measurements in simple porous systems

3.1 Introduction

In the previous chapter we considered the various methods for relating echo attenuation with diffusion in the case of free isotropic diffusion for a single diffusing species. It was observed that the echo signal attenuation was single exponential with respect to q^2 and the correct value of the diffusion coefficient was determined irrespective of the measuring time (i.e., Δ). Due to the relatively long timescale of the diffusion measurement (i.e., Δ), gradient-based measurements are sensitive to the enclosing geometry (or pore) in which the diffusion occurs (i.e., restricted diffusion) and an appropriate model must be used to account for the effects of restricted diffusion when analysing the data. The effects of the restriction can be used to provide structural information for pores with characterisitc distances (a) in the range of 0.01–100 μm. Thus, gradient methods are especially suited to studying the physics of restricted diffusion and transport in porous materials.

Non-single-exponential decays can arise in a number of ways including multi-component systems, anisotropic or restricted diffusion. These effects are the subject of the next two chapters (more complex models are studied in Chapter 4). The relevant analytical formulae for diffusion between planes and inside spheres are presented (diffusion in cylinders is presented in the following chapter). It is remarked that these are the commonly used models for benchmarking numerical approaches.[1] We also mention that Grebenkov has recently presented a review of NMR studies of restricted Brownian motion.[2]

Diffusion measurements in restricted geometries are typically characterised using four length scales: a and MSD as noted in Section 1.7; the pitch of the magnetisation helix (i.e., q^{-1}; see Section 2.2.2); and, in the case of a pore with surface relaxation, the relaxation length,[2]

$$a_M = \frac{D}{M}, \tag{3.1}$$

which corresponds to the distance a particle can travel near the boundary before relaxation significantly reduces its magnetisation.

Ideally, an appropriate model (i.e., diffusion propagator) to the system at hand will be used to analyse the attenuation data; however, in general there is no prior knowledge as to what the appropriate geometry is and even if there were, a suitable model may not exist. Thus, we are in a situation analogous to that in Section 2.5 where we considered the analysis of data from multicomponent freely diffusing systems, where we have to deduce physical information from the data instead of merely applying what a priori is known to be the appropriate model ('forward modelling'). Thus in working with porous media we are often concerned with 'inverse methods'.[3] One can always calculate an apparent diffusion coefficient, $D^{\text{app}}(\Delta)$, using the measured signal with the free diffusion formula for the particular pulse sequence being used. For example, without any knowledge of the form of a restricted geometry, measurements conducted on a restricted system using the Hahn spin-echo or STE sequence could be analysed using Eq. (2.49) to yield a value for $D^{\text{app}}(\Delta)$ irrespective of whether the experiment is conducted within the SGP limit.[4,5]

If the experiment is within the SGP limit considerable analysis of the attenuation data can be performed in a model-free manner starting with only the realisation that $E(\mathbf{q}, \Delta)$ and $\bar{P}(\mathbf{R}, \Delta)$ are Fourier conjugates (i.e., Eq. (2.35)), namely

$$E(\mathbf{q}, \Delta) = \int \bar{P}(\mathbf{R}, \Delta) e^{i2\pi \mathbf{q}\cdot\mathbf{R}} d\mathbf{R}. \tag{3.2}$$

The two model-free approaches are:[6] (i) experimental determination of $\bar{P}(\mathbf{R}, \Delta)$ from the inverse Fourier Transform of $E(\mathbf{q}, \Delta)$ and (ii) experimental determination of the moments of $\bar{P}(\mathbf{R}, \Delta)$ from analysis of the form of $E(\mathbf{q}, \Delta)$, which at low q directly leads to information on the (ensemble averaged) MSD and consequently the time-dependent diffusion coefficient (i.e., $D(\Delta)$). Some of these effects are illustrated below using the diffusion between planes model.

While our coverage of the theory is mainly limited to Hahn- and STE-based PGSE sequences, theory has been developed for CPMG-based PGSE sequences in the presence of restriction.[7] Indeed the effects of restricted diffusion were first noted by Woessner in SGSE experiments on benzene in rubber.[8] Further, the relevant theory has been developed for obtaining information on porous systems using (purposely applied) constant magnetic field gradients[9–11] and using the internal magnetic field gradients (IMFG) present in the system as a result of the susceptibility difference between the pore-filling fluid (or gas) and the solid matrix. Some authors have termed this latter approach 'diffusion decay in the internal field' (DDIF). This is given brief coverage in Section 4.8.

To avert later confusion: in the following chapters we will often be concerned with diffusion coefficients measured on different timescales (i.e., D^0, $D(t)$ and D^∞) as discussed in the previous chapter. We reserve $D^{\mathrm{app}}(t)$ (i.e., an apparent diffusion coefficient) for those occasions in which theory is used to analyse data outside its range of validity (e.g., analysing data from a restricted geometry using theory related to free diffusion or using theory derived with the SGP approximation to analyse data acquired outside the SGP limit). Where it is clear from the context D may be used to refer to any one of D^0, $D(t)$, D^∞ or $D^{\mathrm{app}}(t)$.

3.2 Experimental determination of $\bar{P}(\mathbf{R}, \Delta)$

As can be seen from Eq. (3.2) inverse Fourier transformation of $E(\mathbf{q}, \Delta)$ with respect to $\mathbf{q}$ returns an image of $\bar{P}(\mathbf{R}, \Delta)$, namely

$$\bar{P}(\mathbf{R}, \Delta) = \frac{1}{2\pi} \int E(\mathbf{q}, \Delta) e^{-i2\pi \mathbf{q} \cdot \mathbf{R}} d\mathbf{q} \tag{3.3}$$

And, in the case of free diffusion, the width at half-height, $\Delta\bar{P}(\mathbf{R}, \Delta)_{\frac{1}{2}}$ (nb the first Δ denotes difference), is directly related to the root mean square of $\mathbf{R}$,[12] namely

$$\Delta\bar{P}(\mathbf{R}, \Delta)_{\frac{1}{2}} = 4\sqrt{D(\Delta)\Delta \ln(2)}. \tag{3.4}$$

Thus, providing there is sufficient signal-to-noise and $E(\mathbf{q}, \Delta)$ has been well sampled with respect to q, Fourier inversion of $E(\mathbf{q}, \Delta)$ enables determination of the form of $\bar{P}(\mathbf{R}, \Delta)$.[12,13]

3.3 Experimental determination of the moments of $\bar{P}(\mathbf{R}, \Delta)$

For simplicity, assuming that the gradient is directed along z, $Z = z_1 - z_0$, and by performing a Taylor expansion of the exponential in Eq. (3.2),[6,14–16]

$$\begin{aligned} E(\mathbf{q}, \Delta) &= \int \bar{P}(Z, \Delta) \sum_{n=0}^{\infty} \frac{(i2\pi qZ)^n}{n!} dZ \\ &= 1 + \sum_{n=1}^{\infty} \frac{(i2\pi q)^n}{n!} \int \bar{P}(Z, \Delta) Z^n dZ. \end{aligned} \tag{3.5}$$

The $n = 0$ term is equal to 1 on account of the identity $\int_{-\infty}^{\infty} \bar{P}(Z, \Delta)\, dZ = 1$ (Eq. (1.36)). Noting that the integral represents the moments of $\bar{P}(Z, \Delta)$, i.e. $\langle Z^n(\Delta) \rangle$ (cf. Eq. (1.101)), Eq. (3.5) becomes

$$E(\mathbf{q},\Delta) = 1 + i\underbrace{\left(\frac{(2\pi q)\langle Z(\Delta)\rangle}{1!} - \frac{(2\pi q)^3\langle Z^3(\Delta)\rangle}{3!} + \ldots\right)}_{\text{Coherent Motion}}$$

$$- \underbrace{\left(\frac{(2\pi q)^2\langle Z^2(\Delta)\rangle}{2!} - \frac{(2\pi q)^4\langle Z^4(\Delta)\rangle}{4!} + \ldots\right)}_{\text{Diffusion}}. \qquad (3.6)$$

This shows that the shape of the attenuation curve is completely determined by the moments of the mean propagator. For a purely self-diffusive motion the mean propagator is an even function and thus the odd orders in Eq. (3.5) vanish leaving a real function. In this case analysis could proceed by fitting a polynomial in $(2\pi q)^2$ to $E(\mathbf{q}, \Delta)$. Conversely, for purely coherent motion (e.g., flow or mutual diffusion), only the odd terms are retained, whereas translational motion involving both coherent and diffusive motion retains both the even and odd order, and as noted by Stallmach and Galvosas,[6] in such a case Fourier inversion would be a more prudent approach to data analysis.[17]

When q is small the higher-order moments in Eq. (3.6) can be ignored, leaving

$$E(\mathbf{q}\to 0,\Delta) \approx 1 + i2\pi q Z(\Delta) - \frac{(2\pi q)^2\langle Z^2(\Delta)\rangle}{2!}. \qquad (3.7)$$

Note, that $\langle Z^2(\Delta)\rangle$ is the ensemble-averaged MSD and thus, in the case of purely diffusive motion, and noting Eq. (1.103), this becomes

$$E(\mathbf{q}\to 0,\Delta) \approx 1 - (2\pi q)^2\frac{\langle Z^2(\Delta)\rangle}{2} = 1 - (2\pi q)^2 D(\Delta)\Delta. \qquad (3.8)$$

Equation (3.8) is, neglecting flow, the series expansion to second order of Eq. (2.47) and is very useful since it shows that irrespective of the form of the propagator, the initial decay of $E(\mathbf{q}, \Delta)$ with respect to q yields $\langle Z^2(\Delta)\rangle$ (or equivalently $D(\Delta)$).

In general the reorientational motion of the species being probed has no direct effect on the diffusion measurement. However, it has been pointed out when at least one spatial dimension of the species being studied is larger than the root mean square displacement associated with the diffusion measurement that reorientational motions can lead to large apparent diffusion coefficients which depend on Δ.[18]

Combining Eq. (3.8) with Eq. (1.122) gives[15]

$$E(\mathbf{q}\to 0,\Delta\to 0) = \exp\left\{-(2\pi q)^2\Delta D^0\left[1 - \frac{4}{3d\sqrt{\pi}}\frac{S_{\mathrm{P}}}{V_{\mathrm{P}}}\sqrt{D^0\Delta}\cdots\right]\right\}. \qquad (3.9)$$

In analysing the PGSE data of systems where the propagator is unknown, an apparent diffusion coefficient is often determined by analogy to the free diffusion case from the initial slope of the attenuation plot and thus from Eq. (2.47) (this approach is only valid in the short gradient pulse limit)

$$D(\Delta) = -\frac{1}{4\pi^2\Delta}\frac{\partial \ln[E(\mathbf{q}\to 0,\Delta)]}{\partial [q]^2}. \tag{3.10}$$

As noted in Section 1.8.2, studying $D(\Delta)$ can provide significant information on the surrounding geometry. For example, Szafer *et al.*[19] have presented a model for the apparent diffusion coefficient for water in isotropic and non-isotropic tissue, in which the tissue is modelled as a periodic array of boxes surrounded by partially permeable membranes (cells). Thus, with reference to Eq. (1.121), we obtain

$$D(\Delta) \propto \Delta^{\kappa-1}. \tag{3.11}$$

Generally $\kappa \leq 1$ ($\kappa = 1$ corresponds to normal diffusion). In the case of diffusion within a reflecting pore $\langle Z^2(\Delta)\rangle$ is a constant for long observation times (cf. Figure 1.10) and thus $D(\Delta) \propto \Delta^{-1}$ for $\Delta \to \infty$. When the condition $qa \ll 1$ (a being the pore radius for example) is met the behaviour of $E(\mathbf{q}, \Delta)$ is dominated by the diffusive motion of the spin. If the condition ξ $(=D\Delta/a^2) < 1$ is met then the measured diffusion coefficient will tend to that of the bulk solution. As ξ increases the effects of the restricting geometry will become increasingly important. When ξ is greater than 1, structural information can be obtained directly from the PGSE signal (i.e., diffraction effects, if the restricting geometry has local order) by varying q such that $qa \geq 1$. Indeed, Brownian dynamics simulations have revealed for the case of a sphere[20] that the long-time limit is already applicable at $\xi \approx 1$.

For large q, it has been shown that[10,21–23]

$$E(\mathbf{q},\Delta) \propto \Gamma(\Delta)(S_{\mathrm{P}}/V_{\mathrm{P}})q^{-4} \tag{3.12}$$

where $\Gamma(\Delta)$ is a time-dependent function characteristic of the geometry, in analogy with the Debye–Porod law.

3.4 The spectral function and the return to origin probability

From Eq. (1.80) and the SGP attenuation definition (Eq. (2.34)) and assuming an isotropic medium, the return to origin probability is given by[24,25]

$$\mathrm{RTOP}(\Delta) = \Theta(\Delta) = \frac{4\pi V_{\mathrm{P}}}{(2\pi)^3}\int_0^\infty E(q,\Delta)q^2 dq. \tag{3.13}$$

Thus, RTOP is proportional to the area of the signal attenuation curve and the inclusion of the q^2 term in the integrand weights the probability towards the more highly restricted molecules as it will be these whose signals will be least dephased. In practice, the upper limit of the integral over q is determined by experimental limitations and so ∞ is replaced by q_{max}. In free isotropic solution, RTOP is given by

$$\mathrm{RTOP}^0(\Delta) = \frac{1}{(4\pi D\Delta)^{3/2}} \tag{3.14}$$

And from these two equations a dimensionless RTOP enhancement can be defined by[26]

$$\mathrm{RTOP_e}(\Delta) = \frac{\mathrm{RTOP}(\Delta)}{\mathrm{RTOP}^0(\Delta)}. \tag{3.15}$$

$\mathrm{RTOP_e}(\Delta) = 1$ for a small value of q since neither restricted nor unrestricted species can diffuse out of the sphere characterised by large a_{min} ($= q^{-1}$) within the diffusion time. However, as q increases and a_{min} decreases, species diffusing within a restricted environment are more likely to be found within the sphere at the end of Δ. Consequently, $\mathrm{RTOP_e}(\Delta)$ increases in restricted samples with increasing q or Δ.

3.5 Diffusive diffraction

Equation (2.35) is analogous to the scattering function in neutron scattering with $\mathbf{q}$ corresponding to the scattering wave vector and emphasises the Fourier relationship between attenuation and displacement in the PGSE experiment. However, there are major differences in the temporal and spatial timescales of each type of experiment. Further, $E(\mathbf{q}, \Delta)$ is measured in the time domain in PGSE NMR experiments and in the frequency domain in neutron scattering experiments. Indeed, a direct analogy can be made between PGSE NMR of a spin undergoing restricted diffusion in an enclosed pore and scattering experiments such as optical diffraction by a single slit.[12,13,27–32] Appropriate values of $\mathbf{q}$ are required to return information on the required length scale. For a ^{1}H PGSE experiment taking a maximum value of g to 50 T m^{-1} and $\delta = 2$ ms gives a value of q of 3×10^7 m^{-1} which corresponds to minimum observable displacements of 10 … 100 nm in a PGSE experiment.[33]

From Eq. (2.35) and using the Wiener–Khintchine theorem[34,35] (i.e., the Fourier transform of a time auto-correlation function is the frequency power spectrum), we find that $E(\mathbf{q}, \infty)$ is the power spectrum of $\rho(\mathbf{r}_0)$,[13,28,36]

$$
\begin{aligned}
E(\mathbf{q},\infty) &= \int \bar{P}(\mathbf{R},\infty)e^{i2\pi\mathbf{q}\cdot\mathbf{R}}d\mathbf{R} \\
&= \int\int \rho(\mathbf{r}_0)\,\rho(\mathbf{r}_0+\mathbf{R})d\mathbf{r}_0\,e^{i2\pi\mathbf{q}\cdot\mathbf{R}}d\mathbf{R} \\
&= \int\int \rho(\mathbf{r}_0)\,\rho(\mathbf{r}_1)d\mathbf{r}_0\,e^{i2\pi\mathbf{q}\cdot(\mathbf{r}_1-\mathbf{r}_0)}d\mathbf{R} \\
&= \int \rho(\mathbf{r}_0)\,e^{-i2\pi\mathbf{q}\cdot\mathbf{r}_0}d\mathbf{r}_0 \int \rho(\mathbf{r}_1)\,e^{i2\pi\mathbf{q}\cdot\mathbf{r}_1}d\mathbf{r}_1 \\
&= S^*(\mathbf{q})S(\mathbf{q}) \\
&= |S(\mathbf{q})|^2 = |\bar{P}(q)|^2,
\end{aligned}
\tag{3.16}
$$

where $S(\mathbf{q})$ is the Fourier transform of $\rho(\mathbf{r}_1)$. As will be shown in the next section, this is the origin of diffraction-like effects in PGSE diffusion studies. In quasielastic neutron scattering $|S(\mathbf{q})|^2$ is known as the elastic incoherent structure factor whilst in scattering theory it is referred to as the form factor of the confining volume.[31] $S(\mathbf{q})$ is analogous to the signal measured in conventional NMR imaging (aka k-space or MRI); see Eq. (2.36).[36–38] However, whereas conventional (i.e., k-space) imaging returns the phase-sensitive spatial spectrum of the restricting pore $\rho(\mathbf{r}_0)$, $E(\mathbf{q}, \infty)$ measures the power spectrum, $|S(\mathbf{q})|^2$. Thus $E(\mathbf{q}, \infty)$ is sensitive to average features in local structure, not the motional characteristics. Since $E(\mathbf{q}, \infty)$ measures the power spectrum of $S(\mathbf{q})$, Fourier inversion cannot be used and in consideration of the diffraction analogy, structural information can be obtained in q-space by using the characteristic features of the diffraction pattern such as the position of the nodes. However, as noted in Chapter 2, q-space imaging has the potential to give much higher resolution than conventional k-space imaging since the entire signal from the sample is available to contribute to each pixel in **R**-space (i.e., **R**, the dynamic displacement)[13] rather than from a volume element (i.e., voxel) as in conventional k-space imaging. Typically k-space imaging is limited by sensitivity to voxels with sides of 10 μm or greater; however, the resolution achievable in q-space imaging is limited only by the magnitude of q, with the shortest length scale that can be probed being $\sim q_{\max}^{-1}$, where $q_{\max}$ is the maximum value of q used in an experiment.[26] While q-space imaging cannot return spatial location, in many porous substances, such as suspensions of biological cells and emulsions, it is the spatial details of the small local structures that is of more interest.

It is important to note that the above discussion concerns 'diffusive (i.e., q-space) diffraction' and not 'Mansfield (k-space) diffraction'.[22,39–41] k-space diffraction depends on the relative *positions* (i.e., spatial periodicity of concentrations of nuclei) at fixed time (normal imaging returns an image of $\rho(\mathbf{r})$) whereas q-space diffraction depends on the relative *displacements* from the molecular origin during Δ.

3.6 An example – diffusion between planes

We begin by analysing the case of diffusion between reflecting (i.e., impenetrable) planes with a separation of $2a$. Inserting the expression for the propagator (Eq. (1.84)) and the spin density (Eq. (1.31); $\rho(\mathbf{r}_0) = 1/(2a)$) into Eq. (2.34) to obtain the SGP solution,[42]

$$E(\mathbf{q},\Delta) = \frac{2[1-\cos(2\pi q(2a))]}{(2\pi q(2a))^2} + 4(2\pi q(2a))^2 \sum_{n=1}^{\infty} \exp\left(-\frac{n^2\pi^2 D\Delta}{(2a)^2}\right) \frac{1-(-1)^n\cos(2\pi q(2a))}{\left[(2\pi q(2a))^2-(n\pi)^2\right]^2}. \quad (3.17)$$

This attenuation function is plotted in Figure 3.1.

In the long-time limit ($\xi \gg 1$), only the first term, which results from the zero-eigenvalue, remains, namely

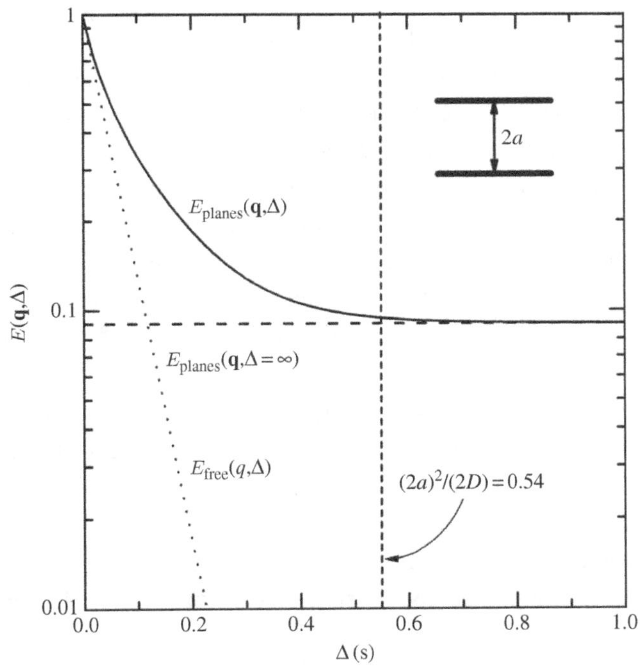

Figure 3.1 A plot of the PGSE attenuation for diffusion between reflecting planes versus Δ. The plot was simulated using the SGP approximation Eq. (3.17) with $D = 2.3 \times 10^{-9}\,\text{m}^2\,\text{s}^{-1}$, $q = 15{,}000\,\text{m}^{-1}$ and $a = 25\,\mu\text{m}$. At very short values of Δ, the attenuation is the same as that for free isotropic diffusion (Eq. (2.46)). As Δ increases the attenuation plateaus out. However, the signal from unrestricted particles is completely attenuated. By the time Δ is large enough that the MSD is equal to the interplanar spacing (vertical line), $E(\mathbf{q}, \Delta)$ has almost decreased to the long-time solution $E(\mathbf{q}, \infty)$ as given by Eq. (3.18).

$$E(\mathbf{q},\infty)=|S(\mathbf{q})|^2=\frac{2[1-\cos(2\pi q(2a))]}{(2\pi q(2a))^2}=|\mathrm{sinc}(\pi q(2a))|^2. \tag{3.18}$$

Importantly, this term is independent of diffusion. At low q ($\sim qa<0.15$) this reduces to

$$E(\mathbf{q}\to 0,\infty)=\exp\left(-(2\pi q)^2\frac{(2a)^2}{12}\right). \tag{3.19}$$

Comparison with Eq. (3.8) and application of Eq. (3.10) yield

$$\langle Z^2\rangle=\frac{(2a)^2}{6} \tag{3.20}$$

and

$$D(\Delta)=\frac{(2a)^2}{12\Delta}, \tag{3.21}$$

respectively. Note the last two equations should be compared with Eqs. (1.106) and (1.107), respectively.

As can be inferred from the attenuation behaviour, as the length of the diffusion experiment is such that the diffusing spins interact with the enclosing geometry (i.e., the RMSD is of the order of the characteristic length of the geometry) the attenuation profile differs significantly from that of the free diffusion model. Indeed, such differences can be put to use for discriminating between species inside and outside such pores (e.g., a suspension of biological cells) since values of q and Δ can be chosen such that only the intrapore signal remains[43] as can be understood from Figure 3.1.

When the interactions with the boundary become significant ($\Delta \gtrsim a^2/D$) an interesting effect is noted if the attenuation is plotted as a function of q as shown in Figure 3.2.

Diffusive diffraction effects are evident in the figure and structural information about the enclosing geometry can be obtained from the characteristics of the diffraction pattern. In analogy with Eq. (2.19), the diffractive minima occur at $q=n/(2a)$ ($n=1, 2, 3,\ldots$) – also as expected from analysis of Eq. (3.18). The effects of surface roughness and plane misalignment have also been considered.[44,45]

The solution for the planar geometry derived using the GPD approximation is quite different:[20]

$$E(\mathbf{g},\Delta)=\exp\left(-\frac{8\gamma^2g^2(2a)^4}{D\pi^6}\sum_{n=0}^{\infty}\frac{1}{(2n+1)^6}\left\{2\delta-\frac{2-2L(\delta)+L(\Delta-\delta)-2L(\Delta)+L(\Delta+\delta)}{(2n+1)^2\pi^2D/(2a)^2}\right\}\right) \tag{3.22}$$

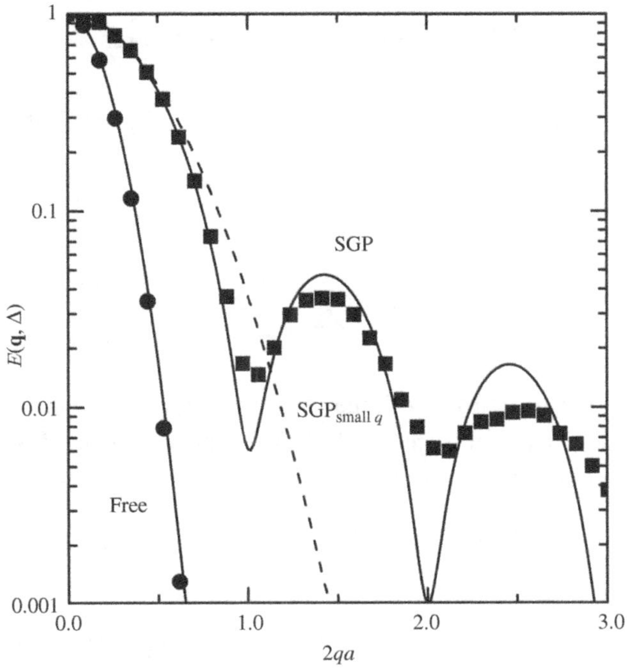

Figure 3.2 Experimental PGSE data (■) for water $D = 3.69 \times 10^{-9}\ \mathrm{m^2\,s^{-1}}$ diffusing between planes separated by a distance ($2a$) of 128.4 μm. The data were acquired with $\Delta = 2$ s and $\delta = 2$ ms, which is very close to SGP conditions with $\xi = 0.4$. The data were simulated (solid line) using Eq. (3.17). The dashed line represents the $E(\mathbf{q}\to 0,\infty)$ solution (Eq. (3.19)). The attenuation of freely diffusing water (●) is also shown for comparison (in this case the dependence on $2a$ as plotted is of course artificial). It is emphasised that diffractive phenomena are only evident at high attenuations (i.e., $E < 0.1$) and plotting the ordinate on a log scale aids their visibility. Modified from Price *et al.*[44]

where $L(t) = \exp\left[-(2n+1)^2\pi^2 Dt/(2a)^2\right]$, which in the steady gradient limit reduces to[46,47]

$$E(\mathbf{g},\Delta) = \exp\left(-\frac{8\gamma^2 g^2 (2a)^4}{D\pi^6}\sum_{n=0}^{\infty}\frac{1}{(2n+1)^6}\left\{2\tau - \frac{3 - 4L(\tau) + L(2\tau)}{(2n+1)^2\pi^2 D/(2a)^2}\right\}\right) \tag{3.23}$$

which reverts to that for the steady gradient experiment for free diffusion (cf. Eq. (2.50)) when $\xi \ll 1$. In the limit $\xi \gg 1$, Eq. (3.23) becomes[47]

$$E(\mathbf{g},\Delta) = -\frac{(2a)^4\gamma^2 g^2}{D}\frac{1}{120}\left(2\tau - \frac{17}{56}\frac{(2a)^2}{D}\right). \tag{3.24}$$

3.7 Diffusion in reflecting spheres

For spins confined to a perfectly reflecting sphere of radius a, the SGP solution obtained by substituting Eq. (1.90) into Eq. (2.34) is given by[20,48]

$$E(\mathbf{q},\Delta) = \left[\frac{3j_1(2\pi qa)}{(2\pi qa)}\right]^2 + 6(2\pi qa)^2 \sum_{n=0}^{\infty} \left[j'_n(2\pi qa)\right]^2 \sum_{m=1}^{\infty} \frac{(2n+1)\alpha_{nm}^2}{\alpha_{nm}^2 - n^2 - n} \frac{\exp\left(-\alpha_{nm}^2 \frac{D\Delta}{a^2}\right)}{\left[\alpha_{nm}^2 - (2\pi qa)^2\right]^2}, \quad (3.25)$$

where α_{nm} is the mth root of the equation $j'_n(\alpha_{nm}) = nJ_{n+1/2}(\alpha_{nm}) - \alpha_{nm} J_{n+3/2}(\alpha_{nm}) = 0$ and j is the spherical Bessel function of the first kind.[49] In the long-time limit, Eq. (3.25) reduces to[42]

$$E(\mathbf{q},\infty) = \left[\frac{3j_1(2\pi qa)}{2\pi qa}\right]^2 = \frac{9[(2\pi qa)\cos(2\pi qa) - \sin(2\pi qa)]^2}{(2\pi qa)^6}, \quad (3.26)$$

and thus the first minima occurs at $q \approx 1.43/(2a)$. At low q this reduces by Taylor expansion to[15]

$$E(\mathbf{q} \to 0,\infty) = \exp\left(-(2\pi q)^2 \frac{(2a)^2}{20}\right). \quad (3.27)$$

Similarly, using the GPD approximation, the attenuation function is[50]

$$E(\mathbf{g},\Delta) = \exp\left(-\frac{2\gamma^2 g^2}{D^2} \sum_{n=1}^{\infty} \frac{2\alpha_n^2 D\delta - 2 + 2L(\delta) - L(\Delta-\delta) + 2L(\Delta) - L(\Delta+\delta)}{\alpha_n^6 (R^2\alpha_n^2 - 2)}\right), \quad (3.28)$$

where $L(t) = \exp(-\alpha_n^2 Dt)$ and the α_n are the roots of $(\alpha_n a)J'_{3/2}(\alpha_n a) - \frac{1}{2} J_{3/2}(\alpha_n a) = 0$. As $\Delta \to \infty$ with $\delta \ll \Delta$, Eq. (3.28) reduces to Eq. (3.27).[47] In the limit $\delta \to \tau$ (i.e., the steady gradient limit, see Figure 2.8) Eq. (3.28) becomes[47]

$$E(\mathbf{g},\Delta) = \exp\left(-\frac{2\gamma^2 g^2}{D^2} \sum_{n=1}^{\infty} \frac{2\alpha_n^2 D\tau - 3 + 4L(\tau) - L(2\tau)}{\alpha_n^6 (a^2\alpha_n^2 - 2)}\right), \quad (3.29)$$

which reverts to that for the steady gradient experiment for free diffusion (cf. Eq. (2.50)) when $\xi \ll 1$. In the limit $\xi \gg 1$ Eq. (3.29) becomes[47]

$$E(\mathbf{g},\Delta) = -\frac{a^4\gamma^2 g^2}{D} \frac{8}{175} \left(2\tau - \frac{581}{840} \frac{a^2}{D}\right). \quad (3.30)$$

3.8 Size distributions of the restricting geometry

Real systems containing restricting geometries generally have a distribution of characteristic length scales. Further, in general, the geometry of the restriction is not known and the systems may even have a range of geometries. Nevertheless, as is commonly found in systems such as emulsions, the form of the restriction is spherical and with an appropriate model, PGSE data can be analysed to give the size distribution; a number of reviews of the applications of such measurements to emulsions have appeared.[51–54] For example, for spins diffusing inside a polydisperse set of isolated spheres (i.e., no exchange) with a distribution of the sphere radii, $P(a)$, as might be found in an emulsion, the signal attenuation is given by a volume average of the PGSE attenuation function[55]

$$E_{\mathrm{poly}}(\mathbf{q},\Delta) = \frac{\int_0^\infty a^3 P(a) E(\mathbf{q},\Delta) da}{\int_0^\infty a^3 P(a) da}, \tag{3.31}$$

where $E(\mathbf{q},\Delta)$ is the attenuation function for a reflecting sphere of a particular radius (e.g., Eq. (3.28)). The a^3 term is included to account for the increase in the number of spins as the radius increases. The distribution function is often taken to be a log-normal function (cf. Eq. (2.114)) although other distributions such as Gaussians have been used.[56] The use of reflective boundary conditions as against absorbing boundary conditions contains the implicit assumption that the spin relaxation is independent of the size of the restriction.[57] Similar to the case of analysing a mixture of freely diffusing components (see Section 2.5), the inversion of Eq. (3.31), a Fredholm equation of the first kind, to provide $P(a)$ is non-trivial.[54,57–59]

An improvement based on the recognition that the echo-attenuation data obtained in the long-time limit contain all the necessary information on the radii distribution has been presented.[60] Fourel *et al.*[61] supplemented Eq. (3.31) by adding an unrestricted diffusion component. The unrestricted component is to either account for the case where the distribution of radii is not perfectly log-normal or to account for a freely diffusing component. The coupling of diffusion measurements with CPMG measurements of transverse relaxation to provide enhanced characterisation of emulsions has also been suggested.[62]

In PGSE data of real systems it can be difficult to discern coherence features in q-space plots. To facilitate the discernment of such phenomena Kuchel *et al.*[63] have demonstrated the use of a numerical procedure in which the coherence features are accentuated. Specifically, the raw data are weighted with a bell-shaped window function, interpolated with a shifting cubic spline, and then the second derivative is taken prior to Fourier transformation. The window function provides apodisation of the noisy data at high q-values, then together with taking the second derivative, this functions as a high-pass filter removing zero and low frequency components in the echo-signal attenuation.

3.9 Finite gradient pulses and the validity of the SGP and GPD approximations

The SGP and GPD approaches are the most commonly used methods for relating PGSE attenuation to the underlying diffusion processes, and consequently it is important to understand the implications and limitations of these two approximations. A conceptual and rather simplistic representation of q-space and the GPD and SGP approximations are given in Figure 3.3, and an example of the validity of the two approaches for simulating PGSE attenuation for diffusion between planes is compared in Figure 3.4.

As its name suggests, the GPD approach requires a Gaussian distribution of spin phases (i.e., Eq. (2.77)); this condition can be met only when the spins being studied undergo free (i.e., Gaussian) diffusion on the timescale of Δ. Consequently, the GPD

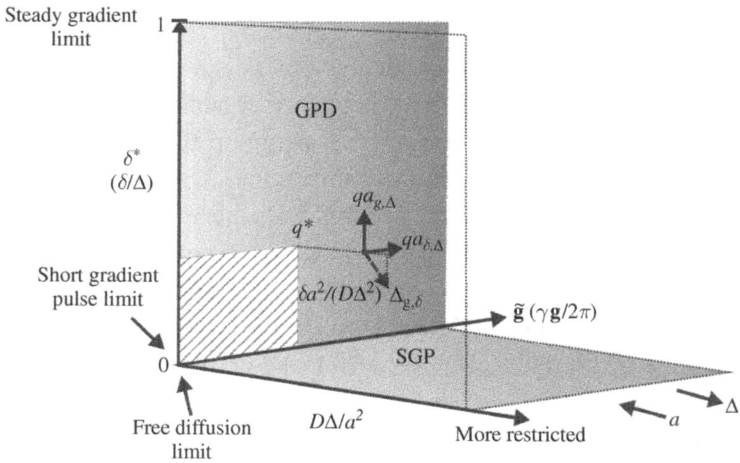

Figure 3.3 A conceptual representation of 'q-space' for a static experimental system (i.e., the characteristic distance of the restricting geometry, a, is unchanged). The shaded planes GPD and SGP signify the limits where these two approximations hold. $\delta/\Delta = 1$ denotes the steady gradient limit. The value of q^* (i.e., $q^* = q/\Delta$) at a particular point corresponds to the striped area on the GPD plane as marked. In a PGSE experiment q-space is probed in one of the three directions indicated on the inset axes by varying either δ, Δ or g (the subscripts denote the experimental variables are held constant). A PGSE experiment in which Δ is varied (dashed line) corresponds to sampling q-space in a line orthogonal to $\tilde{\mathbf{g}} = \gamma\mathbf{g}/2\pi$ and connecting the GPD and the SGP planes with slope $\delta a^2/(D\Delta^2)$. To allow for a more easily intuitive interpretation of data from restricted systems, it is preferable to plot the data using qa instead of $\tilde{g}$. Adapted from Price and Söderman.[64]

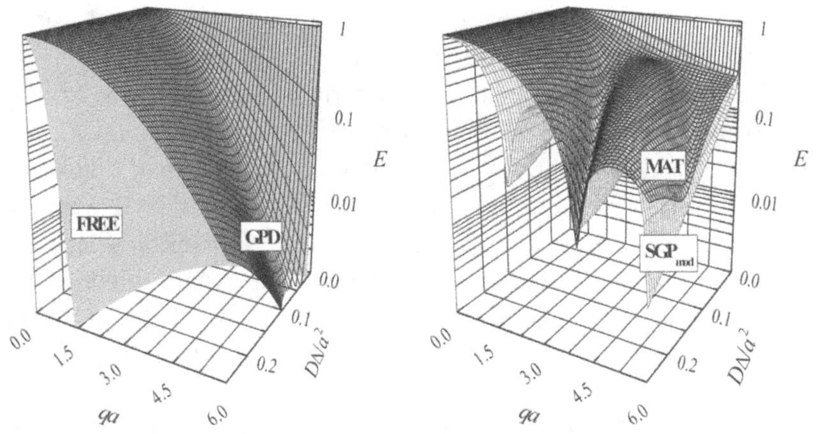

Figure 3.4 PGSE profiles for diffusion between planes simulated using the GPD (Eq. (3.22)), SGP_{mod} (Eq. (3.17) with Δ substituted by $\Delta-\delta/3$), matrix (see Section 3.10.4) and free diffusion formulae (Eq. (2.75)) for $\delta/\Delta = 0.5$.

approach is also applicable for spins diffusing within a restricted geometry providing Δ is so short that very few of the spins are affected by the boundary (i.e., ξ $(=D\Delta/a^2) \ll 1$) as previously explained in Section 1.7. When Δ becomes so long that the probability of being at any position is independent of the starting position, the change in phase becomes independent of the phase distribution.[47] From the central limit theorem the distribution of the sums of the phase changes becomes Gaussian. It has been shown that the GPD approximation solution for diffusion within a reflecting sphere (Eq. (3.22)) emulated Brownian motion simulations very well for $\xi < 1$, fairly well for $\xi \approx 1$ and well for $\xi > 1$.[20] However, as shown in Figure 3.4, the enforcing of a Gaussian phase condition leads to the loss of interference effects.[65,66] Whilst most studies on the validity of the GPD approach have been limited to closed geometries, Zielinski and Sen have considered its validity in open geometries.[5]

In contrast to the GPD approach, the SGP approximation is strictly valid for infinitely small δ such that motion during the gradient pulse can be neglected. Infinitely narrow gradient pulses are of course a technical impossibility and the condition is even less closely approximated in sequences designed to remove the effects of background gradients (see Section 7.6) or where the applied gradients are weak as in MRI-based studies where δ is typically of the order of Δ.[67] The SGP limit can in general be reasonably met in a freely diffusing system by ensuring $\delta \ll \Delta$ as is evident by comparison of the analytical solution for PGSE attenuation (Eq. (2.75)) and corresponding SGP solution (Eq. (2.44)). However, for diffusion within restricted geometries the situation is more complicated since even though it may be possible to ensure that $\delta \ll \Delta$, depending on the size of the restricted geometry, the diffusing spins may have significant interactions with the boundaries on the timescale of δ. In reality, the uncertainty in the initial and final positions due to motion during the gradient pulses is characterised by a sphere of radius $a_{\min}=q^{-1}$ and it is this that determines the lower limit of length scales that can be probed.[26,36,68]

Due to the difficulty in meeting the rigorous criteria required for the validity of the SGP approximation in experiments on many real porous systems (e.g., biological cells), there is considerable interest in understanding the effects of finite δ on PGSE measurements in restricted systems and also in extending the SGP approximation approach to encompass finite values of δ.[44,64,69] While with infinitely short gradient pulses a spin position will be accurately labelled anywhere within a restricted geometry, with increasing pulse lengths the particle will be labelled as being closer and closer to the centre of the pore since the entire pore volume has been sampled during the pulse.[70] Consequently, as δ increases the diffraction peaks shift towards higher q-values (that is to make the pore size appear smaller than the actual size) with the higher-order minima being more affected than the first minimum as shown in Figure 3.5.[15,65,71] Indeed at finite gradient pulse lengths, the echo amplitude is the

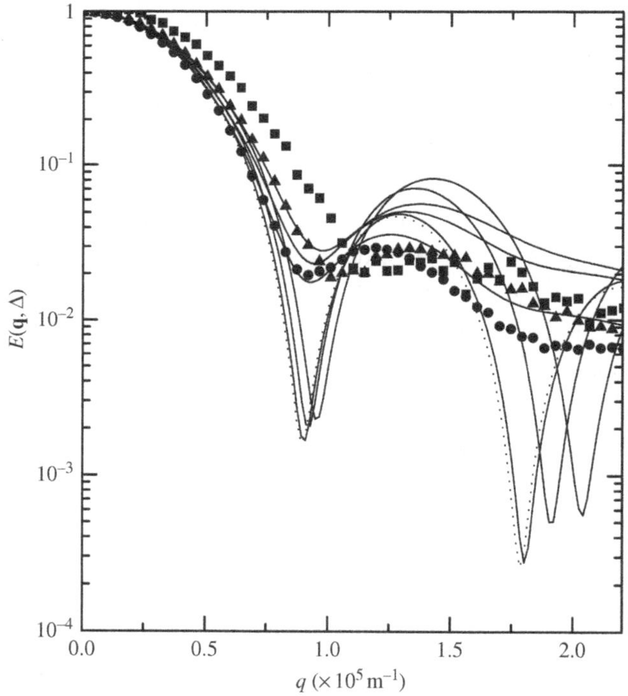

Figure 3.5 Effects of finite gradient pulse length on ^{1}H PGSE coherence patterns of water diffusing between planes. The data were acquired with $\Delta=2$ s, and three different values of δ (●: 2 ms, ▲: 100 ms, and ■: 200 ms). The data were fit using $2a=112$ μm. The dotted line denotes the SGP fit (Eq. (3.17)) to the data. The solid lines denote the matrix fits (see Section 3.10.4) and the three solid lines with shallower minima are Gaussian distributions ($\sigma=20$ μm and number of standard deviations $=1.4$) using the matrix formalism. From Price *et al.*[44]

spatial Fourier transform of a 'centre of mass' propagator.[72] As expected, as $\delta \rightarrow 0$, the centre of mass propagator reduces to the usual diffusion propagator. Indeed, from analysis of diffractive phenomena it has been realised that the rather stringent condition of $D\delta/a^2 \ll 0.02$ needs to be met for the SGP approximation to be valid in which case less than 10% of the signal attenuation will occur during δ and there will be less than 5% shift of the second minimum in the diffraction pattern.[73] Due to the rapid diffusion it is especially difficult to satisfy the SGP approximation in measurements of gas diffusion.[74]

Another point of interest with gradient pulses of finite width is the applicability of Eq. (3.10) to determine $D(\Delta)$ since strictly it is valid only in the SGP limit; outside this limit only an apparent time-diffusion coefficient, $D^{app}(\Delta)$, is determined. Whilst in closed geometries $D^{app}(\Delta)$ and $D(\Delta)$ can differ significantly,[73] in open restricted geometries the two quantities deviate by less than 10%.[5] Indeed, it was found that S_P/V_P and $\Im$ could be calculated from $D^{app}(\Delta)$ in open geometries even outside the SGP limit.[5]

As noted above, analytic solutions starting from the Bloch–Torrey equations account for finite gradient pulse effects; however, this approach is in general mathematically intractable. A simplistic approach that has been used to account for the finite length of δ has been, by analogy to Eqs. (2.44) and (2.75), to replace Δ by $\Delta-\delta/3$ in the SGP solutions for restricting geometries. However, this substitution is without physical basis and fails (see Figure 3.4).[64] Indeed, if this 'correction' is attempted on molecules undergoing restricted diffusion on the timescale of δ, the measured MSD will decrease with increasing δ.[72,75] Lori *et al.*[67] have presented a formalism using a modified q-space index for correcting predictions of the displacement distribution if the diffusion being studied is multi-Gaussian free diffusion. As a result of these limitations more sophisticated approximations and numerical approaches have been developed for analysing PGSE data and these are overviewed in the next section.

3.10 Other methods for determining $E(\mathbf{q}, \Delta)$

From the above discussion it can be understood that the commonly used approaches for relating the echo attenuation to the diffusive phenomena (i.e., solution of the Bloch–Torrey equations and the SGP and GPD approximations) have severe limitations: solutions to the Bloch–Torrey equations for realistic geometries become exceedingly difficult if not intractable, while the SGP and GPD solutions do not adequately represent the experimental reality (e.g., finite δ) or the physical reality (e.g., non-Gaussian diffusion), respectively. Numerous other theoretical approaches have been developed in attempts to account for some or all these limitations and they can be broadly classified into the following categories: solutions to the Bloch–Torrey equations, extensions to the SGP approach and Monte Carlo simulations of Brownian trajectories. Note some of the characterisation in the following subsections is rather imprecise since some studies involve multiple approaches.

3.10.1 Solution of the Bloch–Torrey equations

In this approach the Bloch–Torrey equations are solved with analytic generalisations[76–79] or numerically using finite difference, finite element and lattice methods.[46,71,80–86] Amongst these efforts, as noted by Grebenkov,[2] the work of Robertson,[46] where a quantum mechanical operator formalism was used, is related to later efficient numerical techniques (see Section 3.10.4).

3.10.2 Propagator and SGP approach

In an attempt to alleviate the mathematical complexity involved in the SGP approach, Hagslätt *et al.*[1] have used finite element calculations to generate the appropriate propagator which is then used to calculate the spin-echo intensity using the SGP equation (Eq. (2.34)).

3.10.3 Monte Carlo approaches

In this approach, Monte Carlo simulations of Brownian trajectories, summed over simulations of a large number of particles and a sufficient number of steps of appropriately sized jump distances, are used[87] to determine the phase distribution for the sample, which can then be used with Eq. (2.2) to determine the signal attenuation. There have been many attempts using this approach.[15,20,29,65,72,88–97]

3.10.4 Multiple narrow pulse and matrix formalism

A general, and given appropriate numerical precision, analytical solution to modelling PGSE attenuation data is given by the matrix formulation of Callaghan[69] based on the multiple propagator approach originally suggested by Caprihan *et al.*[98] In this approach involving the eigenvalues of the Laplace operator, the Hahn echo-based PGSE sequence for example is discretised into $2N+1$ intervals of length τ such that the total length of the sequence is $(2N+1)\tau$ as illustrated in Figure 3.6.

Thus at time $t = n\tau$ the gradient impulse will be

$$\mathbf{q}_n = m_n \mathbf{q} \tag{3.32}$$

where

$$\mathbf{q} = (2\pi)^{-1} \gamma \delta \mathbf{g}_{\text{step}} \tag{3.33}$$

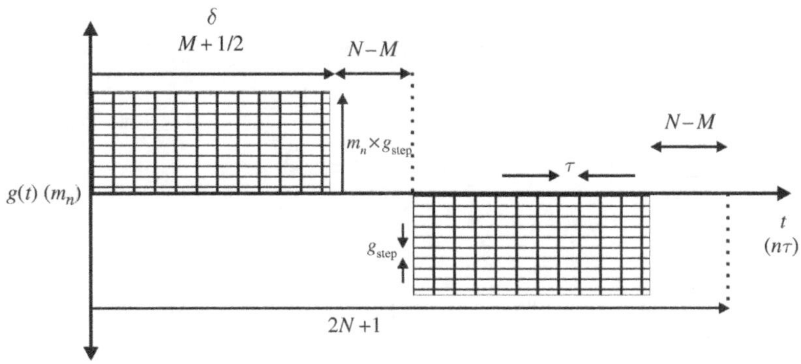

Figure 3.6 Schematic diagram of the discretisation of the Hahn echo-based PGSE sequence.

and m_n is some positive or negative integer multiple of the gradient amplitude discretising unit $\mathbf{g}_{\text{step}}$, viz.

$$m_n = \text{int}\left(\frac{\mathbf{g}(n\tau)}{\mathbf{g}_{\text{step}}}\right). \tag{3.34}$$

The SGP master equation (Eq. (2.34)) can be rewritten as

$$E(\mathbf{q}, \Delta) = \int\int \rho(\mathbf{r}_0) e^{-i2\pi\,\mathbf{q}\cdot\mathbf{r}_0} P(\mathbf{r}_0, \mathbf{r}_1, \Delta) e^{i2\pi\,\mathbf{q}\cdot\mathbf{r}_1} d\mathbf{r}_0\, d\mathbf{r}_1. \tag{3.35}$$

For brevity we write

$$\text{e}(\mathbf{r}_i) = \exp(i2\pi\mathbf{q}_i \cdot \mathbf{r}_i), \tag{3.36}$$

and thus, we can rewrite the SGP equation as

$$E(\mathbf{q}, \Delta) = \int\int \rho(\mathbf{r}_0)\text{e}(\mathbf{r}_0)P(\mathbf{r}_0, \mathbf{r}_1, \Delta)\text{e}(\mathbf{r}_1)d\mathbf{r}_0 d\mathbf{r}_1. \tag{3.37}$$

Eq. (3.37) can be expanded to account for a generalised gradient waveform discretised into N time intervals τ each bounded by impulses, $\mathbf{q}_n$, $\mathbf{q}_{n+1}$, etc. to obtain

$$\begin{aligned} E(\mathbf{q}, \Delta) = \int &\rho(\mathbf{r}_0)\, \text{e}(\mathbf{r}_0)\, P(\mathbf{r}_0, \mathbf{r}_1, \tau)\, \text{e}(\mathbf{r}_1)\, P(\mathbf{r}_1, \mathbf{r}_2, \tau)\, \text{e}(\mathbf{r}_2) \\ &\cdots \text{e}(\mathbf{r}_{N-1})\, P(\mathbf{r}_{N-1}, \mathbf{r}_N, \tau)\, \text{e}(\mathbf{r}_N)\, d\mathbf{r}_0\, d\mathbf{r}_1 \cdots d\mathbf{r}_N, \end{aligned} \tag{3.38}$$

and obtaining a solution in terms of an eigenmode expansion (see Eq. (1.78))

$$P(\mathbf{r}_0, \mathbf{r}_1, t) = \sum_k K_k(\mathbf{r}_0)K_k^*(\mathbf{r}_1) \exp(-\lambda_k t). \tag{3.39}$$

As noted in Section 1.5.3, the eigenfunctions are real and so the complex conjugate is in reality superfluous. Inserting Eq. (3.39) into (3.38) we obtain

$$E(\mathbf{q}, \Delta) = \int \rho(\mathbf{r}_0)\mathrm{e}(\mathbf{r}_0) \left[\sum_{k_1=0}^{\infty} K_{k_1}(\mathbf{r}_0) K_{k_1}^*(\mathbf{r}_1) \exp\left(-\lambda_{k_1} t\right) \right] \mathrm{e}(\mathbf{r}_1)$$
$$\times \left[\sum_{k_2=0}^{\infty} K_{k_2}(\mathbf{r}_1) K_{k_2}^*(\mathbf{r}_2) \exp\left(-\lambda_{k_2} t\right) \right] \mathrm{e}(\mathbf{r}_2)$$
$$\cdots$$
$$\times \left[\sum_{k_N=0}^{\infty} K_{k_N}(\mathbf{r}_{N-1}) K_{k_N}^*(\mathbf{r}_N) \exp\left(-\lambda_{k_N} t\right) \right] \mathrm{e}(\mathbf{r}_N)\, d\mathbf{r}_0\, d\mathbf{r}_1 \cdots d\mathbf{r}_N. \quad (3.40)$$

Putting all the summation indices together this becomes

$$E(\mathbf{q}, \Delta) = \int \rho(\mathbf{r}_0)\mathrm{e}(\mathbf{r}_0) \sum_{k_1, k_2 \cdots k_N} K_{k_1}(\mathbf{r}_0) K_{k_1}^*(\mathbf{r}_1) \exp\left(-\lambda_{k_1} t\right) \mathrm{e}(\mathbf{r}_1)$$
$$\times K_{k_2}(\mathbf{r}_1) K_{k_2}^*(\mathbf{r}_2) \exp\left(-\lambda_{k_2} t\right) \mathrm{e}(\mathbf{r}_2)$$
$$\cdots$$
$$\times K_{k_N}(\mathbf{r}_{N-1}) K_{k_N}^*(\mathbf{r}_N) \exp\left(-\lambda_{k_N} t\right) \mathrm{e}(\mathbf{r}_N)\, d\mathbf{r}_0\, d\mathbf{r}_1 \cdots d\mathbf{r}_N. \quad (3.41)$$

Setting $\rho(\mathbf{r}_0) = V^{-1}$ and rearranging the 'pieces' so that those that depend on the same $\mathbf{r}_n$ are grouped together and defining

$$A_{k_i k_{i+1}}(\mathbf{q}_i) = \int K_{k_i}^*(\mathbf{r}_i)\, K_{k_{i+1}}(\mathbf{r}_i) \mathrm{e}(\mathbf{r}_i) d\mathbf{r}_i, \quad (3.42)$$

$$R_{kk} = \exp(-\lambda_k \tau), \quad (3.43)$$

and

$$S_k(\mathbf{q}_i) = V^{-1/2} \int K_k(\mathbf{r}_i)\, e(\mathbf{r}_i) d\mathbf{r}_i \quad (3.44)$$

and noting that $\exp\left(i2\pi q\right) = \exp\left(-i2\pi q\right)^*$ and after some rearrangement Eq. (3.41) becomes

$$E(q) = \sum_{k_1, k_2 \cdots k_N} S_{k_1}(\mathbf{q}_0) R_{k_1 k_1} A_{k_1 k_2}(\mathbf{q}_1) R_{k_2 k_2} A_{k_2 k_3}(\mathbf{q}_2) \ldots R_{k_N k_N} S_{k_N}^*(-\mathbf{q}_N). \quad (3.45)$$

The delta function identity (Eq. (1.79)) can be used to obtain

$$\sum_{k_{i+1}} A_{k_i k_{i+1}}(\mathbf{q}_i) A_{k_{i+1} k_{i+2}}\left(\mathbf{q}_{i+1}\right) = A_{k_i k_{i+2}}\left(\mathbf{q}_i + \mathbf{q}_{i+1}\right), \quad (3.46)$$

which leads to the result

$$A(n\mathbf{q}) = A(\mathbf{q})^n. \tag{3.47}$$

This is a very important result since it allows us to determine only $A(\mathbf{q})$ for the smallest value used to digitise the waveform. Hence Eq. (3.45) becomes

$$E(\mathbf{q}, \Delta) = S(\mathbf{q}_0)RA(\mathbf{q}_1)RA(\mathbf{q}_2)\ldots RA(\mathbf{q}_{N-1})RS^{\dagger}(-\mathbf{q}_N), \tag{3.48}$$

where the first S is a row vector and the last $S^{\dagger}$ is a column vector ($S^{\dagger} = S^{\mathrm{T}*} = S^{*\mathrm{T}}$).

Thus in the case of a waveform $\mathbf{g}(t)$ that begins and ends with zero amplitude and which has zero time integral at the sampling time in order to satisfy the echo condition (Eq. (2.26)), setting the initial and final impulses to the minimum value of $\mathbf{q}$ (not a serious error when N is large) gives

$$E(\mathbf{q}, \Delta) = S(\mathbf{q})R[A(\mathbf{q})]^{m_2}\cdots R[A(\mathbf{q})]^{m_n}\ldots R[A(\mathbf{q})]^{m_N}RS^{\dagger}(-\mathbf{q}). \tag{3.49}$$

Importantly Eq. (3.49) can be evaluated providing the three matrices, $A(\mathbf{q})$, $R(\tau)$ and $S(\mathbf{q})$ are calculated, where $\mathbf{q}$ is the smallest impulse to digitise the waveform.

We illustrate the use of this formalism for the case of diffusion between reflecting planes. For this formulation to be valid, τ must be less than the characteristic timescales, that is, the following two conditions must be met,[99]

$$\tau \ll \frac{(2a)^2}{D} \tag{3.50}$$

(i.e., the MSD during τ is much less than the planar separation) and

$$\tau \ll \left(\gamma^2 g^2 D\right)^{-1/3} \tag{3.51}$$

(i.e., during τ the spins must travel a distance much less than that required for them to dephase by 2π radians).

Using the above discretisation we have

$$\Delta = (N + 1/2)\tau \tag{3.52}$$

$$\delta = (M + 1/2)\tau, \tag{3.53}$$

hence the total effective scattering wave-vector amplitude is

$$q_{\mathrm{net}} = (M+1)q_\tau = (M+1)(2\pi)^{-1}\gamma g\tau \tag{3.54}$$

and from Eq. (3.49) the matrix equation for the attenuation in the Hahn echo-based PGSE sequence as depicted in Figure 3.6 is

$$E(\mathbf{q}, \Delta) = S(\mathbf{q})[RA(\mathbf{q})]^M R^{N-M}\left[RA^{\dagger}(\mathbf{q})\right]^M RS^{\dagger}(\mathbf{q}) \tag{3.55}$$

Noting the eigenfunctions (Eq. (1.82)) and eigenvalues (Eq. (1.83)) for diffusion between planes, the component matrices are given by

$$S = BS' \tag{3.56}$$

$$A = C^{\dagger} A' C \tag{3.57}$$

and

$$R = \exp\left(\frac{-k^2\pi^2 D\tau}{(2a)^2}\right), \tag{3.58}$$

where B and C are diagonal matrices with

$$B = \begin{bmatrix} \frac{1}{2a} & & & \\ & \frac{\sqrt{2}}{2a} & & \\ & & \ddots & \\ & & & \frac{\sqrt{2}}{2a} \end{bmatrix} \tag{3.59}$$

$$C = \begin{bmatrix} \sqrt{\frac{1}{2a}} & & & \\ & \sqrt{\frac{1}{a}} & & \\ & & \ddots & \\ & & & \sqrt{\frac{1}{a}} \end{bmatrix} \tag{3.60}$$

$$S'_k = \begin{cases} \dfrac{i4a\exp(i2\pi qa)(4\pi qa)\cos(2\pi qa)}{(4\pi qa)^2-(k\pi)^2} & k \text{ odd} \\ \dfrac{4a\exp(i2\pi qa)(4\pi qa)\sin(2\pi qa)}{(4\pi qa)^2-(k\pi)^2} & k \text{ even} \end{cases} \tag{3.61}$$

and

$$A'_{kk'} = \frac{1}{2}\left[S'_{|k-k'|} + S'_{k+k'}\right]. \tag{3.62}$$

Examples of the use of this formulation are given in Figures 3.4 and 3.5.

Solutions using this formalism have also been given for planar, cylindrical and spherical pores including surface relaxation.[99–101] Later, this matrix approach was reformulated by Sukstanskii and Yablonskiy[102] in terms of a random walk.

Barzykin proposed an alternative and highly efficient matrix formalism by considering a stepwise approximation of the temporal gradient profile.[103,104] In his approach, the linear magnetic field operator in the Bloch–Torrey equation is expanded in terms of eigenmodes of the diffusion propagator for the relevant geometry. As a result, the two matrices determining the macroscopic signal depend solely on the confining geometry and have to be calculated only once. All simple geometries were solved analytically by this method with and without surface relaxation. The memory function approximation of Sheltraw and Kenkre[66] (see Section 3.10.5) was obtained in the first order of the general magnetic field operator expansion. It was proved that the stepwise gradient formulation is mathematically equivalent to the multiple propagator formulation in appropriate limits. Barzykin's approach is computationally superior in situations where the temporal gradient profile is naturally stepwise, such as for Hahn or Stejskal–Tanner pulse sequences.

3.10.5 Other approaches

Sheltraw and Kenkre[66] presented a memory function approximation based on application of the projection operator method[105,106] to the evolution equation of the system density matrix. Importantly this approach, whilst related to the GPD approximation, does predict diffraction effects. Other approaches include a lattice Boltzmann procedure to numerically simulate time evolution of magnetisation in a porous medium.[107] Grebenkov[2,108] has recently presented a multiple correlation function approach which is related to that of Axelrod and Sen.[77] We also mention that artificial neural networks have been used to treat complex geometries.[109,110]

References

1. H. Hagslätt, B. Jönsson, M. Nydén, and O. Söderman, Predictions of Pulsed Field Gradient NMR Echo-Decays for Molecules Diffusing in Various Restrictive Geometries. Simulations of Diffusion Propagators Based on a Finite Element Method. *J. Magn. Reson.* **161** (2003), 138–47.
2. D. S. Grebenkov, NMR Survey of the Reflected Brownian Motion. *Rev. Mod. Phys.* **79** (2007), 1077–136.
3. P. N. Sen, Time-Dependent Diffusion Coefficient as a Probe of Geometry. *Concepts Magn. Reson.* **23A** (2004), 1–21.
4. J. E. Tanner, Transient Diffusion in a System Partitioned by Permeable Barriers. Application to NMR Measurements with a Pulsed Field Gradient. *J. Chem. Phys.* **69** (1978), 1748–54.
5. L. J. Zielinski and P. N. Sen, Effects of Finite Width Pulses in the Pulsed Field Gradient Measurement of the Diffusion Coefficient in Connected Porous Media. *J. Magn. Reson.* **165** (2003), 153–61.
6. F. Stallmach and P. Galvosas, Spin Echo NMR Diffusion Studies. In *Annual Reports on NMR Spectroscopy*, ed. G. A. Webb. vol. 61. (London: Elsevier, 2007), pp. 51–131.
7. D. S. Grebenkov, Multiexponential Attenuation of the CPMG Spin Echoes Due to a Geometrical Confinement. *J. Magn. Reson.* **180** (2006), 118–26.
8. D. E. Woessner, N. M. R. Spin-Echo Self-Diffusion Measurements on Fluids Undergoing Restricted Diffusion. *J. Phys. Chem.* **67** (1963), 1365–7.

9. T. M. De Swiet and P. N. Sen, Decay of Nuclear Magnetization by Bounded Diffusion in a Constant Field Gradient. *J. Chem. Phys.* **100** (1994), 5597–604.
10. M. D. Hürlimann, K. G. Helmer, T. M. De Swiet, P. N. Sen, and C. H. Sotak, Spin Echoes in a Constant Gradient and in the Presence of Simple Restriction. *J. Magn. Reson. A* **113** (1995), 260–4.
11. K. G. Helmer, M. D. Hürlimann, T. M. De Swiet, P. N. Sen, and C. H. Sotak, Determination of Ratio of Surface Area to Pore Volume from Restricted Diffusion in a Constant Field Gradient. *J. Magn. Reson. A* **115** (1995), 257–69.
12. D. G. Cory and A. N. Garroway, Measurement of Translational Displacement Probabilities by NMR: An Indicator of Compartmentation. *Magn. Reson. Med.* **14** (1990), 435–44.
13. P. T. Callaghan, D. MacGowan, K. J. Packer, and F. O. Zelaya, High-Resolution q-Space Imaging in Porous Structures. *J. Magn. Reson.* **90** (1990), 177–82.
14. P. T. Callaghan and A. Coy, PGSE NMR and Molecular Translational Motion in Porous Media. In *NMR Probes and Molecular Dynamics*, ed. R. Tycko. (Dordrecht: Kluwer, 1994), pp. 489–523.
15. P. Linse and O. Söderman, The Validity of the Short-Gradient-Pulse Approximation in NMR studies of Restricted Diffusion. Simulations of Molecules Diffusing between Planes, in Cylinders and Spheres. *J. Magn. Reson. A* **116** (1995), 77–86.
16. K. J. Packer, Oil Reservoir Rocks Examined by MRI. In *Encyclopedia of Nuclear Magnetic Resonance*, ed. D. M. Grant and R. K. Harris. vol. 5. (New York: Wiley, 1996), pp. 3365–75.
17. P. T. Callaghan, C. D. Eccles, and Y. Xia, NMR Microscopy of Dynamic Displacements: $\boldsymbol{k}$-Space and $\boldsymbol{q}$-Space Imaging. *J. Phys. E: Sci. Instrum.* **21** (1988), 820–2.
18. A. J. Baldwin, J. Christodolou, P. D. Barker, and C. M. Dobson, Contribution of Rotational Diffusion to Pulsed Field Gradient Diffusion Measurements. *J. Chem. Phys.* **127** (2007), 114505-1–114505-8.
19. A. Szafer, J. Zhong, and J. C. Gore, Theoretical Model for Water Diffusion in Tissues. *Magn. Reson. Med.* **33** (1995), 697–712.
20. B. Balinov, B. Jönsson, P. Linse, and O. Söderman, The NMR Self-Diffusion Method Applied to Restricted Diffusion. Simulation of Echo Attenuation from Molecules in Spheres and Between Planes. *J. Magn. Reson. A* **104** (1993), 17–25 and *J. Magn. Reson. A* **108** (1994), 130.
21. P. N. Sen and M. D. Hürlimann, Analysis of Nuclear Magnetic Resonance Spin Echoes Using Simple Structure Factors. *J. Chem. Phys.* **101** (1994), 5423–30.
22. M. D. Hürlimann, T. M. De Swiet, and P. N. Sen, Comparison of Diffraction and Diffusion Measurements in Porous Media. *J. Non-Cryst. Solids* **182** (1995), 198–205.
23. P. N. Sen, M. D. Hürlimann, and T. M. De Swiet, Debye-Porod Law of Diffraction for Diffusion in Porous Media. *Phys. Rev. B* **51** (1995), 601–4.
24. M. D. Hürlimann, L. M. Schwartz, and P. N. Sen, Probability of Return to the Origin at Short Times: A Probe of Microstructure in Porous Media. *Phys. Rev. B* **51** (1995), 14936–40.
25. P. P. Mitra, Diffusion in Porous Materials as Probed by Pulsed Gradient NMR Measurements. *Physica A* **241** (1997), 122–7.
26. P. P. Mitra, L. L. Latour, R. L. Kleinberg, and C. H. Sotak, Pulsed-Field-Gradient NMR Measurements of Restricted Diffusion and the Return-to-the Origin Probability. *J. Magn. Reson. A* **114** (1995), 47–58.
27. H. G. Hertz, Translational Motions as Studied by Nuclear Magnetic Resonance. In *Molecular Motions in Liquids*, ed. J. Lascombe. (Dordrecht: Reidel, 1974), pp. 337–57.
28. P. T. Callaghan, A. Coy, D. MacGowan, K. J. Packer, and F. O. Zelaya, Diffraction-Like Effects in NMR Diffusion of Fluids in Porous Solids. *Nature* **351** (1991), 467–9.

29. P. T. Callaghan, A. Coy, D. MacGowan, and K. J. Packer, Diffusion of Fluids in Porous Solids Probed by Pulsed Field Gradient Spin Echo NMR. *J. Mol. Liquids* **54** (1992), 239–51.
30. R. M. Cotts, Diffusion and Diffraction. *Nature* **351** (1991), 443–4.
31. G. Fleischer and F. Fujara, NMR as a Generalized Incoherent Scattering Experiment. *NMR Basic Princ. Progr.* **30** (1994), 157–207.
32. M. Appel, G. Fleischer, D. Geschke, J. Kärger, and M. Winkler, Pulsed-Field-Gradient NMR Analogue of the Single-Slit Diffraction Pattern. *J. Magn. Reson. A* **122** (1996), 248–50.
33. J. Kärger and F. Stallmach, PFG NMR Studies of Anomalous Diffusion. In *Diffusion in Condensed Matter*, ed. P. Heitjans and J. Kärger. (Berlin: Springer, 2006), pp. 417–59.
34. D. A. McQuarrie, *Statistical Mechanics*. (New York: Harper & Row, 1976).
35. A. G. Marshall and F. R. Verdun, *Fourier Transforms in NMR, Optical, and Mass Spectroscopy. A User's Handbook*. (Amsterdam: Elsevier, 1990).
36. P. T. Callaghan, *Principles of Nuclear Magnetic Resonance Microscopy*. (Oxford: Clarendon Press, 1991).
37. S. L. Talagala and I. J. Lowe, Introduction to Magnetic Resonance Imaging. *Concepts Magn. Reson.* **3** (1991), 145–59.
38. Y. Xia, Contrast in NMR Imaging and Microscopy. *Concepts Magn. Reson.* **8** (1996), 205–25.
39. P. Mansfield and P. K. Grannell, 'Diffraction' and Microscopy in Solids and Liquids by NMR. *Phys. Rev. B* **12** (1975), 3618–34.
40. P. T. Callaghan, NMR Imaging, NMR Diffraction and Applications of Pulsed Gradient Spin Echoes in Porous Media. *Magn. Reson. Imaging* **14** (1996), 701–9.
41. G. A. Barrall, L. Frydman, and G. C. Chingas, NMR Diffraction and Spatial Statistics of Stationary Systems. *Science* **255** (1992), 714–17.
42. J. E. Tanner and E. O. Stejskal, Restricted Self-Diffusion of Protons in Colloidal Systems by the Pulsed-Gradient, Spin-Echo Method. *J. Chem. Phys.* **49** (1968), 1768–77.
43. J. Andrasko, Measurement of Membrane Permeability to Slowly Penetrating Molecules by a Pulse Gradient NMR Method. *J. Magn. Reson.* **21** (1976), 479–84.
44. W. S. Price, P. Stilbs, and O. Söderman, Determination of Pore Space Shape and Size in Porous Systems Using NMR Diffusometry. Beyond the Short Gradient Pulse Approximation. *J. Magn. Reson.* **160** (2003), 139–43.
45. D. Topgaard and O. Söderman, Experimental Determination of Pore Shape Using *q*-Space NMR Microscopy in the Long Diffusion-Time Limit. *Magn. Reson. Imaging* **21** (2003), 69–76.
46. B. Robertson, Spin-Echo Decay of Spins Diffusing in a Bounded Region. *Phys. Rev.* **151** (1966), 273–7.
47. C. H. Neuman, Spin Echo of Spins Diffusing in a Bounded Medium. *J. Chem. Phys.* **60** (1974), 4508–11.
48. W. S. Veeman, Diffusion in a Closed Sphere. In *Annual Reports on NMR Spectroscopy*, ed. G. A. Webb. vol. 50. (London: Elsevier, 2003), pp. 201–16.
49. M. Abramowitz and I. A. Stegun, *Handbook of Mathematical Functions*. (New York: Dover, 1970).
50. J. S. Murday and R. M. Cotts, Self-Diffusion Coefficient of Liquid Lithium. *J. Chem. Phys.* **48** (1968), 4938–45.
51. O. Söderman and P. Stilbs, NMR Studies of Complex Surfactant Systems. *Prog. NMR Spectrosc.* **26** (1994), 445–82.
52. O. Söderman, Pulsed-Field-Gradient NMR Studies of Emulsions. Droplet Sizes and Concentrated Emulsions. *Progr. Colloid Polym. Sci.* **106** (1997), 34–41.

53. O. Söderman, P. Stilbs, and W. S. Price, NMR Studies of Surfactants. *Concepts Magn. Reson.* **23A** (2004), 121–35.
54. M. L. Johns and K. G. Hollingsworth, Characterisation of Emulsion Systems Using NMR and MRI. *Prog. NMR Spectrosc.* **50** (2007), 51–70.
55. K. J. Packer and C. Rees, Pulsed NMR Studies of Restricted Diffusion. 1. Droplet Size Distributions in Emulsions. *J. Colloid Interface Sci.* **40** (1972), 206–18.
56. P. Callaghan, K. W. Jolley, and R. S. Humphrey, Diffusion of Fat and Water in Cheese as Studied by Pulsed Field Gradient Nuclear Magnetic Resonance. *J. Colloid Interface Sci.* **93** (1983), 521–9.
57. L. Ambrosone, S. Murgia, G. Cinelli, M. Monduzzi, and A. Ceglie, Size Polydispersity Determination in Emulsion Systems by Free Diffusion Measurements via PFG-NMR. *J. Phys. Chem. B* **108** (2004), 18472–8.
58. L. Ambrosone, A. Ceglie, G. Colafemmina, and G. Palazzo, A Novel Approach for Determining the Droplet Size Distribution in Emulsion Systems by Generating Function. *J. Chem. Phys.* **107** (1997), 10756–63.
59. L. Ambrosone, A. Ceglie, G. Colafemmina, and G. Palazzo, General Methods for Determining the Droplet Size Distribution in Emulsion Systems. *J. Chem. Phys.* **110** (1999), 797–854.
60. J. C. Van Den Enden, D. Waddington, H. Van Aalst, C. G. Van Kralingen, and K. J. Packer, Rapid Determination of Water Droplet Size Distributions by PFG-NMR. *J. Colloid Interface Sci.* **140** (1990), 105–13.
61. I. Fourel, J. P. Guillement, and D. Le Botlan, Determination of Water Droplet Size Distributions by Low Resolution PFG-NMR. I. Liquid Emulsions. *J. Colloid Interface Sci.* **164** (1994), 48–53.
62. A. A. Peña and G. J. Hirasaki, Enhanced Characterizations of Oil Field Emulsions via NMR Diffusion and Transverse Relaxation Experiments. *Adv. Colloid Interface Sci.* **105** (2003), 103–50.
63. P. W. Kuchel, T. R. Eykyn, and D. G. Regan, Measurement of Compartment Size in *q*-Space Experiments: Fourier Transform of the Second Derivative. *Magn. Reson. Med.* **52** (2004), 907–12.
64. W. S. Price and O. Söderman, Some 'Reflections' on the Effects of Finite Gradient Pulse Lengths in PGSE NMR Experiments in Restricted Systems. *Isr. J. Chem.* **43** (2003), 25–32.
65. A. Coy and P. T. Callaghan, Pulsed Gradient Spin Echo Nuclear Magnetic Resonance for Molecules Diffusing Between Partially Reflecting Rectangular Barriers. *J. Chem. Phys.* **101** (1994), 4599–609.
66. D. Sheltraw and V. M. Kenkre, The Memory-Function Technique for the Calculation of Pulsed-Gradient NMR Signals in Confined Geometries. *J. Magn. Reson. A* **122** (1996), 126–36.
67. N. F. Lori, T. E. Conturo, and D. Le Bihan, Definition of Displacement Probability and Diffusion Time in *q*-Space Magnetic Resonance Measurements That Use Finite-Duration Diffusion-Encoding Gradients. *J. Magn. Reson.* **165** (2003), 185–95.
68. K. G. Helmer, M. R. Meiler, C. H. Sotak, and J. D. Petrucelli, Comparison of the Return-to-the-Origin Probability and the Apparent Diffusion Coefficient of Water as Indicators of Necrosis in RIF-1 Tumors. *Magn. Reson. Med.* **49** (2003), 468–78.
69. P. T. Callaghan, A Simple Matrix Formalism for Spin Echo Analysis of Restricted Diffusion under Generalized Gradient Waveforms. *J. Magn. Reson.* **129** (1997), 74–84.
70. C. Malmborg, M. Sjöbeck, S. Brockstedt, E. Englund, O. Söderman, and D. Topgaard, Mapping the Intracellular Fraction of Water by Varying the Gradient Pulse Length in *q*-Space Diffusion MRI. *J. Magn. Reson.* **180** (2006), 280–5.

71. M. H. Blees, The Effect of Finite Duration of Gradient Pulses on the Pulsed-Field-Gradient NMR Method for Studying Restricted Diffusion. *J. Magn. Reson. A* **109** (1994), 203–9.
72. P. P. Mitra and B. I. Halperin, Effects of Finite Gradient Pulse Widths in Pulsed Field Gradient Diffusion Measurements. *J. Magn. Reson. A* **113** (1995), 94–101.
73. L. Z. Wang, A. Caprihan, and E. Fukushima, The Narrow-Pulse Criterion for Pulsed-Gradient Spin-Echo Diffusion Measurements. *J. Magn. Reson. A* **117** (1995), 209–19.
74. R. W. Mair, P. N. Sen, M. D. Hürlimann, S. Patz, D. G. Cory, and R. L. Walsworth, The Narrow Pulse Approximation and Long Length Scale Determination in Xenon Gas Diffusion NMR Studies of Model Porous Media. *J. Magn. Reson.* **156** (2002), 202–12.
75. C. Malmborg, D. Topgaard, and O. Söderman, NMR Diffusometry and the Short Gradient Pulse Limit Approximation. *J. Magn. Reson.* **169** (2004), 85–91.
76. V. M. Kenkre, E. Fukushima, and D. Sheltraw, Simple Solution of the Torrey–Bloch Equations in the NMR Study of Molecular Diffusion. *J. Magn. Reson.* **128** (1997), 62–9.
77. S. Axelrod and P. N. Sen, Nuclear Magnetic Resonance Spin Echoes for Restricted Diffusion in an Inhomogeneous Field: Methods and Asymptotic Regimes. *J. Chem. Phys.* **114** (2001), 6878–95.
78. V. M. Kenkre and F. J. Sevilla, Analytic Considerations in the Theory of NMR Microscopy. *Physica A* **371** (2006), 139–43.
79. F. J. Sevilla and V. M. Kenkre, Theory of the Spin Echo Signal in NMR Microscopy: Analytic Solutions of a Generalized Torrey–Bloch Equation. *J. Phys. Condens. Matter* **19** (2007), 065113-1–065113-14.
80. R. C. Wayne and R. M. Cotts, Nuclear-Magnetic-Resonance Study of Self-Diffusion in a Bounded Medium. *Phys. Rev.* **151** (1966), 264–72.
81. G. P. Zientara and J. H. Freed, Spin-Echoes for Diffusion in Bounded, Heterogeneous Media: A Numerical Study. *J. Chem. Phys.* **72** (1980), 1285–92.
82. V. A. Daragan and E. E. Il'ina, Pulsed Field Gradient NMR for the Study of the Structure of Membrane Systems. *Chem. Phys.* **158** (1991), 105–11.
83. E. G. Novikov, D. van Dusschoten, and H. Van As, Modeling of Self-Diffusion and Relaxation Time NMR in Multi-Compartment Systems. *J. Magn. Reson.* **135** (1998), 522–8.
84. L. J. Zielinski and P. N. Sen, Relaxation of Nuclear Magnetization in a Nonuniform Magnetic Field Gradient and in a Restricted Geometry. *J. Magn. Reson.* **147** (2000), 95–103.
85. P. N. Sen, A. André, and S. Axelrod, Spin Echoes of Nuclear Magnetization Diffusing in a Constant Magnetic Field Gradient and in a Restricted Geometry. *J. Chem. Phys.* **111** (1999), 6548–55.
86. S. N. Hwang, C.-L. Chin, F. W. Wehrli, and D. B. Hackney, An Image-Based Finite Difference Model for Simulating Restricted Diffusion. *Magn. Reson. Med.* **50** (2003), 373–82.
87. M. Leibig, Random Walks and NMR Measurements in Porous Media. *J. Phys. A: Math. Gen.* **26** (1993), 3349–67.
88. P. T. Callaghan, A. Coy, T. P. J. Halpin, D. MacGowan, K. J. Packer, and F. O. Zelaya, Diffusion in Porous Systems and the Influence of Pore Morphology in Pulsed Field Gradient Spin-Echo Nuclear Magnetic Resonance Studies. *J. Chem. Phys.* **97** (1992), 651–62.
89. G. Celebre, L. Coppola, and G. A. Raineri, Water Self-Diffusion in Lyotropic Systems by Simulation of Pulsed Field Gradient Spin Echo Nuclear Magnetic Resonance Experiments. *J. Chem. Phys.* **97** (1992), 7781–5.
90. P. P. Mitra, P. N. Sen, and L. M. Schwartz, Short-Time Behaviour of the Diffusion Coefficient as a Geometrical Probe of Porous Media. *Phys. Rev. B* **47** (1993), 8565–74.

91. A. J. Lennon and P. W. Kuchel, Enhancement of the 'Diffraction-Like' Effect in NMR Diffusion Experiments. *J. Magn. Reson. A* **111** (1994), 208–11.
92. P. N. Sen, L. M. Schwartz, P. P. Mitra, and B. I. Halperin, Surface Relaxation and the Long-Time Diffusion Coefficient in Porous Media: Periodic Geometries. *Phys. Rev. B* **49** (1994), 215–25.
93. B. Håkansson, B. Jönsson, P. Linse, and O. Söderman, The Influence of a Nonconstant Magnetic-Field Gradient on PFG NMR Diffusion Experiments. A Brownian-Dynamics Computer Simulation Study. *J. Magn. Reson.* **124** (1997), 343–51.
94. A. Duh, A. Mohorič, and J. Stepišnik, Computer Simulation of the Spin-Echo Spatial Distribution in the Case of Restricted Self-Diffusion. *J. Magn. Reson.* **148** (2001), 257–66.
95. R. M. E. Valckenborg, H. P. Huinink, J. J. van der Sande, and K. Kopinga, Random Walk Simulations of NMR Dephasing Effects Due to Uniform Magnetic Field Gradients in a Pore. *Phys. Rev. E* **65** (2002), 021306-1–021306-8.
96. D. S. Grebenkov, G. Guillot, and B. Sapoval, Restricted Diffusion in a Model Acinar Labyrinth by NMR: Theoretical and Numerical Results. *J. Magn. Reson.* **184** (2007), 143–56.
97. D. S. Grebenkov, Nuclear Magnetic Resonance Restricted Diffusion Between Parallel Planes in a Cosine Magnetic Field: An Exactly Solvable Model. *J. Chem. Phys.* **126** (2007), 104706-1–104706-15.
98. A. Caprihan, L. Z. Wang, and E. Fukushima, A Multiple-Narrow-Pulse Approximation for Restricted Diffusion in a Time-Varying Field Gradient. *J. Magn. Reson. A* **118** (1996), 94–102.
99. S. L. Codd and P. T. Callaghan, Spin Echo Analysis of Restricted Diffusion Under Generalized Gradient Waveforms: Planar, Cylindrical, and Spherical Pores with Wall Relaxivity. *J. Magn. Reson.* **137** (1999), 358–72.
100. K. Hayamizu, E. Akiba, T. Bando, Y. Aihara, and W. S. Price, NMR Studies on Poly (ethylene oxide)-Based Polymer Electrolytes with Different Cross-Linking Doped with $LiN(SO_2CF_3)_2$. Restricted Diffusion of the Polymer and Lithium Ion and Time-Dependent Diffusion of the Anion. *Macromolecules* **36** (2003), 2785–92.
101. B. N. Ryland and P. T. Callaghan, Spin Echo Analysis of Restricted Diffusion under Generalized Gradient Waveforms for Spherical Pores with Relaxivity and Interconnections. *Isr. J. Chem.* **43** (2003), 1–7.
102. A. L. Sukstanskii and D. A. Yablonskiy, Effects of Restricted Diffusion on MR Signal Formation. *J. Magn. Reson.* **157** (2002), 92–105.
103. A. V. Barzykin, Exact Solution of the Torrey–Bloch Equation for a Spin Echo in Restricted Geometries. *Phys. Rev. B* **58** (1998), 14171–4.
104. A. V. Barzykin, Theory of Spin Echo in Restricted Geometries Under a Step-Wise Gradient Pulse Sequence. *J. Magn. Reson.* **139** (1999), 342–53.
105. S. Nakajima, On Quantum Theory of Transport Phenomena. *Prog. Theor. Phys.* **20** (1958), 948–59.
106. R. Zwanzig, *Nonequilibrium Statistical Mechanics*. (Oxford: Oxford, 2001).
107. R. A. Guyer and K. R. McCall, Lattice Boltzmann Description of Magnetization in Porous Media. *Phys. Rev. B* **62** (2000), 3674–88.
108. D. S. Grebenkov, Multiple Correlation Function Approach: Rigorous Results for Simple Geometries. *Diffusion Fundamentals* **5** (2007), 1.1–1.34.
109. A. J. Lennon and P. W. Kuchel, Neural Networks Used to Interpret Pulsed-Gradient Restricted-Diffusion Data. *J. Magn. Reson. A* **107** (1994), 229–35.
110. P. W. Kuchel and C. J. Durrant, Permeability Coefficients from NMR *q*-Space Data: Models with Unevenly Spaced Semi-Permeable Parallel Membranes. *J. Magn. Reson.* **139** (1999), 258–72.

4

PGSE measurements in complex and exchanging systems

4.1 Introduction

A requirement in measuring transport (e.g., transmembrane) or exchange (e.g., ligand binding) is to be able to identify a measurable NMR parameter that has a different value in each state. Modulation of this parameter by the transport or exchange process is examined to characterise the process.[1] Traditionally, NMR chemical shifts or relaxation times have been used for this purpose.[2–4] With the advent of PGSE methods, a difference in diffusion properties (i.e., a difference in diffusion coefficient between sites or a difference in motional restriction) becomes another measurable NMR parameter that can be used to probe transport or exchange.

In the simplest case the exchange will occur between two freely diffusing sites (e.g., a ligand binding to a macromolecule; Figure 4.1); however, in many real systems (e.g., a suspension of biological cells) one site, or both sites if at higher cellular volume fractions, may be restricted. In contrast to the previous chapter where only simple restricting systems with reflecting boundary conditions were considered and the diffusing species did not interact with other restricting geometries, in real systems (e.g., biological cells, porous systems) it may also be necessary to consider the effects of a combination of exchange, restriction, obstruction and polydispersity in addition to surface and bulk relaxation as well as different bulk diffusion coefficients in each medium (e.g., Figure 4.2). As a consequence, modelling such systems can be very complicated and various approximations are necessarily used.

This chapter begins by examining the relatively simple case of exchange between two freely diffusing regions which are modelled with the well-known Kärger equations. Then, in the order of increasing complexity, a system in which particles diffusing within a sphere subject to the absorbing boundary conditions is considered (Figure 4.2A). The absorbing boundary conditions are equivalent to implying infinitely fast relaxation in the outer domain and effectively turn a two-site system

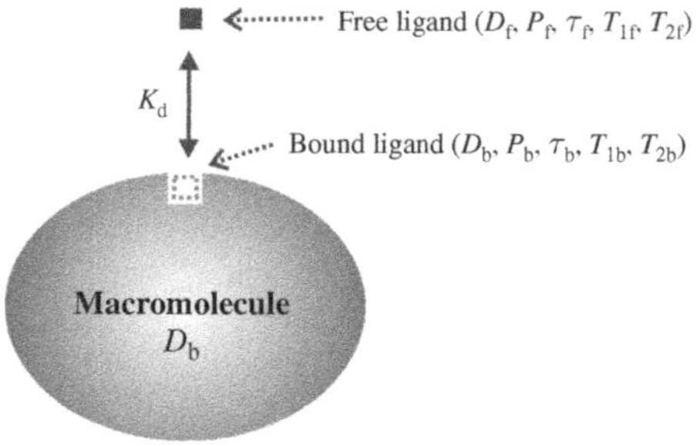

Figure 4.1 A schematic diagram of a ligand (square) exchanging between free solution and a macromolecule binding site on a macromolecule with dissociation constant, K_d. In the free state the ligand has a diffusion coefficient D_f, a population P_f, a lifetime τ_f, a spin–lattice relaxation time T_{1f}, and spin–spin relaxation time, T_{2f}. The corresponding parameters for the ligand in the bound state are denoted with a subscript 'b'. We note that D_b is assumed to be equal to the macromolecule diffusion coefficient.

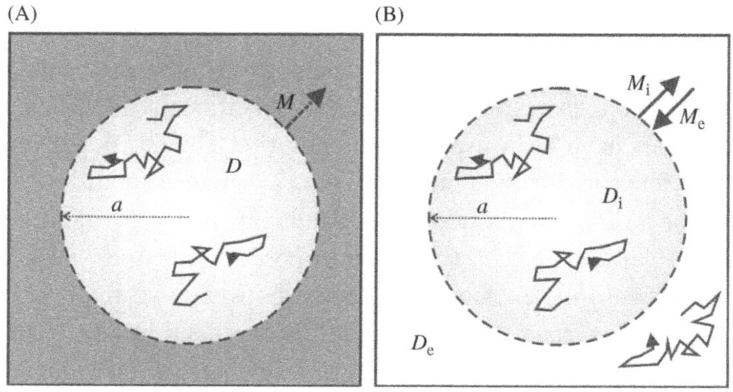

Figure 4.2 Schematic representations of diffusive transport through a spherical interface of radius a. The outer domain is taken as freely diffusing. M denotes the permeability coefficients ($m\,s^{-1}$), it is sometimes convenient to write this as a reduced permeability coefficient, $h = Ma/D$. The subscripts i and e refer to the internal and external medium respectively. (A) When either $D_e \gg D_i$ or $T_{2(i)} \gg T_{2(e)}$ (where the number in braces denotes the medium) this model becomes equivalent to the partially absorbing wall condition. In this model, when a spin reaches the external medium, its signal is instantly annihilated. Thus, the absorbing wall conditions result in a pseudo-one-site system. (B) The full two-site system where the populations and diffusion coefficients in both domains are considered.

into a mathematically simpler pseudo-one-site system. Finally, the full two-site system is examined in which the populations in both the (restricted) internal domain and (freely diffusing) external domain and the exchange between them must be considered (Figure 4.2B).

Some experimental data for hypophosphite ($H_2PO_2^-$) transport in human red blood cells are presented in Figure 4.3 and Figure 4.4.

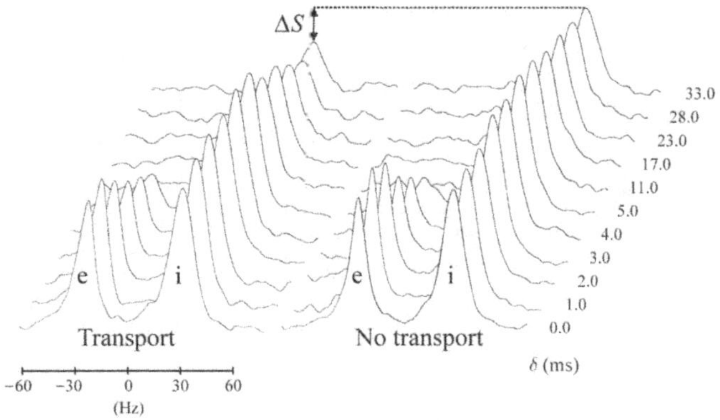

Figure 4.3 ^{31}P PGSE NMR of hypophosphite transport in a suspension of human red blood cells at 310 K. The intra- and extra-cellular species give separate resonances and the transport can be blocked by the addition of dinitrostilbene. The experimental parameters were $\Delta = 100$ ms and $g = 0.28$ T m^{-1}. The haematocrit of the cells was ~0.7 and the cell volume was about 70 fl, which, assuming the cells to be spherical, gives an effective radius of about 2.5 μm. ΔS denotes the difference in signal attenuation of the intracellular species due to transport. The signals of the extracellular species, which have little obstruction attenuate very rapidly. Adapted from Price *et al.*[5]

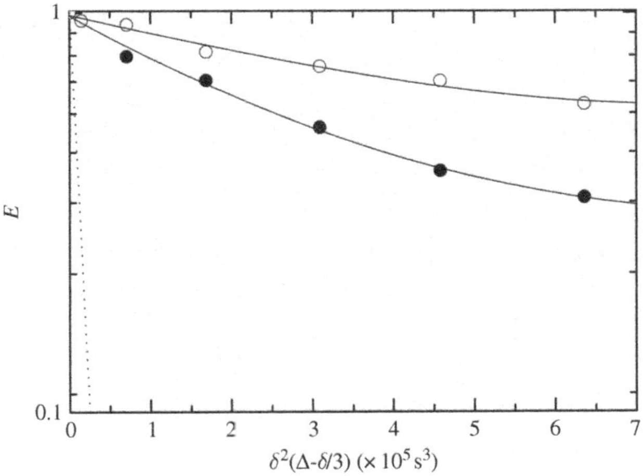

Figure 4.4 Plots of the hypophosphite attenuation data shown in Figure 4.3 versus $\delta^2(\Delta\text{-}\delta/3)$. The intracellular signals from the cells with (○) and without (•) transport inhibition. The solid lines are second-order polynomial interpolations. The difference in attenuation between the two data sets is a result of the transmembrane transport. Compared to the intracellular medium, the hypophosphite in the extracellular medium diffused much more rapidly in both samples giving an apparent diffusion coefficient of $D^{app} = 5.9 \times 10^{-10}$ m^{2}s^{-1} (the approximate attenuation of this signal is depicted by the dotted line). The true diffusion coefficient, as measured in cell-free supernatant was 1.60×10^{-9} m^{2}s^{-1}. This indicates the presence of obstruction effects for the extracellular species. Adapted from Price *et al.*[5]

4.2 Simple binding and exchange: the Kärger equations

The simplest case of exchange between two domains where the exchanging species are both undergoing free diffusion as depicted for a ligand (L) exchangeably binding with a macromolecule receptor (P) to form a complex (LP) in Figure 4.1. The ligand has diffusion coefficients D_{f} and D_{b}, when it is in the free and bound states respectively. The system is described by the equilibrium reaction[6]

$$\mathrm{L}+\mathrm{P} \underset{k_{-1}}{\overset{k_{+1}}{\rightleftharpoons}} \mathrm{LP}. \tag{4.1}$$

The dissociation constant of the complex, K_{d} (M), may be defined in terms of this stoichiometric equation by the ratio of the reverse first-order rate constant (k_{-1} (s^{-1})) and the forward rate constant (k_{+1} ($\mathrm{M}^{-1}\mathrm{s}^{-1}$)) by

$$K_{\mathrm{d}}=\frac{[\mathrm{L}][\mathrm{P}]}{[\mathrm{LP}]}=\frac{k_{-1}}{k_{+1}}. \tag{4.2}$$

Exchange on the timescale of Δ will be reflected in the magnitude of the observed diffusion coefficient of the ligand. Although K_{d} is only related to the *ratio* of k_{-1} and k_{+1} as shown in Eq. (4.2), it is sometimes found that a value of K_{d} less than 1 μM, which would constitute tight binding, is correlated with slow exchange.[7] In the case of slow exchange there would be insufficient exchange to modulate the measured diffusion coefficient and assuming that the bound and free sites had distinct chemical shifts the experiment would reveal the diffusion coefficients of both sites and, after accounting for relaxation differences, the relative populations of the two sites.

The fractional populations (i.e. mole fractions) in the free and bound states are given by

$$P_{\mathrm{f}}=\frac{[\mathrm{L}]}{[\mathrm{L}]_{\mathrm{Total}}},\; P_{\mathrm{b}}=\frac{[\mathrm{LP}]}{[\mathrm{L}]_{\mathrm{Total}}}, \tag{4.3}$$

where [L] is the concentration of the free ligand, [LP] is the concentration of the bound ligand and $[\mathrm{L}]_{\mathrm{Total}}$ is the total ligand concentration. Hence,

$$P_{\mathrm{f}}+P_{\mathrm{b}}=1. \tag{4.4}$$

For the above model, taking $[\mathrm{P}]_{\mathrm{Total}}$ to be the total macromolecule concentration and assuming the possibility of n equivalent binding sites the bound population is given by

$$P_{\mathrm{b}}=\alpha-\sqrt{\alpha^{2}-\beta}, \tag{4.5}$$

where

$$\alpha = \frac{\left([\mathrm{L}]_{\mathrm{Total}} + n[\mathrm{P}]_{\mathrm{Total}} + K_{\mathrm{d}}\right)}{2[\mathrm{L}]_{\mathrm{Total}}} \text{ and } \beta = \frac{n[\mathrm{P}]_{\mathrm{Total}}}{[\mathrm{L}]_{\mathrm{Total}}}. \tag{4.6}$$

The lifetimes of a ligand molecule in the free and bound states are

$$\tau_{\mathrm{f}} = \frac{1}{k_{+1}[\mathrm{P}]} \text{ and } \tau_{\mathrm{b}} = \frac{1}{k_{-1}}, \tag{4.7}$$

respectively. A single lifetime is often used to characterise such exchange processes

$$\frac{1}{\tau} = \frac{1}{\tau_{\mathrm{f}}} + \frac{1}{\tau_{\mathrm{b}}} = k_{-1}\left(1 + \frac{P_{\mathrm{b}}}{P_{\mathrm{f}}}\right) = \frac{k_{-1}}{[L]_{\mathrm{Total}}}. \tag{4.8}$$

The coupled differential equations describing the echo signal intensities of the species free and bound sites are, for the Hahn spin-echo based PGSE sequence,[8–12]

$$\frac{dS_{\mathrm{f}}}{dt} = -(2\pi q)^2 D_{\mathrm{f}} S_{\mathrm{f}} - \frac{S_{\mathrm{f}}}{\tau_{\mathrm{f}}} - \frac{S_{\mathrm{f}}}{T_{2\mathrm{f}}} + \frac{S_{\mathrm{b}}}{\tau_{\mathrm{b}}} \tag{4.9}$$

and

$$\frac{dS_{\mathrm{b}}}{dt} = -(2\pi q)^2 D_{\mathrm{b}} S_{\mathrm{b}} - \frac{S_{\mathrm{b}}}{\tau_{\mathrm{b}}} - \frac{S_{\mathrm{b}}}{T_{2\mathrm{b}}} + \frac{S_{\mathrm{f}}}{\tau_{\mathrm{f}}}. \tag{4.10}$$

The initial conditions are given by $S_{\mathrm{f}}|_{t=0} = P_{\mathrm{f}} = (1 - P_{\mathrm{b}})$ and $S_{\mathrm{b}}|_{t=0} = P_{\mathrm{b}}$. These equations are analogous to those for nuclear relaxation in multiphase systems.[2]

Ignoring the relaxation time differences between the two domains, the equation describing the echo attenuation (in the short gradient pulse limit) is given by a sum of exponentials.[8–11] The individual solutions to Eqs. (4.9) and (4.10) are, respectively,

$$\begin{aligned} S_{\mathrm{b}}(q,\Delta) &= (P_{\mathrm{b}} - C_2)\exp\left(-(2\pi q)^2 D_1 \Delta\right) + C_2 \exp\left(-(2\pi q)^2 D_2 \Delta\right) \\ \text{and} \quad S_{\mathrm{f}}(q,\Delta) &= \left(P_{\mathrm{f}} - C_2'\right)\exp\left(-(2\pi q)^2 D_1 \Delta\right) + C_2' \exp\left(-(2\pi q)^2 D_2 \Delta\right) \end{aligned} \tag{4.11}$$

where

$$C_2 = \frac{P_{\mathrm{f}} D_{\mathrm{f}} - P_{\mathrm{f}} D_1}{D_2 - D_1}$$

and

$$C'_2 = \frac{P_{\mathrm{b}} D_{\mathrm{b}} - P_{\mathrm{b}} D_1}{D_2 - D_1}.$$

D_1 and D_2 are the apparent self-diffusion coefficients defined by

$$D_{1(2)} = \frac{1}{2}\left\{ D_b + D_f + \frac{1}{(2\pi q)^2}\left(\frac{1}{\tau_b} + \frac{1}{\tau_f}\right) \right.$$
$$\left. \mp \left\{ \left[D_f - D_b + \frac{1}{(2\pi q)^2}\left(\frac{1}{\tau_f} - \frac{1}{\tau_b}\right)\right]^2 + \frac{4}{(2\pi q)^4 \tau_b \tau_f} \right\}^{1/2} \right\}.$$

Often individual signals cannot be observed for the free and bound sites and thus the solution becomes

$$E(q, \Delta) = S_b + S_f = P_1 \exp\left(-(2\pi q)^2 D_1 \Delta\right) + P_2 \exp\left(-(2\pi q)^2 D_2 \Delta\right), \quad (4.12)$$

where P_1 and P_2 are the population fractions (relative signal intensities) given by

$$P_1 = 1 - P_2 \quad (4.13)$$

and

$$P_2 = \frac{P_b D_b + P_f D_f - D_1}{D_2 - D_1}. \quad (4.14)$$

Generally, D_f can be determined by measuring the diffusion of the ligand alone. If the free and bound populations have separate chemical shifts, P_f and P_b can be determined directly by integrating the spectral resonances and since

$$\frac{P_f}{\tau_f} = \frac{P_b}{\tau_b}, \quad (4.15)$$

either τ_b or τ_f can be eliminated, reducing the number of unknowns to just D_b and τ_b (or τ_f).

Despite wide application, this approach involves a serious approximation: transport between different subregions is introduced through the mean residence times and conditional hopping probabilities, by combining Fick's second law with the Chapman–Kolmogorov equations.[9] Thus, the space coordinate is applied in a macroscopic sense, leaving the space unit much larger than the diameters of the individual subregions. However, this approximation considerably simplifies the solution of the underlying diffusion problem and can easily be extended to more than two sites – although without extremely precise experimental data the application of such models would be problematic. There are also a number of other important assumptions whose validity may be questionable including: (a) that the spin–spin relaxation times are the same in both sites or very much longer than Δ, (b) that there is no self-association of the macromolecule or the ligand and that the changes in ligand concentration (NB normally the macromolecule concentration is constant) do not perturb the solution viscosity and thus the self-diffusion coefficients of the free and bound species, and (c) that the above derivation is performed in the short gradient pulse limit. As discussed in detail elsewhere these assumptions may not be justified, and corrections for relaxation may need to be included.[9,13–16]

Usually, rates of chemical exchange are defined relative to the NMR timescale based on chemical shift differences.[17] For example, in a slowly exchanging system the individual sites can be observed with different chemical shifts (assuming that there is a difference) as the exchange rate is much less than the frequency difference between the sites. As the exchange rate (i.e., the inverse lifetimes) increases line broadening is first observed, followed by coalescence and finally in fast exchange, where the mean lifetimes in each site are now very short, a motionally narrowed time-averaged spectrum of the different sites is observed. Ordinarily, the transition between fast and slow exchange can only be made by changing the concentration of the participating species or their environment (e.g., temperature). In diffusion measurements of an exchanging system, however, we are primarily concerned with events on the timescale of Δ and thus by cogent choice of Δ it may be possible to change from one regime to another. In the case of slow exchange (i.e., τ_{f}, $\tau_{\mathrm{b}} \rightarrow \infty$) Eq. (4.12) simplifies to

$$E(q, \Delta) = P_{\mathrm{f}} \exp\left(-(2\pi q)^2 D_{\mathrm{f}} \Delta\right) + P_{\mathrm{b}} \exp\left(-(2\pi q)^2 D_{\mathrm{b}} \Delta\right), \tag{4.16}$$

whereas in the case of fast exchange (i.e., τ_{f}, $\tau_{\mathrm{b}} \rightarrow 0$ and thus $D_2 \rightarrow \infty$) and so Eq. (4.12) reduces to the particularly simple single exponential form

$$E(q, \Delta) = S_{\mathrm{b}} + S_{\mathrm{f}} = \exp\left(-(2\pi q)^2 \langle D \rangle_P \Delta\right), \tag{4.17}$$

where

$$\langle D \rangle_P = (1 - P_{\mathrm{b}}) D_{\mathrm{f}} + P_{\mathrm{b}} D_{\mathrm{b}} \tag{4.18}$$

is the population-weighted average diffusion coefficient. Clearly, under conditions of slow or fast exchange it is not possible to determine lifetimes directly from diffusion measurements. The solutions to the Kärger equations are illustrated in Figure 4.5.

Normally, a series of diffusion measurements at constant macromolecule concentration with various ligand concentrations is performed (theoretically the situation could be reversed, but the change in solution viscosity with macromolecule concentration would be far greater). In the systems normally studied, the ligand is in fast exchange between the free and the bound site and so Eq. (4.18) holds. D_{b} and D_{f} can be determined by measuring the diffusion of pure macromolecule and pure ligand solutions, respectively. Equation (4.18) in conjunction with Eq. (4.5) is then regressed on to the data to obtain estimates of K_{d} and n. An example of such an analysis for the salicylate-albumin system is given in Figure 4.6.

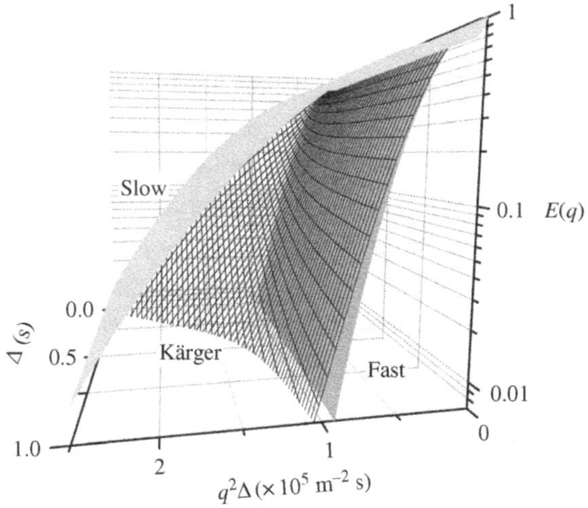

Figure 4.5 A surface plot of the solutions to the Kärger Equations as a function of Δ (black mesh, Eq. (4.12)). The plots were calculated with $D_b = 1.0 \times 10^{-10}$ m^{2}s^{-1}, $D_f = 2.0 \times 10^{-9}$ m^{2}s^{-1}, $P_b = 0.6$ and $\tau_b = 0.04$ s. For comparison, the solutions to the slow (solid grey, Eq. (4.16)) and fast exchange (dark grey, Eq. (4.17)) limits are also shown. As can be clearly seen from the figure, the system can be in either fast exchange or slow exchange (or somewhere in between) on the Δ timescale by the choice of Δ. The slow exchange condition is clearly biexponential, whereas the fast exchange condition is effectively single exponential.

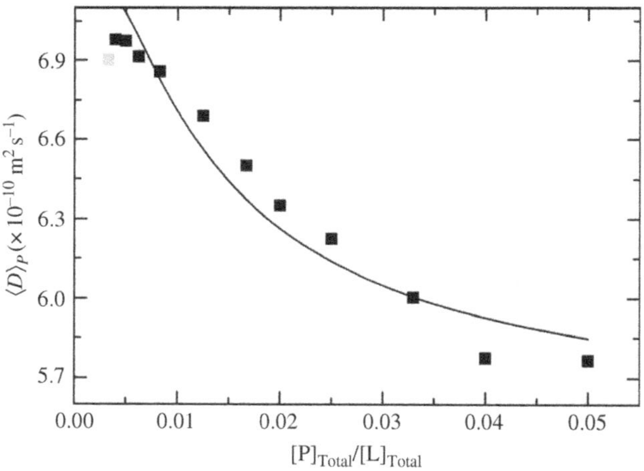

Figure 4.6 Population-weighted average diffusion of salicylate in the presence of bovine serum albumin (0.5 mM) as a function of $[\mathrm{P}]_{\mathrm{Total}}/[\mathrm{L}]_{\mathrm{Total}}$ at 298 K. The curve was obtained by regressing the two-site model (i.e., Eqs. (4.18) and (4.5)) onto the data. From the systematically poor fit of the model to the data, it is evident that this model is too simplistic and does not truly describe the molecular reality of salicylate binding to BSA. The sample with the highest salicylate concentration was excluded from the analysis. Analysis of the diffusion data gave $K_d = 0.030 \pm 0.004$ M with $n = 33 \pm 3$. Modified from Price *et al.*[18]

In Section 2.5 we considered diffusion measurements as a means of separating a mixture of non-exchanging species. However, exchange on the timescale of Δ will influence the outcome of such a measurement giving a different distribution of diffusion coefficients. Using analysis based on the Kärger relations, several studies have examined how exchange will affect the distributions in the case of a two-site system.[19,20]

4.3 Exchange between free and restricted sites

4.3.1 The modified Kärger equations

It is easy to imagine that the conceptually appealing Kärger model (Section 4.2) could be adapted to form an approximate model for transport between two (or more) domains such as transmembrane transport in biological cells in accordance with Figure 4.2A. And a number of variations have been attempted.[11,21–23] Jespersen *et al.*[24] have considered the influence of cellular size distribution on modelling diffusion data with the Kärger equations.

The Kärger model assumes free diffusion in both sites, and thus, if applied to transmembrane transport, the effects of the restriction must be accounted for. A first attempt at such a correction is to replace the intracellular diffusion coefficient with an apparent one using Eq. (3.10). However, as Δ increases such that $\xi \gg 1$ the amplitude of the internal site becomes independent of q. Here we consider one approach of modifying the Kärger equations to model transmembrane exchange:[11] Relabelling Eqs. (4.9) and (4.10) to reflect the internal (i) and external domains (e), and ignoring relaxation time differences to obtain

$$\frac{dS_e}{dt} = -(2\pi q)^2 D_e S_e - \frac{S_e}{\tau_e} + \frac{S_i}{\tau_i} \tag{4.19}$$

$$\frac{dS_i}{dt} = -\frac{S_i}{\tau_i} + \frac{S_e}{\tau_e}, \tag{4.20}$$

with the initial conditions given by $S_e|_{t=0} = P_e = (1 - P_i)$ and, assuming a spherical restriction of radius a in the long-time limit (Eq. (3.26)), $S_i|_{t=0} = \widehat{P}_i = P_i\left[\frac{3j_1(2\pi qa)}{(2\pi qa)}\right]$. Thus, the attenuation at time $t = \Delta$ takes the same form as Eq. (4.12) but with

$$P_1 = P_e - \widehat{P}_i - \frac{D_e P_e - D_i P_e - D_1 \widehat{P}_i}{(D_2 - D_1)} \tag{4.21}$$

$$P_2 = \frac{D_e P_e - D_1 P_e - D_1 \widehat{P}_i}{(D_2 - D_1)} \tag{4.22}$$

and

$$D_{1(2)} = \frac{1}{2}\left\{ D_e + \frac{1}{(2\pi q)^2}\left(\frac{1}{\tau_i} + \frac{1}{\tau_e}\right) \mp \left\{ \left[D_e - \frac{1}{(2\pi q)^2}\left(\frac{1}{\tau_e} - \frac{1}{\tau_i}\right)\right]^2 + \frac{4}{(2\pi q)^4 \tau_i \tau_e} \right\}^{1/2} \right\}. \tag{4.23}$$

Equation (4.23) can be further simplified. In the slow exchange limit (i.e., τ_i, $\tau_e \to \infty$)

$$D_1 \approx \frac{1}{(2\pi q)^2 \tau_e} \tag{4.24}$$

and

$$D_2 \approx D_i + \frac{1}{(2\pi q)^2 \tau_i}. \tag{4.25}$$

In the fast exchange limit (i.e., τ_i, $\tau_e \to 0$) we only need to consider D_1 (since $D_2 \to \infty$ and thus the second term in Eq. (4.12) $\to 0$) and

$$D_1 \approx \frac{\tau_e}{\tau_e + \tau_i}. \tag{4.26}$$

Even allowing for this modification to better account for the reality of diffusion in one domain being subject to restricted diffusion, the Kärger approach still involves a serious approximation. In particular, there is the implicit assumption that exchanging spins sample all points of the sphere before being transported as there is no consideration of the starting point inside the sphere. Hence, this approach is strictly only applicable in the case of very slow exchange. However, this approximation considerably simplifies the solution of the underlying diffusion problem, which is otherwise extremely complicated even for a two-site system. Further, this theoretical approach fails to predict the PGSE diffraction patterns observed in real permeable media such as erythrocyte suspensions,[25] and consequently more complete theoretical approaches must be adopted.

4.3.2 *Absorbing boundaries*

4.3.2.1 *Absorbing planes*

Various authors have presented the SGP solution for diffusion between a pair of absorbing planes with the gradient being applied perpendicular to the planes.[26–29] Starting from the propagator defined in Eq. (1.86) the signal attenuation is[28,29]

$$E(\mathbf{q},\Delta) = 2\sum_{n=0}^{\infty}\left(1+\frac{\sin(2\alpha_n)}{2\alpha_n}\right)^{-1} \times \frac{[(2\pi qa)\sin(2\pi qa)\cos\alpha_n - \alpha_n\cos(2\pi qa)\sin\alpha_n]^2}{\left[(2\pi qa)^2-\alpha_n^2\right]^2} e^{-\alpha_n^2\frac{D\Delta}{a^2}} + 2\sum_{m=0}^{\infty}\left(1-\frac{\sin(2\beta_m)}{2\beta_m}\right)^{-1} \times \frac{[(2\pi qa)\cos(2\pi qa)\sin\beta_m - \beta_m\sin(2\pi qa)\cos\beta_m]^2}{\left[(2\pi qa)^2-\beta_m^2\right]^2} e^{-\beta_m^2\frac{D\Delta}{a^2}}, \quad (4.27)$$

where the eigenvalues α_n and β_m are given by Eqs. (1.87) and (1.88).

Wall relaxation effects do not significantly perturb the diffraction pattern and the major effect is to shift the position of the diffraction minima resulting in a reduced value of a.[28] An extension of Eq. (4.27) which includes relaxation before and after the gradient pulse has also been presented.[28]

4.3.2.2 Absorbing spheres

For the case of spheres with relaxing walls, the SGP solution is[26,29–31]

$$E(\mathbf{q},\Delta) = 6\sum_{n=0}^{\infty}\sum_{m=0}^{\infty}\frac{(2n+1)}{h(h-1)+\alpha_{nm}^2-n(n+1)} \times \left\{\frac{\alpha_{nm}[(n+h)j_n(2\pi qa)-2\pi qa j_{n+1}(2\pi qa)]}{\alpha_{nm}^2-(2\pi q)^2a^2}\right\}^2 e^{-\alpha_{nm}^2\frac{D\Delta}{a^2}}, \quad (4.28)$$

where $h = Ma/D$ is the reduced permeability and α_{nm} are the positive roots of

$$(n+h)\, j_n(\alpha_{nm}) = \alpha_{nm}\, j_{n+1}(\alpha_{nm}). \quad (4.29)$$

The microscopic first-order rate constant for transport through the spherical interface will be given by $k_1 = 3M/a$.[31]

In a typical experiment the condition $D\Delta/a^2 > 1$ holds and only the lowest eigenvalue $\alpha = \alpha_{00}$ is important. Thus, from Eq. (4.29), α satisfies the following equation

$$1 - h = \alpha\cot\alpha. \quad (4.30)$$

In this long-time limit the normalised echo amplitude is independent of $D\Delta/a^2$ and is given by

$$E(\mathbf{q},\infty) = \left\{\frac{1}{h}\frac{\alpha^2}{\alpha^2-(2\pi qa)^2}\left[(h-1)\frac{\sin(2\pi qa)}{2\pi qa}+\cos(2\pi qa)\right]\right\}^2. \quad (4.31)$$

In the limit of weak permeability (i.e., $h \to 0$, reflecting boundary), the lowest positive solution of Eq. (4.30) is $\alpha^2 = 3h$, and when $2\pi qa \ll 1$ this reduces to Eq. (3.26). And in the limit of infinite permeability (i.e., $h \to \infty$, perfectly absorbing boundary)

$$E(\mathbf{q}, \infty) = \left[\frac{\pi^2}{\pi^2 - (2\pi qa)^2}\frac{\sin(2\pi qa)}{2\pi qa}\right]^2 \approx \exp\left[-(2\pi qa)^2(\pi^2 - 6)\Big/(3\pi)^2\right]. \tag{4.32}$$

The attenuation profiles for a range of permeability values are plotted in Figure 4.7. It can be seen that there is only a weak dependence on permeability. Further, there is a weak decrease in the measured diffusion coefficient as given by the initial slope (see Eq. (3.10)) as h is increased. The reason for the apparent slowing of diffusion is that as the permeability increases, more spins leave the sphere and no longer contribute to the signal. At high permeability values, only those spins survive which do not reach the surface by time Δ.

In cell or vesicle systems the partially absorbing wall experiment is, in principle, possible as the relaxation properties of the exterior medium may be manipulated by

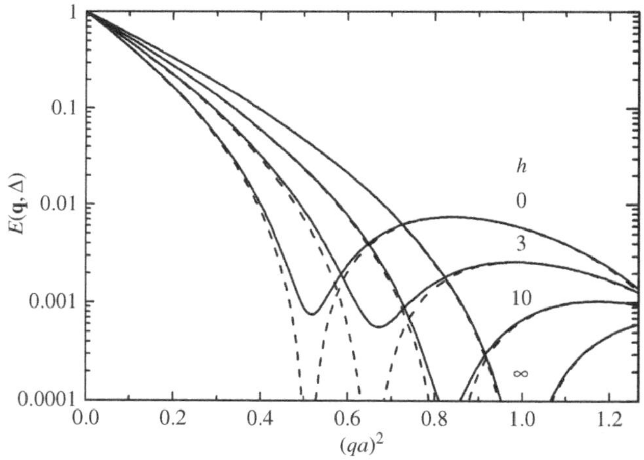

Figure 4.7 A plot of the normalised echo amplitude versus $(qa)^2$ for various values of the reduced permeability coefficient h for diffusion in a partially-absorbing sphere. The calculations were performed using $\xi = 0.765$ which, for instance, corresponds to $D = 1 \times 10^{-10}\,\mathrm{m^2 s^{-1}}$, $\Delta = 60$ ms and $a = 2.8$ μm, according to the values used in an experimental study of bicarbonate diffusion in red blood cells.[5] The solid lines were calculated from Eq. (4.28) whereas the dotted lines were plotted using the long-time limit formula Eq. (4.31). The plot clearly shows that the positions of the minima are h-dependent. In reality the values of h are always small. However, we have included a large range of h values so that the effects may be easily visualised. Adapted from Barzykin *et al.*[31]

addition of relaxation agents. However, it is experimentally difficult to attain the instant quenching condition,[5,31] and the spins in the exterior medium contribute considerably to the observed echo signal. As shown above, even if the true partially absorbing boundary condition is achieved, it is only weakly sensitive to transport. An experimentally more realisable system and similarly a model one-step closer to the full two-site system is where the spins that transport through the interface are rapidly relaxed but not instantly quenched.[31] In this formulation, the finite lifetime of the spins in both domains is included. Accordingly, the diffusion equation must be complemented with a decay term and is given by (here P represents propagator)

$$\frac{\partial P}{\partial t} = D\nabla^2 P - \frac{1}{T_2}P. \tag{4.33}$$

T_2 and D are different in each domain (see Figure 4.2) and are denoted below by the subscripts i and e corresponding to the interior and the exterior, respectively. The boundary conditions (of the fourth kind) for the Green function are given by

$$D_\mathrm{i}\frac{\partial P_\mathrm{i}}{\partial r}\bigg|_{r=a} = D_\mathrm{e}\frac{\partial P_\mathrm{e}}{\partial r}\bigg|_{r=a} \tag{4.34}$$

and

$$D_\mathrm{i}\frac{\partial P_\mathrm{i}}{\partial r} + M_\mathrm{i}P_\mathrm{i} - M_\mathrm{e}P_\mathrm{e}\big|_{r=a} = 0. \tag{4.35}$$

At equilibrium, there is no flux across the interface and $[P_\mathrm{i}/P_\mathrm{e}]_{eq} = M_\mathrm{e}/M_\mathrm{i}$. The contribution to the echo signal from the internal domain is

$$E_\mathrm{i}(\mathbf{q}, \Delta) = 6\sum_{n=0}^{\infty}\sum_{m=0}^{\infty}(2n+1)\exp(-\xi_\mathrm{i}\alpha_{nm}^2)f_n(\alpha_{nm}) \times \left\{\frac{\alpha_{nm}}{\alpha_{nm}^2 - (2\pi qa)^2}\left[\frac{\alpha_{nm}\, j_{n+1}(\alpha_{nm})}{j_n(\alpha_{nm})}j_n(2\pi qa) - 2\pi qa\, j_{n+1}(2\pi qa)\right]\right\}^2, \tag{4.36}$$

where α_{nm} are the positive real roots of the equation

$$\alpha_{nm}\, j_{n+1}(\alpha_{nm})\left[\beta_{nm}K_{n+\frac{3}{2}}(\beta_{nm}) - (n-h_\mathrm{e})K_{n+\frac{1}{2}}(\beta_{nm})\right] = j_n(\alpha_{nm})\left[(n+h_\mathrm{i})\beta_{nm}K_{n+\frac{3}{2}}(\beta_{nm}) - n(n+h_\mathrm{i}-h_\mathrm{e})K_{n+\frac{1}{2}}(\beta_{nm})\right], \tag{4.37}$$

$\beta_{nm} = \left(\varepsilon^{-1} - \zeta\alpha_{nm}^2\right)^{1/2}$ with $\varepsilon = D_\mathrm{e}T_{2\mathrm{e}}/a^2$ and $\zeta = D_\mathrm{i}/D_\mathrm{e}$; $h_\mathrm{i} = M_\mathrm{i}a/D_\mathrm{i}$, $h_\mathrm{e} = M_\mathrm{e}a/D_\mathrm{e}$, K is the modified Bessel function of the third kind,[32] and $f_n(\alpha_{nm})$ is a constant and is further considered below. It is assumed that at least one root exists; this requires small T_{2e} and sufficiently small permeability. The latter condition is normally fulfilled experimentally for most systems of interest. As the lifetime in the outer domain is decreased (i.e., $T_{2e} \to 0$) β_{nm} tends to infinity, and the inner component of the echo reduces to Eq. (4.28) as expected. The contribution to the echo signal from the spins found at time Δ outside the sphere is[31]

$$E_e(\mathbf{q}, \Delta) = 6h_i\zeta \sum_{n=0}^{\infty}\sum_{m=0}^{\infty} (2n+1)\exp(-\xi_i\alpha_{nm}^2)\frac{\alpha_{nm}^2 \tilde{f}_n(\alpha_{nm})}{(\alpha_{nm}^2 - (2\pi qa)^2)(\beta_{nm}^2 + (2\pi qa)^2)}$$
$$\times\left[\frac{\alpha_{nm}\, j_{n+1}(\alpha_{nm})}{j_n(\alpha_{nm})} j_n(2\pi qa) - 2\pi qa\, j_{n+1}(2\pi qa)\right]$$
$$\times\left[\frac{\beta_{nm}K_{n+\frac{3}{2}}(\beta_{nm})}{K_{n+\frac{1}{2}}(\beta_{nm})} j_n(2\pi qa) - 2\pi qa\, j_{n+1}(2\pi qa)\right]. \tag{4.38}$$

The roots are defined as above. For $n=0$, the functions f and $\tilde{f}$ are related via

$$\tilde{f}_0^{-1}(\alpha_{0m}) = (1 + \beta_{0m} + h_e)\, f_0^{-1}(\alpha_{0m}). \tag{4.39}$$

The outer component vanishes as $T_{2e} \to 0$.

In the long-time limit, only the lowest eigenvalue $\alpha = \alpha_{00}$ is important which, from Eq. (4.37), satisfies the following equation

$$\alpha\cot\alpha = 1 - h_i\left(1 + \frac{h_e}{1+\beta}\right)^{-1}, \tag{4.40}$$

where $\beta = \beta_{00}$. The corresponding constant $f_0(\alpha)$ in Eq. (4.36) is given by

$$f_0^{-1}(\alpha) = \alpha^2 + h_i\left(h_i - 1 - \frac{h_e}{1+\beta}\right)\left(1 + \frac{h_e}{1+\beta}\right)^{-2} + \frac{\alpha^2 h_e^2}{\beta(1+\beta+h_e)^2}. \tag{4.41}$$

The overall normalised long-time echo amplitude is again independent of Δ. The components are given by

$$E_i(\mathbf{q}, \infty) = \left(\frac{1}{\alpha^2} + \frac{\zeta}{\beta^2}\right)^{-1}\left\{\frac{1}{h}\frac{\alpha}{\alpha^2 - (2\pi qa)^2}\left[(h-1)\frac{\sin(2\pi qa)}{2\pi qa} + \cos(2\pi qa)\right]\right\}^2 \tag{4.42}$$

and

$$E_e(\mathbf{q},\infty) = \left(\frac{1}{\alpha^2}+\frac{\zeta}{\beta^2}\right)^{-1}\left\{\frac{1}{h}\frac{\alpha}{\alpha^2-(2\pi qa)^2}\left[(h-1)\frac{\sin(2\pi qa)}{2\pi qa}+\cos(2\pi qa)\right]\right\}$$
$$\times\left\{\frac{\beta}{1+\beta}\frac{\alpha}{\beta^2+(2\pi qa)^2}\left[\frac{\sin(2\pi qa)}{2\pi qa}+\frac{1}{\beta}\cos(2\pi qa)\right]\right\}. \quad (4.43)$$

where $h=h_i/(1+h_e/(1+\beta))$. PGSE profiles calculated using these two equations are given in Figure 4.8 for a range of permeability values. In contrast to the partially absorbing wall case, finite relaxation rates in the external medium lead to an increase in the apparent diffusion coefficient given by the initial slope as the permeability increases and this was experimentally observed in a study of bicarbonate ions diffusing through red blood cell membranes into a Mn^{2+} doped extracellular medium.[5] Even though the external relaxation rate is very fast (but not infinite), a small population of spins survive at the end of the PGSE experiment and result in a considerable enhancement in the observed apparent diffusion coefficient. Typically, two stages are observed: a rapid initial attenuation due to the external component followed by a slower attenuation due to the internal component. When the ratio of the

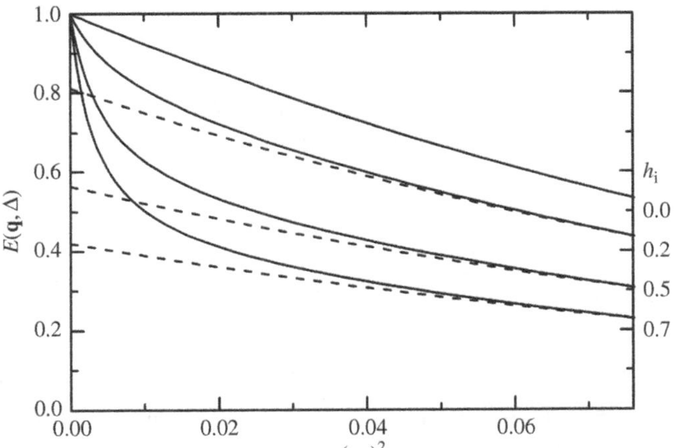

Figure 4.8. PGSE profiles in the long-time limit for spins diffusing through a spherical interface into a rapidly (but not instantly) relaxing external medium calculated using Eqs. (4.42) and (4.43). The calculations were performed with $\xi=0.765$, $\Delta=60$ ms, $a=2.8$ μm and $T_{2e}=26$ ms, $D_i=1\times10^{-10}\,m^2s^{-1}$, $D_e=1\times10^{-9}\,m^2s^{-1}$. In contrast to the case of a perfectly absorbing external medium (see Figure 4.7), an increase in h_i is accompanied by an increase in the apparent diffusion coefficient. Smaller values of h_i were considered than in Figure 4.7 so that β in Eq. (4.40) was real. For comparison the corresponding plots of the contribution to the echo signal from inside the interface (Eq. (4.42)) are also included as dashed lines. For $h_i=0$ the lines are coincident. From Barzykin *et al.*[31]

internal and external diffusion coefficients is very small (i.e., $\zeta \ll 1$), the initial stage is too fast and may not be detected.

The GPD solutions for the spin-echo attenuation for spins diffusing within a sphere including surface and bulk relaxation have also been derived.[33]

4.3.3 Permeable boundaries

An important model for analysing many systems is that of regularly spaced planar or spherical barriers. Useful solutions must take into account the possibility that a diffusing spin may transport through several such barriers. Tanner derived an analytical eigenfunction solution in the SGP limit for an infinite number of regularly spaced permeable barriers (see Figure 1.13) using Eq. (2.34) and the propagator given in Eq. (1.136) as,[34]

$$E(\mathbf{q},\Delta) = \sum_{n=1}^{\infty} \frac{1}{A_n} \begin{bmatrix} (Z+Y)^2 c_{0,n} \sum_i c_{i,n} \cos(2\pi qai) \\ +(Z+Y)(Z-Y) \sum_i \left(c_{0,n} d_{i,n} - c_{i,n} d_{0,n}\right) \sin(2\pi qai) \\ +(Z-Y)^2 d_{0,n} \sum_i d_{i,n} \cos(2\pi qai) \end{bmatrix} \exp\left(\frac{-\beta_n^2 \Delta D}{a^2}\right), \tag{4.44}$$

where

$$A_n = \frac{1}{2} \sum_i \left(c_{i,n}^2 + d_{i,n}^2\right) + \left(\frac{1}{\beta_n}\right) \sin\left(\frac{\beta_n}{2}\right) \cos\left(\frac{\beta_n}{2}\right) \sum_i \left(c_{i,n}^2 - d_{i,n}^2\right) \tag{4.45}$$

and

$$Y = \frac{\sin(\pi qa - \beta_n/2)}{2\pi qa - \beta_n} \text{ and } Z = \frac{\sin(\pi qa + \beta_n/2)}{2\pi qa + \beta_n}. \tag{4.46}$$

Zientara and Freed[35] have presented a finite difference solution of a density matrix/stochastic Liouville approach to model a similar permeable membrane system. In an attempt to find an efficient approximate solution, von Meerwall and Ferguson started with the SGP model for reflecting planes (Eq. (3.17)) and developed an ansatz-like approximation for barriers of finite permeability and also for permeable spherical boundaries.[36,37] An SGP model for unevenly spaced semi-permeable membranes has also been presented.[38] Van der Weerd *et al.*[39] have presented finite difference solutions to a model including diffusion and relaxation in multicompartment systems with cylindrical geometry.

Price and Barzykin have presented a two-site model in which the internal domain is undergoing restricted diffusion within a sphere within the SGP limit and the outside domain is undergoing free diffusion such as might be used to model diffusion data from a cell suspension.[11] Similar to the model presented in Section 4.3.2 in which the extracellular spins have a short but finite lifetime, this two-site model in which there is no restriction on the extracellular lifetime is very sensitive to transport. In a later work related to the pore-hopping model (see Sections 1.8 and 4.5) Jiang *et al.*[40] developed an approximate model in which the external pores between the cells in addition to the size and arrangements of the cells was considered.

Some further aspects of diffusion in complex porous media is considered under 'General Porous Media' in Section 4.5 and in imaging based studies in Chapter 9.

4.4 Anisotropic and low dimensional diffusion

In an anisotropic system the PGSE signal attenuation will have an orientational dependence with respect to **g** since diffusion is now represented by a tensor (i.e., **D**, see Eq. (1.64)). Whereas we have previously assumed that the gradient was oriented along the z axis, now we consider the possibility that it is oriented along any direction. We also need to consider the possibility that the sample is not homogeneously oriented and thus even though we only measure in the direction of the gradient, the result of the measurement is some form of average over all of the orientations.

Similar to isotropic restricted system, we note that the amount of information that can be obtained will depend on whether the system being studied is undergoing anisotropic Gaussian diffusion or anisotropic restricted diffusion. In the former case diffusion measurements can lead to the determination of a true diffusion ellipsoid (see Section 1.6); in the latter case unless the correct propagator is known only an apparent diffusion ellipsoid can be determined (i.e., Eq. (1.115)). Here we will consider both possibilities.

4.4.1 Anisotropic Gaussian diffusion

Had we attempted the macroscopic solution for free diffusion as in Section 2.3.2 but considering anisotropic diffusion, we would have obtained

$$E(\mathbf{q}) = \exp\left(-(2\pi)^2 \int_0^t \mathbf{q}(t')^{\mathrm{T}} \cdot \mathbf{D} \cdot \mathbf{q}(t')\, dt' \right), \tag{4.47}$$

where $\mathbf{D}$ is $\mathbf{D}^{\mathrm{lab}}$ and $\mathbf{q}(t)$ is defined by Eq. (2.15). Assuming that $\mathbf{D}$ is time-independent, we can rewrite this as[41,42]

$$\begin{aligned} E(\mathbf{q}) &= \exp\left(-(2\pi)^2 \int_0^t \mathbf{q}(t')\mathbf{q}(t')^{\mathrm{T}} dt' : \mathbf{D}\right) \\ &= \exp(-\mathbf{b} : \mathbf{D}) \\ &= \exp\left(-\sum_i \sum_j b_{ij} D_{ij}\right), \end{aligned} \tag{4.48}$$

where ':' is the generalised dot product (i.e., the scalar product for rank two tensors). Note, Eq. (4.48) defines $\mathbf{b}$ as a symmetric matrix

$$\mathbf{b} = (2\pi)^2 \int_0^t \mathbf{q}(t')\mathbf{q}(t')^{\mathrm{T}} dt'. \tag{4.49}$$

If, for example, the gradients were applied along all three Cartesian directions then the echo attenuation would be described by

$$E = \exp\left(-\left[b_{xx}D_{xx} + b_{yy}D_{yy} + b_{zz}D_{zz} + 2b_{xy}D_{xy} + 2b_{xz}D_{xz} + 2b_{yz}D_{yz}\right]\right). \tag{4.50}$$

Thus, were we to probe the diffusion tensor for anisotropic Gaussian diffusion using the Hahn- or STE-based sequences the attenuation would be given by

$$E(\mathbf{q}) = \exp\left(-4\pi^2 \mathbf{q}^{\mathrm{T}} \cdot \mathbf{D} \cdot \mathbf{q}\,(\Delta - \delta/3)\right), \tag{4.51}$$

and in particular if diffusion were measured along the z axis using only a z-gradient this could be written in tensor notation as

$$\begin{aligned} E &= \exp(-b_{zz}D_{zz}) \\ &= \exp\left(-4\pi^2 q_z^2 D_{zz}(\Delta - \delta/3)\right). \end{aligned} \tag{4.52}$$

In general, diffusion measurements would be performed in at least six directions which will be linear combinations of the matrix elements of $\mathbf{D}^{\mathrm{lab}}$ with coefficients corresponding to gradient strength and direction (i.e., the b values) as in Eq. (4.50). The choice of gradient directions has been considered in detail.[43] The resulting data is in the form of a set of linear equations that can then be solved to obtain $\mathbf{D}^{\mathrm{lab}}$ which can then be diagonalised to provide the principal diffusivities and the corresponding eigenvectors which define the principal axes frame.[41,44,45]

4.4.2 1D and 2D diffusion and the powder average

Often anisotropic systems are not homogeneously oriented and it is necessary to perform a 'powder average' over the various orientations, or equivalently, assume that there is only a single domain with a defined direction and that it is the field gradient randomly oriented, thus Eq. (2.34) becomes[9,46–48]

$$E(\mathbf{q},\Delta) = \frac{1}{4\pi}\int \rho(\mathbf{r}_0)P(\mathbf{r}_0,\mathbf{r}_1,\Delta)\cos[2\pi\mathbf{q}\cdot(\mathbf{r}_1-\mathbf{r}_0)]\sin\theta\, d\theta\, d\phi\, d\mathbf{r}_0\, d\mathbf{r}_1, \quad (4.53)$$

where θ and ϕ are the polar and azimuthal angles, respectively (see Figure 1.4), and $1/(4\pi)\sin\theta\, d\theta\, d\phi$ is the probability of $\mathbf{q}$ being in the direction defined by θ and ϕ. Interestingly, if a system contains anisotropic couplings (e.g., quadrupolar or chemical shift anisotropy) it may be possible to resolve the powder average into the principal components by investigating the decay of the full spectral shape.[49]

For polycrystalline systems Eq. (4.53) may be further simplified to[48]

$$E(\mathbf{q},\Delta) = \int \rho(\mathbf{r}_0)P(\mathbf{r}_0,\mathbf{r}_1,\Delta)\frac{\sin(2\pi q|\mathbf{r}_1-\mathbf{r}_0|)}{2\pi q|\mathbf{r}_1-\mathbf{r}_0|}d\mathbf{r}_0\, d\mathbf{r}_1. \quad (4.54)$$

Importantly, whilst the echo attenuations from the spins in the individual elements in such anisotropic systems may be Gaussian, the powder average will not be. For example, in a system such as a liquid crystal with axial symmetry it is possible to show that[50,51]

$$E(\mathbf{q},\Delta) = \exp(-4\pi^2q^2D_\perp\Delta)\int_0^1 \exp(-4\pi^2q^2(D_\parallel - D_\perp)\Delta\, x^2)dx, \quad (4.55)$$

where $x=\cos(\theta)$ and θ define the angle between the lamellae normal and the axis of the field gradient. Equation (4.55) can be expressed in terms of the error function (3.321.1 and .2 in ref. 52)

$$E(\mathbf{q},\Delta) = \frac{1}{2}\exp(-4\pi^2q^2D_\perp\Delta)\left(\frac{\pi}{4\pi^2q^2(D_\parallel-D_\perp)\Delta}\right)^{1/2}\mathrm{erf}\left(\sqrt{4\pi^2q^2(D_\parallel-D_\perp)\Delta}\right). \quad (4.56)$$

For one-dimensional (or capillary) diffusion (i.e., $D_\parallel=D$ and $D_\perp=0$), this becomes[48,50,51]

$$\begin{aligned} E(\mathbf{q},\Delta) &= \int_0^1 \exp(-4\pi^2q^2D\Delta\, x^2)dx \\ &= \frac{1}{2}\left(\frac{\pi}{4\pi^2q^2D\Delta}\right)^{1/2}\mathrm{erf}\left(\sqrt{4\pi^2q^2D\Delta}\right) \\ &= 1+\sum_{n=1}^{\infty}(-1)^n\frac{1}{n!(2n+1)!}(2\pi q\Delta D)^n, \end{aligned} \quad (4.57)$$

where the integral has been evaluated in terms of a power series expansion of the error function. Similarly, for two-dimensional (or lamellar) diffusion (i.e., $D_\parallel=0$ and $D_\perp=D$)[48,50,51]

$$E(\mathbf{q},\Delta) = \exp\left(-4\pi^2 q^2 D\Delta\right) \int_0^1 \exp\left(4\pi^2 q^2 D\Delta\, x^2\right) dx$$

$$= \frac{1}{2i} \exp\left(-4\pi^2 q^2 D\Delta\right) \left(\frac{\pi}{-4\pi^2 q^2 D\Delta}\right)^{1/2} \operatorname{erf}\left(i\sqrt{4\pi^2 q^2 D\Delta}\right). \tag{4.58}$$

The three-dimensional case is, of course, free diffusion (see Eq. (2.44)). Thus, diffusion in capillary and lamellar systems exhibit characteristic attenuation profiles. As is often the case for more complicated systems, exact analytical solutions are not available and Eq. (4.53) must be evaluated numerically (see Section 3.10).[47,48]

The situation is much more complicated, though, if in the timescale of Δ the diffusing molecules change from one domain, specified by a unique local director orientation, into another.[48,51] As noted by Callaghan and Söderman[51] this problem is analogous to the curvilinear diffusion of polymers. The echo attenuation for spins changing N times between differently oriented lamellar domains with a total rms curvilinear path length of Λ, will for large values of N, reflect Gaussian diffusion with an apparent diffusion coefficient of[51]

$$D(\Delta) = \frac{2}{3} \frac{D^0 \lambda}{\sqrt{4D^0 \Delta}}, \tag{4.59}$$

where $\lambda = \Lambda/N$. We shall see in Section 8.7 that experiments that correlate diffusion over two time-periods may help to better characterise such systems.

4.4.3 A model of anisotropic diffusion – diffusion in a cylinder

As noted previously, anisotropic diffusion can result from restriction in anisotropic pores such as muscle fibre cells. Also, many porous materials are isotropic on the macroscopic scale but are locally anisotropic. This local anisotropy can give rise to a spread of macroscopically observed diffusion coefficients. We begin our consideration of diffusional anisotropy due to restriction by considering diffusion in a cylinder of radius a and length l and with a (polar) angle, θ, between the symmetry axis of the cylinder and the static magnetic field (which is also the direction of the gradient; see Figure 4.9). As we shall see the θ dependence of the attenuation provides an additional structural probe.

Using the propagator given in Eq. (1.96) the GPD approximation for diffusion perpendicular to the cylinder axis ($\theta = \pi/2$) is[53]

$$E(\mathbf{q},\Delta) = \exp\left(-\frac{2\gamma^2 g^2}{D^2} \sum_{n=1}^{\infty} \frac{2\alpha_n^2 D\delta - 2 + 2L(\delta) - L(\Delta-\delta) + 2L(\Delta) - L(\Delta+\delta)}{\alpha_n^6 \left(a^2 \alpha_n^2 - 1\right)}\right), \tag{4.60}$$

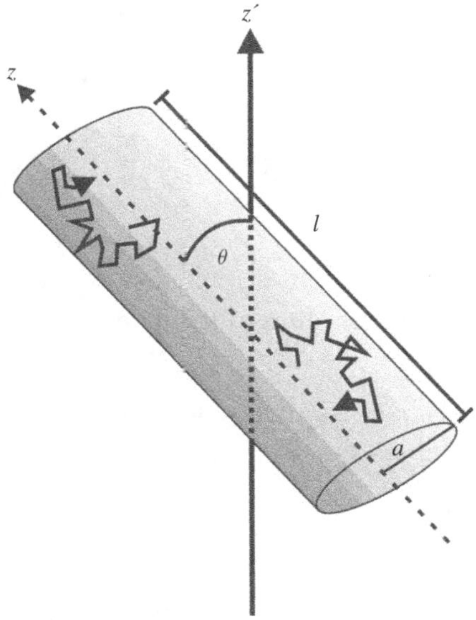

Figure 4.9. A schematic diagram of diffusion in a cylinder of length l and radius a with the symmetry axis of the cylinder subtending an angle θ with the direction of the gradient and static field which is taken to be z. The laboratory or gradient frame is given by (x', y', z') where z' coincides with the gradient direction. The director frame for the cylinder is given by (x, y, z) where z coincides with the symmetry axis of the cylinder. If this were an elliptic cylinder for example, the director frame would be uniquely determined. Clearly if $\theta = 0$ the two reference frames coincide. If $\theta = 0$ and a PGSE diffusion measurement is performed the spin-echo attenuation will be described by diffusion between planar boundaries (i.e., Eq. (3.17)). Conversely if $\theta = \pi/2$ the spin-echo attenuation will be described by diffusion within a cylinder (Eq. (4.66)).

where $L(t) = \exp(-\alpha_n^2 Dt)$ and the α_n are the roots of $J_1'(\alpha_n a) = 0$. In the limit $\xi \gg 1$ and $\delta \ll \Delta$, Eq. (4.60) becomes[53]

$$E(q, \infty) = e^{-(2\pi q a)^2}. \tag{4.61}$$

In the steady gradient limit (i.e., $\delta \to \tau$) Eq. (4.60) becomes[54]

$$E(\mathbf{q}, \Delta) = \exp\left(-\frac{2\gamma^2 g^2}{D^2}\sum_{n=1}^{\infty}\frac{\alpha_n^{-4}}{\alpha_n^2 a^2 - 1}\left(\frac{2\tau\alpha_n^2 D - 3 - 4L(\tau) + L(2\tau)}{\alpha_n^2}\right)\right). \tag{4.62}$$

In the limit $\xi \gg 1$ Eq. (4.60) becomes[54]

$$E(\mathbf{q}, \infty) = -\frac{a^4\gamma^2 g^2}{D}\frac{7}{296}\left(2\tau - \frac{99}{112}\frac{a^2}{D}\right). \tag{4.63}$$

The SGP solution for this geometry is given by[55]

$$E(\mathbf{q},\Delta)=\sum_{n=0}^{\infty}\sum_{k=1}^{\infty}\sum_{m=0}^{\infty}\frac{2K_{nm}a^2(2\pi qa)^4\sin^2(2\theta)[1-(-1)^n\cos(2\pi ql\cos\theta)]\alpha_{km}^2}{l^2\left[(n\pi a/l)^2-(2\pi qa)^2\cos^2\theta\right]^2\left[\alpha_{km}^2-(2\pi qa)^2\sin^2\theta\right]^2\left(\alpha_{km}^2-m^2\right)}$$
$$\times\left[J'_m(2\pi qa\sin\theta)\right]^2\exp\left\{-\left[\left(\frac{\alpha_{km}}{a}\right)^2+\left(\frac{n\pi}{l}\right)^2\right]D\Delta\right\},\tag{4.64}$$

where α_{km} is the kth non-zero root of the equation $J'_m(\alpha_{km}) = 0$ (with $\alpha_{10}=0$) and the constant K_{nm} depends on the values of the indexes n and m according to

$$\begin{aligned}K_{nm} &= 1\,\text{if}\,n=m=0\\ K_{nm} &= 2\,\text{if}\,n\neq 0\,\text{and}\,m=0\,\text{or}\,m\neq 0\,\text{and}\,n=0\\ K_{nm} &= 4\,\text{if}\,n,m\neq 0.\end{aligned}\tag{4.65}$$

If $\theta = 0$ then Eq. (4.64) reduces to the solution for diffusion between planes (i.e., Eq. (3.17)). Similarly, if $\theta=\pi/2$, Eq. (4.64) reduces to[55]

$$E(\mathbf{q},\Delta) = 4(2\pi qa)^2\sum_{k=1}^{\infty}\sum_{m=0}^{\infty}\frac{K_{0m}\alpha_{km}^2\left[J'_m(2\pi qa)\right]^2\exp\left\{-(\alpha_{km}/a)^2D\Delta\right\}}{\left[\alpha_{km}^2-(2\pi qa)^2\right]^2\left(\alpha_{km}^2-m^2\right)}.\tag{4.66}$$

Ref. 56 contains an alternate but equivalent expression. The echo-attenuation curves for diffusion in a cylinder versus Δ are plotted for three different values of θ in Figure 4.10.

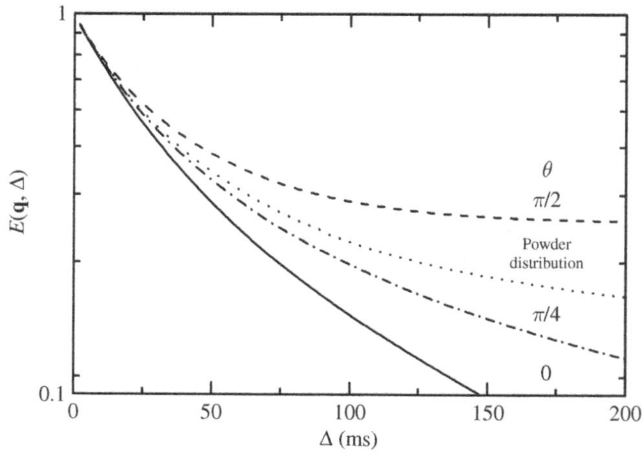

Figure 4.10 Plots of the simulated echo attenuation for PGSE diffusion measurements in a cylinder with the cylinder oriented at three different polar angles with respect to the gradient, i.e., $\theta=0$ (——), $\theta=\pi/4$ (-·-·-) and $\theta=\pi/2$ (- - - -) versus Δ calculated using Eqs. (3.17), (4.64) and (4.66), respectively. Also shown is the result of a powder distribution of polar angles (·······) versus Δ. The parameters used in the simulation are $\delta=1$ ms, $D=5\times10^{-10}\,\text{m}^2\text{s}^{-1}$, $g=1\,\text{T m}^{-1}$, $a=8\,\mu\text{m}$, $l=24\,\mu\text{m}$ and $\gamma(^1\text{H})=2.6571\times10^8\,\text{rad s}^{-1}\,\text{T}^{-1}$. As would be expected, since it is an average over all possible polar angles, the powder average echo attenuation curve is between the limits of the attenuation curves for $\theta=0$ and $\theta=\pi/2$. From Price.[57]

The corresponding long-time limiting formula is[55]

$$E(\mathbf{q},\infty)=\frac{8a^2[1-\cos(2\pi ql\cos\theta)][J_1(2\pi qa\sin\theta)]^2}{(2\pi qa)^4l^2(\cos\theta\sin\theta)^2}. \tag{4.67}$$

When $\theta=0$ this, of course, reduces to the long-time limit for diffusion between planes as given by Eq. (3.18) and when $\theta=\pi/2$ this reduces to[55]

$$E(\mathbf{q},\infty)=\frac{[2J_1(2\pi qa)]^2}{(2\pi qa)^2}, \tag{4.68}$$

and thus the first minima occurs at $q\approx 1.22/(2a)$. At low q this reduces to[56]

$$E(\mathbf{q}\rightarrow 0,\infty)=\exp\left(-(2\pi q)^2\frac{(2a)^2}{16}\right). \tag{4.69}$$

The echo-attenuation curves for diffusion in a cylinder are plotted for three different values of θ in Figure 4.11. It has been noted that in real systems the visibility of such diffraction patterns is very dependent on θ.[58]

Using the propagator given in Eq. (1.98) the SGP solution for a relaxing boundary at a radial distance a from the cylinder centre with the gradient applied across the polar direction is[29]

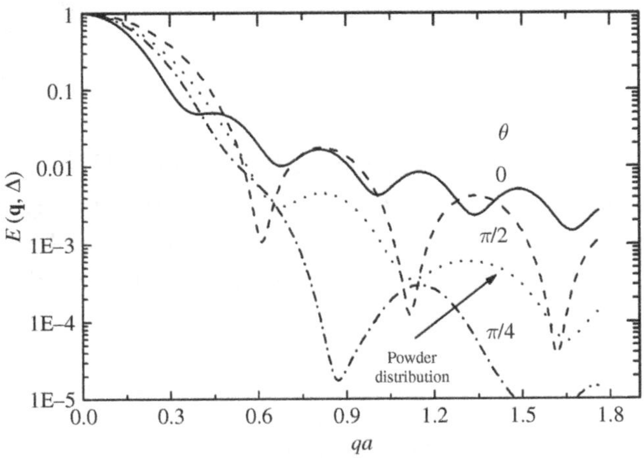

Figure 4.11 Simulated echo attenuation for PGSE diffusion in a cylinder with the cylinder oriented at three different polar angles with respect to the gradient, i.e., $\theta=0$ (——), $\theta=\pi/4$ (- · · - · -) and $\theta=\pi/2$ (- - - -) versus qa calculated using Eqs. (3.17), (4.64) and (4.66), respectively. Also shown is the result of a powder distribution of polar angles (········) versus qa. The parameters used in the simulation were $\delta=1$ ms, $\Delta=100$ ms, $D=1\times10^{-9}$ m^{2}s^{-1}, $g=1$ T m^{-1}, $a=8$ μm, $l=24$ μm and $\gamma(^1\mathrm{H})=2.6571\times10^8$ rad s^{-1} T^{-1}. From Price.[57]

$$E(\mathbf{q},\Delta) = 4\sum_{m=0}^{\infty}\frac{\alpha_{0m}^2}{\left[\left(\frac{Ma}{D}\right)^2+\alpha_{0m}^2\right]}\frac{\left\{(2\pi qa)J_0'(2\pi qa)+\left(\frac{Ma}{D}\right)J_0(2\pi qa)\right\}^2}{\left[(2\pi qa)^2-\alpha_{0m}^2\right]^2}e^{-\alpha_{0m}^2\frac{D\Delta}{a^2}}$$
$$+8\sum_{n=0}^{\infty}\sum_{m=0}^{\infty}\frac{\alpha_{nm}^2}{\left[\left(\frac{Ma}{D}\right)^2+\alpha_{nm}^2-n^2\right]}\frac{\left\{(2\pi qa)J_n'(2\pi qa)+\left(\frac{Ma}{D}\right)J_n(2\pi qa)\right\}^2}{\left[(2\pi qa)^2-\alpha_{nm}^2\right]^2}e^{-\alpha_{nm}^2\frac{D\Delta}{a^2}},$$
(4.70)

where the eigenvalues are given by Eq. (1.98). Diffusion measurements on water in 100 μm i.d. capillaries were well described using this formula.[59] We note that Eq. (4.70) reduces to Eq. (4.66) when M = 0.

4.4.4 Diffusion tensor

Without any information on the underlying geometry it will always be possible to determine the apparent diffusion tensor and thus an apparent diffusion ellipsoid by conducting diffusion measurements with six different gradient directions. If the short gradient pulse limit can be achieved it will be possible to determine the time-dependent diffusion tensor (see Eq. (1.115)) by determining the width of the diffusion propagator, quantified by **H**, by evaluating the low-q part of the data (cf. Eqs. (3.10) and (4.51))[60]

$$E(\mathbf{q}\rightarrow 0,\Delta) = \exp\left(-2\pi^2\mathbf{q}^{\mathrm{T}}\cdot\mathbf{H}\cdot\mathbf{q}\right). \tag{4.71}$$

When a single apparent diffusion tensor is estimated for the entire sample it is sometimes referred to as 'diffusion tensor MR spectroscopy' and when, as is commonly the case in imaging studies, the estimation is performed for each voxel it is referred to as 'diffusion tensor MR imaging'.

From the discussion on time-dependent diffusion ellipsoids in Section 1.6, it was understood that at very short Δ values $\mathbf{D}(\Delta)$ will become isotropic since the diffusion of the probe species is unaffected by the boundaries and this can be easily visualised from the convergence of the echo signal attenuation plots for different values of θ versus Δ in Figure 4.10.

4.5 General porous media

Porous media, as in rocks, constitute the most general case of restricted diffusion and, as noted in Chapter 1, analytical solutions are only available for simple cases such as periodic geometries.[61–63] In general, approximations such as the pore-hopping model or numerical methods are used. We also note that fractal diffusion can be viewed as a generalisation of diffusive behaviour in porous media.

4.5.1 Interconnected pores

For a porous medium consisting of isolated pores of radius a and spacing b (see Figure 1.11), the connectivity matrix is diagonal and Eqs. (1.139) and (3.16), in analogy to Section 3.5 leads to[64]

$$E(\mathbf{q},\infty) = \frac{1}{N}\sum_{i=1}^{N} |S_{0i}(\mathbf{q})|^2. \tag{4.72}$$

Thus, in a porous structure a diffusing spin can migrate between pores and this has the effect of changing the single-slit 'diffraction' into a 'diffusion grating'.[65] Information about the pore spacing and connectivity is contained in $D(t)$ and the tortuosity. The elements of the 'diffraction grating' are weighted according to a diffusion envelope $C(Z, \Delta)$ (a Gaussian, Eq. (1.60)), which describes the probability that a spin will move to another pore a displacement Z parallel to $\mathbf{q}$ from the starting pore after a time Δ. Hence, provided $4a^2/D^0 \ll b^2/D^\infty$ (i.e., the pore-equilibration assumption), $P(\mathbf{r}_0, \mathbf{r}_1, \Delta)$ becomes a product of the pore density and probability of jumps between pores. Thus, the attenuation is given by the Fourier transform of Eq. (1.141)[64–66]

$$E(\mathbf{q},\Delta) = |S_0(\mathbf{q})|^2[\mathrm{FT}\{L(Z)\} \otimes \mathrm{FT}\{C(Z,\Delta)\}], \tag{4.73}$$

where $|S_0(\mathbf{q})|^2$ is the average pore structure factor. $\mathrm{FT}\{L(Z)\}$ is the reciprocal lattice which is broadened by $\mathrm{FT}\{C(Z, \Delta)\}$. This broadening arises because the number of 'scattering centres' in the lattice increases with time as the spins diffuse to more distant pores. If the pore spacing b is irregular the maximum $E(\mathbf{q}, \Delta)$ will occur at $|q| = b^{-1}$. Thus, using small values of q ($< b^{-1}$), $E(q, \Delta)$ will probe D^∞, while for larger values of q, $|S_0(\mathbf{q})|^2$ will be observable as a modulation.

For a model system of monodisperse polystyrene spheres, it was found that Eq. (4.73) gave a qualitative fit to the data but a more precise fit was obtained when a pore-hopping formalism was adopted such that[65,66]

$$E(\mathbf{q},\Delta) = |S_0(q)|^2 \exp\left[-\frac{6D^\infty\Delta}{b^2+3\sigma^2}\left(1-\exp(-2\pi^2q^2\sigma^2)\right)\mathrm{sinc}(2\pi qb)\right], \tag{4.74}$$

where σ is the standard deviation of the pore spacing. This model predicts coherence peaks in $E(\mathbf{q}, \Delta)$ when qb is an integer. It was noted that D^∞ is already effective at times when a single spin would have only diffused across one pore width, while pore structure effects are not visible until a spin has had time to visit a neighbouring pore.[66] By performing inverse Fourier transforms of the experimental data, Callaghan *et al.*[65] constructed an 'image' of the spherically averaged auto-correlation function. In a later study it was found that the coherence peaks were strongly influenced by the sphere size and polydispersity in a more polydisperse set of polystyrene spheres.[67] It has been realised that the length of the gradient pulse can be set to emphasise pore-hopping maxima.[68,69] Using computer simulations Bhattacharya and Mahanti have examined diffraction in two-dimensional disordered media that can be characterised by a single length scale.[70] Their results showed that the diffraction pattern picks up the dominant mean length scale despite their being a distribution of pore sizes about the mean.

4.5.2 Diffusion in fractals

Using the GPD approximation Banavar *et al.*[71] derived the attenuation for the SGP limit of the Hahn-echo-based PGSE sequence to be (note there appears to be a misprint in this reference as 12 should have been removed from the denominator in the exponential)

$$E(\mathbf{q},\Delta) = \exp\left(-(2\pi q)^2 D\Delta^{\kappa}\right), \tag{4.75}$$

where $\kappa = 2/d_w$ and d_w is the random walk dimension. This equation reverts to that for the solution for normal free diffusion when $d_w = 2$ (i.e., $\kappa = 1$).

Based on the Bloch equations modified to include a term for describing diffusion in fractal systems based on a scaling law, the validity of which has been questioned (see refs. 72–74 and references therein) Jug[75] derived an expression for attenuation in the PGSE experiment due to fractal diffusion in agreement with Eq. (4.75) – after the removal of some inconsistencies.[73] In a later paper, Kärger *et al.*,[74] using the GPD approximation and noting the time scaling of the MSD in fractal objects (see Eq. (1.121)) derived the PGSE spin-echo attenuation for diffusion in a fractal network to be

$$E(\mathbf{q},\Delta) = \exp\left\{\frac{\gamma^2 g^2 \alpha}{3(\kappa+1)(\kappa+2)}\left[\delta^{\kappa+2} - \tfrac{1}{2}(\Delta-\delta)^{\kappa+2} + \Delta^{\kappa+2} - \tfrac{1}{2}(\Delta+\delta)^{\kappa+2}\right]\right\}, \tag{4.76}$$

which in the SGP limit returns Eq. (4.75) when the time-independent scaling constant $\alpha = 6D$. And the solution for the steady field gradient case has been derived[74]

$$E(\mathbf{g},2\tau) = \exp\left[-\frac{1-1/2^{\kappa}}{(\kappa+1)(\kappa+2)}\gamma^2 g^2 \frac{\alpha}{6}(2\tau)^{\kappa+2}\right], \tag{4.77}$$

which reduces to the solution for the steady gradient experiment for ordinary diffusion when $\kappa = 1$ and $\alpha = 6D$ (cf. Eq. (2.50)). We also note that solutions for the steady gradient case were also derived by Banavar *et al.*[71] and Widom and Chen,[76] but the three solutions only agree in the case of ordinary diffusion (i.e., $\kappa = 1$).

All of the previous studies[71,73–76] have shown the signal attenuation due to fractional Brownian motion is of the form

$$E(\mathbf{q}, \Delta) \sim \exp\left(-cq^2\Delta^\kappa\right), \tag{4.78}$$

where c is a constant. However, starting with the fractal propagator given by Eq. (1.132) and substituting into Eq. (2.35), Damion and Packer[77] obtained the expression for the echo attenuation in fractal geometries as

$$E(\mathbf{q}, \Delta) = \frac{\sqrt{\pi}}{\Gamma\left(1 - \frac{(1-\kappa)d_f}{2}\right)\Gamma\left(\frac{d_f}{2} - 1\right)} \sum_{n=0}^{\infty} \frac{\Gamma\left(n + 1 - \frac{(1-\kappa)d_f}{2}\right)\Gamma\left(n - 1 + \frac{d_f}{2}\right)(-1)^n}{\Gamma(1 + \kappa n)\Gamma\left(n + \frac{1}{2}\right)n!} \times \left(2\pi q D_\kappa^{1/2}\Delta^{\kappa/2}\right)^{2n} \tag{4.79}$$

where D_κ is the fractional diffusion coefficient and d_f is the dimension of the fractal space. For normal Brownian (i.e., $d_f = 3$ and $\kappa = 1$), Eq. (4.79) reduces to Eq. (2.44) as expected. In the asymptotic limit, $4\pi q D_k^{1/2}\Delta^{\kappa/2} \gg 1$, Eq. (4.79) becomes

$$E(\mathbf{q}, \Delta) \sim \left(q\kappa^{\frac{1}{2}}\Delta^{\frac{\kappa}{2}}\right)^{2-d_f}, \; 2 < d_f < 3. \tag{4.80}$$

The power law behaviour of this equation is in contrast to Eq. (4.78).

4.6 Polymer diffusion

Models for analysing polymer diffusion have been reviewed by Ardelean and Kimmich.[78] As noted above, polymer diffusion is complicated as it can contain both elements of free diffusion and, due to the effects of entanglement, 'soft' restriction. Diffusion in a harmonic well is a model of hindered diffusion in the presence of a 'soft' wall with relevance to diffusion in entangled polymer systems.[79,80] The SGP solution for a particle that is diffusing in an isotropic medium but subject to an additional force pulling it towards the origin in proportion to its distance from the origin (i.e., $\mathbf{F} = -\beta f \mathbf{r}$; where f is the friction coefficient and β is defined through the velocity of the particle, $\mathbf{v} = -\beta\mathbf{r}$) was determined by Stejskal to be[81]

$$E(\mathbf{q}, \Delta) = \exp\left(-4\pi^2 q^2 D\left(1 - e^{-\beta\Delta}\right)/\beta\right). \tag{4.81}$$

Interestingly, attenuation data in accordance with Eq. (4.81) but analysed assuming normal isotropic diffusion (i.e., Eq. (2.44)) results in a time-dependent diffusion coefficient

$$D^{\mathrm{app}}(t) = D\left(1 - e^{-\beta\Delta}\right)/\beta\Delta. \tag{4.82}$$

More recently, the exact solution has been determined for the echo attenuation for the damped oscillator problem at all timescales.[80] This comprehensive solution has the above SGP solution as one of its limits.

Models have been presented for segment diffusion in the reptation tube model.[82,83]

4.7 Flow and velocity distributions

If the Bloch–Torrey equation (Eq. (2.56)) includes a term reflecting plug flow (i.e., $-\nabla \cdot \mathbf{v}_0\mathbf{M}$), where $\mathbf{v}_0$ is the velocity of the medium in which the spins are in and the flow is along the direction of the gradient, a solution of the form[81,84]

$$S(\mathbf{q},\Delta) = S(0)\exp\left(-(2\pi)^2\int_0^t \mathbf{q}(t')^2 dt' D\right)\exp\left(i2\pi\int_0^t \mathbf{q}(t')dt' \cdot \mathbf{v}_0\right) \tag{4.83}$$

is obtained. This, for the Stejskal and Tanner sequence, gives in the case of plug flow (similar to what was derived in Section 2.2.3)[81,84]

$$S(\mathbf{q},\Delta) = S(0)\exp\left(-\underbrace{(2\pi\mathbf{q})^2 D(\Delta-\delta/3)}_{\text{attenuation}} + \underbrace{i2\pi\mathbf{q}\cdot\mathbf{v}_0\Delta}_{\text{phase shift}}\right). \tag{4.84}$$

More generally, and always in the case of laminar flow, there will be a distribution of velocities, $P(v)$, which depends on the geometry of the flow channel. In this case, ignoring diffusion, the echo signal response will be[85]

$$S(\mathbf{q},\Delta) = S(0)\int P(v)\exp(iq\Delta v)dv. \tag{4.85}$$

In the case of a circular pipe with the gradient directed along the long axis, Eq. (4.85) reduces to[85,86]

$$S(\mathbf{q},\Delta) = S(0)\ \mathrm{sinc}(4\pi q\Delta\langle v\rangle), \tag{4.86}$$

where $\langle v\rangle$ is the average velocity. Thus, in addition to a net phase shift, when there is a distribution of velocities, the flow can lead to signal attenuation as there will be a distribution of phases – similar to the case of diffusion (cf. Eq. (2.3)). Indeed, under the assumptions that either the flow is slow or its velocity field is constant in time, Eq. (4.83) can be extended to cover the case of non-uniform flow including convection[87]

$$S(t) \approx S(0) \exp\left(-(2\pi)^2 \int_0^t \mathbf{q}(t')^2 dt' D\right) \int \exp\left(i2\pi \int_0^t \mathbf{q}(t')dt' \cdot \mathbf{v}(\mathbf{r})\right) d\mathbf{r}. \quad (4.87)$$

Despite having a zero average in a closed sample $\mathbf{v}(\mathbf{r})$ has a non-zero width, consequently the second exponent in Eq. (4.87) results in signal attenuation with increasing q.[87]

For slow flows, diffusion can cause the spins to sample a range of velocities with the distribution present (i.e., $P(v)$), whereas turbulence also introduces a time-dependence. Similarly, capillary microcirculation in tissue perfusion can also lead to such time-dependent velocity distributions and as there is generally no preferred orientation – such diffusion like-processes have been referred to as pseudo-diffusion.[85,88,89] In tissue this effect can lead to pseudo-diffusion coefficients many times greater than the true diffusion coefficient. Interestingly, a time-dependent sampling of $P(v)$ during Δ can result in an effectively narrowed $P(v)$ corresponding to a single averaged velocity.[85] For fluid with an average velocity $\langle v \rangle$ undergoing capillary microcirculation where the paths randomly change direction with a characteristic length scale L_C the pseudo-diffusion coefficient is given by[89]

$$D_P = \frac{\langle v \rangle L_C}{6}. \quad (4.88)$$

A theoretical description of the signal attenuation from a turbulent flowing liquid in a PGSE experiment has also been developed.[90,91]

Whereas we have already met diffraction phenomena due to diffusion, it is also possible to observe them in flow measurements in porous media. Flow diffraction patterns can result when the displacement of nuclei is correlated with the length scale of the pore spacing.[92,93]

Various more complicated correlations with flow are considered in Section 8.7 and Chapter 9.

4.8 Internal magnetic field gradients

In Chapter 2 we learnt that there were a number of sources of background gradients. Here we are concerned with the internal magnetic field gradients ($\mathbf{g}_{\text{int}}$; IMFGs) that arise from magnetic susceptibility (χ) heterogeneities in samples. IMFGs are thus prevalent in porous samples. A very simple example is given in Figure 4.12 for the gradients in an NMR sample (which can be viewed simplistically as a very large pore) which arise in the sample out of the difference in susceptibility inside and outside of the sample. Indeed, for a spherical-shaped pore of radius r and susceptibility difference $\Delta\chi$, the gradient at the pore surface is[94,95]

$$g_{\text{int}} \approx \pi B_0 \Delta\chi / r. \tag{4.89}$$

Thus, the size and homogeneity of the internal gradients are inherently related to the size, shape and distribution of pores within a sample and importantly their magnitude scales with B_0. Non-homogeneous internal gradients are common within many samples. For example, red blood cells have been estimated to have gradients up to $2 \times 10^{-2}\,\mathrm{T\,m^{-1}}$,[97] and internal gradients can be of the order of $0.2\,\mathrm{T\,m^{-1}}$ and $0.5\,\mathrm{T\,m^{-1}}$ in zeolite and hydride samples, respectively.[98,99] Even minute air bubbles in apple can lead to large background gradients[100] and over very short length scales just the interface between air and water can generate gradients in the range of hundreds of $\mathrm{T\,m^{-1}}$.[101] Similarly, internal gradients in iron(III)-doped kaolin clay have been found to be of the order of 1–$10\,\mathrm{T\,m^{-1}}$ inside the pores and may exceed $1000\,\mathrm{T\,m^{-1}}$ at the pore surface.[102] Background gradients can effectively be made time-dependent by the spins diffusing between different regions of differing background gradient strength and are encountered in situations such as spins

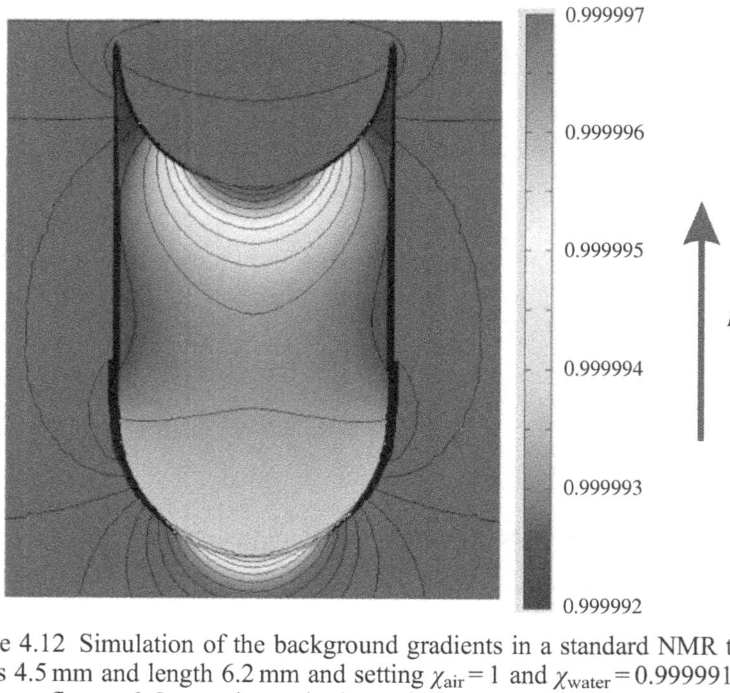

Figure 4.12 Simulation of the background gradients in a standard NMR tube of radius 4.5 mm and length 6.2 mm and setting $\chi_{\text{air}} = 1$ and $\chi_{\text{water}} = 0.999991$. Each contour reflects a 0.5 ppm change in the static magnetic field. From Price *et al.*[96]

diffusing in media containing superparamagnetic particles used as contrast agents in MRI.[103]

In the present work the coverage of diffusion measurements in porous media focuses on cases in which the applied magnetic gradient pulses are ideally of a constant amplitude throughout the sample and IMFGs arising out of magnetic susceptibility differences are viewed as a source of artefacts and methods to remove their effects are considered in Section 7.6. However, given that background gradients exist in all porous media and, in many cases, such as well-logging, it is physically difficult to get the requisite hardware for generating strong applied magnetic gradients into position. There is great interest on extracting information on the porous characteristics of many porous media by studying the diffusive decay due to diffusion in the inherent internal gradients (also known as 'diffusion decay in the internal field' (DDIF)[104,105]). Importantly, because the internal fields are created by the pore structure, they directly reflect the pore geometry.

The importance of diffusion through inhomogeneous gradients has long been realised[106] and a large body of work exists dealing with it,[104,105,107–114] including continuous Gaussian random fields,[115,116] in magnetic fields with parabolic, cosine and sinusoidal profiles,[117–122] restricted diffusion between planes in a cosine magnetic field,[123] in magic angle spinning (MAS)[124] and in grossly inhomogeneous fields where the rf pulses are weak relative to the field inhomogeneity.[125] It is also noted that diffusion measurements performed with single-sided NMR (e.g., NMR-MOUSE) also involve non-constant gradients unless great efforts are made to generate highly uniform gradients.[126,127] The analysis of such systems becomes quite complicated since the effects of diffusion and relaxation are intertwined.

Sukstanskii and Yablonskiy developed the theory of macroscopic signal formation in the presence of mesoscopic structure-specific magnetic field inhomogeneities using the GPD approximation.[128,129]

References

1. B. Lévay, Studies on Self-Association Equilibria by Self-Diffusion Measurements. *J. Phys. Chem.* **77** (1973), 2118–21.
2. J. R. Zimmerman and W. E. Brittin, Nuclear Magnetic Resonance Studies in Multiple Phase Systems: Lifetime of a Water Molecule in an Adsorbing Phase on Silica Gel. *J. Phys. Chem.* **61** (1957), 1328–33.
3. C. S. Johnson, Jr., Chemical Rate Processes and Magnetic Resonance. *Adv. Magn. Reson.* (1965), 33–102.
4. B. Meyer and T. Peters, NMR Spectroscopy Techniques for Screening and Identifying Ligand Binding to Protein Receptors. *Angew. Chemie (Int. Ed. Engl.)* **42** (2003), 864–90.

5. W. S. Price and P. W. Kuchel, Restricted Diffusion of Bicarbonate and Hypophosphite Ions Modulated by Transport in Suspensions of Red Blood Cells. *J. Magn. Reson.* **90** (1990), 100–10.
6. G. C. K. Roberts, The Determination of Equilibrium Dissociation Constants of Protein-Ligand Complexes by NMR. In *BioNMR in Drug Research*, ed. O. Zerbe and R. Mannhold. (New York: Wiley, 2003), pp. 309–19.
7. F. S. Dukhovich, E. N. Gorbatova, M. B. Darkhovskii, and V. K. Kurochkin, Relationship Between the Dissociation Constant and the Lifetime for Complexes of Biologically Active Substances with Receptors and Enzymes. *Pharm. Chem. J.* **36** (2002), 248–54.
8. J. Kärger, NMR Self-Diffusion Studies in Heterogeneous Systems. *Adv. Colloid Interface Sci.* **23** (1985), 129–48.
9. J. Kärger, H. Pfeifer, and W. Heink, Principles and Applications of Self-Diffusion Measurements by Nuclear Magnetic Resonance. *Adv. Magn. Reson.* **12** (1988), 1–89.
10. A. R. Waldeck, P. W. Kuchel, A. J. Lennon, and B. E. Chapman, NMR Diffusion Measurements to Characterise Membrane Transport and Solute Binding. *Prog. NMR Spectrosc.* **30** (1997), 39–68.
11. W. S. Price, A. V. Barzykin, K. Hayamizu, and M. Tachiya, A Model for Diffusive Transport Through a Spherical Interface Probed by Pulsed-Field Gradient NMR. *Biophys. J.* **74** (1998), 2259–71.
12. C. S. Johnson, Jr., Diffusion Ordered Nuclear Magnetic Resonance Spectroscopy: Principles and Applications. *Prog. NMR Spectrosc.* **34** (1999), 203–56.
13. A. J. Lennon, B. E. Chapman, and P. W. Kuchel, T_2 Effects in PFG NMR Analysis of Ligand Binding to Macromolecules. *Bull. Magn. Reson.* **17** (1996), 224–5.
14. W. S. Price, Recent Advances in NMR Diffusion Techniques for Studying Drug Binding. *Aust. J. Chem.* **56** (2003), 855–60.
15. T. Adalsteinsson, W.-F. Dong, and M. Schönhoff, Diffusion of 77,000 g/mol Dextran in Sub-Micron Polyelectrolyte Capsule Dispersion Measured Using PFG-NMR. *J. Phys. Chem. B* **108** (2004), 20056–63.
16. K. I. Momot and P. W. Kuchel, Pulsed Field Gradient Nuclear Magnetic Resonance as a Tool for Studying Drug Delivery Systems. *Concepts Magn. Reson.* **19A** (2003), 51–64.
17. A. D. Bain, Chemical Exchange in NMR. *Prog. NMR Spectrosc.* **43** (2003), 63–103.
18. W. S. Price, F. Elwinger, C. Vigouroux, and P. Stilbs, PGSE-WATERGATE, a New Tool for NMR Diffusion-Based Studies of Ligand-Macromolecule Binding. *Magn. Reson. Chem.* **40** (2002), 391–95.
19. C. S. Johnson, Jr., Effects of Chemical Exchange in Diffusion-Ordered 2D NMR Spectra. *J. Magn. Reson. A* **102** (1993), 214–18.
20. E. J. Cabrita, S. Berger, P. Bräuer, and J. Kärger, High-Resolution DOSY NMR with Spins in Different Chemical Surroundings: Influence of Particle Exchange. *J. Magn. Reson.* **157** (2002), 124–31.
21. J. Andrasko, Water Diffusion Permeability of Human Erythrocytes Studied by a Pulsed Field Gradient NMR Technique. *Biochim. Biophys. Acta* **428** (1976), 304–11.
22. J. Pfeuffer, U. Flögel, W. Dreher, and D. Leibfritz, Restricted Diffusion and Exchange of Intracellular Water: Theoretical Modelling and Diffusion Time Dependence of ^{1}H NMR Measurements on Perfused Glial Cells. *NMR Biomed.* **11** (1998), 19–31.
23. G. J. Stanisz, J. G. Li, G. A. Wright, and R. M. Henkelman, Water Dynamics in Human Blood via Combined Measurements of T_2 Relaxation and Diffusion in the Presence of Gadolinium. *Magn. Reson. Med.* **39** (1998), 223–33.
24. S. N. Jespersen, M. Pedersen, and H. Stodkilde-Jorgensen, The Influence of a Cellular Size Distribution on NMR Diffusion Measurements. *Eur. Biophys. J.* **34** (2005), 890–8.

25. P. W. Kuchel, A. Coy, and P. Stilbs, NMR 'Diffusion-Diffraction' of Water Revealing Alignment of Erythrocytes in a Magnetic Field and Their Dimensions and Membrane Transport Characteristics. *Magn. Reson. Med.* **37** (1997), 637–43.
26. P. P. Mitra and P. N. Sen, Effects of Microgeometry and Surface Relaxation on NMR Pulsed-Field-Gradient Experiments: Simple Pore Geometries. *Phys. Rev. B* **45** (1992), 143–56.
27. J. E. M. Snaar and H. Van As, NMR Self-Diffusion Measurements in a Bounded System with Loss of Magnetization at the Walls. *J. Magn. Reson. A* **102** (1993), 318–26.
28. A. Coy and P. T. Callaghan, Pulsed Gradient Spin Echo Nuclear Magnetic Resonance for Molecules Diffusing Between Partially Reflecting Rectangular Barriers. *J. Chem. Phys.* **101** (1994), 4599–609.
29. P. T. Callaghan, Pulsed Gradient Spin Echo NMR for Planar, Cylindrical and Spherical Pores under Conditions of Wall Relaxation. *J. Magn. Reson. A* **113** (1995), 53–9.
30. P. P. Mitra, P. N. Sen, and L. M. Schwartz, Short-Time Behaviour of the Diffusion Coefficient as a Geometrical Probe of Porous Media. *Phys. Rev. B* **47** (1993), 8565–74.
31. A. V. Barzykin, W. S. Price, K. Hayamizu, and M. Tachiya, Pulsed Field Gradient NMR of Diffusive Transport Through a Spherical Interface into an External Medium Containing a Relaxation Agent. *J. Magn. Reson. A* **114** (1995), 39–46.
32. M. Abramowitz and I. A. Stegun, *Handbook of Mathematical Functions* (New York: Dover, 1970).
33. P. W. Kuchel, A. J. Lennon, and C. J. Durrant, Analytical Solutions and Simulations for Spin-Echo Measurements of Diffusion of Spins in a Sphere with Surface and Bulk Relaxation. *J. Magn. Reson. B* **112** (1996), 1–17.
34. J. E. Tanner, Transient Diffusion in a System Partitioned by Permeable Barriers. Application to NMR Measurements with a Pulsed Field Gradient. *J. Chem. Phys.* **69** (1978), 1748–54.
35. G. P. Zientara and J. H. Freed, Spin-Echoes for Diffusion in Bounded, Heterogeneous Media: A Numerical Study. *J. Chem. Phys.* **72** (1980), 1285–92.
36. E. von Meerwall and R. D. Ferguson, Interpreting Pulsed-Gradient Spin-Echo Diffusion Experiments with Permeable Membranes. *J. Chem. Phys.* **74** (1981), 6956–9.
37. E. D. von Meerwall and R. D. Ferguson, A Fortran Program to Fit Diffusion Models to Field-Gradient Spin Echo Data. *Comput. Phys. Commun.* **21** (1981), 421–9.
38. P. W. Kuchel and C. J. Durrant, Permeability Coefficients from NMR q-Space Data: Models with Unevenly Spaced Semi-Permeable Parallel Membranes. *J. Magn. Reson.* **139** (1999), 258–72.
39. L. van der Weerd, S. M. Melnikov, F. J. Vergeldt, E. G. Novikov, and H. Van As, Modelling of Self-diffusion and Relaxation Time NMR in Multicompartment Systems with Cylindrical Geometry. *J. Magn. Reson.* **156** (2002), 213–21.
40. P.-C. Jiang, T.-Y. Yu, W.-C. Perng, and L.-P. Hwang, Pore-to-Pore Hopping Model for the Interpretation of the Pulsed Gradient Spin Echo Attenuation of Water Diffusion in Cell Suspension Systems. *Biophys. J.* **80** (2001), 2493–504.
41. P. J. Basser, J. Mattiello, and D. Le Bihan, Estimation of the Effective Self-Diffusion Tensor from the NMR Spin Echo. *J. Magn. Reson. B* **103** (1994), 247–54.
42. D. Güllmar, J. Haueisen, and J. R. Reichenbach, Analysis of b-Value Calculations in Diffusion Weighted and Diffusion Tensor Imaging. *Concepts Magn. Reson.* **25A** (2005), 53–66.
43. A. Özcan, (Mathematical) Necessary Conditions for the Selection of Gradient Vectors in DTI. *J. Magn. Reson.* **172** (2005), 238–41.

44. P. J. Basser, J. Mattiello, and D. Le Bihan, MR Diffusion Tensor Spectroscopy and Imaging. *Biophys. J.* **66** (1994), 259–67.
45. D. Le Bihan, J.-F. Mangin, C. Poupon, C. A. Clark, S. Pappata, N. Molko, and H. Chabriat, Diffusion Tensor Imaging: Concepts and Applications. *J. Mag. Res. Imaging* **13** (2001), 534–46.
46. F. D. Blum, A. S. Padmanabhan, and R. Mohebbi, Self-Diffusion of Water in Polycrystalline Smectic Liquid Crystals. *Langmuir* **1** (1985), 127–31.
47. G. Celebre, G. Coppola, and G. A. Raineri, Water Self-Diffusion in Lyotropic Systems by Simulation of Pulsed Field Gradient Spin Echo Nuclear Magnetic Resonance Experiments. *J. Chem. Phys.* **97** (1992), 7781–5.
48. G. Celebre, G. Chidichimo, L. Coppola, C. La Mesa, R. Muzzalupo, L. Pogliani, G. A. Ranieri, and M. Terenzi, Water Self-Diffusion in Lyotropic Liquid Crystals: Pulsed Gradient Spin-Echo NMR and Simulation Techniques. *Gazz. Chim. Ital.* **126** (1996), 489–503.
49. S. V. Dvinskikh and I. Furó, Measurement of the Principal Values of the Anisotropic Diffusion Tensor in an Unoriented Sample by Exploiting the Chemical Shift Aniostropy: ^{19}F PGSE NMR with Homonuclear Decoupling. *J. Magn. Reson.* **148** (2001), 73–7.
50. P. T. Callaghan, K. W. Jolley, and J. Lelievre, Diffusion of Water in the Endosperm Tissue of Wheat Grains as Studied by Pulsed Field Gradient Nuclear Magnetic Resonance. *Biophys. J.* **28** (1979), 133–42.
51. P. T. Callaghan and O. Söderman, Examination of the Lamellar Phase of Aerosol OT/Water Using Pulsed Field Gradient Nuclear Magnetic Resonance. *J. Phys. Chem.* **87** (1983), 1737–44.
52. I. S. Gradshteyn and I. M. Ryzhik, *Table of Integrals, Series, and Products*, 7th edn. (New York: Academic Press, 2007).
53. P. van Gelderen, D. DesPres, P. C. M. Van Zijl, and C. T. W. Moonen, Evaluation of Restricted Diffusion in Cylinders. Phosphocreatine in Rabbit Leg Muscle. *J. Magn. Reson. B* **103** (1994), 255–60.
54. C. H. Neuman, Spin Echo of Spins Diffusing in a Bounded Medium. *J. Chem. Phys.* **60** (1974), 4508–11.
55. O. Söderman and B. Jönsson, Restricted Diffusion in Cylindrical Geometry. *J. Magn. Reson. A* **117** (1995), 94–7.
56. P. Linse and O. Söderman, The Validity of the Short-Gradient-Pulse Approximation in NMR Studies of Restricted Diffusion. Simulations of Molecules Diffusing between Planes, in Cylinders and Spheres. *J. Magn. Reson. A* **116** (1995), 77–86.
57. W. S. Price, Pulsed Field Gradient NMR as a Tool for Studying Translational Diffusion, Part I. Basic Theory. *Concepts Magn. Reson.* **9** (1997), 299–336.
58. L. Avram, Y. Assaf, and Y. Cohen, The Effect of Rotational Angle and Experimental Parameters on the Diffraction Patterns and Micro-Structural Information Obtained from *q*-Space Diffusion NMR: Implication for Diffusion in White Matter Fibers. *J. Magn. Reson.* **169** (2004), 30–8.
59. S. J. Gibbs, Observations of Diffusive Diffraction in a Cylindrical Pore by PFG NMR. *J. Magn. Reson.* **124** (1997), 223–6.
60. P. J. Basser, Relationships Between Diffusion Tensor and *q*-Space MRI. *Magn. Reson. Med.* **47** (2002), 392–7.
61. D. J. Bergman and K.-J. Dunn, Self-Diffusion in a Periodic Porous Medium with Interface Absorption. *Phys. Rev. E* **51** (1995), 3401–16.
62. D. J. Bergman, K.-J. Dunn, L. M. Schwartz, and P. P. Mitra, Self-Diffusion in a Periodic Porous Medium: A Comparison of Different Approaches. *Phys. Rev. E* **51** (1995), 3393–400.

63. K.-J. Dunn and D. J. Bergman, Self Diffusion of Nuclear Spins in a Porous Medium with a Periodic Microstructure. *J. Chem. Phys.* **102** (1995), 3041–54.
64. P. T. Callaghan, D. MacGowan, K. J. Packer, and F. O. Zelaya, High-Resolution q-Space Imaging in Porous Structures. *J. Magn. Reson.* **90** (1990), 177–82.
65. P. T. Callaghan, A. Coy, D. MacGowan, K. J. Packer, and F. O. Zelaya, Diffraction-Like Effects in NMR Diffusion of Fluids in Porous Solids. *Nature* **351** (1991), 467–9.
66. P. T. Callaghan, A. Coy, T. P. J. Halpin, D. MacGowan, K. J. Packer, and F. O. Zelaya, Diffusion in Porous Systems and the Influence of Pore Morphology in Pulsed Field Gradient Spin-Echo Nuclear Magnetic Resonance Studies. *J. Chem. Phys.* **97** (1992), 651–62.
67. A. Coy and P. T. Callaghan, Pulsed Gradient Spin-Echo NMR 'Diffusive Diffraction' Experiments on Water Surrounding Close-Packed Polymer Spheres. *J. Colloid Interface Sci.* **168** (1994), 373–9.
68. P. P. Mitra and B. I. Halperin, Effects of Finite Gradient Pulse Widths in Pulsed Field Gradient Diffusion Measurements. *J. Magn. Reson. A* **113** (1995), 94–101.
69. C. Malmborg, D. Topgaard, and O. Söderman, NMR Diffusometry and the Short Gradient Pulse Limit Approximation. *J. Magn. Reson.* **169** (2004), 85–91.
70. A. Bhattacharya and S. D. Mahanti, Spin Echo Diffraction in Disordered Media with Single Length Scales. *Phys. Rev. B* **55** (1997), 11230–5.
71. J. R. Banavar, M. Lipsicas, and J. F. Willemsen, Determination of the Random-Walk Dimension of Fractals by Means of NMR. *Phys. Rev. B* **32** (1985), 6066.
72. J. R. Banavar and J. F. Willemsen, Probability Density for Diffusion on Fractals. *Phys. Rev. B* **30** (1984), 6778–9.
73. J. Kärger and G. Vojta, On the Use of NMR Pulsed Field-Gradient Spectroscopy for the Study of Anomalous Diffusion in Fractal Networks. *Chem. Phys. Lett.* **141** (1987), 411–13.
74. J. Kärger, H. Pfeifer, and G. Vojta, Time Correlation During Anomalous Diffusion in Fractal Systems and Signal Attenuation in NMR Field-Gradient Spectroscopy. *Phys. Rev. A* **37** (1988), 4514–17.
75. G. Jug, Theory of NMR Field-Gradient-Spectroscopy for Anomalous Diffusion in Fractal Networks. *Chem. Phys. Lett.* **131** (1986), 94–7.
76. A. Widom and H. J. Chen, Fractal Brownian Motion and Nuclear Spins. *J. Phys. A Math. Gen.* **28** (1995), 1243–7.
77. R. A. Damion and K. J. Packer, Predictions for Pulsed-Field-Gradient NMR Experiments of Diffusion in Fractal Spaces. *Proc. R. Soc. London A* **453** (1997), 205–11.
78. I. Ardelean and R. Kimmich, Principles and Unconventional Aspects of NMR Diffusometry. In *Annual Reports on NMR Spectroscopy*. ed. G. A. Webb. vol. 49. (London: Elsevier, 2003), pp. 43–115.
79. M. Doi and S. F. Edwards, *The Theory of Polymer Dynamics*. (Oxford: Clarendon Press, 1986).
80. E. M. Terentjev, P. T. Callaghan, and M. Warner, Pulsed Gradient Spin-Echo Nuclear Magnetic Resonance of Confined Brownian Particles. *J. Chem. Phys.* **102** (1995), 4619–24.
81. E. O. Stejskal, Use of Spin Echoes in a Pulsed Magnetic-Field Gradient to Study Anisotropic Restricted Diffusion and Flow. *J. Chem. Phys.* **43** (1965), 3597–603.
82. N. Fatkullin and R. Kimmich, Theory of Field-Gradient NMR Diffusometry of Polymer Segment Displacements in the Tube-Reptation Model. *Phys. Rev. E* **52** (1995), 3273–6.
83. N. Fatkullin, E. Fischer, C. Mattea, U. Beginn, and R. Kimmich, Polymer Dynamics under Nanoscopic Constraints: The Corset Effect as Revealed by NMR Relaxometry and Diffusometry. *Chem. Phys. Chem.* **5** (2004), 884–94.

84. P. T. Callaghan, *Principle of Nuclear Magnetic Resonance Microscopy.* (Oxford: Clarendon Press, 1991).
85. K. J. Packer, Diffusion & Flow in Liquids. In *Encyclopedia of Nuclear Magnetic Resonance*, ed. D. M. Grant and R. K. Harris. vol. 3. (New York: Wiley, 1996), pp. 1615–26.
86. R. J. Hayward, K. J. Packer, and D. J. Tomlinson, Pulsed Field-Gradient Spin-Echo NMR Studies of Flow in Fluids. *Mol. Phys.* **23** (1972), 1083–102.
87. K. I. Momot and P. W. Kuchel, PFG NMR Diffusion Experiments for Complex Systems. *Concepts Magn. Reson.* **28A** (2006), 249–69.
88. D. Le Bihan, E. Breton, D. Lallemand, P. Grenier, E. Cabanis and M. Laval-Jeantet, MR Imaging of Intravoxel Incoherent Motions: Application to Diffusion and Perfusion in Neurologic Disorders. *Radiology* **161** (1986), 401–7.
89. R. Turner and P. Keller, Angiography and Perfusion Measurements by NMR. *Prog. NMR Spectrosc.* **23** (1991), 93–133.
90. J.-H. Gao and J. C. Gore, Turbulent Flow Effects on NMR Imaging: Measurement of Turbulent Intensity. *Med. Phys.* **18** (1991), 1045–51.
91. J. C. Gatenby and J. C. Gore, Characterization of Turbulent Flows by NMR Measurements with Pulsed Field Gradients. *J. Magn. Reson. A* **110** (1994), 26–32.
92. J. D. Seymour and P. T. Callaghan, 'Flow-Diffraction' Structural Characterization and Measurement of Hydrodynamic Dispersion in Porous Media by PGSE NMR. *J. Magn. Reson. A* **122** (1996), 90–3.
93. J. D. Seymour and P. T. Callaghan, Generalized Approach to NMR Analysis of Flow and Dispersion in Porous Media. *AIChE J.* **43** (1997), 2096–111.
94. J. A. Glasel and K. H. Lee, On the Interpretation of Water Nuclear Magnetic Resonance Relaxation Times in Heterogeneous Systems. *J. Am. Chem. Soc.* **96** (1974), 970–8.
95. R. Blinc, J. Dolinšek, G. Lahajnar, A. Sepe, I. Zupančič, S. Žumer, F. Milia, and M. M. Pintar, Spin-Lattice Relaxation of Water in Cement Gels. *Z. Naturforsch.* **43A** (1988), 1026–38.
96. W. S. Price, P. Stilbs, B. Jönsson, and O. Söderman, Macroscopic Background Gradient and Radiation Damping Effects on High-Field PGSE NMR Diffusion Measurements. *J. Magn. Reson.* **150** (2001), 49–56.
97. Z. H. Endre, P. W. Kuchel, and B. E. Chapman, Cell-Volume Dependence of ^{1}H Spin-Echo NMR Signals in Human Erythrocyte Suspensions. The Influence of *In Situ* Field Gradients. *Biochim. Biophys. Acta* **803** (1984), 137–44.
98. J. Kärger, H. Pfeifer, and S. Rudtsch, The Influence of Internal Magnetic Field Gradients on NMR Self-Diffusion Measurements of Molecules Adsorbed on Microporous Crystallites. *J. Magn. Reson.* **85** (1989), 381–7.
99. W. D. Williams, E. F. W. Seymour, and R. M. Cotts, A Pulsed-Gradient Multiple-Spin-Echo NMR Technique for Measuring Diffusion in the Presence of Background Magnetic Field Gradients. *J. Magn. Reson.* **31** (1978), 271–82.
100. J. Lian, D. S. Williams, and I. J. Lowe, Magnetic Resonance Imaging of Diffusion in the Presence of Background Gradients and Imaging of Background Gradients. *J. Magn. Reson. A* **106** (1994), 65–74.
101. P. J. McDonald, Stray Field Magnetic Resonance Imaging. *Prog. NMR Spectrosc.* **30** (1997), 69–99.
102. R. M. E. Valckenborg, L. Pel, and K. Kopinga, NMR Relaxation and Diffusion Measurements on Iron(III)-Doped Kaolin Clay. *J. Magn. Reson.* **151** (2001), 291–7.
103. H. Lee, H. Shao, Y. Huang, and B. Kwak, Synthesis of MRI Contrast Agent by Coating Superparamagnetic Iron Oxide with Chitosan. *IEEE Trans. Magn.* **41** (2005), 4102–4.

104. Y.-Q. Song, S. Ryu, and P. N. Sen, Determining Multiple Length Scales in Rocks. *Nature* **406** (2000), 178–81.
105. Y.-Q. Song, Using Internal Magnetic Fields to Obtain Pore Size Distributions of Porous Media. *Concepts Magn. Reson.* **18A** (2003), 97–110.
106. K. J. Packer, The Effects of Diffusion through Locally Inhomogenous Magnetic Fields in Transverse Nuclear Spin Relaxation in Heterogenous Systems. Proton Transverse Relaxation in Striated Muscle Tissue. *J. Magn. Reson.* **9** (1973), 438–43.
107. J. C. Tarczon and W. P. Halperin, Interpretation of NMR Diffusion Measurements in Uniform- and Nonuniform-Field Profiles. *Phys. Rev. B* **32** (1985), 2798–807.
108. C. H. Recchia, K. Gorny, and C. H. Pennington, Gaussian-Approximation Formalism for Evaluating Decay of NMR Spin Echoes. *Phys. Rev. B* **54** (1996), 4207–17.
109. M. D. Hürlimann, Effective Gradients in Porous Media Due to Susceptibility Differences. *J. Magn. Reson.* **131** (1998), 232–40.
110. M. D. Hürlimann, K. G. Helmer, and C. H. Sotak, Dephasing of Hahn Echo in Rocks by Diffusion in Susceptibility-Induced Field Inhomogeneities. *Magn. Reson. Imaging* **16** (1998), 535–9.
111. Y.-Q. Song, Determining Pore Sizes Using an Internal Magnetic Field. *J. Magn. Reson.* **143** (2000), 397–401.
112. Y.-Q. Song, Detection of the High Eigenmodes of Spin Diffusion in Porous Media. *Phys. Rev. Lett.* **85** (2000), 3878–81.
113. G. Q. Zhang and G. J. Hirasaki, CPMG Relaxation by Diffusion with Constant Magnetic Field Gradient in a Restricted Geometry: Numerical Simulation and Application. *J. Magn. Reson.* **163** (2003), 81–91.
114. A. Valfouskaya, P. M. Adler, J.-F. Thovert, and M. Fleury, Nuclear Magnetic Resonance Diffusion with Surface Relaxation in Porous Media. *J. Colloid Interface Sci.* **295** (2006), 188–201.
115. P. P. Mitra and P. Le Doussal, Long-Time Magnetization Relaxation of Spins Diffusing in a Random Field. *Phys. Rev. B* **44** (1991), 12035–8.
116. N. Fatkullin, Spin Relaxation and Diffusion Damping of the Spin Echo Amplitude of a Particle Moving in a Random Gaussian Magnetic Field. *Sov. Phys. JETP* **74** (1992), 833–8.
117. P. Bendel, Spin-Echo Attenuation by Diffusion in Nonuniform Field Gradients. *J. Magn. Reson.* **86** (1990), 509–15.
118. P. Le Doussal and P. N. Sen, Decay of Nuclear Magnetization by Diffusion in a Parabolic Magnetic Field: An Exactly Solvable Model. *Phys. Rev. B* **46** (1992), 3465–85.
119. P. M. Joseph, An Analytical Model for Diffusion of Spins in an Inhomogeneous Field Gradient. *J. Magn. Reson. B* **105** (1994), 95–7.
120. B. Håkansson, B. Jönsson, P. Linse, and O. Söderman, The Influence of a Nonconstant Magnetic-Field Gradient on PFG NMR Diffusion Experiments. A Brownian-Dynamics Computer Simulation Study. *J. Magn. Reson.* **124** (1997), 343–51.
121. L. J. Zielinski and P. N. Sen, Relaxation of Nuclear Magnetization in a Nonuniform Magnetic Field Gradient and in a Restricted Geometry. *J. Magn. Reson.* **147** (2000), 95–103.
122. G. Lin, Z. Chen, J. Zhong, D. Lin, and X. Liao, A Novel Propagator Approach for NMR Signal Attenuation due to Anisotropic Diffusion Under Various Magnetic Field Gradients. *Chem. Phys. Lett.* **335** (2001), 249–56.
123. D. S. Grebenkov, Nuclear Magnetic Resonance Restricted Diffusion Between Parallel Planes in a Cosine Magnetic Field: An Exactly Solvable Model. *J. Chem. Phys.* **126** (2007), 104706-1–104706-15.

124. G. Leu, X.-W. Tang, S. Peled, W. E. Maas, S. Singer, D. G. Cory, and P. N. Sen, Amplitude Modulation and Relaxation Due to Diffusion in NMR Experiments with a Rotating Sample. *Chem. Phys. Lett.* **332** (2000), 344–50.
125. L. J. Zielinski and P. N. Sen, Restricted Diffusion in Grossly Inhomogeneous Fields. *J. Magn. Reson.* **164** (2003), 145–53.
126. C. Casieri, S. Bubici, and F. De Luca, Self-Diffusion Coefficient by Single-Sided NMR. *J. Magn. Reson.* **162** (2003), 348–55.
127. D. G. Rata, F. Casanova, J. Perlo, D. E. Demco, and B. Blümich, Self-Diffusion Measurements by a Mobile Single-Sided NMR Sensor with Improved Magnetic Field Gradient. *J. Magn. Reson.* **180** (2006), 229–35.
128. A. L. Sukstanskii and D. A. Yablonskiy, Gaussian Approximation in the Theory of MR Signal Formation in the Presence of Structure-Specific Magnetic Field Inhomogeneities. *J. Magn. Reson.* **163** (2003), 236–43.
129. A. L. Sukstanskii and D. A. Yablonskiy, Gaussian Approximation in the Theory of MR Signal Formation in the Presence of Structure-Specific Magnetic Field Inhomogeneities. Effects of Impermeable Susceptibility Inclusions. *J. Magn. Reson.* **167** (2004), 56–7.

5
PGSE hardware

5.1 Introduction

This chapter details the instrumentation for generating magnetic gradients and related technical issues. A basic understanding of gradient pulse generation provides insight into spectrometer limitations and related problems. The basic considerations and components of NMR probes[1–3] and of the generation of high-intensity pulsed field gradients[4,5] have been reviewed elsewhere. Many of the complications that affect PGSE measurements also apply to imaging experiments, consequently some of the solutions to the technical problems were developed with imaging in mind.[6] Indeed, the design of a B_0 gradient probe for diffusion measurements is essentially similar to that of an NMR imaging or microscopy probe[6,7] except that the gradients used for the B_0 gradient probe are often larger and greater precision is required in gradient pulse generation (i.e., pairs of gradient pulses need to be matched to the ppm level). Many high-resolution NMR probes come equipped with gradient coils capable of generating magnetic gradients in the range of 0.5 T m^{-1}, whereas modern high-gradient diffusion probes are capable of generating gradients in excess of 20 T m^{-1} (Figure 5.1). There is also an interest in making probes capable of performing measurements on samples at high temperature and pressure[8,9] and for use in solid-state studies.[10]

To perform PGSE measurements, the spectrometer must be equipped with a current amplifier under the control of the acquisition computer which can send current pulses to a gradient coil placed around the sample. The hardware aspects of pulsed field gradient NMR have been discussed by numerous authors.[6,11–15] The additional hardware that must be added to a spectrometer to generate gradient pulses is summarised in Figure 5.2. Specifically, the spectrometer, in accordance with the pulse sequence, needs to output either a logic pulse (if only rectangular pulses are required) or a shaped voltage pulse (thereby affording the possibility of shaped gradient pulses) to an amplifier. Ideally, the polarity of the gradient pulse will also able to be specified. In turn, the amplifier outputs a corresponding current pulse to the gradient coil.

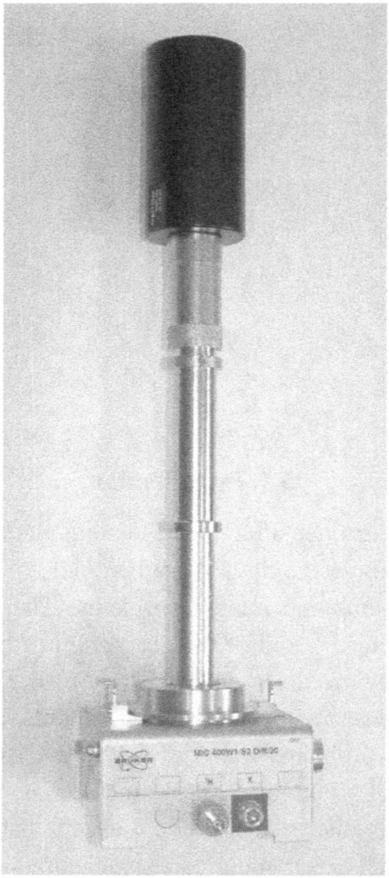

Figure 5.1 A specialised high z-gradient diffusion 400 MHz heteronuclear probe capable of generating gradients up to 30 T m^{-1} when coupled to a suitable gradient amplifier (i.e., providing a current of 60 A). The entire gradient coil consists of various individual cylindrical coils in a multilayer design with channels for water cooling. The inner diameter of the gradient system (black cylinder at the top of the probe) is 19 mm. The probe can be configured with fixed or with exchangeable rf-coils up to 10 mm inner diameter. By changing the gradient set to one containing three orthogonal x–y–z-gradient coils, the same probe can be used for investigating anisotropic diffusion and for imaging studies. In this case the maximum gradient strength gets reduced to 3 T m^{-1} at currents of 60 A. Photo courtesy of Bruker Biospin.

5.2 Gradient coils and fringe fields

Many gradient coil designs exist,[14,17–24] including quadrupolar (for g_x and g_y; see Figure 5.3),[4,25–30] planar array (for g_x and g_y) coils,[7] saddle coils (for g_x and g_y) and at the magic angle.[31,32]

The simplest commonly used geometry for producing gradients along the z-direction in superconducting magnets is the Maxwell pair of coils (i.e., anti-Helmholtz) (see

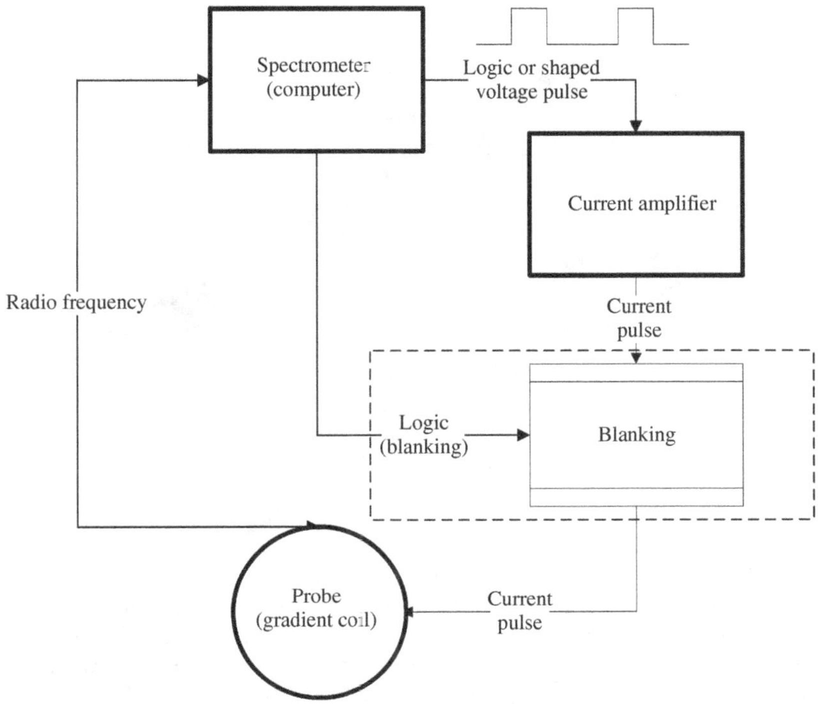

Figure 5.2 A schematic diagram of the instrumentation for performing PGSE NMR measurements. At the appropriate points in the pulse sequence the spectrometer sends logic pulses or, on more sophisticated machines, shaped voltage pulses (waveforms) such as trapezoidal and bipolar pulses or pulses with preemphasis to the current amplifier. The current amplifier is in turn connected to the gradient coils placed around the sample in the probe head (see Figure 5.6). More advanced spectrometers also include current blanking circuitry which prevents ground-loops (see also Section 5.4). Modified from Price.[16]

Figure 5.4A). The magnetic field strength at a point $P = (r_p, z_p)$ (see Figure 5.4) from a single winding can be estimated from the Biot-Savart law[33]

$$B(r_p, z_p) = \frac{\mu_0 I}{2\pi} \frac{1}{\left((r_c + r_p)^2 + z_p^2\right)^{1/2}} \times \left\{ K\left(\sqrt{\frac{4 r_c r_p}{(r_c + r_p)^2 + z_p^2}}\right) + \frac{r_c^2 - r_p^2 - z_p^2}{(r_c - r_p)^2 + z_p^2} E\left(\sqrt{\frac{4 r_c r_p}{(r_c + r_p)^2 + z_p^2}}\right) \right\}, \quad (5.1)$$

where K and E are the elliptic integrals of the first and second kinds, respectively.[34] μ_0 is the permittivity constant, r_p is the radius of the point at which the gradient is calculated, r_c is the radius of the gradient coil and z_p is the displacement along the z-axis from the coil (see Figure 5.4). The gradient at P can then be computed by

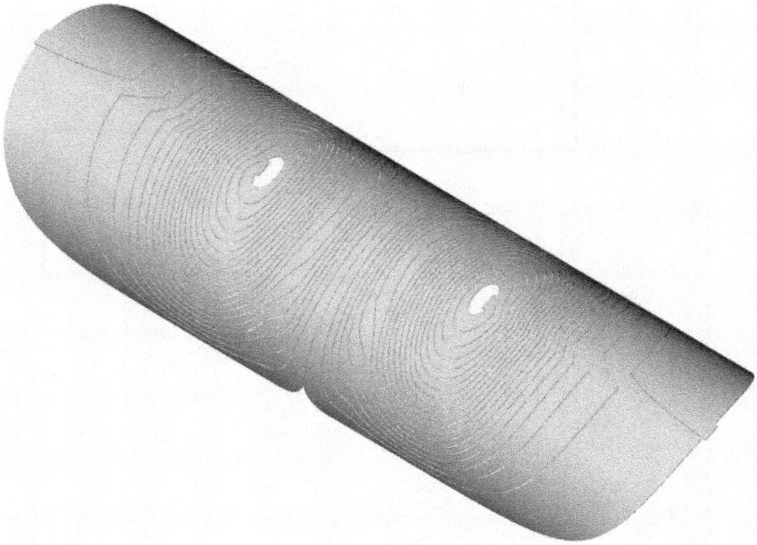

Figure 5.3 One section of a quadrupolar gradient coil suitable for generating transverse gradients (i.e. normal to the static field) in a high-resolution diffusion probe. Two such sections are combined to form a complete cylinder for the generation of the gradient field. Additional cylinders are usually arranged in a concentric way to increase the gradient strength and to realise a perfectly constant gradient in a large volume with perfect shielding properties. The cylinders are positioned relative to each other with very high precision and potted in epoxy for a rigid fixation. Diagram courtesy of Bruker Biospin.

calculating the magnetic field strength at two points separated by a distance along the z axis, zd, (i.e., $P_1 = (r_\mathrm{p}, z_\mathrm{p} + zd/2)$ and $P_2 = (r_\mathrm{p}, z_\mathrm{p} - zd/2)$) and dividing by the distance between the two points

$$g_{z,P} = \frac{\sum\limits_{\substack{\text{coil}\\ \text{windings}}} \left[B\left(r_\mathrm{p}, z_\mathrm{p} + \frac{zd}{2}\right) - B\left(r_\mathrm{p}, z_\mathrm{p} - \frac{zd}{2}\right)\right]}{zd}. \tag{5.2}$$

The sum in Eq. (5.2) runs over both windings of the Maxwell pair and, due to the opposite polarity, one coil winding is taken as negative. Ideally, the gradient coils should produce a perfectly constant gradient, but due to the space constraints inside the probe and inherent limitations in construction, such as attempting to produce a continuous magnetic field distribution from a finite number of turns, the gradient coils never produce a perfectly constant gradient. A field plot for the gradient coils depicted in Figure 5.4A is given in Figure 5.4B.

The limiting factor in achieving very large magnetic field gradients is the rapid increase in coil resistance with efficiency since the maximum usable wire diameter must decrease as the number of coils is increased. Accordingly, much effort has

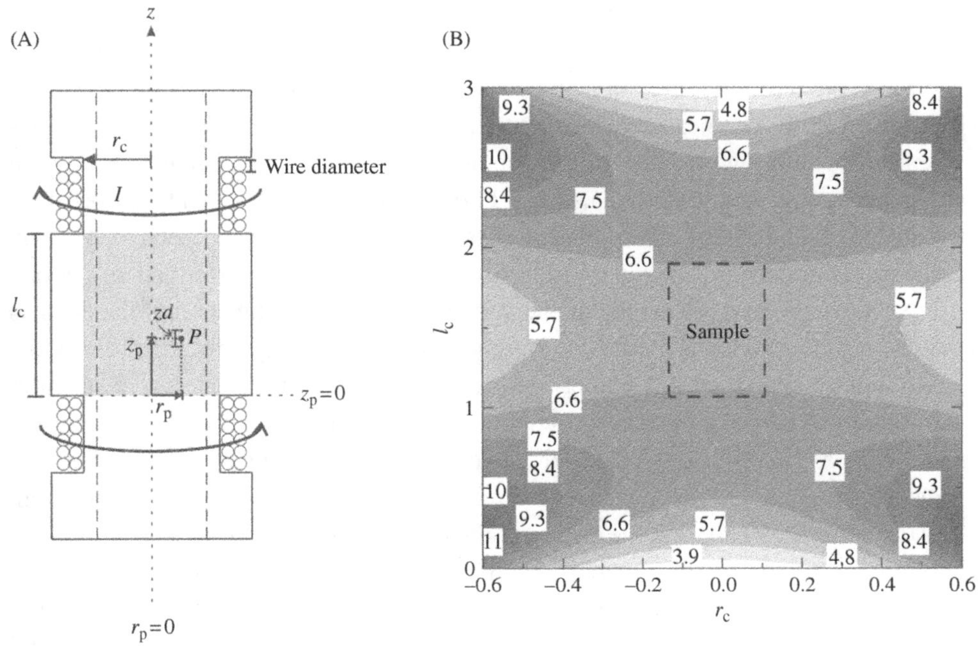

Figure 5.4 (A) A schematic depiction of a cross-section through a Maxwell pair. In computing the gradient using Eqs. (5.1) and (5.2), the coil radius r_c is adjusted according to the actual winding being calculated. The gradient, g_z at a point P is calculated by computing the magnetic field at two points separated by a distance along the z axis, zd, (i.e., $P_1 = (r_p, z_p + zd/2)$ and $P_2 = (r_p, z_p - zd/2)$, denoted by the smaller solid circles) and dividing by the distance between them. (B) The corresponding contour plot of the gradient in the shaded region of the gradient coil taking r_c to be 0.6 cm, l_c to be 3 cm, the wire diameter to be 0.5 mm and $I = 1$ A. The numbers on the contours denote the gradient strength in G cm^{-1}. Ideally, the sample would be restricted to a volume of constant gradient (e.g., the dashed box). Modified from Price.[16]

been spent in designing efficient coils such as a four-layer z-gradient coil of 8 mm i.d. capable of producing gradients up to 52 T m^{-1} with an efficiency of 1.73 T m^{-1} A^{-1}, a resistance of 1.8 Ω and an inductance of 50 μH.[35] Currently, the strongest gradients achieved on commercially available coils are in the range of 30 T m^{-1} (see Figure 5.1). In comparison, the maximum rate of change of the field (i.e., the stray-field) below a 9.4 T wide bore (89 mm) superconducting magnet is around 60 T m^{-1}.[36,37] Special superconducting anti-Helmholtz coils for stray-field measurements have also been constructed that are capable of generating gradients to more than 184 T m^{-1}.[38,39] Stray-field gradients, although not switchable, have the advantage of being very stable and quite homogeneous over reasonably large sample volumes. Analytical models for the field profile of superconducting magnets have been developed.[40]

Ideally, the gradient coils should produce a perfectly constant gradient. However, deviations are typically in the range of a fraction of a per cent to 5% or more.[35,41–44] A simple, but tedious, experimental means of testing the constancy of the gradient is to perform diffusion measurements using a very small sample at different positions within the volume where the sample would normally lie. An alternative is to map the constancy of the gradients using a PGSE sequence with a read gradient (see Figure 5.5).[45] In this approach, which is related to the 1D imaging approach for gradient calibration discussed in Section 6.4, a diffusion measurement is performed on a strong sample such as water and a read gradient is applied during acquisition of the echo so that the resulting signal is a diffusion weighted 1D image along the gradient direction.

As will be explained in Chapter 7, serious spectral disturbances can result from the induction of eddy currents in conducting materials surrounding the gradient coils (such as the probe outer shell) due to the rapidly changing magnetic fields. The most direct solution is to place a shield gradient coil outside the (primary) gradient coil (see Figure 5.6) to limit the effects of the gradient pulse to the sample volume by suppressing the (outward directed) magnetic fields from the primary gradient coil. In this way, no or at least greatly reduced eddy currents are generated, typically to less than 1%.[46] Shielded gradient coils were originally proposed by Mansfield *et al.*[47–51] and many of the theoretical aspects of shielded gradient coil design and technical considerations have been summarised elsewhere.[6,23,43,52–54] Numerical optimisation procedures for designing coils have been presented.[3,55–60] A negative aspect of shielded gradient coils is that the shield coils decrease the strength and constancy of the gradient produced by the primary gradient coil.[26] Carlson *et al.*[52] discussed techniques to design shielded gradient-coil systems; specifically they considered the design compromises between gradient homogeneity, construction complexity, accessible bore and coil efficiency. Eddy-current effects rapidly

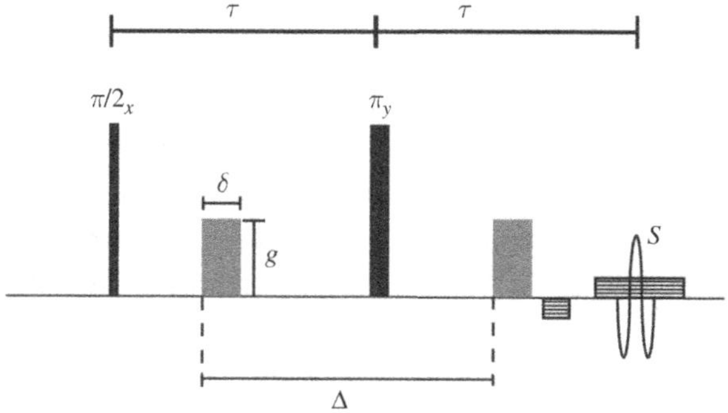

Figure 5.5 Hahn-echo-based PGSE sequence incorporating a read gradient (striped gradient pulses) to map the gradient constancy.[45] By using a strong pair of diffusion gradients, the resulting 1D image of the NMR sample is now heavily diffusion weighted.

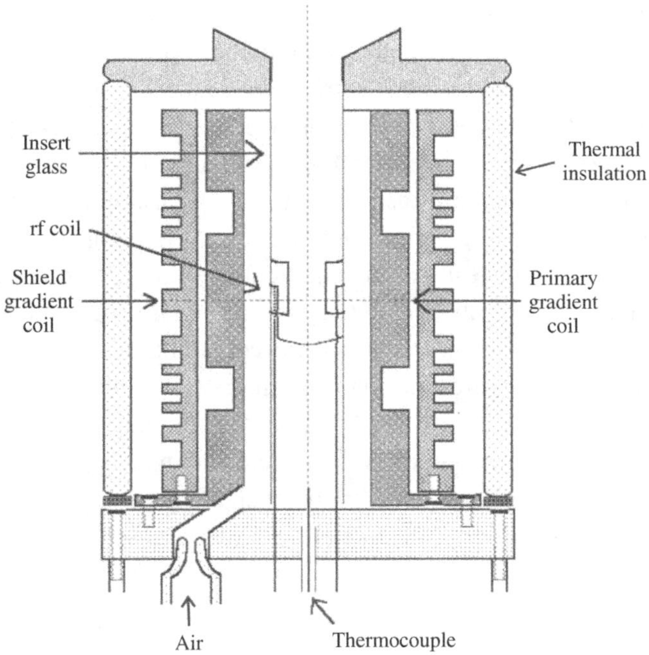

Figure 5.6 An example of a shielded magnetic gradient coil system in an NMR probe head. Only the coil formers are shown. Modified from Price *et al.*[14]

attenuate with increasing distance and thus, there is considerable advantage in using small shielded gradient coils in a wide-bore magnet.

5.3 Current amplifiers

Ideally, we desire infinitely fast rise and fall times of the gradient pulses. In practice, there are two factors which limit the maximum current switching speed; the first is that the power supply voltage must equal $RI + L\, dI/dt$, where I is the current and L and R are the load (i.e., gradient coils + leads) inductance and resistance respectively, and the second is the 'slew rate' (i.e., the maximum rate of change of the output voltage) of the power supply. Thus, the amplifier used must have suitable current and voltage parameters to drive the gradient coil used. Typical rise and fall times of the gradient pulses in a high-resolution probe are on the order of 50 μs. Various techniques have been used to determine the temporal evolution of the gradient waveform including using pickup coils,[61] Hall effect current probes,[62] and more recently imaging-based techniques.[63–66] Simply measuring current flow through gradient coils is not necessarily sufficiently accurate to monitor the field as it does not account for the effects of eddy currents.

Since the current through a gradient coil induces heating (which can be considerable[54]), which in turn results in a change in gradient coil resistance, the amplitude of the gradient pulses might change during the sequence or even cause physical damage to the coil. Consequently, especially when used with large currents or duty cycles, gradient coils need to be air and/or water cooled. Similarly, in conducting variable temperature diffusion measurements, the gas used for heating and cooling the sample will also have some effect on the gradient coil temperature. The use of a constant current supply, instead of a constant voltage amplifier, obviates the need to calibrate the gradient for each sample temperature or particular experimental parameters. The negative aspects of a constant current amplifier are that it is difficult to achieve very low noise figures and rapid settling. A number of amplifier designs have been discussed.[62,67–70] Doty has discussed the various technical considerations of amplifiers for driving gradient coils.[43] We note that in the last decade, largely due to the investment in MRI, commercially available units generally outperform home-built amplifiers and thus most PGSE experiments are now conducted using commercial units.

5.4 Connecting to the spectrometer

In the absence of gradient pulses, there should be zero current flowing through the gradient coils; however, in practice slight differences in potential difference between different parts of the spectrometer (e.g., the amplifier and the input line may not have the same zero voltage) result in a ground-loop[71]-infected input which results in currents flowing through the gradient coils between pulses resulting in non-random gradients. Similarly, the amplifier will also have a noise level resulting in small currents through the coils. Although very small, such ‘ground loop’ and noise currents result in troublesome background gradients and can completely thwart high-resolution diffusion experiments since they will be present during signal acquisition (similar to bad shimming) as well as attenuating the signal. Ground-loop currents can be detected by physically disconnecting the gradient circuit and looking at the effect on the lineshape or shift of the signal in the observed spectrum or, if available, by the effects on the lock signal.

To prevent the effects of ground-loops and noise, all of the components in the spectrometer and current amplifier should be earthed to the same point and ideally the gradient coil should be blanked (i.e., disconnected) from the current circuit between gradient pulses (see Figure 5.2). Blanking, however, will not prevent the effects of noise during the gradient pulses. Very small ground-loop effects can be ‘shimmed out’ if the ground-loop currents result in a steady gradient.

5.5 Field-frequency locking

Normally, the ^{2}H (or other suitable nucleus) lock is coupled to a z_0 shim coil to counteract the natural drift of the magnet. A gradient pulse will affect this mechanism since after a gradient pulse the deuterium signal will be dephased (and thus unlockable) and the time required for there to be sufficient deuterium signal to regain locking will be determined by the longitudinal relaxation rate. Further, the 'jolt' given to the lock by the gradient pulse can result in a shift of the field. A partial solution is to turn the lock off and with spectrometers based on superconducting magnets, after resolution is achieved by shimming, running unlocked generally has almost no effect on resolution. Even experiments requiring high degrees of resolution can normally be performed so long as the duration of the experiment is not long (~ hours, but strongly dependent on the magnet type) with respect to the drift rate of the magnet. A better solution is to gate the lock off before a gradient pulse and then to gate it on at the end of the pulse (after the dissipation of any eddy-current effects). It is important to centre the B_0 field and adjust the lock phase to minimise field disturbances due to lock gating. The situation is improved further with the use of bipolar gradient pulses (BPP) (see Section 7.7).

5.6 Temperature control and calibration

The sample temperature can differ significantly from that set on a spectrometer's temperature control unit and the deviation between the two values can itself be a function of temperature. Reasons for this inaccuracy include insufficient air flow around the sample and the thermocouple being somewhat distant from the sample (they cannot be placed too close to an rf coil since they can work like an antenna thereby severely decreasing S/N). Typically, if the probe temperature is changed, at least 10 min must be allowed for the probe and sample temperature to reach equilibrium. The exact time required will be dependent on the probe, airflow, sample and sample size. The calibration becomes more problematic when the sample has a high ionic strength and high-power proton decoupling is used.[72] Common methods of calibrating the sample temperature are to use some compound with temperature-dependent chemical shifts such as ethylene glycol or methanol[73–75] or piezoelectric thermometers.[76] We note that the various methods for measuring temperature using NMR have recently been reviewed.[77]

References

1. F. D. Doty, Probe Design & Construction. In *Encyclopedia of Nuclear Magnetic Resonance*, ed. D. M. Grant and R. K. Harris. vol. 6. (New York: Wiley, 1996), pp. 3753–62.

2. H. D. Hill, Probes for High Resolution. In *Encyclopedia of Nuclear Magnetic Resonance*, ed. D. M. Grant and R. K. Harris. vol. 6. (New York: Wiley, 1996), pp. 3762–8.
3. F. D. Doty, G. Entziminger, and Y. A. Yang, Magnetism in High-Resolution NMR Probe Design. I: General Methods. *Concepts Magn. Reson.* **10** (1998), 133–56.
4. P. T. Callaghan, M. E. Komlosh, and M. Nydén, High Magnetic Field Gradient PGSE NMR in the Presence of a Large Polarizing Field. *J. Magn. Reson.* **133** (1998), 177–82.
5. F. Stallmach and P. Galvosas, Spin Echo NMR Diffusion Studies. In *Annual Reports on NMR Spectroscopy*, ed. G. A. Webb. vol. 61. (New York: Elsevier, 2007), pp. 51–131.
6. P. T. Callaghan, *Principles of Nuclear Magnetic Resonance Microscopy*. (Oxford: Clarendon Press, 1991).
7. Y. Xia, K. R. Jeffrey, and P. T. Callaghan, Purpose-Designed Probes and Their Applications for Dynamic NMR Spectroscopy in an Electromagnet. *Magn. Reson. Imaging* **10** (1992), 411–26.
8. U. Matenaar, J. Richter, and M. D. Zeidler, High-Temperature-High-Pressure NMR Probe for Self-Diffusion Measurements in Molten Salts. *J. Magn. Reson. A* **122** (1996), 72–5.
9. K. Yoshida, C. Wakai, N. Matubayasi, and M. Nakahara, A New High-Temperature Multinuclear-Magnetic-Resonance Probe and the Self-Diffusion of Light and Heavy Water in Sub- and Supercritical Conditions. *J. Chem. Phys.* **123** (2005), 164506-1–164506-10.
10. W. Zhang and D. G. Cory, Pulsed Gradient NMR Probes for Solid State Studies. *J. Magn. Reson.* **132** (1998), 144–9.
11. P. T. Callaghan, C. M. Trotter, and K. W. Jolley, A Pulsed Field Gradient System for a Fourier Transform Spectrometer. *J. Magn. Reson.* **37** (1980), 247–59.
12. P. Stilbs, Fourier Transform Pulsed-Gradient Spin-Echo Studies of Molecular Diffusion. *Prog. NMR Spectrosc.* **19** (1987), 1–45.
13. J. Kärger, H. Pfeifer, and W. Heink, Principles and Applications of Self-Diffusion Measurements by Nuclear Magnetic Resonance. *Adv. Magn. Reson.* **12** (1988), 1–89.
14. W. S. Price, W.-T. Chang, W.-M. Kwok, and L.-P. Hwang, Design and Construction of a Pulsed Field-Gradient NMR Probe for a High-Field Superconducting Magnet. *J. Chin. Chem. Soc. (Taipei)* **41** (1994), 119–27.
15. W. S. Price, Gradient NMR. In *Annual Reports on NMR Spectroscopy*, ed. G. A. Webb. vol. 32. (London: Academic Press, 1996), pp. 51–142.
16. W. S. Price, Pulsed Field Gradient NMR as a Tool for Studying Translational Diffusion, Part II. Experimental Aspects. *Concepts Magn. Reson.* **10** (1998), 197–237.
17. A. M. Peters and R. W. Bowtell, Biplanar Gradient Coil Design by Simulated Annealing. *Magn. Reson. Mater. Phys. Bio. Med.* **2** (1994), 387–9.
18. E. R. Andrew and E. Szczesniak, Low Inductance Transverse Gradient System of Restricted Length. *Magn. Reson. Imaging* **13** (1995), 607–13.
19. Z. H. Cho and J. H. Yi, Planar Surface Gradient Coil. *Concepts Magn. Reson.* **7** (1995), 95–114.
20. W. W. Brey, T. H. Mareci, and J. Dougherty, A Field-Gradient Coil Using Concentric Return Paths. *J. Magn. Reson. B* **112** (1996), 124–30.
21. C. F. Maier, K. C. Chu, B. A. Chronik, and B. K. Rutt, A Novel Transverse Gradient Coil Design for High-Resolution MR Imaging. *Magn. Reson. Med.* **34** (1995), 604–11.
22. Q. Liu, D. G. Hughes, and P. S. Allen, Improved, Minimum-Inductance, Elliptic-Cylinder *z*-Gradient Coil Using Axial and Azimuthal Current Flow. *J. Magn. Reson. B* **113** (1996), 228–35.
23. R. Turner, Gradient Coil Systems. In *Encyclopedia of Nuclear Magnetic Resonance*, ed. D. M. Grant and R. K. Harris. vol. 4. (New York: Wiley, 1996), pp. 2223–33.

24. J. Leggett, S. Crozier, S. Blackband, B. Beck, and R. Bowtell, Multilayer Transverse Gradient Coil Design. *Concepts Magn. Reson. B* **16** (2003), 38–46.
25. D. S. Webster and K. H. Marsden, Improved Apparatus for the NMR Measurement of Self-Diffusion Coefficients Using Pulsed Field Gradients. *Rev. Sci. Instrum.* **45** (1974), 1232–4.
26. A. Jasinski, T. Jakubowski, M. Rydzy, P. Morris, I. C. P. Smith, P. Kozlowski, and J. K. Saunders, Shielded Gradient Coils and Radio Frequency Probes for High-Resolution Imaging of Rat Brains. *Magn. Reson. Med.* **24** (1992), 29–41.
27. J. S. Schoeniger and S. J. Blackband, The Design and Construction of a NMR Microscopy Probe. *J. Magn. Reson. B* **104** (1994), 127–34.
28. K. C. Chu and B. K. Rutt, Quadrupole Gradient Coil Design and Optimization: A Printed Circuit Board Approach. *Magn. Reson. Med.* **31** (1994), 652–9.
29. C. J. Rofe, J. Van Noort, P. J. Back, and P. T. Callaghan, NMR Microscopy Using Large, Pulsed Magnetic-Field Gradients. *J. Magn. Reson. B* **108** (1995), 125–36.
30. O. Oishi and S. Miyajima, New PFG NMR Spectrometer with a Rotatable Quadrupole Coil for the Measurement of an Anisotropic Self-Diffusion Coefficient Tensor. *J. Magn. Reson. A* **123** (1996), 64–71.
31. R. Bowtell and A. Peters, Magic-Angle Gradient-Coil Design. *J. Magn. Reson. A* **115** (1995), 55–9.
32. T. M. Barbara and C. E. Bronnimann, Target Field Design for Magic Angle Gradient Coils. *J. Magn. Reson.* **140** (1999), 285–8.
33. W. R. Smythe, *Static and Dynamic Electricity*, 3rd edn. (London: Taylor and Francis, 1989).
34. M. Abramowitz and I. A. Stegun, *Handbook of Mathematical Functions*. (New York: Dover, 1970).
35. R. Bowtell and P. Robyr, Multilayer Gradient Coil Design. *J. Magn. Reson.* **131** (1998), 286–94.
36. P. J. McDonald, Stray Field Magnetic Resonance Imaging. *Prog. NMR Spectrosc.* **30** (1997), 69–99.
37. P. J. McDonald and B. Newling, Stray Field Magnetic Resonance Imaging. *Rep. Prog. Phys.* **61** (1998), 1441–93.
38. F. Fujara, E. Ilyina, H. Nienstadt, H. Sillescu, R. Spohr, and C. Trautmann, Anisotropic Diffusion in Etched Particle Tracks Studied by Field Gradient NMR. *Magn. Reson. Imaging* **12** (1994), 245–6.
39. I. Chang, F. Fujara, B. Geil, G. Hinze, H. Sillescu, and A. Tölle, New Perspectives of NMR in Ultrahigh Static Magnetic Field Gradients. *J. Non-Cryst. Solids* **172–174** (1994), 674–81.
40. P. Jeglič, A. Lebar, T. Apih, and J. Dolinšek, Modeling the Static Fringe Field of Superconducting Magnets. *J. Magn. Reson.* **150** (2001), 39–42.
41. I. Furó and H. Jóhannesson, Accurate Anisotropic Water-Diffusion Measurements in Liquid Crystals. *J. Magn. Reson. A* **119** (1996), 15–21.
42. B. Håkansson, B. Jönsson, P. Linse, and O. Söderman, The Influence of a Nonconstant Magnetic-Field Gradient on PFG NMR Diffusion Experiments. A Brownian-Dynamics Computer Simulation Study. *J. Magn. Reson.* **124** (1997), 343–51.
43. F. D. Doty, MRI Gradient Coil Optimization. In *Spatially Resolved Magnetic Resonance*, ed. P. Blümler, B. Blümich, R. E. Botto, and E. Fukushima. (Weinheim: Wiley-VCH, 1998).
44. A. Jerschow and G. Bodenhausen, Mapping the B_1 Field Distortion with Nonideal Gradients in a High-Resolution NMR Spectrometer. *J. Magn. Reson.* **137** (1999), 108–15.

45. R. E. Hurd, A. Deese, M. O'Neil, S. Sukumar, and P. C. M. Van Zijl, Impact of Differential Linearity in Gradient-Enhanced NMR. *J. Magn. Reson. A* **119** (1996), 285–8.
46. M. Burl and I. R. Young, Eddy Currents & Their Control. In *Encyclopedia of Nuclear Magnetic Resonance*, ed. D. M. Grant and R. K. Harris. vol. 3. (New York: Wiley, 1996), pp. 1841–6.
47. P. Mansfield and B. Chapman, Active Magnetic Screening of Gradient Coils in NMR Imaging. *J. Magn. Reson.* **66** (1986), 573–6.
48. P. Mansfield and B. Chapman, Active Magnetic Screening of Coils for Static and Time-Dependent Magnetic Field Generation. *J. Phys. E: Scientific Instruments* **19** (1986), 540–5.
49. R. Turner, A Target Field Approach to Optimal Coil Design. *J. Phys. D. Appl. Phys.* **19** (1986), L147–51.
50. R. Turner and R. M. Bowley, Passive Screening of Switched Magnetic Field Gradients. *J. Phys. E: Sci. Instrum.* **19** (1986), 876–9.
51. P. Mansfield and B. Chapman, Multishield Active Magnetic Screening of Coil Structures in NMR. *J. Magn. Reson.* **72** (1987), 211–23.
52. J. W. Carlson, K. A. Derby, K. C. Hawryszko, and M. Weideman, Design and Evaluation of Shielded Gradient Coils. *Magn. Reson. Med.* **26** (1992), 191–206.
53. R. Turner, Gradient Coil Design: A Review of Methods. Magn. *Reson. Imaging* **11** (1993), 903–20.
54. K. C. Chu and B. K. Rutt, MR Gradient Coil Heat Dissipation. *Magn. Reson. Med.* **34** (1995), 125–32.
55. S. J. Gibbs, K. F. Morris, and C. S. Johnson, Jr., Design and Implementation of a Shielded Gradient Coil for PFG NMR Diffusion and Flow Studies. *J. Magn. Reson.* **94** (1991), 165–9.
56. S. Crozier and D. M. Doddrell, Gradient-Coil Design by Simulated Annealing. *J. Magn. Reson. A* **103** (1993), 354–7.
57. S. Crozier, L. K. Forbes, and D. M. Doddrell, The Design of Transverse Gradient Coils of Restricted Length by Simulated Annealing. *J. Magn. Reson. A* **107** (1994), 126–8.
58. C. Gasparovic, M. Cabañas, and C. Arús, A Simple Approach to the Design of a Shielded Gradient Probe for High-Resolution in Vivo Spectroscopy. *J. Magn. Reson. B* **109** (1995), 146–52.
59. J. Chládek, P. Konzbul, P. Ošmera, and A. Gottvald, Evolutionary and Genetic Optimization of NMR Gradient and Shim Coils. *IEEE Trans. Magn.* **36** (2000), 1102–5.
60. M. A. Brideson, L. K. Forbes, and S. Crozier, Determining Complicated Winding Patterns for Shim Coils Using Stream Functions and the Target-Field Method. *Concepts Magn. Reson.* **14** (2002), 9–18.
61. D. J. Jensen, W. W. Brey, J. L. Delayre, and P. A. Narayana, Reduction of Pulsed Gradient Settling Time in the Superconducting Magnet of a Magnetic Resonance Instrument. *Med. Phys.* **14** (1987), 859–62.
62. W. S. Price, An NMR Study of Diffusion, Viscosity, and Transport of Small Molecules in Human Erythrocytes. 1990. PhD Thesis. University of Sydney.
63. B. J. Balcom, M. Bogdan, and R. L. Armstrong, Single-Point Imaging of Gradient Rise, Stabilization and Decay. *J. Magn. Reson. A* **118** (1996), 122–5.
64. M. T. Alley, G. H. Glover, and N. J. Pelc, Gradient Characterization Using a Fourier-Transform Technique. *Magn. Reson. Med.* **39** (1998), 581–7.
65. D. J. Goodyear, M. Shea, S. D. Beyea, N. J. Shah, and B. J. Balcom, Single Point Measurements of Magnetic Field Gradient Waveform. *J. Magn. Reson.* **163** (2003), 1–7.

66. V. Jellúš, J. C. Sharp, B. Tomanek, and P. Latta, An NMR Technique for Measurement of Magnetic Field Gradient Waveforms. *J. Magn. Reson.* **162** (2003), 189–97.
67. G. B. Matson, A Precision Current Pulse Generator for NMR Self-Diffusion Measurements by the Pulsed Gradient Technique. *Rev. Sci. Instrum.* **43** (1972), 1504–8.
68. T. R. Saarinen and W. S. Woodward, Computer-Controlled Pulsed Magnetic Field Gradient NMR System for Electrophoretic Mobility Measurements. *Rev. Sci. Instrum.* **59** (1988), 761–3.
69. R. M. Boerner and W. S. Woodward, A Computer-Controlled Bipolar Magnetic-Field-Gradient Driver for NMR Electrophoretic and Self-Diffusion Measurements. *J. Magn. Reson. A* **106** (1994), 195–202.
70. D. A. Seeber, J. H. Hoftiezer, and C. H. Pennington, Pulsed Current Gradient Power Supply for Microcoil Magnetic Resonance Imaging. *Concepts Magn. Reson.* **15** (2002), 189–200.
71. H. W. Ott, *Noise Reduction Techniques in Electronic Systems*, 2nd edn. (New York: Wiley, 1998).
72. W. A. Bubb, K. Kirk, and P. W. Kuchel, Ethylene Glycol as Thermometer for X-nucleus Spectroscopy in Biological Samples. *J. Magn. Reson.* **77** (1988), 363–8.
73. A. L. Van Geet, Calibration of the Methanol and Glycol Nuclear Magnetic Resonance Thermometers with a Static Thermistor Probe. *Anal. Chem.* **40** (1968), 2227–9.
74. A. L. Van Geet, Calibration of Methanol Nuclear Magnetic Resonance Thermometer at Low Temperature. *Anal. Chem.* **42** (1970), 679–80.
75. D. S. Raiford, C. L. Fisk, and E. D. Becker, Calibration of Methanol and Ethylene Glycol Nuclear Magnetic Resonance Thermometers. *Anal. Chem.* **51** (1979), 2050–1.
76. D.-J. Wang and J. S. Leigh, Wireless Precision Piezoelectric Thermometer Using an RF Excitation-Detection Technique with an NMR Probe. *J. Magn. Reson. B* **105** (1994), 25–30.
77. A. G. Webb, Temperature Measurements Using Nuclear Magnetic Resonance. In *Annual Reports on NMR Spectroscopy*, ed. G. A. Webb. vol. 45. (London: Academic Press, 2002), pp. 1–67.

6

Setup and analysis of PGSE experiments

6.1 Introduction

This chapter is concerned with the practical issues and key considerations involved in setting up PGSE experiments and the subsequent data analysis. Selection of PGSE parameters is discussed in Section 6.2 and sample preparation is discussed in Section 6.3. The various methods of gradient calibration are considered in Section 6.4. Finally, PGSE data analysis and display are considered in Section 6.5. Under favourable conditions it is possible to measure diffusion coefficients with greater than 99% accuracy.[1] Indeed simple PGSE experiments have been shown to be reasonably robust with respect to experimental parameters (e.g., rf pulse flip angle).[2] It cannot be overemphasised that the overall accuracy of a diffusion measurement is intimately connected to the accuracy of the gradient calibration. It is too easy to confuse the apparent precision of a diffusion measurement obtained from analysing the PGSE data with the true overall accuracy. For example, the PGSE data obtained from an experiment may be highly single exponential, but the gradient calibration or temperature control may have been inaccurate such that the analysis of the PGSE data leads to a highly precise but unfortunately a highly inaccurate diffusion coefficient.

Irrespective of the aim of the PGSE experiment, the analysis is always simplified by starting with a distortion-free data set with good signal-to-noise and, especially when the system has multiple components, good resolution. Consequently, it can be advantageous to work at higher field strengths since, in addition to the greater S/N, the greater chemical shift dispersion may lead to resonances that do not overlap. It can be beneficial to try and remove the effects of spectral overlap by spectral editing or the use of multi-dimensional PGSE approaches in which PGSE is combined with another experiment such as TOCSY or multiple quantum selection as covered in Chapter 8. However, in polydisperse systems (e.g., synthetic polymers and proteins) and mixtures of different species, the spectra are crowded and spectral overlap is generally inevitable. In some circumstances, there may be benefit in using heteronuclear PGSE experiments since the increased spectral dispersion available may allow the resolution of otherwise overlapped resonances. Although most PGSE measurements are performed using ^{1}H, apart from NMR sensitivity, there is no difference in working with heteronuclei. However, lower γ nuclei require greater applied gradient strengths. In addition to averaging sufficient scans simply to obtain sufficient S/N, it is also important to use purge pulses and an appropriate number of scans to complete the relevant phase cycle to ensure clean coherence selection.[3]

The symptoms and physical basis of some of the technical problems that can be alleviated by prudent experimental setup are covered in detail in Chapter 7. Specialised PGSE sequences are discussed in Chapter 8.

6.2 Selection of PGSE parameters

Using the Hahn spin-echo based sequence (Fig. 2.6A) as an example, the PGSE experiment is performed by varying one of the PGSE parameters (i.e., δ, Δ or g) while τ is generally kept constant so that, at least in the case of a single diffusing species, relaxation may be factored out (see Eq. (2.49)). Ideally, the PGSE parameters would be set to values to best probe the system under study. In the simplest case of a monodisperse single freely diffusing species, it is desirable to study the attenuation over at least an order of magnitude; however, to probe restricted diffusion it might be necessary to study the attenuation over a much greater range and to pick Δ on consideration of the MSD with respect to the size of the restricting geometry. To set cogent values for Δ, δ and g in the experiment, it is prudent to simulate the experiment using Eq. (2.75) with an approximate value for the diffusion coefficient. The choice of q spacings has been considered and is akin to the problem of picking spacings in relaxation measurements, and logarithmic and geometric spacing have been suggested as better alternatives to linear spacings.[2,4–6] Song *et al.* have also considered experimental design from the perspective of maximising the information obtained from performing Laplace inversion of the data.[7] Clearly, the choice of gradient spacings becomes even more important when only few q values are used or when the sample contains a mixture of species with different diffusion coefficients. In theory, a diffusion measurement can be performed by measuring the echo attenuation at only two q values – but this requires certainty of the type of diffusion being measured and, of course, exceptional accuracy in determining the echo amplitudes and consideration must be given to the balance between sufficient attenuation and not losing the signal in the noise.[8] For free diffusion the ratio of the signal amplitudes obtained for two different q values should lie between 0.2 and 0.6.[9] Problems with determining diffusion values from a small number of q values is more commonly encountered in experiments involving MRI.

In practice, the applicable range of Δ depends on the spectrometer hardware and the relaxation properties of the sample. For example, neglecting the length of the rf pulses – which are normally insignificant compared to the length of δ and Δ, inspection of the Hahn PGSE sequence (Figure 2.6A) reveals that the minimum possible value of Δ is given by

$$\Delta_{\min} = \delta. \tag{6.1}$$

However, such a short Δ would be unusable as it would not allow sufficient time for the dissipation of eddy currents arising from the first gradient pulse (see Section 7.7) prior to the π rf pulse. Similarly, for a given echo time, τ, the maximum possible Δ would be given by

$$\Delta_{\max} = 2\tau - \delta, \tag{6.2}$$

which would generally be impracticable as there is no time for the eddy currents arising from the second pulse to dissipate prior to acquisition. Ultimately, the largest usable Δ is determined by the maximum usable value of τ which is, in turn, related to the spin–spin relaxation time (i.e., T_2) of the species being measured. In practice, Δ is normally between several ms to several seconds.

In general, the analysis of PGSE data is simplified when each scan starts from thermal equilibrium. Thus, setting a recycle delay $\geq 5 \times T_1$ is safe, albeit a time-consuming practice. However, for many experimental systems, setting the recycle delay shorter than this will still lead to usable results especially if a crusher gradient is included in the prescan delay; however this needs to be assessed on a system-by-system basis.[10]

When studying species for which $T_1 \gg T_2$ (e.g., macromolecules) it is advantageous to use the stimulated echo sequence (Figure 2.6B) since the delays in the pulse sequence can be chosen so that the magnetisation is 'stored' along the z-axis for most of Δ. A comparison of the relative signal amplitude from the Hahn and STE sequences is given in Figure 6.1.

There can also be complications from cross-relaxation effects (see Section 7.9) and scalar coupling (see Section 7.10) and there is a chemical shift dependence of the amplitude of the first point when the gradient is zero.[12]

6.2.1 Heteronuclear decoupling

Especially in isotropic systems with large coupling constants, applying heteronuclear decoupling during the PGSE sequence can result in seemingly anomalous changes in signal intensity from incomplete decoupling during gradient pulses. The reason being that heteronuclear J-couplings behave as chemical shifts and therefore any echo sequence will refocus heteronuclear effects. Thus, generally, decoupling should only be applied during the recycle delay to obtain NOE enhancement and

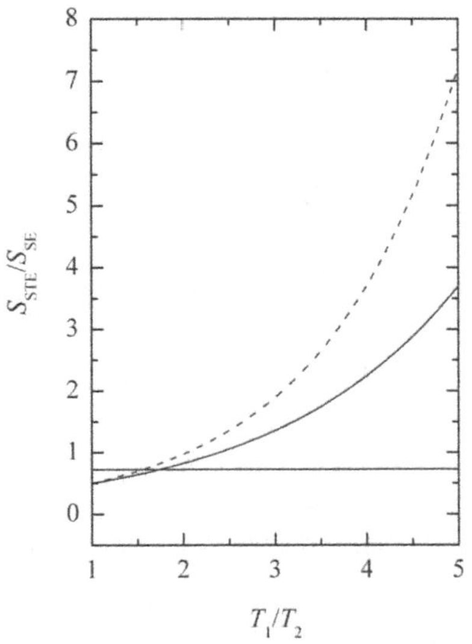

Figure 6.1 The ratio of the signal obtained from the stimulated echo (STE; Eq. (2.52)) sequence to that obtained from the Hahn spin-echo (SE; Eq. (2.48)) sequence versus T_1/T_2 in the absence of gradients. The calculations were performed assuming that $2\tau = T_1$ in the SE sequence and $2\tau_1 + \tau_2 = T_1$ in the STE sequence. The simulations were performed for the cases of $\tau_1 = \tau_2/2$ (——) and $\tau_1 = \tau_2/4$ (- - - -). The solid horizontal line indicates the boundary above which the STE sequence gives better signal-to-noise than the SE sequence. Adapted from Price.[11]

acquisition to remove the couplings. S/N problems with heteronuclei may also be alleviated by using inverse detection and the like (see Section 8.3).

In systems where non-vanishing static dipolar interactions are significant (e.g., in solids and oriented systems such as liquid crystals), specialised pulse sequences that reduce the effect of dipolar coupling are required.[13,14] Diffusion measurements involving liquid crystals have recently been reviewed by Furó and Dvinskikh.[15,16]

6.2.2 Solvent suppression

Many solvents are protonated and give rise to very strong resonances resulting in numerous difficulties including poor spectral baselines, inadequate digitisation of small signals, obscuring peaks near solvent resonances and artefacts arising from radiation damping and the distant dipolar fields as will be discussed in Section 7.3. Switching to NMR inert solvents (e.g., ^2H_2O) is not always practicable. In general, so long as the species being measured are considerably larger than the solvent, the solvent signal becomes less of a problem as the value of q increases in a PGSE experiment as its intensity is rapidly suppressed relative to that of the (more slowly moving) solute.[17–19] Neglecting relaxation effects, the relative attenuation of the solvent to the solute in a PGSE experiment is given by

$$\frac{\ln(E_{\text{solvent}})}{\ln(E_{\text{solute}})} = \frac{D_{\text{solvent}}}{D_{\text{solute}}}. \tag{6.3}$$

Many non-diffusion-based solvent suppression techniques exist[20] but only some of these are suitable for use with PGSE sequences. The simple presaturation approach militates against the observation of solute resonances near the solvent frequency and is a poor choice for use with very high molecular weight solutes due to the effects of spin diffusion. Gradient-based suppression sequences have proven to be very useful but it has been noted that these sequences, especially those based on spatial encoding (see Figure 2.1),[21–23] are deleteriously affected by the inadvertent generation of gradient-recalled echoes due to background gradients and other gradient pulses in the sequences.[24] The generation of gradient-recalled echoes can be prevented by changing the gradient strength and direction (ideally orthogonally) of successive gradient episodes.[25,26] However, the joint action of the two feedback fields, radiation damping and the long range dipolar interaction (see Section 7.3), gives rise to chaotic spin dynamics and the resurrection of previously crushed (i.e., completely dephased) solvent magnetisation.[27,28] It is likely that this feedback mechanism is sometimes confused with gradient recalled echoes in thwarting suppression. Due to the complexity in working with non-linear spin dynamics it is simpler to obviate the problem in PGSE measurements by using *Q*-switching[29] as discussed in Section 7.3.2.

Since PGSE measurements are often conducted on rapidly relaxing samples or polydisperse samples, the best suppression techniques will not lengthen the sequence and thus the inclusion of WATERGATE-like elements into Hahn spin-echo and STE sequences has been very successful (see Figures 6.2 and 6.3).[30–32]

The echo-attenuation function for the PGSTE–WATERGATE sequence is identical to that of the standard Hahn-echo-based sequence. However, the attenuation of the PGSTE–WATERGATE sequence is, for the non-solvent resonances,[32]

$$\ln(E) = -\gamma^2 D\delta^2\left[\left(\Delta - \frac{4}{3}\delta - 2\delta_2\right)(g_2 - g_1)^2 - \frac{2}{3}\delta(g_2 - g_1)g_1 + \left(4\delta_2 + \frac{4}{3}\delta\right)g_1^2\right], \tag{6.4}$$

and for the solvent resonances,

$$\ln(E) = -\gamma^2 D\delta^2\left[\left(\Delta - \frac{4}{3}\delta - 2\delta_2\right)(g_1 + g_2)^2 + \frac{2}{3}\delta(g_1 + g_2)g_1 + \left(4\delta_2 + \frac{4}{3}\delta\right)g_1^2\right], \tag{6.5}$$

and the delays are defined in Figure 6.2B. Importantly, the use of asymmetric gradients provides coherence selection, allowing only a four-step phase cycle.

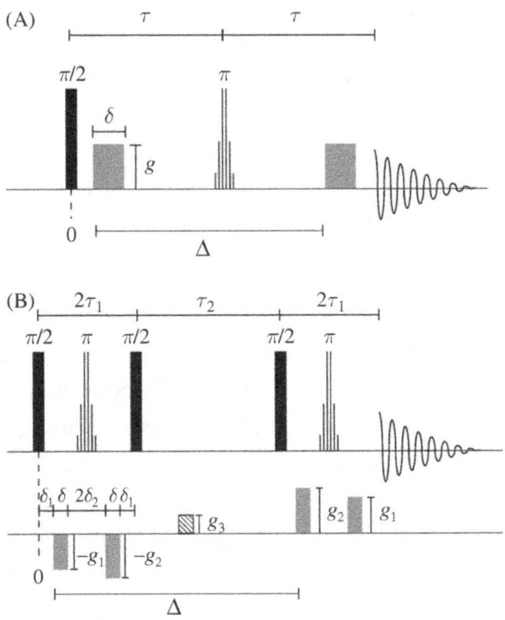

Figure 6.2. The (A) PGSE–WATERGATE and (B) PGSTE–WATERGATE sequences. The phase cycling and details of the binomial pulses (striped rf pulses) are given elsewhere.[31,32]

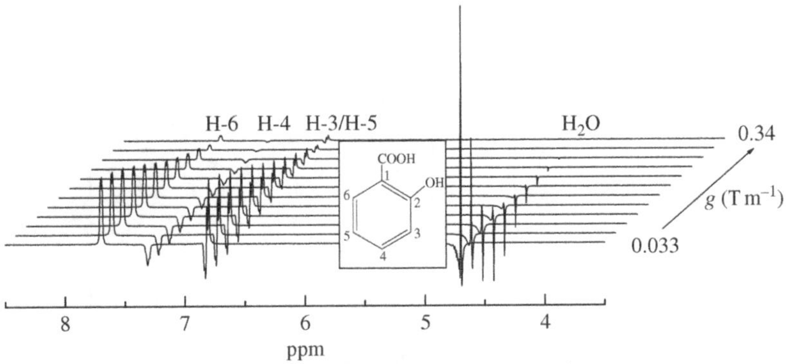

Figure 6.3 A series of PGSE–WATERGATE spectra obtained using the sequence given in Figure 6.2A of a sample containing 80 mM salicylate and 0.5 mM bovine serum albumin in water (10:90 ^2H_2O:^1H_2O) at 298 K. The water resonance gives rise to the peak at 4.7 ppm and the three peaks to the left originate from salicylate (from left to right: H-6, H-4, H-3/H-5; also see inset). The water resonance has been suppressed by more than four orders of magnitude. The effects of *J*-modulation are evident on the H-4 and H-3/H-5 resonances. From Price *et al.*[31]

Another alternative for solvent suppression is through the use of multiple quantum experiments (see Section 8.3); however, these are not generally applicable. Sequences that include both solvent suppression and convection compensation have been presented (see Section 7.4).

6.3 Sample preparation

To prevent motional artefacts the sample should be firmly held inside the magnet so that the sample maintains a constant position with respect to the gradient coil former (see Section 7.8). The sample should be wholly contained inside the constant region of the gradient coils and thus the sample is typically contained in a volume less than 1 cm high in a high-resolution probe. Such a sample, though, has large changes in magnetic susceptibility close to the rf coils and consequently it can be very difficult to achieve good resolution. A solution is depicted in Figure 6.4. This method, compared to just coaxially inserting a bulb into an NMR tube, has the advantage in that it is easy to clean the sample tube or to work with viscous substances. It also gives a precise shape with no meniscus effect. Susceptibility-matched tubes and inserts are also commercially available (e.g., Shigemi NMR tubes). An alternative to physically restricting the size of the sample is to use a slice-selective pulse in the PGSE sequence.[33,34] If the sample contains internal magnetic gradients it is necessary to use one of the sequences presented in Section 7.6.2.

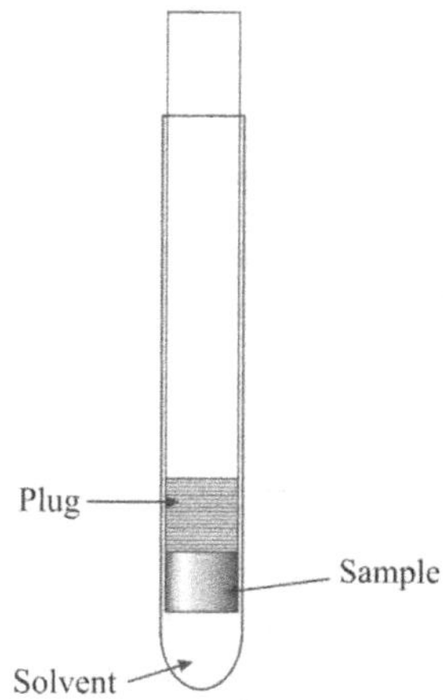

Figure 6.4 A sample geometry for containing the sample within the constant region of the magnetic field gradient. The sample is placed in a cylindrical sample tube and then capped with a vortex plug ideally of the same magnetic susceptibility. This tube is then coaxially inserted into a tube containing either an NMR inert solvent with a similar magnetic susceptibility or the same solvent but without the solute of interest. Thus the 'NMR active' part of the sample is short whilst the sample is still magnetically long and allows easier shimming (cf. Figure 4.12). This sample arrangement also confines the sample to a region of highly homogeneous rf.

High B_0 homogeneity is always desirable since background gradients have deleterious effects on diffusion measurements. If it is necessary to use a short sample to stay within the constant region of the gradient and susceptibility-matched NMR tubes or the like are unavailable, the process of shimming becomes very difficult. Further, the initial lineshape may be so poor that it is impossible to shim the sample using the lock signal if available. In such circumstances, it is easier to first shim the probe on a normal long sample and then iteratively shim and gradually reduce the volume of the sample to the (short) sample to be measured. After the first shimming of the long sample, the non-spinning shims (i.e., those without axial symmetry) should be largely correct, but due to the shortness of the sample the z and z^2 shims will require particular attention. It is not uncommon to have to use very large values for the z^2 shim and it is generally easier to shim using the FID.[35,36] In some cases, gradient shimming may also be of use; but some degree of parameter optimisation is required to work well with the short sample length.[37] Although spinning the sample helps to average out the effect of background gradients allowing a higher resolution spectrum to be obtained, the spinning may cause motion along the direction of the magnetic gradient in the sample. However, a 'stop-and-go' sample spinner suitable for use in PGSE experiments has been developed that allows for the spinning to be arrested during the motion-sensitive part of the experiment and yet spun, to achieve higher resolution, during acquisition.[38]

6.4 Gradient calibration

Numerous methods exist for calibrating gradient coils and these are summarised in Table 6.1 and the methods are considered in detail in the following sections. In theory, the applied gradient can be calculated from the known dimensions, geometry and the number of turns of wire in the coil and the current applied (see Section 5.2). In practice, however, coil calculation gives an estimate with an error of $\leq 10\%$. The major reason for this inaccuracy being interaction with nearby metal in the probe and non-ideal gradient pulse generation problems. Similarly, manufacturer's specifications are rarely sufficient to conduct accurate diffusion measurements. A 1D image is a quick method of experimentally verifying that the calibration is approximately correct. Typically, if a suitable reference compound exists, this should be used for 'fine tuning' the calibration. Ultimately, the gradient should be calibrated to within 1% of its true value. The final form of the calibration is a coil constant with unit $\mathrm{T\,m^{-1}\,A^{-1}}$ or as a function of the spectrometer input variable defining the amplitude of the current pulse to the gradient coil.

Table 6.1 *Summary of the gradient calibration methods*

Method	Range of application	Comments
Coil calculation	Unlimited	Can be complicated to perform Not very accurate
Echo shape[a]	$\gamma g l$ < receiver bandwidth High S/N	Numerous systematic errors including misalignment of sample geometry and gradient
1D image	$\gamma g l$ < receiver bandwidth High S/N	Information on gradient constancy
Gradient pulse mismatch	Similar to echo shape	Similar to echo shape
Standard sample	Need to have a relevant standard	Simple Includes gradient nonconstancy Few suitable and accurate standards Need accurate temperature control

[a] l is the dimension of the sample in the direction of the gradient. The echo shape and 1D image methods are essentially the same due to the Fourier relationship between the two.
Source: Modified from Price[11]

It is important to consider the relationship between the accuracy of the determination of the gradient and the accuracy of a subsequent diffusion measurement. For example, if a calibration using a 1D image provides a gradient calibration with an x% error, the inherent error (i.e., best possible result) in a subsequent diffusion measurement would be $2x$% of the diffusion coefficient determined. Calibration using a standard sample has an inherent advantage over calibration based on a 1D image since in using a standard sample the calibration is effectively being performed on g^2 and not g itself – although the accuracy of a gradient calibration ultimately rests on the accuracy of the original experiment from which the standard diffusion coefficient was taken. Thus, the overall accuracy of the diffusion measurement will depend on the accuracy of the gradient calibration and other factors including S/N. Hence, in the simplest case and assuming the other experimental parameters are ideal, if the gradient calibration had an error of y% and analysis of the PGSE data gave an error of z% the total accuracy of the diffusion measurement would be the propagation of y% and z%.[39]

Another important consideration is that the experimental parameters for controlling the gradient pulses, the pulse duration (δ) and the amplitude (g), are linear functions of the input current pulses, for example the gradient amplifier may be unable to generate accurately shaped large amplitude pulses as easily as low amplitude pulses and similarly for various reasons (e.g., increased resistance due to coil heating) the amplitude of the gradient pulses may not increase linearly with input current. Hence, calibrations at low gradient amplitudes may not extrapolate to higher gradient amplitudes.

The situation can become more complicated when PGSE experiments are incorporated into MRI sequences as 'cross-terms' between imaging and diffusion gradients need to be considered, otherwise they will lead to incorrect estimates of the diffusion coefficient.[40–42] However, even in straight PGSE experiments, it is possible to have orthogonal gradients and these have the effect of mixing the diffusion in various directions proportional to the squares of the respective gradient components and thus the effect is scaled by the diffusional anisotropy.[1,43]

For completeness, we note that if the diffusion probe is equipped with more than one gradient, the measured diffusion anisotropy in isotropic media can be used as a basis for calibrating and aligning magnetic field gradients.[44]

6.4.1 A standard sample with known diffusion coefficient

The most common way of calibrating a gradient is to use a 'standard sample' of known diffusion coefficient and suitable standard samples are listed in Table 6.2. More comprehensive listings can be found elsewhere.[45,46] Also, the diffusion coefficients of a range of alkanes and alcohols with diffusion properties suitable for cross-checking clinical MRI diffusion sequences have also been published.[47] For lower diffusion coefficients, suitable reference compounds become scarce, and for very low diffusion coefficients monodisperse polymers are generally used.[48] Glycerol has often been used as a reference but its diffusion coefficient is highly temperature dependent and greatly affected by water content.[18,45]

In this method, an initial guess at the gradient strength, g^{app}, is used to measure the diffusion coefficient, D^{app}, of the standard sample. These two values are then

Table 6.2 *Some selected reference compounds and their diffusion coefficients at 298 K useful for calibrating PGSE experiments*

Observed nucleus	Compound	Diffusion coefficient (m^2s^{-1})	Reference
^1H	H_2O	2.30×10^{-9}	49,50
^2H	HO^2H in ^2H_2O	1.90×10^{-9}	50
	^2H_2O	1.87×10^{-9}	50
^7Li	LiCl (0.25 M) in H_2O	9.60×10^{-10}	45
^{13}C	C_6H_6	2.21×10^{-9}	51
^{19}F	C_6H_6F	2.40×10^{-9}	45
^{21}Ne	Ne (4 MPa) in ^2H_2O	4.18×10^{-9}	52
^{23}Na	NaCl (2 M) in H_2O	1.14×10^{-9}	53
^{31}P	$(C_6H_5)_3P$ (3 M) in C_6D_6	3.65×10^{-10}	45
^{129}Xe	Xe (3 MPa) in H_2O	1.90×10^{-9}	54
^{133}Cs	CsCl (2 M) in H_2O	1.90×10^{-9}	45

Source: Modified from Price[11]

used to calculate the true gradient strength, g, from the true diffusion coefficient, D^0, namely

$$g = g^{\mathrm{app}} \sqrt{\frac{D^{\mathrm{app}}}{D^0}} \tag{6.6}$$

Importantly, the same experimental conditions (i.e., sample shape, delays, pulse lengths and gradient strengths etc.) should be used in a subsequent experiment so that the calibration automatically includes, to some extent, non-ideal gradient behaviour including background gradients. However, because eddy-current effects increase with gradient strength, a calibration at one current may not extrapolate to another value of the applied current. Further, because the calibration is very often performed with a pure liquid (leading to a 'strong' NMR signal), radiation damping and long-range dipolar interactions can thwart calibration measurements (see Section 7.3).

6.4.2 Shape analysis of the spin-echo and one-dimensional images

Relating back to the idea of magnetisation gratings in Section 2.2.2, an FID acquired in the presence of a read gradient, g_r, when Fourier transformed, gives a 1D projection (i.e., 1D image) of the sample in the direction of the gradient that reflects both the gradient amplitude and the shape of the sample (see Figure 6.5; note the example is given using a Hahn-echo with a read gradient, of course the method could equally well have been performed with an STE-base sequence). Consequently, the gradient strength can be calculated from the echo shape from a sample of known geometry,[55–59] or indeed from performing a 1D image of a sample of known length,[11] or moving the sample by a precise amount between images.[60] But this method is susceptible to systematic errors.[58,61–64] To illustrate this method, detailed derivations are given of both the FID and the corresponding Fourier-transformed spectrum (i.e., 1D images) for a gradient directed transverse and longitudinal to a right cylindrical sample of length l and radius a (see Figure 6.5) in the Appendix. To avoid problems with insufficient rf pulse power (as in the steady gradient PGSE experiment, see Section 2.2.4), the echo sequence given in Figure 6.5 is used; nevertheless, even modest gradients require large receiver bandwidths in order to acquire the signal and a correspondingly large number of scans are required to obtain sufficient S/N. For example, simple reasoning with the Larmor equation reveals that for the gradient directed along the cylinder as in Figure 6.5, the spectrum must be rectangular with a linewidth, $\Delta\nu$ (Hz), given by

$$\Delta\nu = \frac{\gamma g_r l}{2\pi}. \tag{6.7}$$

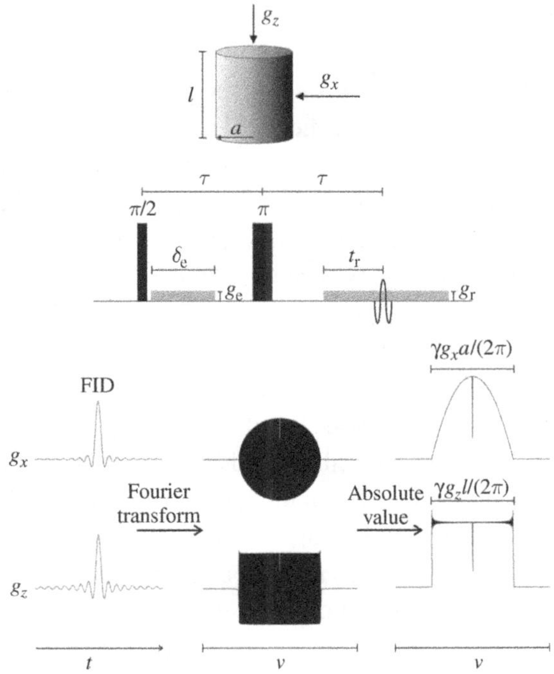

Figure 6.5 A cylindrical sample of length l and radius a used for determining the gradient strength, and of the Hahn spin-echo sequence incorporating an encoding gradient of strength g_e and read gradient of strength g_r used for obtaining a one-dimensional image of the sample. The echo occurs when $t_r = \delta_e g_e / g_r$ (i.e., according to Eq. (2.26)). In the lower half of the figure, two simulated FIDs are given for the case of $g_e, g_r = g_x$ (upper) or g_z (lower). The FID for the g_x case has the characteristic Bessel function profile (see Eq. (A.22)) whilst the FID acquired in the presence of g_z has a sinc function profile (see Eq. (A.33)). In theory, the gradient strength can be determined by analysing the FID shape (e.g., from the zeros of the Bessel function in the case of g_x), but in practice the Fourier transforms, which return images of the sample cross-sections with respect to the gradient directions, are easier to analyse. The Fourier transforms are rapidly oscillating functions with sharp frequency cutoffs in both cases. The power spectrum makes the cutoff easier to visualise (right hand spectra). The width ($\Delta\nu$; Hz) of the spectra are $\gamma g_r a/(2\pi)$ and $\gamma g_r l/(2\pi)$ for the spectra acquired with g_r (and g_e) replaced by g_x and g_z, respectively. The power spectra were calculated numerically from the simulated FID (not from Eqs. (A.29) and (A.44)) which was slightly truncated at the ends, hence the oscillations at the top of the absolute value of the transform of the sinc function and also the dip in the middle of both. Experimentally, the FIDs are often more seriously truncated and as a consequence the oscillation artefacts are more pronounced. As the number of points used in the transform decrease, the edges of the absolute value spectra are not so sharp. Further, if the echo is not quite in the middle of the acquisition, the transformed spectrum appears to have strange phasing; however, the absolute value spectrum solves the problem. Modified from Price.[11]

The shape of the corresponding FID is given by sinc($\gamma g_r(2\tau - t)l$) (see Eq. (A.33)) and in theory this can be used to determine the gradient strength. This method is most useful in the case where the gradient is along the cylinder axis since the shape of the image should be rectangular as depicted in Figure 6.5. The FID is recorded in the presence of the gradient for a number of different applied currents, and then by plotting the width of the spectrum versus the current, the gradient strength can be calibrated using Eq. (6.7).[11] Alternatively, as noted above, the sample could be imaged once and then imaged again after moving by a precise amount.[60] If the transmitter offset is placed at the resonance frequency of the sample (i.e., in the absence of the gradient) and if the sample is correctly centred in the gradient, the gradient broadened spectrum will expand symmetrically around the transmitter offset as the gradient strength is increased. This method requires that the length of the sample containing cell be known accurately and the final calibration will have an error of less than 5%. This method can be performed without any knowledge of the sample diffusion coefficient and, so long as the dimensions of the container holding the sample are insensitive to temperature, it can be used to cross-check that the gradient strength is not temperature-dependent. This method also allows some indication of the gradient constancy from the 1D image.

Saarinen and Johnson[60] also noted that the period of rapid oscillations in the frequency spectrum (see Figure 6.5) is given by

$$v = \frac{g_r}{g_e \delta} \tag{6.8}$$

(note g_r and g_e are collinear) thereby allowing g_e to be determined.

6.4.3 Intentional gradient pulse mismatch

Hrovat and Wade[8,59,65] suggested using the time displacement of the echo maximum caused by the intentional mismatch of gradient pulses with sequence conducted in the presence of a small background gradient (Figure 6.6). In their procedure several echoes are collected as a function of an intentional mismatch in

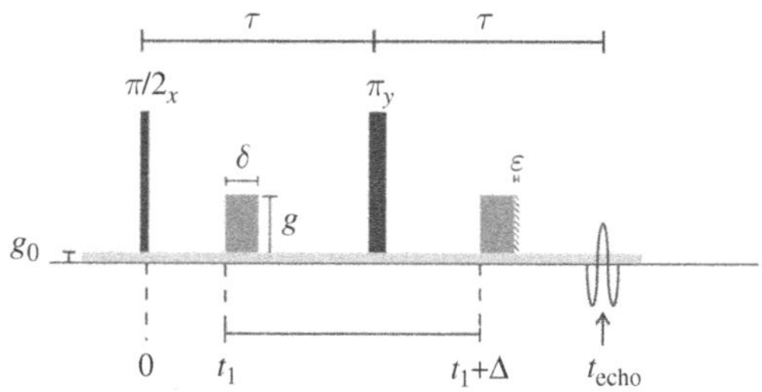

Figure 6.6 The mismatched gradient pulse Hahn-echo sequence of Hrovat and Wade.[59] The second gradient pulse is slightly longer by a duration ε.

duration, ε, of the gradient pulses. Assuming that $\mathbf{g}$ is directed along a cylindrical sample, the echo occurs at,

$$t_{\text{echo}} = 2\tau - \frac{g\varepsilon\cos(\alpha)}{g_0} \tag{6.9}$$

where α is the angle between $\mathbf{g}_0$ and $\mathbf{g}$. The echo attenuation is given by (cf. Eq. (A.22))

$$E = \frac{2J_1[\gamma r g\varepsilon\sin(\alpha)]}{\gamma r g\varepsilon\sin(\alpha)}, \tag{6.10}$$

which is a maximum as expected for $\varepsilon=0$. The background gradient g_0 can be determined from the lineshape of the spin-echoes (see Section 6.4.2) and then from Eqs. (6.9) and (6.10) g and α can be determined. From Eq. (6.9) it can be seen that the sensitivity of t_{echo} to ε increases as g_0 becomes smaller.

6.4.4 Calibration of very high gradients

The calibration of very high amplitude gradient pulses presents special problems due to the difficulty in finding reliable diffusion standards and because projection methods cannot be extended to such gradient amplitudes due to lack of receiver bandwidth and/or insufficient signal due to the gradient broadening. Wright *et al.*[66] have presented a two-step method in which the gradient is first calibrated at low gradient strengths ($0-4\ \mathrm{T\,m^{-1}}$) and then, based on this initial calibration, higher amplitude gradients (in their case from 4 to $50\ \mathrm{T\,m^{-1}}$) are calibrated using a gradient echo-based technique in which a longer low-amplitude calibrated gradient pulse is used to calibrate a higher amplitude but shorter gradient pulse (Figure 6.7). An alternative method of calibration is to use a diffusion standard containing two receptive nuclei (i.e., with different γs).[67]

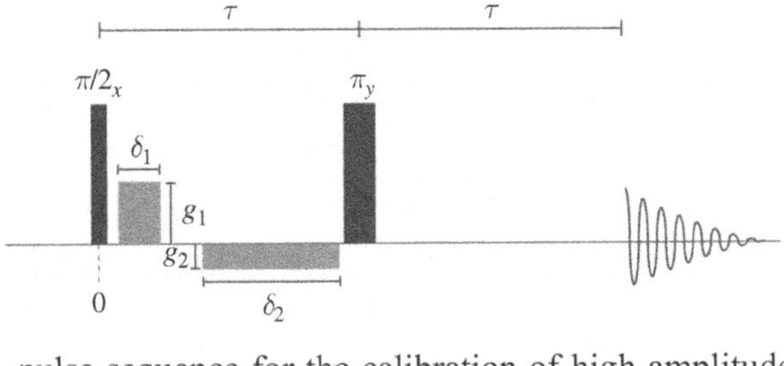

Figure 6.7 A pulse sequence for the calibration of high amplitude gradients.[66] The duration of the second (calibrated lower amplitude) gradient pulse (δ_2) is adjusted until a maximum signal is acquired (i.e., when the two gradient pulses are matched).

6.5 Analysis

6.5.1 Spectrally isolated components

Data analysis begins by transforming all of the data and applying careful phase and baseline corrections and perhaps reference deconvolution (see Section 7.7).[11,68] The data should be presented in phase-sensitive mode as it provides better resolution and, if phase-sensitive PGSE spectra are compared as a function of q, it allows the presence of artefacts such as eddy currents to be identified. Ideally, the same phase correction should be applicable to the whole data set. Although it might seem advantageous to use absolute value (also referred to as power or magnitude) spectra to overcome eddy current–induced phase instability, other gradient generation artefacts and the effects of J-evolution (see Section 7.10), the improvement is mainly artificial and complicates data fitting by making all noise positive.

Baseline corrections can be complicated if, for example, there is a very large solvent peak that attenuates rapidly with respect to other species leading to the need for different baseline corrections as a function of q. In some cases, the 'wings' of the solvent resonances can spread under those of the species of interest. Consequently, it is best to have PGSE data with the solvent peaks well suppressed.

If the spectral lineshapes are Lorentzian and the resonance(s) of interest is (are) isolated, then using peak heights or integration are both valid means of calculating the echo attenuation. Although, in cases of spectral overlap (e.g., ^{1}H NMR spectra of proteins) there may be no choice other than to integrate over the region and to subsequently attempt to deconvolute the data.[69] However, due to the effects of radiation damping (see Section 7.3), the signal linewidth can vary with signal amplitude and thus q values and this can lead to an error in the diffusion coefficient determined if peak heights rather than integrals are used in the analysis.[70]

If it is a purely freely diffusing system then a very obvious way to analyse the data when free diffusion is expected, especially if a 'rough and ready' analysis is acceptable, is simply to plot ln(E) versus $\gamma^2 g^2 \delta^2(\Delta - \delta/3)$ (cf. Eq. (2.75)), in which case the diffusion coefficient can be obtained from the slope (i.e., $-D$). Todica and Pop[71] have proposed other representations which may have benefits for data including effects of background gradients (see Section 7.6). However, this approach gives unequal weighting to the noise, particularly as the signal approaches zero. For this reason, non-linear least squares regression (e.g., the Levenberg–Marquadt algorithm[72]) of the relevant attenuation equation onto the experimental data is preferred. The final result should be an estimate of the diffusion coefficient and its standard error (i.e., $D \pm \sigma$).

Further, depending upon the system, the presence of restricted diffusion or exchange may add another layer of complexity to analysing a multicomponent data set and knowledge of the type of restriction would be greatly beneficial (e.g., differences in motional restriction could also be used to separate different species (see Chapter 3)). Neural networks have been used to analyse data from restricted systems.[73] A superposition of Fredholm integrals with different kernels has been proposed as an alternative to working with the Laplace transform in cases where the shape of the NMR decay is non-exponential (e.g., due to restricted diffusion).[74]

We comment that the analysis of many diffusion studies in the chemical and biological fields is performed on the basis of the Einstein-Sutherland equation (Eq. (1.13)). And when attempting to obtain information on hydrodynamic radii from diffusion measurements it can be difficult to separate changes in hydrodynamic radii from changes in viscosity. Apart from directly measuring the viscosity (e.g. with a viscometer), some have suggested the use of an internal, and hopefully non-interacting, reference standard of known hydrodynamic radius (and similar to the species of interest) as a means of determining the viscosity.[75] Nevertheless, due to the complexity of the solvent dynamics as discussed in Chapter 1, this is only an approximate solution to this problem and should be used with caution.

6.5.2 Resolution of complex systems

The various techniques to analyse multicomponent PGSE data sets (see Sections 2.5 (free diffusion) and 3.8 (restricted diffusion)) are detailed below. Amongst these approaches, CONTIN and maximum entropy are the most suited to analysing polydisperse systems. It should be recalled that whilst these methods will in general give a 'result', this result may not be the physical reality as often unjustified assumptions are made, for example, about the nature of the distributions of diffusion coefficients. And so the results of such analysis should be treated with caution and a degree of scepticism. The approaches can be roughly separated as follows.

6.5.2.1 Univariate

In univariate methods the signal at each frequency is analysed separately. This class of methods include biexponential fitting.[76] For a discrete number of components the inversion program DISCRETE can be used.[77,78]

If there is a distribution of diffusion coefficients the available methods include the coupled inverse Laplace transform approach (SPLine MODel 'SPLMOD'),[79,80] the (linear regularisation) program CONTIN,[80–82] and non-linear regularisation programs NLREG[83] and GENEREG.[83,84] Other approaches include maximum entropy processing,[85,86] linear prediction using the concept of Hankel diagonalisation[87] and Hopfield neural networks.[88]

6.5.2.2 Curve (multivariate) resolution methods

These methods, also known as complete bandshape methods, exploit the covariance between the signals from a given species and specified constraints on the number of species present to try to resolve the entire spectrum of each component even when the diffusion coefficients of the component species are close.

Analysing the diffusion data using the whole absorption bandshape with the global least-squares approach (e.g., Levenberg–Marquardt non-linear least squares fitting) is known by the acronym CORE (COmponent REsolved spectroscopy).[89–92] Analysis is performed at two levels: optimisation for each spectral component and of the global self-diffusion coefficient. The basic premise is to use prior knowledge on (or systematically assume and vary to best fit) the number of diffusion coefficients in the PGSE-multi-component data set and cyclically make a global least squares fit to all of its frequency and PGSE timing parameters. Its principal advantage is that it can use knowledge gained from regions of the spectrum that do not overlap to assist in the interpretation of regions of the spectrum that do overlap. Corrections to the CORE fitting procedure for gradient non-constancy and B_0 instability have been considered.[92]

A multivariate statistical approach (originally based on tools (NIPALS and Procrustes rotation) developed for optical spectroscopy by Kubista[93]) was adapted for PGSE studies by Schulze and Stilbs,[94] and later improved into the DECRA family of procedures by Antalek *et al.*[95–97] Antalek[98,99] has attempted in a method termed 'q-DECRA' to account for spin relaxation in PGSE experiments in order to obtain resolved quantitative spectra from mixtures. In this method, based on the STE pulse sequence, the τ_1 and τ_2 delays were varied simultaneously and proportionally thereby allowing extrapolation of the spectrum without the effects of relaxation. However, chemical exchange, cross-relaxation and J-modulation lead to non-quantitative conditions. The effects of non-exponential and multi-exponential decay behaviour were examined and found to have rather deleterious effects on the performance of DECRA.[100]

Multivariate curve resolution, which is based on principal component analysis, by itself and/or in combination with non-linear least square regression[68,101–104] unlike DECRA, which assumes exponential decay, makes no such assumption and thus provide more robust possibilities for analysing multicomponent PGSE data sets. Positive matrix factorisation,[105] information entropy minimisation[106] and other statistical approaches[107] have also been tried.

The regularised resolvent transform (RRT)[108] has been adapted to solve the inverse Laplace transform (iRRT)[109] and has been used to generate DOSY spectra of data sets that have poor chemical shift resolution.

References

1. I. Furó and H. Jóhannesson, Accurate Anisotropic Water-Diffusion Measurements in Liquid Crystals. *J. Magn. Reson A.* **119** (1996), 15–21.
2. S. Augé, B. Amblard-Blondel, and M.-A. Delsuc, Investigation of the Diffusion Measurement Using PFG and Test of Robustness Against Experimental Conditions and Parameters. *J. Chim. Phys.* **96** (1999), 1559–65.
3. G. A. Morris, Diffusion-Ordered Spectroscopy (DOSY). In *Encyclopedia of Nuclear Magnetic Resonance*, ed. D. M. Grant and R. K. Harris. vol. 9. (New York: Wiley, 2002), pp. 35–44.
4. D. P. Hinton and C. S. Johnson, Jr., Diffusion Ordered 2D NMR Spectroscopy of Phospholipid Vesicles: Determination of Vesicle Size Distributions. *J. Phys. Chem.* **97** (1993), 9064–72.
5. C. Labadie, D. Gounot, Y. Mauss, and B. Dumitresco, Data Sampling in MR Relaxation. *Magn. Reson. Mater. Phys. Bio. Med.* **2** (1994), 383–5.
6. C. S. Johnson, Jr., Diffusion Ordered Nuclear Magnetic Resonance Spectroscopy: Principles and Applications. *Prog. NMR Spectrosc.* **34** (1999), 203–56.
7. Y.-Q. Song, L. Venkataramanan, and L. Burcaw, Determining the Resolution of Laplace Inversion Spectrum. *J. Chem. Phys.* **122** (2005), 104104-1–104104-10.
8. M. I. Hrovat and C. G. Wade, NMR Pulsed Gradient Diffusion Measurements. II. Residual Gradients and Lineshape Distortions. *J. Magn. Reson.* **45** (1981), 67–80.
9. J. Bedet, D. Canet, S. Leclerc, P. Mutzenhardt, D. Stemmelen, and G. Trausch, Optimal Conditions for Two-Point Estimation of Self-Diffusion Coefficients Through rf Gradient NMR Experiments. *Chem. Phys. Lett.* **408** (2005), 237–40.
10. T. Stait-Gardner, P. G. Anil Kumar and W. S. Price, Steady State Effects in PGSE NMR Diffusion Studies. *Chem. Phys. Lett.* **462** (2008), 331–6.
11. W. S. Price, Pulsed Field Gradient NMR as a Tool for Studying Translational Diffusion, Part II. Experimental Aspects. *Concepts Magn. Reson.* **10** (1998), 197–237.
12. A. Chen, C. S. Johnson, Jr., M. Lin, and M. J. Shapiro, Chemical Exchange in Diffusion NMR Experiments. *J. Am. Chem. Soc.* **120** (1998), 9094–5.
13. M. Silva-Crawford, B. C. Gerstein, A.-L. Kuo, and C. G. Wade, Diffusion in Rigid Bilayer Membranes. Use of Combined Multiple Pulse and Multiple Pulse Gradient Technqiues in Nuclear Magnetic Resonance. *J. Am. Chem. Soc.* **102** (1980), 3728–32.
14. S. V. Dvinskikh, R. Sitnikov, and I. Furó, ^{13}C PGSE NMR Experiment with Heteronuclear Dipolar Decoupling to Measure Diffusion in Liquid Crystals and Solids. *J. Magn. Reson.* **142** (2000), 102–10.
15. I. Furó and S. V. Dvinskikh, NMR Methods Applied to Anisotropic Diffusion. *Magn. Reson. Chem.* **40** (2002), S3–14.
16. S. V. Dvinskikh and I. Furó, Nuclear Magnetic Resonance Studies of Translational Diffusion in Thermotropic Liquid Crystals. *Russ. Chem. Rev.* **75** (2006), 497–506.
17. C. H. Everhart and C. S. Johnson, Jr., The Determination of Tracer Diffusion Coefficients for Proteins by Means of Pulsed Field Gradient NMR with Applications to Hemoglobin. *J. Magn. Reson.* **48** (1982), 466–74.
18. P. Stilbs, Fourier Transform Pulsed-Gradient Spin-Echo Studies of Molecular Diffusion. *Prog. NMR Spectrosc.* **19** (1987), 1–45.
19. P. C. M. Van Zijl and C. T. W. Moonen, Complete Water Suppression for Solutions of Large Molecules Based on Diffusional Differences between Solute and Solvent (DRYCLEAN). *J. Magn. Reson.* **87** (1990), 18–25.
20. W. S. Price, Water Signal Suppression in NMR Spectroscopy. In *Annual Reports on NMR Spectroscopy*, ed. G. A. Webb. vol. 38. (London: Academic Press, 1999), pp. 289–354.

21. A. Haase, J. Frahm, W. Hänicke, and D. Matthei, ^{1}H NMR Chemical Shift Selective (CHESS) Imaging. *Phys. Med. Biol.* **30** (1985), 341–4.
22. D. M. Doddrell, G. J. Galloway, W. M. Brooks, J. Field, J. M. Bulsing, M. G. Irving, and H. Baddeley, Water Signal Elimination in Vivo, Using 'Suppression by Mistimed Echo and Repetitive Gradient Episodes'. *J. Magn. Reson.* **70** (1986), 176–80.
23. A. S. Altieri and R. A. Byrd, Randomization Approach to Water Suppression in Multidimensional NMR Using Pulsed Field Gradients. *J. Magn. Reson. B* **107** (1995), 260–6.
24. M. Czisch, A. Ross, C. Cieslar, and T. A. Holak, Some Practical Aspects of B_0 Gradient Pulses. *J. Biomol. NMR* **7** (1996), 121–30.
25. C. T. W. Moonen and P. C. M. Van Zijl, Highly Effective Water Suppression for In Vivo Proton NMR Spectroscopy (DRYSTEAM). *J. Magn. Reson.* **88** (1990), 28–41.
26. S. H. Smallcombe, S. L. Patt, and P. A. Keifer, WET Solvent Suppression and its Applications to LC NMR and High-Resolution NMR Spectroscopy. *J. Magn. Reson. A* **117** (1995), 295–303.
27. Y.-Y. Lin, N. Lisitza, S. Ahn, and W. S. Warren, Resurrection of Crushed Magnetization and Chaotic Dynamics in Solution NMR Spectroscopy. *Science* **290** (2000), 118–21.
28. S. Datta, S. Y. Huang, and Y.-Y. Lin, Understanding Spin Turbulence in Solution Magnetic Resonance Through Phase Space Dynamics and Instability. *Concepts Magn. Reson.* **28A** (2006), 410–21.
29. W. S. Price and M. Wälchli, NMR Diffusion Measurements of Strong Signals: The PGSE-*Q*-Switch Experiment. *Magn. Reson. Chem.* **40** (2002), S128–32.
30. M. Liu, J. K. Nicholson, and J. C. Lindon, High-Resolution Diffusion and Relaxation Edited One and Two-Dimensional ^{1}H NMR Spectroscopy of Biological Fluids. *Anal. Chem.* **68** (1996), 3370–6.
31. W. S. Price, F. Elwinger, C. Vigouroux, and P. Stilbs, PGSE-WATERGATE, a New Tool for NMR Diffusion-Based Studies of Ligand-Macromolecule Binding. *Magn. Reson. Chem.* **40** (2002), 391–5.
32. G. Zheng, T. Stait-Gardner, P. G. Anil Kumar, A. M. Torres, and W. S. Price, PGSTE-WATERGATE: A Stimulated-Echo-Based PGSE NMR Sequence with Excellent Solvent Suppression. *J. Magn. Reson.* **191** (2008), 159–63.
33. Y. Xia, Contrast in NMR Imaging and Microscopy. *Concepts Magn. Reson.* **8** (1996), 205–25.
34. B. Antalek, Using Pulsed Gradient Spin Echo NMR for Chemical Mixture Analysis: How to Obtain Optimum Results. *Concepts Magn. Reson.* **14** (2002), 225–58.
35. G. N. Chmurny and D. I. Hoult, The Ancient and Honourable Art of Shimming. *Concepts Magn. Reson.* **2** (1990), 131–49.
36. V. W. Miner and W. W. Conover, Shimming of Superconducting Magnets. In *Encyclopedia of Nuclear Magnetic Resonance*, ed. D. M. Grant and R. K. Harris. vol. 7. (New York: Wiley, 1996), pp. 4340–56.
37. M. Weiger, T. Speck, and M. Fey, Gradient Shimming with Spectrum Optimisation. *J. Magn. Reson.* **182** (2006), 38–48.
38. D.-H. Wu, W. S. Woodward, and C. S. Johnson, Jr., A Sample Spinner for Vibration-Sensitive Liquid-State Experiments with Application to Diffusion-Ordered 2D NMR. *J. Magn. Reson. A* **104** (1993), 231–3.
39. J. R. Taylor, *An Introduction to Error Analysis*, 2nd edn. (New York: University Science Books, 1997).
40. M. Neeman, J. P. Freyer, and L. O. Sillerud, Pulsed-Gradient Spin-Echo Diffusion Studies in NMR Imaging. Effects of the Imaging Gradients on the Determination of the Diffusion Coefficients. *J. Magn. Reson.* **90** (1990), 303–12.

41. J. Mattiello, P. J. Basser, and D. Le Bihan, Analytical Expressions for the *b* Matrix in NMR Diffusion Imaging and Spectroscopy. *J. Magn. Reson. A* **108** (1994), 131–41.
42. M. Eis and M. Hoehn-Berlage, Correction of Gradient Crosstalk and Optimization of Measurement Parameters in Diffusion MR Imaging. *J. Magn. Reson. B* **107** (1995), 222–34.
43. J. Kärger, H. Pfeifer, and W. Heink, Principles and Applications of Self-Diffusion Measurements by Nuclear Magnetic Resonance. *Adv. Magn. Reson.* **12** (1988), 1–89.
44. P. J. Basser, J. Mattiello, and D. Le Bihan, Estimation of the Effective Self-Diffusion Tensor from the NMR Spin Echo. *J. Magn. Reson. B* **103** (1994), 247–54.
45. M. Holz and H. Weingärtner, Calibration in Accurate Spin-Echo Self-Diffusion Measurements Using ^{1}H and Less-Common Nuclei. *J. Magn. Reson.* **92** (1991), 115–25.
46. M. Holz, S. R. Heil, and A. Sacco, Temperature-Dependent Self-Diffusion Coefficients of Water and Six Selected Molecular Liquids for Calibration in Accurate ^{1}H NMR PFG Measurements. *Phys. Chem. Chem. Phys.* **2** (2000), 4740–2.
47. P. S. Tofts, D. Lloyd, C. A. Clark, G. J. Barker, G. J. M. Parker, P. McConville, C. Baldock, and J. M. Pope, Test Liquids for Quantitative MRI Measurements of Self-Diffusion Coefficient In Vivo. *Magn. Reson. Med.* **43** (2000), 368–74.
48. P. T. Callaghan and A. Coy, Evidence for Reptational Motion and the Entanglement Tube in Semidilute Polymer Solutions. *Phys. Rev. Lett.* **68** (1992), 3176–9.
49. H. Weingärtner, Self-Diffusion in Liquid Water. A Reassessment. *Z. Phys. Chem.* **132** (1982), 129–49.
50. R. Mills, Self-Diffusion in Normal and Heavy Water in the Range 1–45°. *J. Phys. Chem.* **77** (1973), 685–8.
51. A. F. Collings and R. Mills, Temperature-Dependence of Self-Diffusion for Benzene and Carbon Tetrachloride. *J. Chem. Soc., Faraday Trans.* **66** (1970), 2761–6.
52. M. Holz, R. Haselmeier, R. K. Mazitov, and H. Weingärtner, Self-Diffusion of Neon in Water by ^{21}Ne NMR. *J. Am. Chem. Soc.* **116** (1994), 801–2.
53. R. Mills and V. V. M. Lobo, *Self Diffusion in Electrolyte Solutions*. (Amsterdam: Elsevier, 1989).
54. H. Weingärtner, R. Haselmeier, and M. Holz, ^{129}Xe NMR as a New Tool for Studying Gas Diffusion in Liquids: Self-Diffusion of Xenon in Water. *Chem. Phys. Lett.* **195** (1992), 596–601.
55. H. Y. Carr and E. M. Purcell, Effects of Diffusion on Free Precession in Nuclear Magnetic Resonance Experiments. *Phys. Rev.* **94** (1954), 630–8.
56. D. C. Douglass and D. W. McCall, Diffusion in Paraffin Hydrocarbons. *J. Phys. Chem.* **62** (1958), 1102–7.
57. D. W. McCall, D. C. Douglass, and E. W. Anderson, Self-Diffusion Studies by Means of Nuclear Magnetic Resonance Spin-Echo Techniques. *Ber. Bunsenges. Phys. Chem.* **67** (1963), 336–40.
58. E. Fukushima and S. B. W. Roeder, *Experimental Pulse NMR A Nuts and Bolts Approach*. (London: Addison-Wesley, 1981).
59. M. I. Hrovat and C. G. Wade, NMR Pulsed-Gradient Diffusion Measurements I. Spin-Echo Stability and Gradient Calibration. *J. Magn. Reson.* **44** (1981), 62–75.
60. T. R. Saarinen and C. S. Johnson, Jr., Imaging of Transient Magnetization Gratings in NMR. Analogies with Laser-Induced Gratings and Applications to Diffusion and Flow. *J. Magn. Reson.* **78** (1988), 257–70.
61. J. S. Murday, Measurement of Magnetic Field Gradient by Its Effect on the NMR Free Induction Decay. *J. Magn. Reson.* **10** (1973), 111–20.

62. R. L. Vold, R. R. Vold, and H. E. Simon, Errors in Measurements of Transverse Relaxation Rates. *J. Magn. Reson.* **11** (1973), 283–98.
63. E. Fukushima, A. A. V. Gibson, and T. A. Scott, Carbon-13 NMR of Carbon Monoxide. II. Molecular Diffusion and Spin-Rotation Interaction in Liquid CO. *J. Chem. Phys.* **71** (1979), 1531–6.
64. D. M. Lamb, P. J. Grandinetti, and J. Jonas, Fixed Field Gradient NMR Diffusion Measurements Using Bessel Function Fits to the Spin-Echo Signal. *J. Magn. Reson.* **72** (1987), 532–9.
65. M. I. Hrovat and C. G. Wade, Absolute Measurements of Diffusion Coefficients by Pulsed Nuclear Magnetic Resonance. *J. Chem. Phys.* **73** (1980), 2509–10.
66. A. C. Wright, H. Bataille, H. H. Ong, S. L. Wehrli, H. K. Song, and F. W. Wehrli, Construction and Calibration of a 50 T/m *z*-Gradient Coil for Quantitative Diffusion Microimaging. *J. Magn. Reson.* **186** (2007), 17–25.
67. N. Yadav, A. M. Torres, and W. S. Price, An Improved Approach to Calibrating High Magnetic Field Gradients for Pulsed Field Gradient Experiments. *J. Magn. Reson.* **194** (2008), 25–8.
68. R. Huo, R. Wehrens, and L. M. C. Buydens, Improved DOSY NMR Data Processing by Data Enhancement and Combination of Multivariate Curve Resolution with Non-Linear Least Square Fitting. *J. Magn. Reson.* **169** (2004), 257–69.
69. W. S. Price, F. Tsuchiya, and Y. Arata, Lysozyme Aggregation and Solution Properties Studied Using PGSE NMR Diffusion Measurements. *J. Am. Chem. Soc.* **121** (1999), 11503–12.
70. M. A. Connell, A. L. Davis, A. M. Kenwright, and G. A. Morris, NMR Measurements of Diffusion in Concentrated Samples: Avoiding Problems with Radiation Damping. *Anal. Bioanal. Chem.* **378** (2004), 1568–73.
71. M. Todica and A. Pop, Algorithm for the Evaluation of the Self-Diffusion Coefficient of Small Molecules, Using the NMR Magnetic Field Gradient Method. *Int. J. Mod. Phys. B* **16** (2002), 2875–84.
72. W. H. Press, S. A. Teukolsky, W. T. Vetterling, and B. P. Flannery, *Numerical Recipes*, 3rd edn. (Cambridge: Cambridge University Press, 2007).
73. A. J. Lennon, N. R. Scott, B. E. Chapman, and P. W. Kuchel, Hemoglobin Affinity for 2,3-Bisphosphoglycerate in Solutions and Intact Erythrocytes: Studies Using Pulsed-Field Gradient Nuclear Magnetic Resonance and Monte Carlo Simulations. *Biophys. J.* **67** (1994), 2096–109.
74. R. Lamanna, On the Inversion of Multicomponent NMR Relaxation and Diffusion Decays in Heterogeneous Systems. *Concepts Magn. Reson.* **26A** (2005), 78–90.
75. T. Brand, E. J. Cabrita, and S. Berger, Intermolecular Interaction as Investigated by NOE and Diffusion Studies. *Prog. NMR Spectrosc.* **46** (2005), 159–96.
76. M. Nilsson, M. A. Connell, A. L. Davis, and G. A. Morris, Biexponential Fitting of Diffusion-Ordered NMR Data: Practicalities and Limitations. *Anal. Chem.* **78** (2006), 3040–5.
77. S. W. Provencher, An Eigenfunction Expansion Method for the Analysis of Exponential Recovery Decay Curves. *J. Chem. Phys.* **64** (1976), 2772–7.
78. S. W. Provencher, A Fourier Method for the Analysis of Exponential Decay Curves. *Biophys. J.* **16** (1976), 27–41.
79. S. W. Provencher and R. H. Vogel, *Regularization Techniques for Inverse Problems in Molecular Biology.* In *Numerical Treatment of Inverse Problems in Differential and Integral Equations*, ed. P. Deuflhard and E. Hairer. (Boston: Birkhäuser, 1983), pp. 304–19.

80. K. F. Morris and C. S. Johnson, Jr., Resolution of Discrete and Continuous Molecular Size Distributions by Means of Diffusion-Ordered 2D NMR Spectroscopy. *J. Am. Chem. Soc.* **115** (1993), 4291–9.
81. S. W. Provencher, A Constrained Regularization Method for Inverting Data Represented by Linear Algebraic or Integral Equations. *Comput. Phys. Commun.* **27** (1982), 213–27.
82. S. W. Provencher, Contin: A General Purpose Constrained Regularization Program for Inverting Noisy Linear Algebraic and Integral Equations. *Comput. Phys. Commun.* **27** (1982), 229–42.
83. J. Weese, A Regularization Method for Nonlinear Ill-Posed Problems. *Comput. Phys. Commun.* **77** (1993), 429–40.
84. T. Roths, M. Marth, J. Weese, and J. Honerkamp, A Generalized Regularization Method for Nonlinear Ill-Posed Problems Enhanced for Nonlinear Regularization Terms. *Comput. Phys. Commun.* **139** (2001), 279–96.
85. M. A. Delsuc and T. E. Malliavin, Maximum Entropy Processing of DOSY NMR Spectra. *Anal. Chem.* **70** (1998), 2146–8.
86. T. E. Malliavin, V. Louis, and M. A. Delsuc, The DOSY Experiment Provides Insights into the Protein–Lipid Interactions. *J. Chim. Phys.* **95** (1998), 178–86.
87. C. Mouro, P. Mutzenhardt, B. Diter, and D. Canet, HR-DOSY Experiments with Radiofrequency Field Gradients (RFG) and Their Processing According to the HD Method. *Magn. Reson. Chem.* **40** (2002), S133–8.
88. R. C. Sebastião, C. N. Pacheco, J. P. Braga, and D. Piló-Veloso, Diffusion Coefficient Distribution from NMR-DOSY Experiments Using Hopfield Neural Network. *J. Magn. Reson.* **182** (2006), 22–8.
89. P. Stilbs and K. Paulsen, Global Least-Squares Analysis of Large, Correlated Spectral Data Sets and Application to Chemical Kinetics and Time-Resolved Fluoresence. *Rev. Sci. Instrum.* **67** (1996), 4380–6.
90. P. Stilbs, K. Paulsen, and P. C. Griffiths, Global Least-Squares Analysis of Large, Correlated Spectral Data Sets: Application to Component-Resolved FT-PGSE NMR Spectroscopy. *J. Phys. Chem.* **100** (1996), 8180–9.
91. P. Stilbs, Component Separation in NMR Imaging and Multidimensional Spectroscopy through Global Least-Squares Analysis, Based on Prior Knowledge. *J. Magn. Reson.* **135** (1998), 236–41.
92. M. Nilsson and G. A. Morris, Correction of Systematic Errors in CORE Processing of DOSY Data. *Magn. Reson. Chem.* **44** (2006), 655–60.
93. M. Kubista, A New Method for the Analysis of Correlated Data Using Procrustes Rotation Which is Suitable for Spectral Analysis. *Chemom. Intell. Lab. Syst.* **7** (1990), 273–9.
94. D. Schulze and P. Stilbs, Analysis of Multicomponent FT-PGSE Experiments by Multivariate Statistical Methods Applied to the Complete Bandshapes. *J. Magn. Reson. A.* **105** (1993), 54–8.
95. B. Antalek and W. Windig, Generalized Rank Annihilation Method Applied to a Single Multicomponent Pulsed Gradient Spin Echo NMR Data Set. *J. Am. Chem. Soc.* **118** (1996), 10331–2.
96. W. Windig and B. Antalek, Direct Exponential Curve Resolution Algorithm (DECRA): A Novel Application of the Generalized Rank Annihilation Method for a Single Spectral Mixture Data Set with Exponentially Decaying Contribution Profiles. *Chemom. Intell. Lab. Sys.* **37** (1997), 241–54.
97. B. Antalek, J. M. Hewitt, W. Windig, P. D. Yacobucci, T. Mourey, and K. Le, The Use of PGSE NMR and DECRA for Determining Polymer Composition. *Magn. Reson. Chem.* **40** (2002), S60–71.

98. B. J. Antalek, Accounting for Spin Relaxation in Quantitative Pulse Gradient Spin Echo NMR Mixture Analysis. *J. Am. Chem. Soc.* **128** (2006), 8402–3.
99. B. Antalek, Using PGSE NMR for Chemical Mixture Analysis: Quantitative Aspects. *Concepts Magn. Reson.* **30A** (2007), 219–35.
100. T. M. Alam and M. K. Alam, Effect of Non-Exponential and Multi-Exponential Decay Behavior on the Performance of the Direct Exponential Curve Resolution Algorithm (DECRA) in NMR Investigations. *J. Chemom.* **17** (2003), 583–93.
101. L. C. M. Van Gorkom and T. M. Hancewicz, Analysis of DOSY and GPC-NMR Experiments on Polymers by Multivariate Curve Resolution. *J. Magn. Reson.* **130** (1998), 125–30.
102. R. Huo, R. Wehrens, J. van Duynhoven, and L. M. C. Buydens, Assessment of Techniques for DOSY NMR Data Processing. *Anal. Chim. Acta* **490** (2003), 231–51.
103. R. Huo, R. A. van der Molengraaf, J. A. Pikkemaat, R. Wehrens, and L. M. C. Buydens, Diagnostic Analysis of Experimental Artefacts in DOSY NMR Data by Covariance Matrix of the Residuals. *J. Magn. Reson.* **172** (2005), 346–8.
104. R. Huo, R. Wehrens, and L. M. C. Buydens, Robust DOSY NMR Data Analysis. *Chemom. Intell. Lab. Syst.* **85** (2007), 9–19.
105. Y.-L. Xie and P. K. Hopke, Positive Matrix Factorization Applied to a Curve Resolution Problem. *J. Chemom.* **12** (1998), 357–64.
106. E. Widjaja and M. Garland, Entropy Minimization and Spectral Dissimilarity Curve Resolution Technique Applied to Nuclear Magnetic Resonance Data Sets. *J. Magn. Reson.* **173** (2005), 175–82.
107. L. M. Smith, A. D. Maher, O. Cloarec, M. Rantaleinin, H. Tang, P. Elliott, J. Stamler, J. C. Lindon, E. Holmes, and J. K. Nicholson, Statistical Correlation and Projection Methods for Improved Information Recovery from Diffusion-Edited NMR Spectra of Biological Samples. *Anal. Chem.* **79** (2007), 5682–9.
108. J. Chen, A. J. Shaka, and V. A. Mandelshtam, RRT: The Regularized Resolvent Transform for High-Resolution Spectral Estimation. *J. Magn. Reson.* **147** (2000), 129–37.
109. G. S. Armstrong, N. M. Loening, J. E. Curtis, A. J. Shaka, and V. A. Mandelshtam, Processing DOSY Spectra Using the Regularized Resolvent Transform. *J. Magn. Reson.* **163** (2003), 139–48.

7

PGSE hardware and sample problems

7.1 Introduction

There are a number of potential problems that must be addressed in PGSE measurements if high quality data is to be obtained. The problems include: (i) rf interference, (ii) radiation damping and long-range dipolar interactions, (iii) convection, (iv) homogeneity of the applied magnetic field gradient, (v) background magnetic gradients, (vi) eddy currents and static magnetic field disturbances generated by the gradient pulses, and lastly (vii) gradient pulse mismatch and sample movement. Almost invariably these problems lead to increased signal attenuation and thus overestimates of the diffusion coefficient and misinterpretation of the experimental data, and it has been noted that all PGSE systems have thresholds below which artefactual attenuation exceeds diffusive attenuation.[1] Here, we consider the origins of these problems, their symptoms and some methods to alleviate them.

7.2 RF problems

The addition of gradient coils to an NMR probe can have deleterious effects on probe performance. Due to the proximity of the gradient coils to the sample region, the gradient coils and leads can, without appropriate precautions, act as antennae and introduce rf interference. A related problem is the possible strong mutual inductance between the gradient and the rf coils. Thus, the quality factor Q ($= \omega L/R$ where ω, L and R are the resonance frequency, inductance and resistance, respectively) of the rf coil(s) are diminished resulting in longer pulses for the same flip angle, poorer decoupling efficiency and S/N.

The effects of non-ideal B_1 pulses and B_1 inhomogeneity are well-known on spin-echoes,[2–4] but have not been widely considered with respect to NMR diffusion measurements. This may in part stem from the samples used in diffusion measurements generally being small and thus the sample being contained within more homogeneous B_1 fields.

7.3 Radiation damping and long-range dipolar field effects

7.3.1 Origin and symptoms

The origin and effects of the demagnetising field were discussed in Section 2.6. The radiation damping[5–11] and long-range dipolar fields[11–17] are complicated effects that are intrinsic in all NMR experiments; however, they only become significant when there is strong net magnetisation. Thus their effects are amplified at higher magnetic fields and in high-sensitivity probes[18,19] and, similarly, narrower resonances (and better shimmed samples) are more susceptible. Both of these fields cause the magnetic field during a pulse sequence to become time-dependent which complicates pulse sequences and can result in artefacts such as frequency shifts. For example, a strong (solvent) resonance will effectively see a different pulse sequence than the solute.[20–24] The combination of the two feedback fields, radiation damping and the long-range dipolar interaction, results in chaotic spin dynamics with very deleterious effects on diffusion measurements.[25–27]

Radiation damping is the result of coupling between the transverse magnetisation and the rf coil. The oscillating current that is generated in the receiver coil by the precessing transverse spin magnetisation in turn generates an oscillating magnetic field, the radiation damping field. This field, which points in a direction perpendicular to both $\mathbf{B}_0$ and the magnetisation vector,[8] acts as a torque on the magnetisation and tends to rotate the same magnetisation back to its equilibrium position. Bloch equations modified to include the effects of radiation damping have been presented.[28,29] For a single resonance, in the limit of small flip angles of M_0, the exponential damping due to radiation damping can be characterised by a time constant (in S.I. units)[6,10]

$$T_{\mathrm{rd}} = \frac{2}{\gamma \mu_0 \eta_{\mathrm{F}} Q M_0}, \tag{7.1}$$

where μ_0 is the magnetic permeability and η_{F} is the filling factor of the probe. While the nutation induced by an applied rf pulse has a constant angular velocity, the nutation induced by radiation damping is not a constant and is given by

$$\left(\frac{d\theta}{dt}\right)_{\mathrm{rd}} = -\frac{\sin\theta}{T_{\mathrm{rd}}}, \tag{7.2}$$

where θ is the angle between the magnetisation and the $\mathbf{B}_0$ field.

Radiation damping is a well-known source of problems in NMR experiments[8,30,31] and in a PGSE experiment it can effectively relax a strong solvent signal back to equilibrium many times faster than the intrinsic spin–lattice relaxation time ($T_{\mathrm{rd}} < T_1$; see Figure 7.1) as well as broaden the resonances (Figure 7.2). In a PGSE sequence, radiation damping can result in the echo signal initially increasing before decreasing with gradient strength (see Figure 7.3).

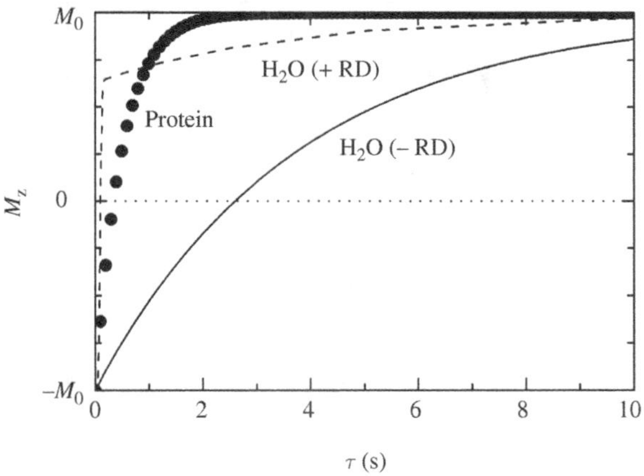

Figure 7.1 The effects of radiation damping on an inversion recovery measurement (i.e., π–τ–$\pi/2$–Acq.) of the ^{1}H longitudinal relaxation of water in a sample of 10 mM lysozyme at 300 MHz. In the absence of radiation damping (—), the water has a relaxation time of about 3.8 s, whereas in the presence of radiation damping (RD), the effective longitudinal relaxation of the water (- - -) is considerably more rapid with the null-point of the signal (i.e., $M_z = 0$) occurring before 0.1 s (i.e., apparent $T_1 \sim 140$ ms). Radiation damping has an insignificant effect on the relaxation of the (much weaker) protein resonances (●●●●). Modified from Price *et al.*[23]

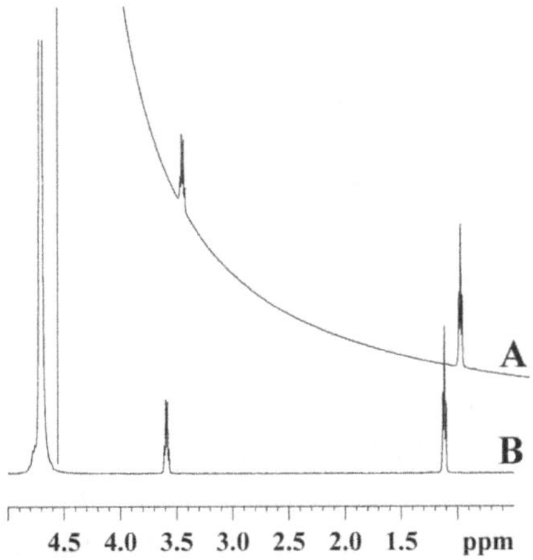

Figure 7.2 PGSE spectrum of aqueous ethanol (90% H_2O, 9% ^1H_2O and 1% ethanol) in the (A) presence and (B) absence of radiation damping. The spectra were obtained using the Hahn spin-echo pulse sequence (Fig. 2.6A) with $\tau = 141.2$ ms, $t_1 = 125.0$ ms, $\Delta = 30$ ms, $\delta = 2$ ms and $g = 0.475$ T m^{-1}. Spectrum B was obtained with Q-switching. From Price and Wälchli.[32]

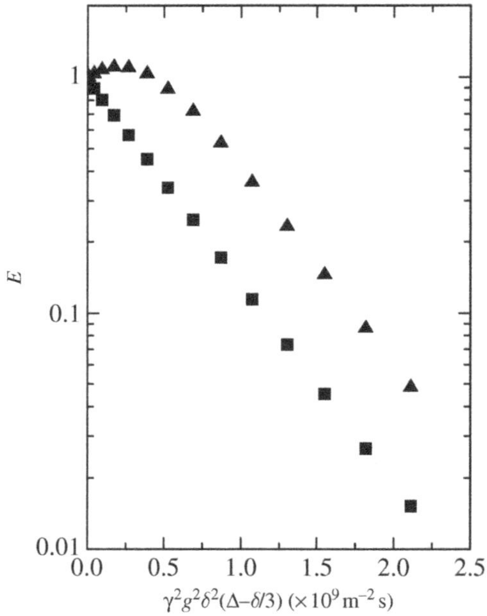

Figure 7.3 Plots of the PGSE attenuation curves of a strong ^1H_2O resonance acquired at 300 MHz using the Hahn spin-echo sequence (Figure 2.6A) in which the first gradient pulse was placed immediately after the $\pi/2$ pulse and the second immediately: after the π pulse (▲) and immediately before acquisition (■). The data is normalised to the first positive value of $g = 0.037$ T m^{-1} in each case. The former case is susceptible to the effects of radiation damping for nearly the whole second τ period since the magnetisation is not spatially encoded after the second gradient pulse. The rise and fall of the attenuation curve of the data acquired using this sequence is due to the competition between radiation damping and diffusive attenuation. The sequence in which the magnetisation is spatially encoded during the entire sequence is not subject to radiation damping and gives the expected single exponential decay. Adapted from Price *et al.*[33]

7.3.2 Solutions

The keys to solving radiation damping problems lie in Eq. (7.1), specifically to reduce η_F, Q or M_0. One simple solution to reduce η_F, and consequently M_0, is by using a very small sample or to de-tune the probe (i.e., lowering Q). As it is the solvent that is normally the source of the strong signal, another solution is to reconstitute the sample in an inert solvent (e.g., in ^2H_2O if observing ^1H). In STE-based PGSE sequences another possibility is to use an initial rf pulse with a smaller flip angle.[29] Finally, the addition of a species to produce chemical exchange with the solvent peak (e.g., adding glycine to aqueous solutions) has also been suggested.[34] However, none of these methods are generally applicable or even perhaps desirable.

More generally, practical solutions include the use of solvent suppression sequences (see Section 6.2.2). Importantly, when strong signals are suppressed the deleterious effects of both the radiation damping and dipolar fields are removed. Although as noted from Eq. (2.120) the effects of the demagnetising field should also disappear when magnetic field gradients are applied at the magic angle (i.e., $\theta = 54.7°$).[20,21,35]

Gradient-based methods have become the most widely used for removing effects of radiation damping and these rely on reducing M through spatial encoding (see Section 2.2.1). Since a gradient has no effect on magnetisation that is collinear with it, a z-gradient pulse will not totally eliminate the potential for radiation damping (i.e., there could be a large net solvent magnetisation directed along $-z$), but while the gradient is being applied it will prevent its onset.[22,23] Gradient-based methods for suppressing radiation damping during different parts of pulse sequences include gradient echoes inserted between the rf pulses of a DANTE train,[36] during the evolution and mixing periods[37] and between acquisition of FID data points.[38] The effects of the radiation damping field can also be counterbalanced during the FID acquisition by adding small flip-angle rf pulses between the acquisition of each data point.[39]

The effects of radiation damping can be largely overcome by moving the gradient pulses within the PGSE sequence so that the magnetisation is spatially encoded for most of the sequence (i.e., there is no net magnetisation) as shown in Figure 7.3. However, this makes the sequence more susceptible to eddy currents and also disallows Δ to be moved independently of the length of the sequence.[33] If available, the incorporation of Q-switching into the PGSE sequence is an ideal and truly flexible solution (see Figures 7.4 and 7.5).[32]

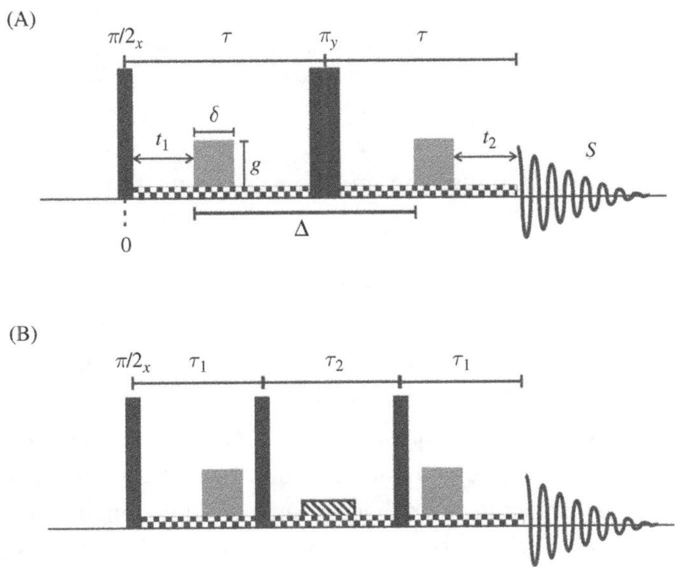

Figure 7.4 Q-switch variants of the (A) Hahn spin-echo and (B) stimulated-echo-based PGSE sequences. The Q of the probe was set to low during the checked periods along the bottom of each sequence. The slanted striped pulse in the STE sequence represents a purge gradient pulse to reduce the phase cycling requirements. Modified from Price and Wälchli.[32]

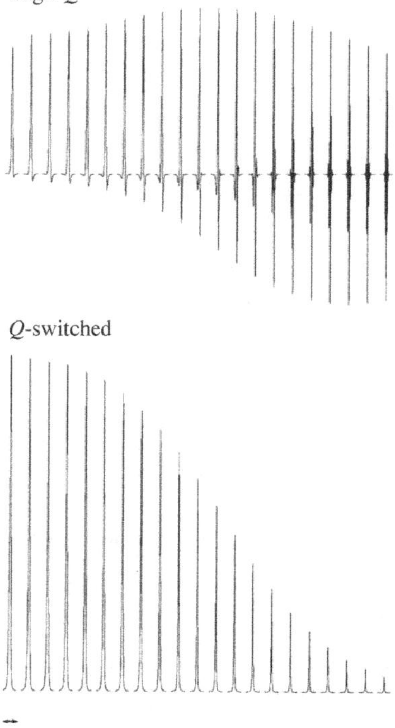

Figure 7.5 A series of Hahn spin-echo-based PGSE spectra of the water resonance in an aqueous ethanol solution acquired with and without *Q*-switching with the timings given in the caption to Figure 7.2. The spectra in each series start from $g = 0$ with increments of 0.068 $\mathrm{T\,m^{-1}}$ (left to right). The spectra acquired without *Q*-switching are severely distorted. Modified from Price and Wälchli.[32]

7.4 Convection and flow compensation

7.4.1 Origin and symptoms

Sample temperature regulation in NMR probes is normally performed by blowing temperature regulated gas over the sample (see Figure 5.6), thus temperature gradients generally exist in the sample when measurements are conducted away from ambient. The severity of the temperature gradient will also depend upon the efficiency of heat transfer and viscosity in the sample as well as the experimental factors (e.g., gas flow rate, geometry and size of the sample and interior dimensions of the probe). If the temperature gradient is large enough convective flows will be induced[40–44] and convective effects have been realised since the earliest days of NMR diffusion measurements.[45] Convection can also be induced by strong rf irradiation as in heteronuclear decoupling.[44,46]

The likelihood of convection is determined from the Rayleigh number[40]

$$R_a = \frac{g\alpha}{\kappa v} a^4 \Delta T, \tag{7.3}$$

where g is the acceleration due to gravity, α is the thermal expansion, κ is the thermal diffusivity of the liquid, v is the kinematic viscosity, a is the tube radius and ΔT is the temperature gradient. For rigid conducting walls, convection results when R_a exceeds a critical value R_c, approximately 1700.

The chemical shift of the ^{59}Co resonance of the (very symmetric) complex $K_3Co(CN)_6$ has an extremely large temperature dependence (1.45 ppm K^{-1}) and is thus a suitable compound for investigating the presence of thermal gradients.[40]

Convection is a source of insidious artefacts in PGSE experiments.[41,42,46–49] While a net flow of spins along the direction of the gradient is clearly indicated by the resulting net phase change (see Section 2.2.3), convective currents,[42,50–52] which typically occur along the long axis of the NMR tube, do not produce a phase change since the flow of the spins along the direction of the gradient is exactly matched by the flow in the anti-parallel direction. However, convection causes a cosine modulation of the PGSE signal attenuation (for a single diffusing species)[47]

$$E(q, \Delta) = \exp\left(-\gamma^2 g^2 D\delta^2(\Delta - \delta/3)\right) \cos(\gamma g \delta v \Delta), \tag{7.4}$$

where v is the velocity of the flow. However, convection is generally not apparent when $v^2\Delta \ll D$ and $\delta \ll \Delta$, due to the similarity between the cosine and Gaussian functions. In such circumstances, the attenuation data appears to be well described by the usual exponential function giving, to first approximation, an apparent diffusion coefficient of

$$D^{\text{app}}(\Delta) = D + \frac{\Delta v^2}{2}. \tag{7.5}$$

Consequently, the presence of convection is suspected when the measured diffusion coefficient of a non-exchanging freely diffusing species increases with Δ. Importantly, whereas diffusion depends on molecular size, convection is mainly controlled by the solvent and all solutes are equally affected regardless of molecular size.

7.4.2 Solutions

Apart from better probe design to reduce temperature gradients, there exist both hardware and pulse sequence methods for minimising the effects of convection. The hardware methods are: (i) using a narrower tube or specially designed tube,[53] (ii) in the case of some samples, glass wool can be placed in the sample (Ernst von Meerwall, private communication), (iii) surrounding the sample with a fluid of higher heat capacity[42] or (iv) spinning the sample.[50–52]

The PGSE pulse sequence solutions are based on gradient moment nulling (see Section 2.2.2).[41] Equation (2.23) provides the basis of the so-called gradient moment nulling methods and flow compensation. Since it is assumed that the convection current has a constant laminar flow in the z-direction, this method does not compensate for turbulent convection. The even echoes in a CPMG sequence are unaffected by the effects of convection since the first moment of the gradient is nulled at these positions.

An example of a flow-compensated double-stimulated echo sequence[41,54] preceded by a WET pulse sequence[55,56] for multiple solvent suppression is shown in Figure 7.6 and the signal attenuation due to diffusion for this sequence is

$$E = \exp\left(-\gamma^2 g^2 D\delta^2 \left(T + \frac{4\delta}{3} + 2\tau\right)\right), \tag{7.6}$$

where T and τ are defined in Figure 7.6. Second-order effects (e.g., flow acceleration) can be eliminated by nulling higher moments.

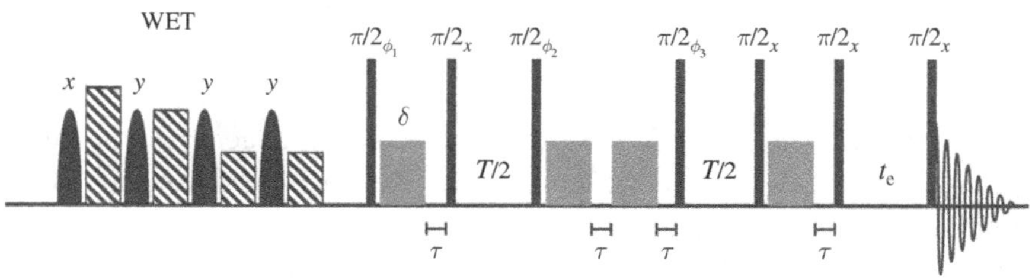

Figure 7.6 A double stimulated echo sequence that is compensated for flow including an LED delay, t_e.[41] The sequence refocusses all constant-velocity effects. The first moment of the effective gradient over the whole sequence is zero. The phase cycling for this modified sequence containing a WET suppression sequence can be found in the paper by Simorellis and Flynn.[54]

A convection compensating sequence is essentially some variation of a double PGSE sequence. Another, and perhaps more intuitive, way of viewing the convection compensating effect of a double PGSE experiment such as that depicted in Figure 7.6 is to realise that (starting from Eq. (2.37)) the echo attenuation will be of the form[57,58]

$$E(\mathbf{q}) = \langle \exp(i2\pi\mathbf{q}_1 \cdot \mathbf{R}_1) \exp(i2\pi\mathbf{q}_2 \cdot \mathbf{R}_2) \rangle. \tag{7.7}$$

Providing the two pairs of gradient pulses are applied with identical Δ, but with the opposite sense, Eq. (7.7) becomes

$$E(\mathbf{q}) = \langle \exp(i2\pi\mathbf{q}_1 \cdot (\mathbf{R}_1 - \mathbf{R}_2)) \rangle, \tag{7.8}$$

and thus the scheme will compensate for any motion common (e.g., constant flow) to both periods.

Various compensating sequences have been presented such as the double (CPMG) multiple spin-echo sequence with bipolar gradients;[59] variations on double spin-echo and double-stimulated echo sequences with eddy-current suppression;[60] CONVEX,[61,62] which combines water suppression with convection compensation and DQDIFF,[62,63] which combines convection compensation with double quantum filtering. OGSE sequences also have inherent convection compensation properties (see Section 2.4).[64]

7.5 Gradient constancy

7.5.1 Origin and symptoms

It has been found in practice that a reasonable deviation from perfect gradient constancy is allowable for many experiments;[65–68] however, it has also been noted that if different regions of the sample volume experience slightly different gradients, the diffusional decay of the resonance of a single freely diffusing species will deviate from exponential behaviour in a manner that is indistinguishable from that of a polydisperse system in the presence of a perfectly constant gradient.[69]

7.5.2 Solutions

A simple approach is to restrict the effective sample volume using one of the methods outlined in Section 6.3. In some NMR probes, the receptive volumes of the rf coil is smaller than the volume of constant gradient, affording the possibility of using standard NMR tubes without requiring sample volume restriction. In some cases, as noted in Section 6.3, it is possible to map the spatial distributions of the gradient strength for a specific probe and account for this in the data fitting.[65–67] With this information it is possible to calculate the theoretical form of the PGSE attenuation as a function of the nominally applied gradient for a specific pulse sequence. For example, the Stejskal and Tanner equation (Eq. (2.49)) is then numerically integrated over the sample and the PGSE decay is represented as an exponential of a power series, namely[66,67]

$$S = S(0)\exp\left(-\sum_{n=1}^{N} c_n\left[\gamma^2 \widehat{g}^2 D\delta^2(\Delta - \delta/3)\right]\right), \tag{7.9}$$

where $\widehat{g}$ is now the nominal gradient strength. Implicit in this equation is the assumption that the MSD of the measured spins is sufficiently small that they stay within regions of constant gradient during Δ. The coefficients c_n are determined experimentally by using a sample with a known diffusion coefficient in the mapping experiment. Typically, $N \leq 4$ is sufficient. The decay function has also been parameterised as the sum of two error functions.[65]

7.6 Background gradients

7.6.1 Origin and symptoms

In PGSE experiments based on the use of applied magnetic gradients, the presence of both homogeneous and inhomogenous (i.e., both in direction and magnitude and perhaps time-dependent) background gradients (Section 4.8) can greatly complicate measurements, both directly through the inclusion of gradient cross-terms and indirectly through decreasing the observed T_2.[33,70–75] The effects of background gradients and methods for suppressing them have recently been reviewed.[76,77]

The salient points of how background gradients affect PGSE measurements can be understood as follows: in the presence of a uniform constant background gradient of strength $\mathbf{g}_0$ (see Figure 2.10) and neglecting relaxation effects, the PGSE signal of the Hahn spin-echo sequence is given by (Eq. (2.69))[78]

$$S(2\tau) = M_0 \exp\left(\underbrace{-\gamma^2 g^2 D\delta^2(\Delta - \delta/3)}_{g\text{ term}} \underbrace{+\gamma^2 \mathbf{g}\cdot\mathbf{g}_0 D\delta\left[t_1^2 + t_2^2 + \delta(t_1+t_2) + \frac{2}{3}\delta^2 - 2\tau^2\right]}_{\mathbf{g}\cdot\mathbf{g}_0\text{ cross-term}} \underbrace{-\gamma^2 g_0^2 D\frac{2}{3}\tau^3}_{g_0\text{ term}}\right), \tag{7.10}$$

where $t_2 = 2\tau - (t_1 + \Delta + \delta)$. Similarly, for the STE sequence[79]

$$S(2\tau) = \frac{M_0}{2} \exp\left(\underbrace{-\gamma^2 g^2 D\delta^2(\Delta - \delta/3)}_{g\text{ term}} \underbrace{+\gamma^2 \mathbf{g}\cdot\mathbf{g}_0 D\delta\left[t_1^2 + t_2^2 + \delta(t_1+t_2) + \frac{2}{3}\delta^2 - 2(\tau_1+\tau_2)\tau_1\right]}_{\mathbf{g}\cdot\mathbf{g}_0\text{ cross-term}} \underbrace{-\gamma^2 g_0^2 D\frac{2}{3}\tau_1(\tau_2 + 2\tau_1/3)}_{g_0\text{ term}}\right), \tag{7.11}$$

where $t_2 = \tau_1 - \delta - t_1$.

Generally, as noted in Section 2.3.2, the g_0 term can be normalised out in Eqs. (7.10) and (7.11). However, unless the condition $\mathbf{g}\delta \gg \mathbf{g}_0\tau$ holds, or if $\mathbf{g}_0$ is non-constant, $\mathbf{g}\delta \gg \langle \mathbf{g}_0^2 \rangle^{1/2}\tau$, the $\mathbf{g}\cdot\mathbf{g}_0$ cross-term cannot be neglected. The cross-term can result in the echo attenuation depending not only on δ and Δ, but also on the other delays in the sequence. As shown in Figure 7.7, the cross-term causes the plot of $\ln(E)$ versus $\gamma^2 g^2 \delta^2(\Delta - \delta/3)$ to be non-linear, thereby complicating any subsequent analysis.[33,75,80] The duration τ in the Hahn-based sequence is generally much longer than τ_1 in the STE sequence, and consequently the effects of the cross-term are smaller in the STE sequence.[81]

Since the sign of the $\mathbf{g}\cdot\mathbf{g}_0$ term depends on the relative direction between the applied and internal gradients ($\mathbf{g}\cdot\mathbf{g}_0 = gg_0 \cos\theta$, where θ is the angle between $\mathbf{g}$ and $\mathbf{g}_0$) as shown in Figure 7.7, the presence of background gradients can be tested for by reversing the polarity of the applied gradient.[33,73,82] However, if there is a distribution of background gradients, reversing the polarity of $\mathbf{g}$ will only affect the signal amplitude if the distribution of $\mathbf{g}_0$ is not symmetric about $g_0 = 0$.[83,84] Further, apparent anisotropic diffusion could result from measurements conducted in the presence of anisotropic background gradients due to the $\mathbf{g}\cdot\mathbf{g}_0$ term.[85–87] This type of problem can also result due to cross-terms between the diffusion and imaging gradients in imaging pulse sequences involving diffusion measurements.[88] Lian and co-workers[89] have demonstrated that an image of D or $\langle D\mathbf{g}_0^2 \rangle$ can be obtained without the corrupting $\mathbf{g}\cdot\mathbf{g}_0$ terms by appending a standard imaging sequence to an alternating pulsed field gradient sequence[90] or a Carr–Purcell sequence.

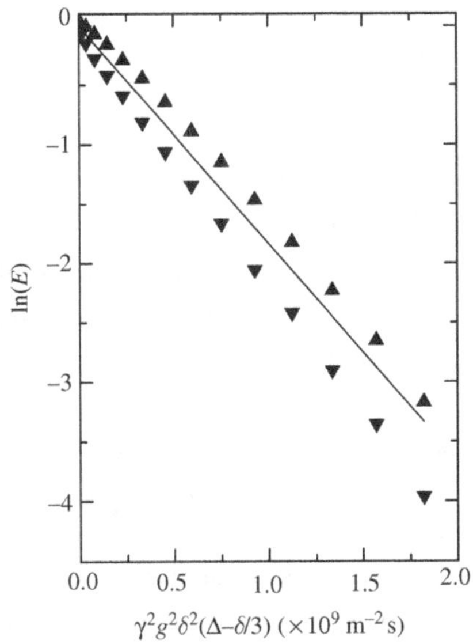

Figure 7.7 ^{1}H PGSE NMR attenuation data acquired at 300 MHz for the residual water in model system designed to give a background gradient through the sample consisting of 50 μl of ^2H_2O in a susceptibility-matched Shigemi tube but without the plunger. The data were acquired using positive gradient polarity (▲) and negative gradient polarity (▼). Regression of Eq. (2.49) onto the average of the normalised positive and negative polarity data (solid line) gives a diffusion coefficient of $1.831 \pm 0.01 \times 10^{-9}$ m^{2}s^{-1}. From Price *et al.*[33]

Finally, we stress that in contrast to our simple treatment above, where only constant background gradients were considered, background gradients will normally be highly inhomogeneous as in porous media. Thus, if a species moves between volume elements of different background gradient slowly, such that it stays within a volume of constant $\mathbf{g}_0$ during the sequence, the acquired signal will be of the form[33]

$$S = \int S(\mathbf{g}, \mathbf{g}_0(\mathbf{r}))\, d\mathbf{r}. \tag{7.12}$$

Thus, it is likely that the measured diffusion will be lower than the actual diffusion coefficient since the measured diffusion is in essence an ensemble average and the internal gradients will weight this distribution by the time of signal acquisition with the apparently faster diffusing spins being more attenuated and most of the signal will be contributed by the apparently slower moving spins.[91] This is analogous to the effect found for spins diffusing in a restricted geometry having an absorbing wall (see Section 4.3.2). The effects of background gradients can also be mistaken for restricted diffusion.[74] Depending upon the diffusion coefficient and the length scales of the magnetic inhomogeneities, the background gradients could also be time-dependent. An obvious example, with relevance to MRI-based studies, is where spins diffuse around in a medium containing a contrast agent consisting of superparamagnetic particles.[92]

7.6.2 Solutions

Where the background gradient arises due to an imperfect B_0 field, one solution is to restrict the sample size either physically using a susceptibility-matched NMR tube or effectively using slice selection.[93] Hong and Dixon have considered the problems resulting from 'background gradients' arising from imaging gradients in MRI-based diffusion measurements.[94]

If the distribution of $\mathbf{g}_0$ is symmetric about $g_0 = 0$ and not too large, a series expansion can be used to correct for the background gradients.[83] In general, however, specialised pulse sequences are required to suppress background gradient problems. The basis of most sequences for the removal of the $\mathbf{g}_0$ term is to add additional π pulses to the PGSE sequence to refocus the dephasing effects of $\mathbf{g}_0$ in a way analogous to the CPMG sequence.[84,95] Clearly, such sequences must be designed with an odd number of π pulses between the gradient pulses (of the same polarity) since an even number of pulses would simply result in the effects of the second gradient pulse adding to the dephasing effects of the first gradient pulse.[84,95] In 1980, Karlicek and Lowe[90] proposed the use of alternating (bipolar) pulsed field gradients in a CPMG sequence (see Figure 7.8A) for which the attenuation function is[90]

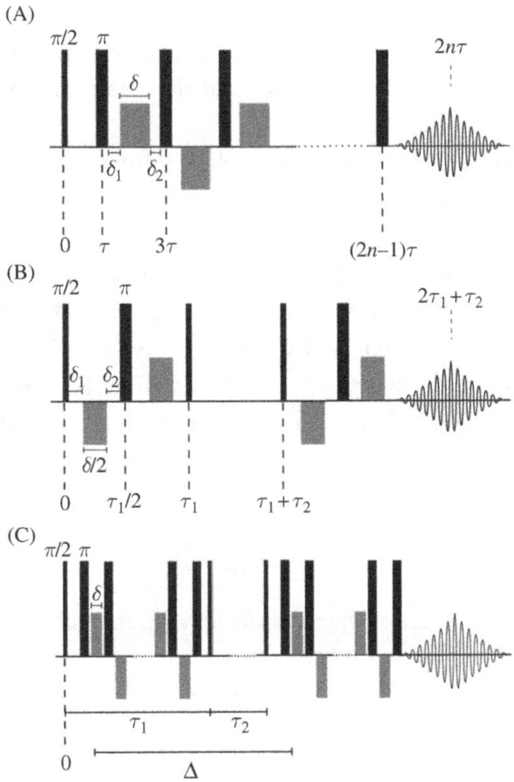

Figure 7.8 Sequences for removal of background gradients. (A) The Karlicek and Lowe sequence,[90] (B) the nine-pulse sequence of Cotts *et al.*,[97] and (C) the improved stimulated-echo sequence of Latour *et al.*[75]

$$E(g, 2n\tau) = \exp\left(-\frac{2}{3}\gamma^2 D\left[ng_0^2\tau^3 + (n-1)^3 g^2\left(\tau - \frac{(\delta_1+\delta_2)}{2}\right)^2 \times\left(\tau + \frac{(\delta_1+\delta_2)}{(n-1)^2}\right)\right]\right), \tag{7.13}$$

where the integer n, δ_1 and δ_2 are defined in Figure 7.8A. In passing, it is noted that the equation relating attenuation of the Karlicek–Lowe alternating PGSE experiment, but neglecting the effects of the background gradient, has been derived for diffusion within a sphere using the GPD approximation.[96]

CPMG-based sequences are limited by T_2 and it is desirable to have STE-based pulse sequences. Cotts and co-workers[97] presented three modified STE sequences incorporating alternating pulsed field gradients (see Figure 7.8B) and later Latour *et al.*[75] presented a pulse sequence that combined features of the Karlicek and Lowe and Cotts pulse sequences in which the gradient pulses in the normal STE echo pulse sequence are replaced by a series of short alternating gradient pulses (see Figure 7.8C).

In later works, Van Dusschoten and co-workers[81] proposed a variation of the Hahn-echo with the (monopolar) gradient pulses sandwiched between π rf pulses termed the PGSE multiple-spin-echo (PGSE MSE) pulse sequence, and Sørland and co-workers[98] devised a 13-interval STE-based sequence using asymmetric bipolar gradient pulses to suppress the effects of constant background gradients. Of particular interest are STE-based PGSE sequences including asymmetric bipolar gradient pulses with the ability to suppress the effects of non-constant background gradients.[99,100] Analysis of the sequence of Sun *et al.*[99] (see Figure 7.9) reveals that if the ratio of the gradient pulses in the asymmetric pairs is set to a 'magic' ratio defined by[77,99,101,102]

$$\eta = \frac{g_1}{g_2} = -\frac{\delta_1^2 + \delta_1\delta + \delta^2/3}{\delta_1^2 + \delta_1\delta + \delta^2/3 - 2\tau_1^2}, \tag{7.14}$$

then the $\mathbf{g}\cdot\mathbf{g}_0$ cross-term can be suppressed leaving the echo attenuation given by

$$E = \exp\left(-\gamma^2 D\delta^2\left\{[\tau_2 + \tau_1 - (\delta_2 - \delta_1)](g_1 - g_2)^2 + 2(\tau_1 + \delta_2 - \delta_1)g_1^2 - \frac{1}{3}\delta\left(g_1^2 + g_2^2\right)\right\}\right). \tag{7.15}$$

Finally, we note that Vasenkov *et al.*[101] have noted that data obtained using a standard PGSTE sequence in the presence of spatially varying IMFGs can be analysed by extrapolation of D^{app} as a function of τ_1 to obtain the true diffusion coefficients.

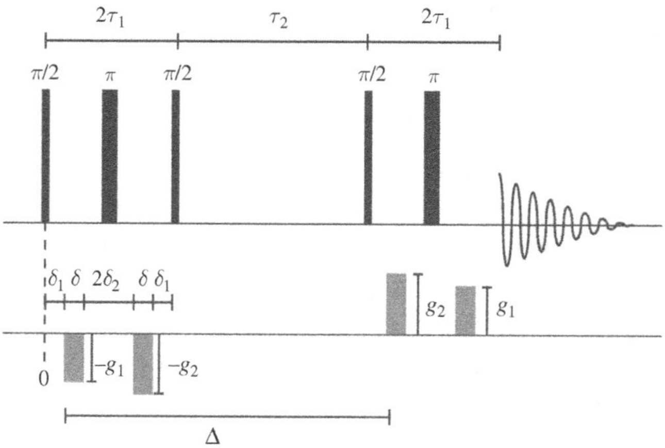

Figure 7.9 The asymmetric bipolar STE-based PGSE sequence of Sun *et al.*[99]

7.7 Eddy currents and perturbation of B_0

7.7.1 Origin and symptoms

Out of mathematical simplicity, most PGSE sequences prescribe 'rectangular' gradient pulses, but the rapid rise of the gradient pulses can generate eddy currents in the surrounding conducting surfaces[103] and the severity of the eddy currents is proportional to the time differential of the current generating the gradient pulses (i.e., dI/dt). Hence, the severity of eddy currents increases with both the intensity and the rapidity of the rise and fall of the gradient pulses. Eddy currents can have very deleterious effects on PGSE- and MRI-based[104] measurements, including: (i) phase changes in the observed spectrum and anomalous changes in the attenuation, (ii) gradient-induced broadening of the observed spectrum, (iii) time-dependent but spatially invariant B_0 (and thus frequency) shift effects, and finally (iv) disturb the field frequency lock.

The minimum delay, t_e, that must be left after a gradient pulse prior to signal acquisition or the application of a refocussing rf pulse is determined by the time required for the dissipation of the eddy currents. If acquisition is begun in the presence of eddy currents, the resulting spectrum will have a combination of phase and amplitude distortions.[105] Further, for example, if the eddy currents from the first gradient pulse extends into the second transverse evolution period in a Hahn-echo sequence, then the integral of magnetic gradients during the second evolution period will not equal, resulting in an effective gradient pulse mismatch (see Section 7.8) and both the echo time and amplitude may be affected.[106–109]

Although there have been attempts to map and characterise eddy currents including MRI-based techniques,[110–114] to determine if eddy-current effects are significant, a simple measurement can be performed on a sample with a diffusion coefficient lower than that which can be detected with the experimental parameters in question and for which true diffractive peaks cannot occur (e.g., a very large monodisperse polymer such as high MW polydimethylsiloxane (MW $\sim$ 700,000) has a diffusion coefficient below $10^{-15}\,\mathrm{m^2s^{-1}}$).[115–117] If no attenuation is observed, the presence of artefacts can be excluded. Another simple way to determine if eddy-current effects are present, and in particular to determine t_e, is to use the pulse sequence shown in Figure 7.10.[107] Some example spectra acquired using this pulse sequence are shown in Figure 7.11.

Simplistically, assuming that the gradient pulses have infinitely fast rise times but with exponential falls that represent the eddy currents, and using the method outlined in Section 2.3.2, the echo signal attenuation in the Hahn-echo PGSE is given by[118]

$$E(g) = \exp\left(-\gamma^2 g^2 D\left\{\delta^2(\Delta - \delta/3) + f(t)\right\}\right), \tag{7.16}$$

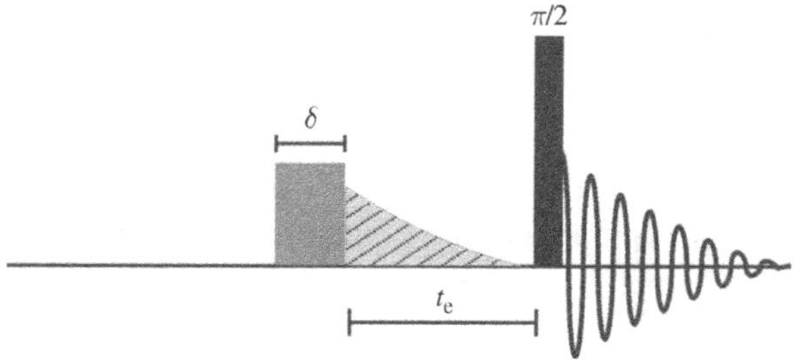

Figure 7.10 A simple pulse sequence for characterising eddy current dissipation. Spectra are acquired with successively shorter t_e delays to determine the minimum time required for the eddy current effects to decay. Some example spectra are shown in Figure 7.11.

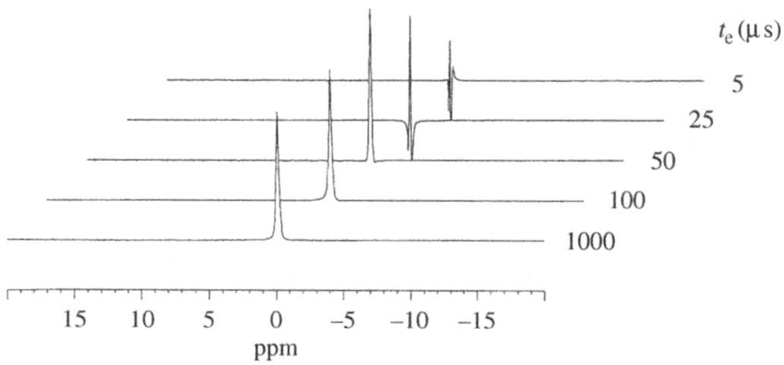

Figure 7.11 Experimental spectra acquired using a sample of $^{13}CCl_4$ and the pulse sequence given in Figure 7.10 for various values of t_e. The gradient pulse used had a duration of 1 ms and a strength of 0.45 T m^{-1}. Eddy-current effects result in the spectra appearing to be badly distorted when $t_e \leq 100$ μs. From Price.[118]

where

$$\begin{aligned} f(t) =& \frac{1}{k}\left(2\delta\Delta - \delta^2 + \delta^2 e^{-k(\Delta-\delta)} + 4\delta e^{-k(\tau-t_1-\delta)}\left[\tau - \Delta - t_1 - \frac{\delta}{2}\right]\right) \\ &+ \frac{1}{k^2}\left(\Delta - 2\delta + 4\tau e^{-2k(\tau-t_1-\delta)} + 4[\tau - \Delta - t_1]e^{-k(\tau-t_1-\delta)}\right. \\ &\left.+[2\tau - \Delta - t_1]\left(e^{-2k(\Delta-\delta)} - 4e^{-k(\Delta+\tau-t_1-2\delta)}\right)\right) \\ &+ \frac{1}{2k^3}\left(8e^{-k(\tau+\Delta-t_1-2\delta)} + 8e^{-k(3\tau-\Delta-2t_1-2\delta)} - 4e^{-k(2\tau-t_1-2\delta)}\right. \\ &\left.-8e^{-2k(\tau-t_1-\delta)} - e^{-2k(\Delta-\delta)} - e^{-2k(2\tau-\Delta-t_1-\delta)} - 1\right), \end{aligned}$$

where k is the exponential rate constant. This analysis is simplistic both in the form of the eddy currents and also as it does not consider the effects of the eddy currents on signal acquisition. It has been noted that the downward curvature in a plot of $\ln(E)$ versus $\gamma^2 g^2 \delta^2(\Delta-\delta/3)$ in a simple freely diffusing system is indicative of eddy-current effects.[108]

In addition to the direct eddy-current effects, gradient pulses can have serious effects on the stability of the main magnetic field. The final result is that the main field may be caused to oscillate or at least shift from its normal value (i.e., a time-dependent but spatially invariant B_0 shift).[119] If such B_0 shifts continue through acquisition, 'ringing' will be apparent in the observed spectrum.

7.7.2 Solutions

7.7.2.1 Hardware solutions

Ideally, well-designed shielded gradient coils (see Section 5.2) will generate negligible eddy currents and require no further experimental adjustments. However, this is usually not sufficient and additional means for dealing with eddy currents are required. Another commonly used approach is termed 'pre-emphasis', in which, according to Lenz's law, the sign of the fields generated by the eddy currents will be opposed to the changes which induced them (see Figure 7.12). Thus, the current at the leading and tailing edges of the gradient pulses is overdriven and in this way the coils self-compensate for the induced eddy currents. This is generally performed by adding small exponential corrections of different amplitude and time constants to the desired current waveform base.[119–122] For example, the out-of-phase components of FIDs detected after a gradient pulse are used to interactively set the pre-emphasis variables[123] or imaging-based mapping of eddy-current fields to determine the eddy-current compensation parameters.[112] But care must be taken in its implementation since additional eddy currents can arise out of the pre-emphasis procedure.[103] In performing pre-emphasis, the difference between the desired and the measured

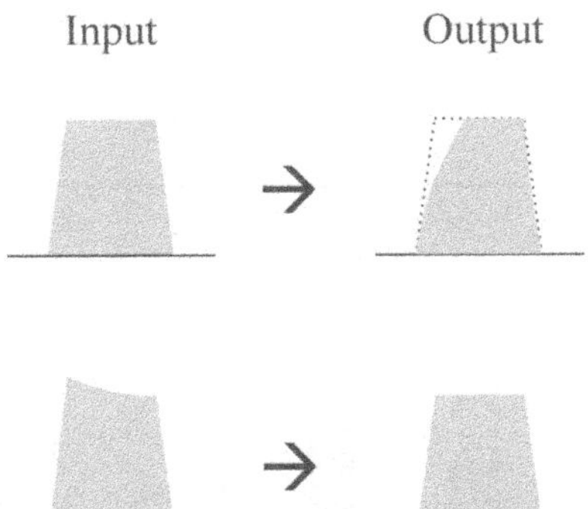

Figure 7.12 A conceptual idea of the pre-emphasis procedure. Ideally the input waveform (i.e., current pulse; top left) into the gradient coil would produce a gradient pulse of the same shape. However, due to the generation of eddy currents the resulting gradient waveform is distorted (top right). A solution is to shape the input waveform to counteract the eddy current effects (bottom left) and thereby produce the desired gradient shape.

gradient waveform indicates the distortion due to the eddy currents. The pre-emphasis time constants are determined using an iterative approach with the sequence shown in Figure 7.10. In practice, pre-emphasis is experimentally complicated and the method is imperfect since the spatial distribution of the fields produced by the eddy currents in the surrounding metal and those produced by the gradient coils are not identical,[124] nor does it solve the B_0 shift problem and can even exacerbate it.[125] Nevertheless, pre-emphasis is commonly used even with shielded coil systems to improve performance.

Since eddy-current severity is proportional to dI/dt, a means of reducing eddy-current effects is to slow the rise and fall times of the gradient pulses.[126,127] The echo-attenuation functions have been calculated for Hahn-echo PGSE sequences using nearly rectangular and sinusoidal gradient pulses (see Figure 7.13) and for the nearly rectangular pulses are found to be of the form[128]

$$E(g) = \exp\left(-\gamma^2 g^2 D\left\{\delta^2(\Delta - \delta/3) + f(t)\right\}\right), \tag{7.17}$$

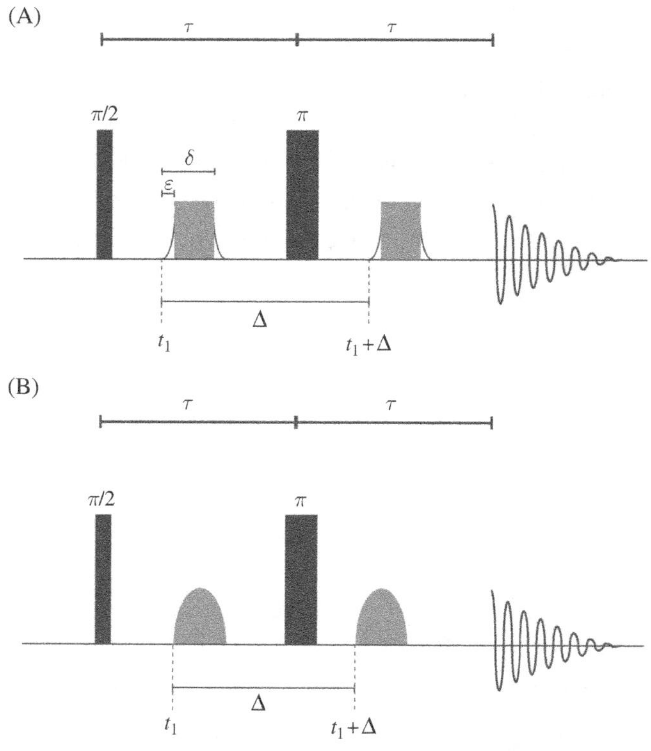

Figure 7.13 The Stejskal and Tanner pulse sequence with (A) approximately rectangular gradient pulses; ε represents the rise and fall times of the gradient pulses and (B) sinusoidal gradient pulses.

where $f(t)$ is defined in Table 7.1. The attenuation functions for sequences containing sinusoidal gradient pulses are given in Table 7.2.

Table 7.1 *The f(t) term in Eq. (7.17) for various nearly rectangular gradient pulse shapes in the Hahn-echo sequence*

Gradient pulse shape	$f(t)$
Identical rectangular	0
Exponential rise and fall	$-2/k^2(1/k-\delta)+4/(ke^{\varepsilon k})(\varepsilon\delta-1/k^2-\varepsilon^2/2)$ $+2/(k^2e^{2\varepsilon k})\times(\delta+1/k)$
Exponential rise and fall with overshoot and undershoot	$(-8\delta/k^2-12/k^3+e^{-\varepsilon k}(2\varepsilon^2-4\varepsilon^2\delta+6\varepsilon^2/k$ $-8\delta\varepsilon/k+12\varepsilon/k^2-8\delta/k^2+12/k^3))/g$ $+4e^{-\varepsilon k}(\varepsilon^2-\varepsilon\delta-\delta/k)/(g^2k)+e^{-2\varepsilon k}(2\varepsilon^2\delta$ $+6\varepsilon^2/k+4\varepsilon\delta/k+6\varepsilon/k^2+2\delta/k^2+3/k^3)/g^2$ $+(2\delta-3/k)/(g^2k^2)$
Sine rise and fall	$4\varepsilon^2(\varepsilon-2\delta)/\pi+8\varepsilon^2(\varepsilon-3\delta)/\pi^2-64\varepsilon^3/\pi^3$
Ramped rise and fall	$\varepsilon^3/30-\delta\varepsilon^2/6$
Mismatched[a]	$\varepsilon^2(2\tau-t_1-\Delta-\delta-2\varepsilon/3)$

Note: k is the time constant of the exponential function describing the gradient shape.
[a] In this sequence only one of the pair of rectangular gradient pulses is extended by ε.
Source: From Price and Kuchel.[128]

Table 7.2 *Attenuation equations for various trigonometric gradient pulse shapes in the Hahn-echo sequence*

Gradient pulse shape	E
$\sin(t)$	$\exp\left(-\gamma^2g^2D\delta^2(4\Delta-\delta)/\pi^2\right)$
$\sin(t)^2$	$\exp\left(-\gamma^2g^2D\delta^2\left(\Delta/4+\delta/12+5\delta/(4\pi)^2\right)\right)$
$1-\cos(2\pi t/\tau)$	$\exp\left(-\gamma^2g^2D\tau^3\left(\frac{2}{3}+\frac{5}{4\pi^2}\right)\right)$

Source: From Stejskal and Tanner,[78] Price and Kuchel,[128] Gross and Kosfeld,[129] and Le Bihan.[130]

The results show that providing the pulses are nearly rectangular, the precise shape is unimportant as long as the 'area' (i.e., δg) of each pulse is equal to that of the ideal rectangular pulse. Shaped pulses result in a loss of gradient intensity in proportion to their integral. As noted by Galvosas and Stallmach,[76] the shape of the ideal gradient pulse depends on the capabilities of the gradient generation system. The optimised shape has exponential rising and falling edges and gives the maximum gradient efficiency available within the bounds of stable feedback, above which the feedback loop of the generation system would no longer be reliable and gradient mismatch may result.

Dynamic shimming and B_0 compensation by pulsing a B_0 coil (typically z and z^2 shim coils) simultaneously with the gradient pulse can be used to reduce B_0 shift effects.[121,125]

7.7.2.2 *Pulse sequences*

Delaying the acquisition until the eddy currents have dissipated. As noted in Section 7.7.1, it is important to delay acquisition and refocussing rf pulses until the eddy currents have dissipated. Early attempts included appending a train of π rf pulses to an STE sequence to refocus the stimulated echo so as to delay the acquisition until after the eddy currents had subsided.[131] However, since the magnetisation is transverse during this period it is susceptible to transverse relaxation, J-modulation and phase distortions from the eddy currents. The well-known STE-based LED (*l*ongitudinal-*e*ddy-current-*d*elay) pulse sequence was proposed in 1991,[132] and its bipolar gradient variant (BPP–LED) sequence in 1995 (see Figure 7.14 and below).[133] The LED approach is useful when the T_1's of the species in question are longer than t_e. The echo attenuation for the BPP–LED is[133]

$$E = \exp\left(-\gamma^2 D g^2 \delta^2 \left(\Delta - \delta/3 - \tau_g/2\right)\right), \tag{7.18}$$

and τ_g is defined in the figure.

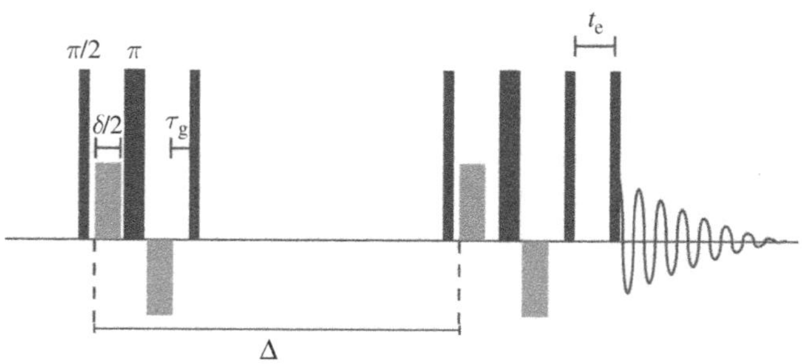

Figure 7.14 The BPP–LED pulse sequence incorporating bipolar gradient pulses.[133] The self-compensating effects of the bipolar gradient pulse 'sandwiches' (i.e., $g - \pi - (-g)$) largely cancel the generation of eddy currents. The two π pulses in the bipolar-gradient pulse sandwiches have the beneficial effect of reducing the active volume of the sample to the region of homogeneous rf. t_e denotes the delay included before acquisition for the eddy currents to dissipate.

A problem common to both the LED sequence and to the STE sequences is that extensive phase cycling is required. For the LED sequence at least 64 steps are required – although this can be reduced by inclusion of a homospoil pulse. However, with modern commercially available actively shielded probes, the time required for eddy-current dissipation when using moderate strength gradient pulses (e.g., 0.5 T m^{-1}) can be quite short (~100 μs) and thus the extra complexity involved in sequences such as LED is often not required.

Bipolar gradients. A more elegant and proactive solution than waiting for the effects of the eddy currents to dissipate, although requiring more sophisticated gradient control, is the use of bipolar (self-compensating) gradient pulses[133,134] as shown in the BPP-LED sequence in Figure 7.14. The two 'halves' of each bipolar gradient pulse are of opposite sign, but the π rf pulse has the effect of negating the phase change by the first 'half' pulse, thus a bipolar gradient is equivalent to a gradient pulse of duration δ with the polarity being equal to that of the second 'half'(i.e., $\delta/2$) of the bipolar pulse pair. Since eddy currents typically have settling times of the order of tens of milliseconds, the eddy currents generated by the first pulse will be approximately cancelled by the effects of the second gradient pulse which is of opposite polarity. An additional advantage in high-resolution studies is that BPP have much less effect on the lock since the lock (e.g., ^{2}H) resonances are unaffected by the π pulse; the two gradient pulses in each BPP behave as a gradient echo and so the lock is functional immediately at the end of the BPP instead of the time required for resumption of the lock function depending on the deuterium relaxation rate.

Prepulses. A means of decreasing eddy-current problems is to place additional, appropriately spaced (i.e., Δ apart), gradient pulses of the same amplitude prior to the 'start' of the rf pulse sequence.[106,108,117,135,136] However, it has been noted that it may be prudent to decrease the amplitude of the prepulses as the amplitude of the (diffusion) gradient pulses are ramped up to keep the net energy dissipated constant in each scan.[66] Which approach is better will depend strongly on the experimental circumstances.

7.7.2.3 Postprocessing

Post-processing methods aim to correct the measured FID for eddy-current effects. The simplest post-processing scheme is to measure the phase of an on-resonance signal from a single component spectrum as it evolves following a gradient pulse.[125] This 'reference' phase angle evolution can then be subtracted from all subsequent spectra obtained under the same conditions to remove the effect of B_0 variation.[137] Other techniques for removing the B_0-based distortions include wavelet transforms[138] and, if a suitable reference signal is available in the spectrum, reference deconvolution[139–141] to produce high-resolution spectra.

Importantly, post-processing techniques cannot correct for the effects of eddy currents during the sequence, which may be a problem in itself (perhaps resulting in anomalous echo intensities), but merely cosmetically correct the phase distortions in the acquired signal.

7.8 Gradient mismatch and sample movement

7.8.1 Origin and symptoms

Stable and perfectly reproducible gradient pulses are crucial for accurate PGSE measurements.[106,107,117,142] The defocussing and refocussing effects of the gradient pulses in the pulse sequence must be very finely matched and the gradient magnitude needed to measure a dynamic displacement n orders of magnitude smaller than the sample dimensions will result in a dephasing of order 10^n cycles along the sample; consequently, the refocussing must be accurate to within a few degrees.[143] Thus, at 10^{-8} m resolution, the matching required in the gradient pulse areas is on the order of 10 ppm.[1] In practice, the gradient amplifier may be incapable of delivering perfectly reproducible noise-free gradient pulses in quick succession and this may be further complicated by heating of the gradient coils and consequent changes in coil resistance during a sequence.[136] Due to the additional demands placed on a current amplifier in generating BPP, it is likely that gradient mismatch will be exacerbated in BPP sequences, especially with large amplitude gradient pulses. As noted above, eddy currents can also result in effectively mismatched gradient pulses. It has also been observed that the refocussing rf pulse (e.g., the π pulse in the Hahn-PGSE sequence) can induce a signal in the gradient coils which in turn elicits a small current pulse from the current amplifier.[144]

The echo attenuation for the Hahn-PGSE sequence, where the second gradient pulse is mismatched by a duration ε longer than the first pulse (see Figure 7.13A), is given by Eq. (7.17) and the expression in Table 7.1. The mismatch introduces a τ and t_1 dependence to the equation and ε enters into the equation in second order.[106] However, whilst the mismatch may have only a small effect on the attenuation due to the diffusion, even extremely small mismatches, too small to measure using conventional techniques (e.g., by oscilloscope), will result in a severe loss in echo signal intensity due to residual phase twist as will be further examined below.

Mechanical stability is also extremely important since movements on the order of 10 nm will restrict diffusion measurements to $D \geq 10^{-15}$ m^{2}s^{-1}.[117,142,145,146] Sample movement is similar to flow in that, in the case of a rigid sample, all spins receive an equal phase shift and it can result in greatly increased echo attenuation.[146]

The possibility of gradient mismatch and vibration artefacts can be checked by performing a measurement under the same conditions to be used experimentally with a very large monodisperse polymer with a diffusion coefficient below the limits of measurability as described above. If no attenuation is observed, the presence of artefacts can be excluded. However, this test does not account for independent movement of the sample with respect to the sample tube as might occur with an incompletely compacted powder sample (e.g., zeolite).[146]

The effects of the gradient pulse mismatch and vibration/sample movement in the PGSE experiment for the case of a freely diffusing species can be theoretically considered by rewriting the SGP attenuation expression (Eq. (2.35)) in terms of $\mathbf{R}$ and including a phase-shift, Φ, due to the effects of a gradient mismatch, $\Delta\mathbf{q}$, and sample movement, $\Delta\mathbf{r}$, between the first and second gradient pulses (in $\Delta\mathbf{q}$ and $\Delta\mathbf{r}$, 'Δ' denotes difference) obtaining[142]

$$E(\mathbf{q}, \Delta) = \int \rho(\mathbf{r}_0) \int \bar{P}(\mathbf{R}, \Delta) e^{i[2\pi\mathbf{q}\cdot\mathbf{R}+\phi]} d\mathbf{R}\, d\mathbf{r}_0. \tag{7.19}$$

Assuming that the gradient is directed along the z-direction and that $\Delta\mathbf{q}$ is parallel to $\mathbf{q}$ (i.e., a magnitude mismatch), the phase term can be expressed as

$$2\pi\mathbf{q}\cdot\mathbf{R} + \phi = 2\pi[qZ + (q + \Delta q)\Delta z + \Delta q z_0], \tag{7.20}$$

where z_0, Z and Δz are the z-components of $\mathbf{r}_0$, $\mathbf{R}$ and $\Delta\mathbf{r}$, respectively. Thus, Eq. (7.19) becomes

$$E(q, \Delta) = \underbrace{\int \bar{P}(Z, \Delta) e^{i2\pi qZ} dZ}_{E_{\mathrm{Diff}}(q,\Delta)} \underbrace{\left\{e^{i2\pi(q+\Delta q)\Delta z}\right\}}_{\Phi(\Delta z)} \underbrace{\int \rho(z_0) e^{i2\pi\Delta q z_0} dz_0}_{E_{\mathrm{Phase}}(\Delta q)} \tag{7.21}$$

The first term, $E_{\mathrm{Diff}}(q, \Delta)$, is merely the usual attenuation due to diffusion (Eq. (2.44)). The second term, $\Phi(\Delta z)$, is the residual phase-shift due to vibration/sample movement and thus has the same form as Φ_{Flow} (Eq. (2.28)). This complex valued function produces net-phase-shifts and is likely to produce complicated phase behaviour through the series of spectra in a PGSE measurement since Δz is likely to be a function of q. The third term (i.e., $E_{\mathrm{Phase}}(\Delta q)$) is the residual phase-twist resulting from the gradient pulse mismatch and is given by Eq. (2.17).

7.8.2 Solutions

The simplest solution to sample vibration problems is to securely locate the sample with respect to the gradient coils and, if the sample is a powder, to completely compact it. Gradient mismatch problems are reduced by using a better current amplifier and to provide temperature control of the gradient coil. It has also been noted that the hum from the mains power supply can be a source of gradient pulse mismatch and consequently it can be beneficial to trigger the pulse sequence to the phase of the power supply.[76]

We now consider what other approaches can be undertaken to minimise the problems associated with gradient mismatch and sample vibration that do not involve changing the spectrometer hardware by considering the SGP model developed in the previous section (Eq. (7.21)). To begin with, the phase shift term, $\Phi(\Delta z)$, is position-independent as it depends on Δz and can be removed by individually phase correcting each spectrum or by computing the absolute value spectrum.[142]

If Δq is taken as a fixed proportion of q (not an unreasonable assumption), then $E_{\text{Phase}}(\Delta q)$ is a damped oscillation with the degree of damping and the periodicity being exquisitely sensitive to the degree of mismatch. If Δq does not become too large, the residual phase effects will appear as 'unphysical' downward curvature in PGSE attenuation plots. In extreme cases, it can even produce artefactual 'diffraction' peaks.[117] $E_{\text{Phase}}(\Delta q)$ cannot be removed by phase correction and computing absolute value spectra is only a cosmetic improvement. Thus, we are left with the possibilities of: (i) empirically matching the gradient pulse pairs; (ii) prefixing the PGSE with a number of Δ-spaced gradient prepulses, as discussed above, which may also alleviate motional disturbances during the encoding and decoding gradient pulses, (iii) using shaped gradient pulses (see above) which are easier for the amplifier electronics to generate reproducibly and are also less likely to produce vibration of the sample/probe,[117] or (iv) to use the imaging-based MASSEY (*m*odulus *a*ddition using *s*patially *s*eparated *e*cho spectroscop*y*) sequence[142] to remove the phase-twist as discussed below.

The empirical matching of gradient pulses is tedious and difficult due to the very high precision needed and the correction time depends on g, δ and Δ.[109] Further, since the eddy-current-generated fields may not be even in the same direction as the applied gradients the empirical matching of gradient pulses may offer no solution.[108]

The MASSEY sequence[142] for minimising phase instability in very high-gradient NMR spectroscopy in which a read gradient, g_r, is incorporated into the Hahn-echo PGSE sequence is depicted in Figure 7.15. The addition of g_r allows the restoration of spatially-dependent phase shifts as caused by a gradient pulse mismatch. Importantly, the same gradient coil is used for both generating the gradient pulses and also the read-gradient.

The requirements for spatially separating the phase shifts using the imaging process with the read gradient are as follows. Suppose that N points in the k-space dimension are sampled with a sampling interval of T. The spectral separation of adjacent pixels in k-space will then be $1/NT$ which corresponds to a spatial separation of $2\pi/(\gamma g_r NT)$. Thus, we require that this be less than or equal to the 'wavelength' of the phase twist so that the phase modulation is well resolved. Thus,

$$g_r > \frac{2\pi\Delta q}{\gamma NT}. \tag{7.22}$$

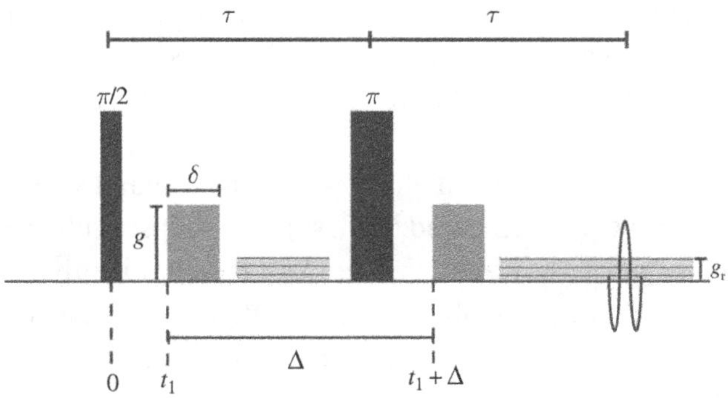

Figure 7.15 The Hahn-echo-based MASSEY sequence which incorporates a read gradient, g_r, for removing phase instability.[142]

It is desirable to keep the acquisition time as long as possible providing that the pixel separation is larger than the homogeneous linewidth ($\Delta\nu_{1/2} = 1/(\pi T_2)$), which sets the lower bandwidth limit as

$$\frac{1}{T} \geq \frac{N}{\pi T_2}. \tag{7.23}$$

The phase twist caused by pulse mismatch is resolved by Fourier transformation of the whole echo with respect to $k\,(=(2\pi)^{-1}\gamma g_r t)$ (k covers both positive and negative values)

$$E(q,\Delta,k) = E_{\mathrm{Diff}}(q,\Delta)e^{i2\pi(q+\Delta q)\Delta z}\int \rho(z_0)e^{i2\pi\Delta q z_0}e^{i2\pi k z_0}dz_0, \tag{7.24}$$

which gives

$$E(q,\Delta,z) = E_{\mathrm{Diff}}(q,\Delta)\underbrace{e^{i2\pi(q+\Delta q)\Delta z}}_{\substack{\text{0th order}\\ \text{phase shift}}}\rho(z)\underbrace{e^{i2\pi\Delta q z_0}}_{\substack{\text{1st order}\\ \text{phase shift}}}. \tag{7.25}$$

This is the 1D projection image of the echo (this is related to the 1D imaging calibration method given in Section 6.4.2). $E_{\mathrm{Diff}}(q,\Delta)$ can be recovered because the phase shifts are resolved in Eq. (7.25). If the S/N is high, the spectrum can be resolved by autophasing; while for poorer S/N, the absolute value of the spectrum must be taken, producing $E_{\mathrm{Diff}}(q,\Delta)\rho(z)$. The signal averaging using absolute value spectra is, however, less efficient due to the coaddition of noise and the absence of phase-cycling.[142]

When $k=-\Delta q$ Eq. (7.24) reduces to

$$E(q,\Delta,k) = E_{\mathrm{Diff}}(q,\Delta)e^{i2\pi(q+\Delta q)\Delta z}, \tag{7.26}$$

and thus at $t=-\Delta\, 2\pi q/(\gamma g_r)$, with respect to the echo centre, the phase-twisted echo will cause a coherent superposition (whether this is before or after the echo centre depends on the sign of the mismatch). Since $E_{\text{Diff}}(q, \Delta)$ can be recovered, it is possible to perform signal averaging even though Δq and Δz may fluctuate between scans. Since t will vary as Δq fluctuates the phase-twist analysis is performed after every scan.

Another negative aspect of this method is that since signal acquisition occurs in the presence of a gradient, this method is not suitable for use with spectra containing more than one resonance; however, the serious gradient disturbances that warrant the use of MASSEY are normally only associated with measurements of large slowly diffusing species (e.g., polymers) and so spectral resolution is less likely to be an issue.

Based on the MASSEY approach, Galvosas *et al.*[76,147] developed a procedure for automatically correcting for gradient pulse mismatch in which the strength of the correcting (i.e., read) gradient is adjusted on the basis of convoluting the spin-echo signals obtained with and without gradients to generate the auto-correlation function of the NMR signal at the original position with that shifted by the mismatched gradient pulses, from which the maximum position can be determined more accurately than directly from the NMR signal. For cases where chemical shift is required, a small read gradient is used earlier in the sequence but not during the acquisition of the FID.

7.9 Cross relaxation

Another complication is that of intermolecular nuclear Overhauser effects when performing diffusion measurements of macromolecular systems using STE-based sequences.[148–150] The problem can be understood as follows. Consider the case of measuring the diffusion of water in a macromolecule solution. After the first $\pi/2$ pulse, both the macromolecule and water magnetisations are in the x-y plane. For simplicity, we assume that the T_2 relaxation time of the macromolecule is much less than that of the water so that by the end of the τ_1 period the macromolecules are fully relaxed whereas the relaxation of the water magnetisation is insignificant. After the application of the second $\pi/2$ pulse the z magnetisation of the macromolecule will be zero since it was entirely aligned along the z-axis prior to the pulse. For the water magnetisation, the situation is entirely different, after the pulse the water magnetisation is proportional to $\cos(qz)$. However, as qz ranges over many periods, the net z magnetisation over the sample is zero. Thus, the local normalised deviation from equilibrium in the macromolecule phase will be -1 and for the water phase $\cos(qz)-1$. Thus, during τ_2 cross-relaxation results from the equilibrium magnetisation differences in both phases. Consequently, the cross-relaxation rate will depend on q. Equations have been derived to account for this cross-relaxation in a two-phase system.[148] If significant cross-relaxation occurs, it can affect the measured signal intensities thereby complicating diffusion measurements. Under limited conditions, it is possible to determine the exchange parameters to allow D to be calculated correctly. It has been pointed out that providing the diffusion experiment is performed by varying the gradient amplitude (as is usually the case), single exponential decays will be observed in the presence of cross-relaxation or exchange – although the signal amplitude may be severely decreased.[150] However, the variation of the signal attenuation with Δ becomes non-exponential which could be misinterpreted for restricted diffusion.

Although such cross-relaxation effects are generally seen as sources of experimental artefacts, under some circumstances it can be put to good use such as epitope mapping of ligand-receptor interactions.[151] Importantly, the problem of cross-relaxation does not apply to Hahn-echo-based sequence.

In a study of the diffusion of hydrogen-bond molecular capsules, it was found that the choice of PGSE sequence (i.e., SE, STE, LED or BPP–LED) can have a dramatic effect on the observed diffusion behaviour of the exchanging water molecules.[152] The SE and STE sequences resulted in single exponential behaviour of the water resonance, but bi-exponential behaviour was obtained with the LED and BPP–LED sequences. Shortening the LED delay in the LED and BPP–LED sequences resulted in the removal of the (additional) slow diffusing component. It was concluded that in systems where there is a possibility of NOE between nuclei with considerably different diffusion coefficients, care should be taken in the choice of sequence.

7.10 Homonuclear scalar couplings

The effects of the evolution of homonuclear scalar couplings[153] can result in poor S/N and peak distortion which considerably complicates the acquisition of good PGSE data and its subsequent analysis.[145] During the echo sequences, the peak phases and thus the refocussing depend on the magnitude of the spin (or scalar) coupling constant, J (Hz). Further, the rf pulses exchange the spin states of the coupled nuclei. For a coupled pair of nuclei, echo maxima occur when (see Figure 2.6, for the STE sequence $\tau = \tau_1$)

$$\tau = n/J, \tag{7.27}$$

and negative maxima occur when

$$\tau = \frac{n}{2J}, \tag{7.28}$$

where n is an integer.

In general, echo spectra from J-coupled spin systems cannot be phased into purely absorption mode spectra. However, if integration is performed it will correctly predict the diffusional decay, but with greatly reduced S/N. Thus, when performing a PGSE experiment, it is important to consider the pulse sequence delays with respect to J to obtain good S/N. Thus, providing τ_1 is kept small ($\ll 1/J$), the STE sequence is preferable to SE for minimising J-modulation problems, although cogent choice of echo timings with respect to the coupling constants in the system may help to minimise J-modulation effects. OGSE sequences (see Section 2.4) allow more flexibility in the setting of the echo time with respect to the couplings to minimise J-modulation effects.[64] But choice of echo delays is a very restrictive and imperfect solution; further, in most real samples there may be more than one coupling constant and thus no single choice of echo time will provide high quality spectra. The most general solution is to incorporate anti-phase magnetisation purging elements at the end of the PGSE sequence such as a spin-lock[154] or a chirp based z-filter.[155]

A 'postprocessing' solution is to try to phase such a spectrum; to resemble absorption mode or take the absolute value of the spectrum; however, as noted in Section 6.5, such an approach is only of cosmetic value and has quite negative consequences.

References

1. P. T. Callaghan, M. E. Komlosh, and M. Nydén, High Magnetic Field Gradient PGSE NMR in the Presence of a Large Polarizing Field. *J. Magn. Reson.* **133** (1998), 177–82.
2. S. Meiboom and D. Gill, Modified Spin-Echo Method for Measuring Nuclear Relaxation Times. *Rev. Sci. Instrum.* **29** (1958), 688–91.
3. J. Simbrunner and G. Zieger, Analysis of Three-Component Composite 180° Pulses. *J. Magn. Reson. B* **106** (1995), 142–6.
4. M. Zweckstetter and T. A. Holak, An Adiabatic Multiple Spin-Echo Pulse Sequence: Removal of Systematic Errors Due to Pulse Imperfections and Off-Resonance Effects. *J. Magn. Reson.* **133** (1998), 134–47.
5. G. Suryan, Nuclear Magnetic Resonance and the Effect of the Methods of Observation. *Curr. Sci.* **6** (1949), 203–4.
6. N. Bloembergen and R. V. Pound, Radiation Damping in Magnetic Resonance Experiments. *Phys. Rev.* **95** (1954), 8–12.
7. A. Abragam, *The Principles of Nuclear Magnetism*. (Oxford: Clarendon Press, 1961).
8. X.-A. Mao and C.-H. Ye, Understanding Radiation Damping in a Simple Way. *Concepts Magn. Reson.* **9** (1997), 173–87.
9. V. V. Krishnan, Radiation Damping: Suryans' Line Broadening Revisited in High Resolution NMR. *Curr. Sci.* **74** (1998), 1049–53.
10. M. P. Augustine, Transient Properties of Radiation Damping. *Prog. NMR Spectrosc.* **40** (2002), 111–50.
11. J. Jeener, Dipolar Field and Radiation Damping: Collective Effects in Liquid-State NMR. In *Encyclopedia of Magnetic Resonance*, ed. D. M. Grant and R. K. Harris. vol. 9. (New York: Wiley, 2002), pp. 642–79.
12. G. Deville, M. Bernier, and J. M. Delrieux, NMR Multiple Echoes Observed in Solid ^{3}He. *Phys. Rev. B* **19** (1979), 5666–88.
13. H. T. Edzes, The Nuclear Magnetization as the Origin of Transient Changes in the Magnetic Field in Pulse NMR Experiments. *J. Magn. Reson.* **86** (1990), 293–303.

14. M. H. Levitt, Demagnetization Field Effects in Two-Dimensional Solution NMR. *Concepts Magn. Reson.* **8** (1996), 77–103.
15. A. Vlassenbroek, J. Jeener, and P. Broekaert, Macroscopic and Microscopic Fields in High-Resolution Liquid NMR. *J. Magn. Reson. A* **118** (1996), 234–46.
16. E. D. Minot, P. T. Callaghan, and N. Kaplan, Multiple Echoes, Multiple Quantum Coherence, and the Dipolar Field: Demonstrating the Significance of Higher Order Terms in the Equilibrium Density Matrix. *J. Magn. Reson.* **140** (1999), 200–5.
17. I. Ardelean and R. Kimmich, Principles and Unconventional Aspects of NMR Diffusometry. In *Annual Reports on NMR Spectroscopy*, ed. G. A. Webb. vol. 49. (London: Academic Press, 2003), pp. 43–115.
18. V. V. Krishnan, Radiation Damping in Microcoil NMR Probes. *J. Magn. Reson.* **179** (2006), 294–8.
19. V. V. Krishnan, Corrigendum to 'Radiation Damping in Micro-Coil NMR Probes'. *J. Magn. Reson.* **179** (2006) 294–8; **182** (2006), 178.
20. P. C. M. Van Zijl, M. O. Johnson, S. Mori, and R. E. Hurd, Magic-Angle-Gradient Double-Quantum-Filtered COSY. *J. Magn. Reson. A* **113** (1995), 265–70.
21. D. L. Mattiello, W. S. Warren, L. Mueller, and B. T. Farmer, II, Minimizing the Water Resonance in Biological NMR: Characterization and Suppression of Intermolecular Dipolar Interactions by Multiple-Axis Gradients. *J. Am. Chem. Soc.* **118** (1996), 3253–61.
22. W. S. Price and Y. Arata, The Manipulation of Water Relaxation and Water Suppression in Biological Systems Using the Water-PRESS Pulse Sequence. *J. Magn. Reson. B* **112** (1996), 190–2.
23. W. S. Price, K. Hayamizu, and Y. Arata, Optimization of the Water-PRESS Pulse Sequence and its Integration into Pulse Sequences for Studying Biological Molecules. *J. Magn. Reson.* **126** (1997), 256–65.
24. S. Y. Huang, C. Anklin, J. D. Walls, and Y.-Y. Lin, Sizable Concentration-Dependent Frequency Shifts in Solution NMR Using Sensitive Probes. *J. Am. Chem. Soc.* **126** (2004), 15936–7.
25. Y.-Y. Lin, N. Lisitza, S. Ahn, and W. S. Warren, Resurrection of Crushed Magnetization and Chaotic Dynamics in Solution NMR Spectroscopy. *Science* **290** (2000), 118–21.
26. S. Datta, S. Y. Huang, and Y.-Y. Lin, Understanding Spin Turbulence in Solution Magnetic Resonance Through Phase Space Dynamics and Instability. *Concepts Magn. Reson.* **28A** (2006), 410–21.
27. S. Y. Huang, Y.-Y. Lin, N. Lisitza, and W. S. Warren, Signal Interferences From Turbulent Spin Dynamics in Solution Nuclear Magnetic Resonance Spectroscopy. *J. Chem. Phys.* **116** (2002), 10325–37.
28. H. Barjat, G. P. Chadwick, G. A. Morris, and A. G. Swanson, The Behaviour of Multiplet Signals under 'Radiation Damping' Conditions. 1. Classical Effects. *J. Magn. Reson. A* **117** (1995), 109–12.
29. M. A. Connell, A. L. Davis, A. M. Kenwright, and G. A. Morris, NMR Measurements of Diffusion in Concentrated Samples: Avoiding Problems with Radiation Damping. *Anal. Bioanal. Chem.* **378** (2004), 1568–73.
30. A. Vlassenbroek, J. Jeener, and P. Broekaert, Radiation Damping in High Resolution Liquid NMR: A Simulation Study. *J. Chem. Phys.* **103** (1995), 5886–97.
31. W. S. Price, Water Signal Suppression in NMR Spectroscopy. In *Annual Reports on NMR Spectroscopy*, ed. G. A. Webb. vol. 38. (London: Academic Press, 1999), pp. 289–354.
32. W. S. Price and M. Wälchli, NMR Diffusion Measurements of Strong Signals: The PGSE-*Q*-Switch Experiment. *Magn. Reson. Chem.* **40** (2002), S128–32.

33. W. S. Price, P. Stilbs, B. Jönsson, and O. Söderman, Macroscopic Background Gradient and Radiation Damping Effects on High-Field PGSE NMR Diffusion Measurements. *J. Magn. Reson.* **150** (2001), 49–56.
34. J. C. Rodríguez, P. A. Jennings, and G. Melacini, Effect of Chemical Exchange on Radiation Damping in Aqueous Solutions of the Osmolyte Glycine. *J. Am. Chem. Soc.* **124** (2002), 6240–1.
35. C. Dalvit and J. M. Böhlen, Simultaneous Suppression of the H_2O Double-Quantum Signal and of the Radiation-Damping Effect in Double-Quantum Experiments. *J. Magn. Reson. B* **113** (1996), 195–200.
36. A. Böckmann and E. Guittet, Suppression of Radiation Damping During Selective Excitation of the Water Signal: The WANTED Sequence. *J. Biomol. NMR* **8** (1996), 87–92.
37. V. Sklenár, Suppression of Radiation Damping in Multidimensional NMR Experiments Using Magnetic Field Gradients. *J. Magn. Reson. A* **114** (1995), 132–5.
38. S. Zhang and D. G. Gorenstein, Suppression of Radiation Damping during Acquisition by Pulsed Field Gradients. *J. Magn. Reson. A* **118** (1996), 291–4.
39. H. Barjat, D. L. Mattiello, and R. Freeman, Suppression of Radiation Damping in High-Resolution NMR. *J. Magn. Reson.* **136** (1999), 114–17.
40. W. J. Goux, L. A. Verkruyse, and S. J. Salter, The Impact of Rayleigh-Benard Convection on NMR Pulsed Field Gradient Diffusion Measurements. *J. Magn. Reson.* **88** (1990), 609–14.
41. A. Jerschow and N. Müller, Suppression of Convection Artifacts in Stimulated-Echo Diffusion Experiments. Double-Stimulated-Echo Experiments. *J. Magn. Reson.* **125** (1997), 372–5.
42. N. Hedin and I. Furó, Temperature Imaging by ^{1}H NMR and Suppression of Convection in NMR Probes. *J. Magn. Reson.* **131** (1998), 126–30.
43. A. Jerschow and N. Müller, Convection Compensation in Gradient Enhanced Nuclear Magnetic Resonance Spectroscopy. *J. Magn. Reson.* **132** (1998), 13–18.
44. N. M. Loening and J. Keeler, Measurement of Convection and Temperature Profiles in Liquid Samples. *J. Magn. Reson.* **139** (1999), 334–41.
45. H. Y. Carr and E. M. Purcell, Effects of Diffusion on Free Precession in Nuclear Magnetic Resonance Experiments. *Phys. Rev.* **94** (1954), 630–8.
46. A. Jerschow, Thermal Convection Currents in NMR: Flow Profiles and Implications for Coherence Pathway Selection. *J. Magn. Reson.* **145** (2000), 125–31.
47. N. Hedin, T. Y. Yu, and I. Furó, Growth of $C_{12}E_{18}$ Micelles with Increasing Temperature. A Convection-Compensated PGSE NMR Study. *Langmuir* **16** (2000), 7548–50.
48. A. Mohorič and J. Stepišnik, Effect of Natural Convection in a Horizontally Oriented Cylinder on NMR Imaging of the Distribution of Diffusivity. *Phys. Rev. E* **62** (2000), 6628–35.
49. M. Nilsson and G. A. Morris, Improving Pulse Sequences for 3D DOSY: Convection Compensation. *J. Magn. Reson.* **177** (2005), 203–11.
50. J. Lounila, K. Oikarinen, P. Ingman, and J. Jokisaari, Effects of Thermal Convection on NMR and their Elimination by Sample Rotation. *J. Magn. Reson. A* **118** (1996), 50–4.
51. S. Augé, B. Amblard-Blondel, and M.-A. Delsuc, Investigation of the Diffusion Measurement Using PFG and Test of Robustness Against Experimental Conditions and Parameters. *J. Chim. Phys.* **96** (1999), 1559–65.
52. N. Esturau, F. Sánchez-Ferrando, J. A. Gavin, C. Roumestand, M.-A. Delsuc, and T. Parella, The Use of Sample Rotation for Minimizing Convection Effects in Self-Diffusion Measurements. *J. Magn. Reson.* **153** (2001), 48–55.

53. K. Hayamizu and W. S. Price, A New Type of Sample Tube for Reducing Convection Effects in PGSE-NMR Measurements of Self-Diffusion Coefficients of Liquid Samples. *J. Magn. Reson.* **167** (2004), 328–33.
54. A. Simorellis and P. F. Flynn, A PFG Experiment for Translational Diffusion Measurements in Low-Viscosity Solvents Containing Multiple Resonances. *J. Magn. Reson.* **170** (2004), 322–8.
55. R. J. Ogg, P. B. Kingsley, and J. S. Taylor, WET, a T_1- and B_1-Insensitive Water-Suppression Method for In Vivo Localized ^{1}H NMR Spectroscopy. *J. Magn. Reson. B* **104** (1994), 1–10.
56. S. H. Smallcombe, S. L. Patt, and P. A. Keifer, WET Solvent Suppression and its Applications to LC NMR and High-Resolution NMR Spectroscopy. *J. Magn. Reson. A* **117** (1995), 295–303.
57. P. T. Callaghan and A. A. Khrapitchev, Time-Dependent Velocities in Porous Media Dispersive Flow. *Magn. Reson. Imaging* **19** (2001), 301–5.
58. P. T. Callaghan and M. E. Komlosh, Locally Anisotropic Motion in a Macroscopically Isotropic System: Displacement Correlations Measured Using Double Pulsed Gradient Spin-Echo NMR. *Magn. Reson. Chem.* **40** (2002), S15–19.
59. Xu. Zhang, C.-G. Li, C.-H. Ye, and M.-L. Liu, Determination of Molecular Self-Diffusion Coefficient Using Multiple Spin-Echo NMR Spectroscopy with Removal of Convection and Background Gradient Artifacts. *Anal. Chem.* **73** (2001), 3528–34.
60. G. H. Sørland, J. G. Seland, J. Krane, and H. W. Anthonsen, Improved Convection Compensating Pulsed Field Gradient Spin-Echo and Stimulated-Echo Methods. *J. Magn. Reson.* **142** (2000), 323–5.
61. K. I. Momot and P. W. Kuchel, Convection-Compensating PGSE Experiment Incorporating Excitation-Sculpting Water Suppression (CONVEX). *J. Magn. Reson.* **169** (2004), 92–101.
62. K. I. Momot and P. W. Kuchel, PFG NMR Diffusion Experiments for Complex Systems. *Concepts Magn. Reson.* **28A** (2006), 249–69.
63. K. I. Momot and P. W. Kuchel, Convection-Compensating Diffusion Experiments with Phase-Sensitive Double-Quantum Filtering. *J. Magn. Reson.* **174** (2005), 229–36.
64. K. I. Momot, P. W. Kuchel, and B. E. Chapman, Acquisition of Pure-Phase Diffusion Spectra Using Oscillating-Gradient Spin Echo. *J. Magn. Reson.* **176** (2005), 151–9.
65. P. Damberg, J. Jarvet, and A. Gräslund, Accurate Measurement of Translational Diffusion Coefficients: A Practical Method to Account for Nonlinear Gradients. *J. Magn. Reson.* **148** (2001), 343–8.
66. G. A. Morris, Diffusion-Ordered Spectroscopy (DOSY). In *Encyclopedia of Nuclear Magnetic Resonance*, ed. D. M. Grant and R. K. Harris. vol. 9. (New York: Wiley, 2002), pp. 35–44.
67. M. Nilsson and G. A. Morris, Correction of Systematic Errors in CORE Processing of DOSY Data. *Magn. Reson. Chem.* **44** (2006), 655–60.
68. M. Nilsson and G. A. Morris, Improved DECRA Processing of DOSY Data: Correcting for Non-Uniform Field Gradients. *Magn. Reson. Chem.* **45** (2007), 656–60.
69. D. L. VanderHart, Magnetic Susceptibility & High Resolution NMR of Liquids & Solids. In *Encyclopedia of NMR*, ed. D. M. Grant and R. K. Harris. vol. 5. (New York: Wiley, 1996), pp. 2938–46.
70. J. A. Glasel and K. H. Lee, On the Interpretation of Water Nuclear Magnetic Resonance Relaxation Times in Heterogeneous Systems. *J. Am. Chem. Soc.* **96** (1974), 970–8.
71. P. Gillis and S. H. Koenig, Transverse Relaxation of Solvent Protons Induced by Magnetized Spheres: Application to Ferritin, Erythrocytes, and Magnetite. *Magn. Reson. Med.* **5** (1987), 323–45.

72. S. Majumdar and J. C. Gore, Studies of Diffusion in Random Fields Produced by Variations in Susceptibility. *J. Magn. Reson.* **78** (1988), 41–55.
73. I. Zupančič, Effect of the Background Gradients on PGSE NMR Diffusion Measurements. *Solid State Commun.* **65** (1988), 199–200.
74. J. Zhong and J. C. Gore, Studies of Restricted Diffusion in Heterogeneous Media Containing Variations in Susceptibility. *Magn. Reson. Med.* **19** (1991), 276–84.
75. L. L. Latour, L. Li, and C. H. Sotak, Improved PFG Stimulated-Echo Method for the Measurement of Diffusion in Inhomogeneous Fields. *J. Magn. Reson. B* **101** (1993), 72–7.
76. F. Stallmach and P. Galvosas, Spin Echo NMR Diffusion Studies. In *Annual Reports on NMR Spectroscopy*, ed. G. A. Webb. vol. 61. (New York: Elsevier, 2007), pp. 51–131.
77. G. Zheng and W. S. Price, Suppression of Background Gradients in (B_0 Gradient-Based) NMR Diffusion Experiments. *Concepts Magn. Reson.* **30A** (2007), 261–77.
78. E. O. Stejskal and J. E. Tanner, Spin Diffusion Measurements: Spin Echoes in the Presence of a Time-Dependent Field Gradient. *J. Chem. Phys.* **42** (1965), 288–92.
79. J. E. Tanner, Use of the Stimulated Echo in NMR Diffusion Studies. *J. Chem. Phys.* **52** (1970), 2523–26.
80. J. Kärger, H. Pfeifer, and S. Rudtsch, The Influence of Internal Magnetic Field Gradients on NMR Self-Diffusion Measurements of Molecules Adsorbed on Microporous Crystallites. *J. Magn. Reson.* **85** (1989), 381–7.
81. D. van Dusschoten, P. A. De Jager, and H. Van As, Flexible PFG NMR Desensitized for Susceptibility Artifacts, Using the PFG Multiple-Spin-Echo Sequence. *J. Magn. Reson. A* **112** (1995), 237–40.
82. I. Furó and H. Jóhannesson, Accurate Anisotropic Water-Diffusion Measurements in Liquid Crystals. *J. Magn. Reson. A* **119** (1996), 15–21.
83. J. S. Murday and R. M. Cotts, Self-Diffusion in Molten Lithium. *Z. Naturforsch.* **26a** (1971), 85–93.
84. W. D. Williams, E. F. W. Seymour, and R. M. Cotts, A Pulsed-Gradient Multiple-Spin-Echo NMR Technique for Measuring Diffusion in the Presence of Background Magnetic Field Gradients. *J. Magn. Reson.* **31** (1978), 271–82.
85. M. E. Moseley, J. Kucharczyk, H. S. Asgari, and D. Norman, Anisotropy in Diffusion-Weighted MRI. *Magn. Reson. Med.* **19** (1991), 321–6.
86. D. Le Bihan, Molecular Diffusion, Tissue Microdynamics and Microstructure. *NMR Biomed.* **8** (1995), 375–86.
87. J. D. Trudeau, D. W. Thomas, and J. Hawkins, The Effect of Inhomogeneous Sample Susceptibility on Measured Diffusion Anisotropy Using MRI Imaging. *J. Magn. Reson. B* **108** (1995), 22–30.
88. M. Neeman, J. P. Freyer, and L. O. Sillerud, Pulsed-Gradient Spin-Echo Diffusion Studies in NMR Imaging. Effects of the Imaging Gradients on the Determination of the Diffusion Coefficients. *J. Magn. Reson.* **90** (1990), 303–12.
89. J. Lian, D. S. Williams, and I. J. Lowe, Magnetic Resonance Imaging of Diffusion in the Presence of Background Gradients and Imaging of Background Gradients. *J. Magn. Reson. A* **106** (1994), 65–74.
90. R. F. Karlicek, Jr. and I. J. Lowe, A Modified Pulsed Gradient Technique for Measuring Diffusion in the Presence of Large Background Gradients. *J. Magn. Reson.* **37** (1980), 75–91.
91. J.-H. Zhong, R. P. Kennan, and J. C. Gore, Effects of Susceptibility Variations on NMR Measurements of Diffusion. *J. Magn. Reson.* **95** (1991), 267–80.
92. H. Lee, H. Shao, Y. Huang, and B. Kwak, Synthesis of MRI Contrast Agent by Coating Superparamagnetic Iron Oxide with Chitosan. *IEEE Trans. Magn.* **41** (2005), 4102–4.

93. Ad. Bax and R. Freeman, Enhanced NMR Resolution by Restricting the Effective Sample Volume. *J. Magn. Reson.* **37** (1980), 177–81.
94. X. Hong and W. T. Dixon, Measuring Diffusion in Inhomogeneous Systems in Imaging Mode Using Antisymmetric Sensitizing Gradients. *J. Magn. Reson.* **99** (1992), 561–70.
95. K. J. Packer, C. Rees, and D. J. Tomlinson, A Modification of the Pulsed Magnetic Field-Gradient Spin Echo Method of Studying Diffusion. *Mol. Phys.* **18** (1970), 421–3.
96. E. H. Sevilla and A. Sevilla, Effects of Restricted Diffusion in the Karlicek-Lowe NMR Alternating Pulsed Field Gradient Experiments. *J. Magn. Reson.* **79** (1988), 534–9.
97. R. M. Cotts, M. J. R. Hoch, T. Sun, and J. T. Markert, Pulsed Field Gradient Stimulated Echo Methods for Improved NMR Diffusion Measurements in Heterogeneous Systems. *J. Magn. Reson.* **83** (1989), 252–66.
98. G. H. Sørland, B. Hafskjold, and O. Herstad, A Stimulated-Echo Method for Diffusion Measurements in Heterogeneous Media Using Pulsed Field Gradients. *J. Magn. Reson.* **124** (1997), 172–6.
99. P. Z. Sun, J. G. Seland, and D. Cory, Background Gradient Suppression in Pulsed Gradient Stimulated Echo Measurements. *J. Magn. Reson.* **161** (2003), 168–73.
100. P. Galvosas, F. Stallmach, and J. Kärger, Background Gradient Suppression in Stimulated Echo NMR Diffusion Studies Using Magic Pulsed Field Gradient Ratios. *J. Magn. Reson.* **166** (2004), 164–73.
101. S. Vasenkov, P. Galvosas, O. Geier, N. Nestle, F. Stallmach, and J. Kärger, Determination of Genuine Diffusivities in Heterogeneous Media Using Stimulated Echo Pulsed Field Gradient NMR. *J. Magn. Reson.* **149** (2001), 228–33.
102. P. Z. Sun, S. A. Smith, and J. Zhou, Analysis of the Magic Asymmetric Gradient Stimulated Echo Sequence with Shaped Gradients. *J. Magn. Reson.* **171** (2004), 324–9.
103. S. Robertson, D. G. Hughes, Q. Liu, and P. S. Allen, Analysis of the Temporal and Spatial Dependence of the Eddy Currents in a 40-cm Bore Magnet. *Magn. Reson. Med.* **25** (1992), 158–66.
104. T. G. Reese, O. Heid, R. M. Weisskopf, and V. J. Wedeen, Reduction of Eddy-Current-Induced Distortion in Diffusion MRI Using a Twice-Refocused Spin Echo. *Magn. Reson. Med.* **49** (2003), 177–82.
105. M. Czisch, A. Ross, C. Cieslar, and T. A. Holak, Some Practical Aspects of B_0 Gradient Pulses. *J. Biomol. NMR* **7** (1996), 121–30.
106. M. I. Hrovat and C. G. Wade, NMR Pulsed-Gradient Diffusion Measurements I. Spin-Echo Stability and Gradient Calibration. *J. Magn. Reson.* **44** (1981), 62–75.
107. M. I. Hrovat and C. G. Wade, NMR Pulsed Gradient Diffusion Measurements. II. Residual Gradients and Lineshape Distortions. *J. Magn. Reson.* **45** (1981), 67–80.
108. E. von Meerwall and M. Kamat, Effect of Residual Field Gradients on Pulsed Gradient NMR Diffusion Measurements. *J. Magn. Reson.* **83** (1989), 309–23.
109. X. X. Zhu and P. M. Macdonald, Empirical Compensation Function for Eddy Current Effects in Pulsed Field Gradient Nuclear Magnetic Resonance Experiments. *Solid State Nucl. Magn. Reson.* **4** (1995), 217–27.
110. M. Burl and I. R. Young, Eddy Currents & Their Control. In *Encyclopedia of Nuclear Magnetic Resonance*, ed. D. M. Grant and R. K. Harris. vol. 3. (New York: Wiley, 1996), pp. 1841–6.
111. J. Schiff, H. Rotem, S. Stokar, and N. Kaplan, New Efficient Eddy-Field-Mapping Procedure (FAME). *J. Magn. Reson. B* **104** (1994), 73–6.
112. M. Terpstra, P. M. Andersen, and R. Gruetter, Localized Eddy Current Compensation Using Quantitative Field Mapping. *J. Magn. Reson.* **131** (1998), 139–43.

113. D. J. Goodyear, M. Shea, S. D. Beyea, N. J. Shah, and B. J. Balcom, Single Point Measurements of Magnetic Field Gradient Waveform. *J. Magn. Reson.* **163** (2003), 1–7.
114. V. Jellúš, J. C. Sharp, B. Tomanek, and P. Latta, An NMR Technique for Measurement of Magnetic Field Gradient Waveforms. *J. Magn. Reson.* **162** (2003), 189–97.
115. M. Appel and G. Fleischer, Investigation of the Chain Length Dependence of Self-Diffusion of Poly(dimethyl siloxane) and Poly(ethylene oxide) in the Melt with Pulsed Field Gradient NMR. *Macromolecules* **26** (1993), 5520–5.
116. W. Heink, J. Kärger, G. Seiffert, G. Fleischer, and J. Rauchfuss, PFG NMR Self-Diffusion Measurements with Large Field Gradients. *J. Magn. Reson. A* **114** (1995), 101–4.
117. W. S. Price, K. Hayamizu, H. Ide, and Y. Arata, Strategies for Diagnosing and Alleviating Artifactual Attenuation Associated with Large Gradient Pulses in PGSE NMR Diffusion Measurements. *J. Magn. Reson.* **139** (1999), 205–12.
118. W. S. Price, Pulsed Field Gradient NMR as a Tool for Studying Translational Diffusion, Part II. Experimental Aspects. *Concepts Magn. Reson.* **10** (1998), 197–237.
119. J. J. Van Vaals and A. H. Bergman, Optimization of Eddy-Current Compensation. *J. Magn. Reson.* **90** (1990), 52–70.
120. P. D. Majors, J. L. Blackley, S. A. Altobelli, A. Caprihan, and E. Fukushima, Eddy Current Compensation by Direct Field Detection and Digital Gradient Modification. *J. Magn. Reson.* **87** (1990), 548–53.
121. P. Jehenson, M. Westphal, and N. Schuff, Analytical Method for the Compensation of Eddy-Current Effects Induced by Pulsed Magnetic Field Gradients in NMR Systems. *J. Magn. Reson.* **90** (1990), 264–78.
122. R. E. Wysong, D. P. Madio, and I. J. Lowe, A Novel Eddy Current Compensation Scheme for Pulsed Gradient Systems. *Magn. Reson. Med.* **31** (1994), 572–5.
123. R. E. Wysong and I. J. Lowe, A Simple Method of Measuring Gradient Induced Eddy Currents to Set Compensation Networks. *Magn. Reson. Med.* **29** (1993), 119–21.
124. M. A. Morich, D. A. Lampman, W. R. Dannels, and F. T. D. Goldie, Exact Temporal Eddy Current Compensation in Magnetic Resonance Imaging Systems. *IEEE Trans. Med. Imaging* **7** (1988), 247–54.
125. P. Jehenson and A. Syrota, Correction of Distortions Due to the Pulsed Magnetic Field Gradient-Induced Shift in B_0 Field by Postprocessing. *Magn. Reson. Med.* **12** (1989), 253–6.
126. P. J. Basser, J. Mattiello, and D. Le Bihan, Estimation of the Effective Self-Diffusion Tensor from the NMR Spin Echo. *J. Magn. Reson. B* **103** (1994), 247–54.
127. O. Söderman and P. Stilbs, NMR Studies of Complex Surfactant Systems. *Prog. NMR Spectrosc.* **26** (1994), 445–82.
128. W. S. Price and P. W. Kuchel, Effect of Nonrectangular Field Gradient Pulses in the Stejskal and Tanner (Diffusion) Pulse Sequence. *J. Magn. Reson.* **94** (1991), 133–9.
129. B. Gross and R. Kosfeld, Anwendung der Spin-Echo-Methode bei der Messung der Selbstdiffusion. *Messtechnik* **7** (1969), 171–7.
130. D. Le Bihan, Molecular Diffusion Nuclear Magnetic Resonance Imaging. *Magn. Reson. Q.* **7** (1991), 1–30.
131. L. Griffiths, R. Horton, and T. Cosgrove, NMR Diffusion Measurements Using Refocused Three-Pulse Stimulated Echoes. *J. Magn. Reson.* **90** (1990), 254–63.
132. S. J. Gibbs and C. S. Johnson, Jr., A PFG NMR Experiment for Accurate Diffusion and Flow Studies in the Presence of Eddy Currents. *J. Magn. Reson.* **93** (1991), 395–402.
133. D. Wu, A. Chen, and C. S. Johnson, Jr., An Improved Diffusion-Ordered Spectroscopy Experiment Incorporating Bipolar-Gradient Pulses. *J. Magn. Reson. A* **115** (1995), 260–4.
134. G. Wider, V. Dötsch, and K. Wüthrich, Self-Compensating Pulsed Magnetic-Field Gradients for Short Recovery Times. *J. Magn. Reson. A* **108** (1994), 255–8.

135. K. F. Morris and C. S. Johnson, Jr., Diffusion-Ordered Two-Dimensional Nuclear Magnetic Resonance Spectroscopy. *J. Am. Chem. Soc.* **114** (1992), 3139–41.
136. R. M. Boerner and W. S. Woodward, A Computer-Controlled Bipolar Magnetic-Field-Gradient Driver for NMR Electrophoretic and Self-Diffusion Measurements. *J. Magn. Reson. A* **106** (1994), 195–202.
137. R. J. Ordridge and I. D. Cresshull, The Correction of Transient B_0 Field Shifts Following the Application of Pulsed Gradients by Phase Correction in the Time Domain. *J. Magn. Reson.* **69** (1986), 151–5.
138. D. Barache, J.-P. Antoine, and J.-M. Dereppe, The Continuous Wavelet Transform, an Analysis Tool for NMR Spectroscopy. *J. Magn. Reson.* **128** (1997), 1–11.
139. G. A. Morris, Compensation of Instrumental Imperfections by Deconvolution Using an Internal Reference Signal. *J. Magn. Reson.* **80** (1988), 547–52.
140. H. Barjat, G. A. Morris, S. Smart, A. G. Swanson, and S. C. R. Williams, High-Resolution Diffusion-Ordered 2D Spectroscopy (HR-DOSY) – A New Tool for the Analysis of Complex Mixtures. *J. Magn. Reson. B* **108** (1995), 170–2.
141. G. A. Morris, H. Barjat, and T. J. Horne, Reference Deconvolution Methods. *Prog. NMR Spectrosc.* **31** (1997), 197–257.
142. P. T. Callaghan, PGSE-MASSEY, a Sequence for Overcoming Phase Instability in Very-High-Gradient Spin-Echo NMR. *J. Magn. Reson.* **88** (1990), 493–500.
143. P. T. Callaghan, *Principles of Nuclear Magnetic Resonance Microscopy.* (Oxford: Clarendon Press, 1991).
144. P. T. Callaghan, K. W. Jolley, and C. M. Trotter, Stable and Accurate Spin Echoes in Pulsed Gradient NMR. *J. Magn. Reson.* **39** (1980), 525–7.
145. P. Stilbs, Fourier Transform Pulsed-Gradient Spin-Echo Studies of Molecular Diffusion. *Prog. NMR Spectrosc.* **19** (1987), 1–45.
146. N. K. Bär, J. Kärger, C. Krause, W. Schmitz, and G. Seiffert, Pitfalls in PFG NMR Self-Diffusion Measurements with Powder Samples. *J. Magn. Reson. A* **113** (1995), 278–80.
147. P. Galvosas, F. Stallmach, G. Seiffert, J. Kärger, U. Kaess, and G. Majer, Generation and Application of Ultra-High-Intensity Magnetic Field Gradient Pulses for NMR Spectroscopy. *J. Magn. Reson.* **151** (2001), 260–8.
148. L. J. C. Peschier, J. A. Bouwstra, J. De Bleyser, H. E. Junginger, and J. C. Leyte, Cross-Relaxation Effects in Pulsed-Field-Gradient Stimulated-Echo Measurements on Water in a Macromolecular Matrix. *J. Magn. Reson. B* **110** (1996), 150–7.
149. A. Chen and M. Shapiro, Nuclear Overhauser Effect on Diffusion Measurements. *J. Am. Chem. Soc.* **121** (1999), 5338–9.
150. S. V. Dvinskikh and I. Furó, Cross-Relaxation Effects in Stimulated-Echo-Type PGSE NMR Experiments by Bipolar and Monopolar Gradient Pulses. *J. Magn. Reson.* **146** (2000), 283–9.
151. J. Yan, A. D. Kline, H. Mo, E. R. Zartler, and M. J. Shapiro, Epitope Mapping of Ligand-Receptor Interactions by Diffusion NMR. *J. Am. Chem. Soc.* **124** (2002), 9984–5.
152. L. Avram and Y. Cohen, Diffusion Measurements for Molecular Capsules: Pulse Sequences Effect on Water Signal Decay. *J. Am. Chem. Soc.* **127** (2005), 5714–19.
153. H. Kessler, M. Gehrke, and C. Griesinger, Two-Dimensional NMR Spectroscopy: Background and Overview of the Experiments. *Angew. Chem.* (*Int. Ed. Engl.*) **27** (1988), 490–536.
154. M. D. Pelta, H. Barjat, G. A. Morris, A. L. Davis, and S. J. Hammond, Pulse Sequences for High-Resolution Diffusion-Ordered Spectroscopy (HR-DOSY). *Magn. Reson. Chem.* **36** (1998), 706–14.
155. A. M. Torres, R. Dela Cruz, and W. S. Price, Removal of *J*-coupling peak distortion in PGSE experiments. *J. Magn. Reson.* **193** (2008), 311–6.

8

Specialised PGSE and related techniques

8.1 Introduction

This chapter primarily deals with specialised NMR pulse sequences for measuring diffusion and flow. Sequences for MRI applications are given in Chapter 9. Steady gradient methods and especially those involving the stray field of superconducting magnets are outside the scope of the present work and so only a brief coverage is given in Section 8.2. Multiple-quantum and heteronuclear measurements are covered in Section 8.3. There has been considerable development of fast diffusion pulse sequences and these are covered in Section 8.4. Methods for handling samples that contain overlapping resonances with differences in relaxation time are considered in Section 8.5. Multi-dimensional methods for mixture separation and diffusion editing are presented in Section 8.6. Double PGSE and multi-dimensional motional correlation experiments are discussed in Section 8.7. Flow and Electrophoretic NMR are covered in Sections 8.8 and 8.9, respectively. Finally, the use of long-range dipolar interactions and miscellaneous sequences are presented in Section 8.10.

8.2 Steady gradient and stray field measurements

The earliest gradient-based diffusion measurements were based on the (technically simple) steady gradient experiments as discussed in Chapter 2. However, due to the limitations mentioned in Section 2.2.4, PGSE has generally overshadowed SGSE. However, SGSE approaches can be preferable since (i) there are no eddy-current problems and when performed in the fringe (or stray) field of superconducting magnets allows the use of very large stable gradients,[1,2] (ii) it may be difficult to generate gradient pulses with particular samples (e.g., in well logging), or (iii) it is desired to study a porous (and magnetically heterogeneous) sample on the basis of internal gradients.

Norwood and Quilter[3] developed a constant time, pulse and gradient amplitude diffusion experiment (CTPG) which avoids some of the problems associated with traditional SGSE experiments. And later several new sequences for measuring diffusion, some of which can be used to measure restricted diffusion and species having coupled spins and short T_2 were presented.[4] However, these approaches have not been widely adopted.

At present, the main interest in SGSE approaches focuses on very large static field gradients of superconducting magnets and purpose-built superconducting coils, this is sometimes referred to as stray-field imaging or STRAFI and they have been summarised in a number of reviews.[5–7] Developments in fringe field approach have allowed the production of relaxation-independent diffusion decays, multislice experiments, experiments using shaped rf pulses and two-dimensional variants.[8–10]

Importantly, the use of such large gradients provides a means of investigating extremely slow diffusive phenomena. Indeed, by conducting experiments in which the whole sample was mechanically oscillated in a stimulated echo experiment in the presence of a 180 T m^{-1} fringe field, Feiweier *et al.* determined that the lower limit of detectable motion was 7 nm.[11] Hole-burning diffusion measurements in high magnetic field gradients (42 T m^{-1}) have been demonstrated that can image diffusion along one dimension on a submicron scale.[12] Nevertheless, the inherent properties of using the fringe field are a major drawback. Because of the large field inhomogeneity, only a thin layer of the sample is on resonance and the S/N is dramatically reduced. The large gradient also means that the echo will become very sharp and so it may be difficult to properly digitise the echo. The linewidths of very slowly diffusing species are often so broad that the loss of chemical shift resolution becomes less of an issue. We also note that the analysis of the evolution of the magnetisation is complicated in highly inhomogeneous fields.[13–20] Also, gradient calibration is not straightforward and a method based on heteronuclear and field-profiling STRAFI techniques has been presented.[21]

Whereas with normal PGSE or SGSE measurements the magnetic field resulting from the applied gradient is much lower in strength than B_0, with stray-field measurements or when working at the Earth's magnetic field (~50 μT),[22,23] this condition no longer applies and thus the simplifying assumption implied in Eq. (2.10) no longer holds and this must be accounted for, else incorrect conclusions will be drawn.[24,25]

In a technique related to both STRAFI and MRI, Callaghan and Stepisnik[26] performed spatially distributed pulsed gradient spin-echo NMR experiments using a single wire through the sample. Since the gradient is produced by a single wire instead of a (more inductive) coil, they were able to achieve very rapid switching times with extremely large gradients (~100 T m^{-1}). It was noted that even larger gradients should be feasible, allowing measurements of molecular displacements smaller than 10 nm. Using this technique, they were able to observe anisotropic diffusion in a liquid crystal.

8.3 Multiple quantum and heteronuclear

8.3.1 General

It is possible to perform multiple quantum PGSE measurements but there are some restrictions on the systems that can be studied since the spectrum of the species in question must exhibit scalar, dipolar or quadrupolar coupling.[27–33]

In multiple quantum experiments, it is the effective sum of the γ values, γ_{eff}, of the nuclei involved in the coherence which is relevant to the attenuation. Thus, for the Stejskal and Tanner pulse sequence, the echo signal attenuation is given by

$$E(q,\Delta) = \exp\left(-\gamma_{\text{eff}}\, g^2 D\delta^2(\Delta - \delta/3)\right). \tag{8.1}$$

For the normal (i.e., single quantum) experiment $\gamma_{\text{eff}} = \gamma^2$. For homonuclear multiple quantum experiments, $\gamma_{\text{eff}} = (n\gamma)^2$.[27,28] For heteronuclear multiple quantum experiments, the definition of γ_{eff} is not so straightforward. For I spin-detected heteronuclear double quantum experiments with an IS spin system, $\gamma_{\text{eff}} = \gamma_I((\gamma_I + \gamma_S)/\gamma_S)^2$.[31]

Thus, for example, for a homonuclear double quantum coherence there is a fourfold decrease in the lower limit of the diffusion coefficient that can be measured with the same gradient strength or conversely the same degree of attenuation can be achieved but with smaller gradients and therefore smaller eddy-current problems. However, a multiple quantum coherence will be more susceptible to eddy-current effects than a single quantum coherence.

8.3.2 Homonuclear multiple quantum

Multiple quantum filtering can also help to remove spectral crowding by removing peaks that are not part of a scalar coupled spin system.[34,35] Further, in the case of homonuclear studies, multiple quantum spectra have the added benefit of providing solvent suppression. Some representative sequences for studying diffusion using multiple quantum spin-echoes are given in Figure 8.1. Multiple quantum steady gradient experiments have also been devised including a multiple quantum version of the CTPG experiment.[36]

^{1}H double quantum PGSE experiments including magic angle gradients have been presented as a means of solvent suppression and for suppressing demagnetising field effects.[34,37] However, the method suppresses only singlet resonances. In the case of quadrupolar nuclei, if the spectrum contains a static quadrupolar splitting the PGSE experiment can be performed using a quadrupolar echo instead of a

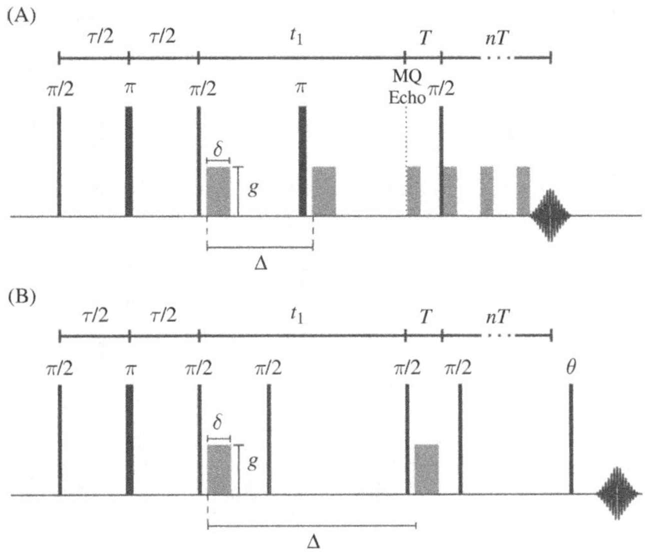

Figure 8.1 Some representative multiple quantum PGSE sequences based on (A) the spin-echo sequence by Zax and Pines[28] and (B) the STE sequence of van Dam *et al.*[32] The flip angle θ is optimised for the particular experiment.

spin-echo (i.e., the π pulse would be replaced by a $\pi/2$ pulse in the Hahn-echo PGSE sequence together with appropriate phase cycling).[38–40]

8.3.3 Heteronuclear multiple quantum

Without isotopic enrichment heteronuclei generally have low sensitivity and because of their low γ values much larger gradients must be used. ^{13}C or ^{15}N labelling is commonplace in macromolecular NMR studies and that large gains in sensitivity can be made through using heteronuclear PGSE sequences which invoke polarisation transfer from protons to the heteronuclei. A number of such sequences have now appeared including heteronuclear multiple-quantum coherence (HMQC).[30,31,41–44]

In particular, the use of heteronuclear inverse DEPT (i.e., inverse detection) and IHETCOR-based sequences have particular advantages.[30,31] The DEPT-based sequence involves polarisation transfer from the I spin (usually ^{1}H) to the S spin. Whereas the IHETCOR sequence is more suited to the observation of protons because of the unfavourable polarisation transfer from the less abundant heteronuclear population to the proton population.

A pulse sequence (see Figure 8.2) has been proposed for measuring multiple quantum coherences involved in the spin operator $8I_xI_yI_zS_y$ of a weakly coupled SI_3 system,[45] and demonstrated its use on the $^{13}CH_3$ group of alanine at natural abundance (Figure 8.3). The pulse sequence was subsequently modified to include bipolar gradient pulses.[46]

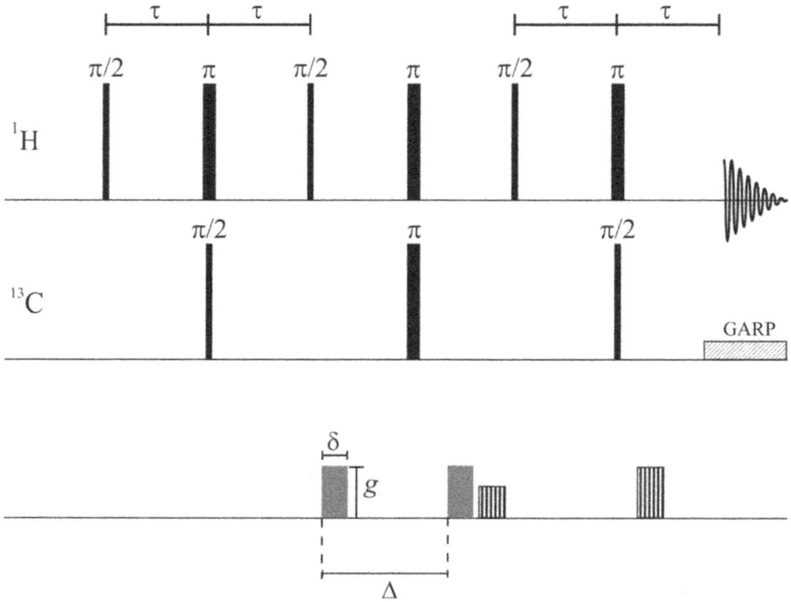

Figure 8.2 The LED-based pulse sequence for MQ diffusion measurements.[45] The delay $\tau = 1/(2J)$ and GARP (globally optimised alternating phase rectangular pulse) is a standard composite pulse method for ^{13}C decoupling. The striped gradient pulses are for coherence selection.

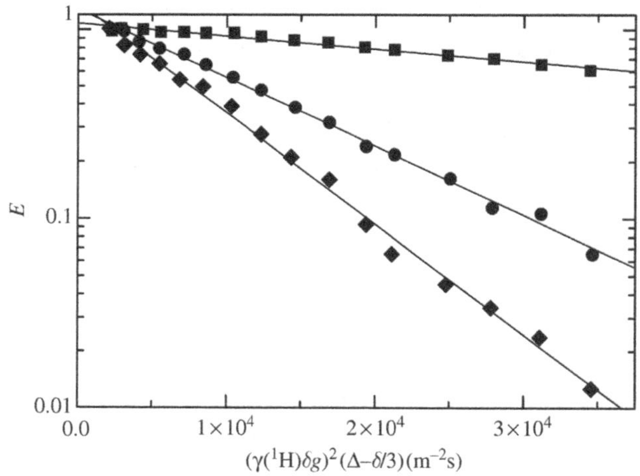

Figure 8.3 Diffusion plots of the multiple quantum coherences ($3I + S$, ♦; $3I - S$, ●; $I + S$, ■) of the $^{13}CH_3$ group of alanine in ^2H_2O measured at 400 MHz and 303 K using the pulse sequence in Figure 8.2. From Liu *et al.*[45]

Finally, we note that by using heteronuclear stimulated echoes in which the magnetisation is stored in a $2I_zS_z$ state, the Δ delay can be made much longer.[47,48] This method has particular application to the study of large proteins.

8.3.4 *Zero quantum*

A zero-quantum coherence can only be formed in a spin system consisting of at least two coupled spins and its precession frequency will be the difference in chemical shifts of the contributing spins but since they are 'spin-forbidden' they can only be excited and observed indirectly.[49] As can be seen from Eq. (8.1), zero-quantum coherences are unaffected by magnetic field gradients and can indirectly be used to study diffusion and a pulse sequence used for observing zero-quantum coherences is shown in Figure 8.4.

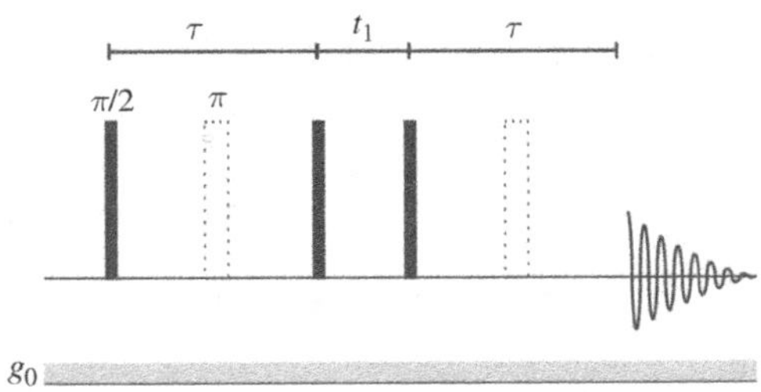

Figure 8.4. A pulse sequence for detecting homonuclear zero-quantum coherence in a constant magnetic field (g_0).[50] The first τ delay corresponds to the preparation period, the evolution period is denoted by t_1 and the second τ delay corresponds to the detection period. The two dotted π pulses are included to enable observation of the zero-quantum coherence without the effects of diffusion.

Diffusion will be encoded with respect to t_1. The signal attenuation is given by[50]

$$E = \exp\left[-\frac{t_1}{T_2^{\mathrm{ZQC}}} - \frac{2\tau}{T_2} - (\gamma g_0 \tau)^2 D[t_1 + (2/3)\tau]\right], \tag{8.2}$$

where T_2^{ZQC} is the transverse relaxation time of the zero-quantum coherence which evolves during t_1. Hence, the decay of the t_1 FID signal due to relaxation and diffusion is given by[50]

$$E(t_1) = \exp\left[-\frac{t_1}{T_2^{\mathrm{ZQC}}} - (\gamma g_0 \tau)^2 D t_1\right]. \tag{8.3}$$

Thus the amplitude and linewidths of zero-quantum coherences are diffusion-dependent. Assuming a Lorentzian lineshape, the linewidth of a zero-quantum coherence at half peak height due to relaxation and diffusion will be[50]

$$\Delta\nu_{1/2} = \frac{1}{\pi}\left(\frac{1}{T_2^{\mathrm{ZQC}}} + (\gamma g_0 \tau)^2 D\right). \tag{8.4}$$

The diffusion-dependent part can be separated by acquiring a zero-quantum coherence by adding a π pulse to the middle of each τ period of the sequence as shown in Figure 8.4. The difference in the measured linewidths will be determined solely by diffusion. There are a few major problems with this method: the method is time-consuming, high resolution in the indirect dimension (i.e., F_1) is required, the S/N decreases as g_0 increases and it has only limited ability to separate individual components.

More recently, it has been demonstrated that singlet states can be excited for any pair of homonuclear coupled spins and, provided that the spins are sufficiently isolated from magnetic interactions that break the symmetry (e.g., intramolecular couplings to remote protons), the singlet relaxation time, T_s, can be considerably longer than T_1. Thus, when used with a modified stimulated-echo sequence Δ can be considerably lengthened.[51,52]

8.4 Fast sequences

Signal averaging is normally required to achieve sufficient S/N. However, if a sample is capable of generating a high S/N in a single scan, the speed of measurement may be limited by the need to complete the minimum phase cycle of a pulse sequence and/or the recycle delay required to regain thermal equilibrium. Further, impetus to develop faster sequences is found in clinical MRI. The development of fast sequences also broadens the potential applications of diffusion measurements such as studying rapid reaction kinetics.

A possibility to rapidly measure diffusion is to perform the experiment using less or even only two q values as noted in Section 6.2. However, apart from needing 'prior knowledge' of the type of diffusion being studied, this requires exceptional accuracy. Similarly, in some cases, such as where a ratio of diffusion coefficients is required as in the determination of the ratio of the apparent to the isotropic diffusion coefficient like in surface-to-volume measurements (Eq. (1.122)) or in determining tortuosity (Eq. (1.126)), considerable timesaving can be had by just comparing the PGSE intensity at two different Δ values, but the same b value (e.g., see Eq. (2.75)).[53] However, it is to be expected that greater error would be involved since the intensity is only measured at two points.

Fast diffusion sequences (i.e., sequences with high S/N efficiency) can be loosely classified into one-shot and single-shot sequences. A one-shot sequence has a reduced phase cycling such that ideally one q-space point is applied in a single scan. In contrast, a single-shot sequence is where a complete diffusion measurement (i.e., multiple q-space points) is conducted in a single scan and is based on a multiecho (e.g., CPMG) approach. We stress that these rapid techniques generally require very good S/N and have other drawbacks such as loss of chemical shift information, difficulty in separation of relaxation and diffusion effects and limited suitability for studying diffusion in restricted systems.

8.4.1 Single-shot sequence (total measurement in one scan)

Packer[54] proposed a modification to the CPMG pulse sequence in which π pulses are inserted at each echo maximum (e.g., at 2τ, 4τ, 6τ,... in Figure 2.9B). These π pulses reflect the phase at the echo maxima causing the phase effects of the flow to be cumulative whilst the effects of diffusion are the same as in the CPMG sequence. However, this sequence is difficult in practice due to the presence of the π pulses in the echoes. Packer then proposed a modified CPMG-like sequence, $\pi/2-\tau-\pi-\tau'-\pi-\tau-\pi-\tau'\ldots$, where $\tau' > \tau$, thereby separating the echoes from the π pulses. In fact, there are two sequences of echoes: one at $(2\tau+2N\tau')$ and the other at $2N\tau'$, where $N=0,1,2,\ldots$ which for large N leads to

$$S(t) = M_0 \exp\left(-\frac{t}{T_2}\right) \exp\left(-\frac{\gamma^2 g^2 D t}{3}\left(\tau'^2 - 3\tau\tau' + 3\tau^2\right)\right) \cos(\gamma g v t(\tau - \tau'/2)), \tag{8.5}$$

where $t=2N\tau'$. Thus, in theory it is possible to determine an estimate for the diffusion coefficient by monitoring the echo amplitudes in a single shot. Song *et al.*[55] reported a method closely related to that of Packer based on a modified CPMG sequence (KCPMG).

A single-shot method based on a CPMG-like sequence with incremented (pair-wise) gradient pulses in each τ period has been proposed by Li and Sotak,[56] and later converted for use in an imaging sequence.[57] It was noted that if the gradient pulses are applied alternatively along three orthogonal directions it can be used to probe anisotropic diffusion. Mair *et al.*[58] have presented a variation of the single-shot technique of Li and Sotak[56,57] for gas diffusion measurements.

Van Gelderen and co-workers[59] have proposed another single-shot diffusion method based on a spin-echo experiment during which a series of gradient echoes are created (Figure 8.5). Since each subsequent echo is attenuated by an additional gradient pair, the total diffusion curve can be acquired within a single experiment. In the presence of a static field gradient (i.e., B_0 inhomogeneity), neglecting g_0^2 terms, the signal attenuation for the nth echo is given by

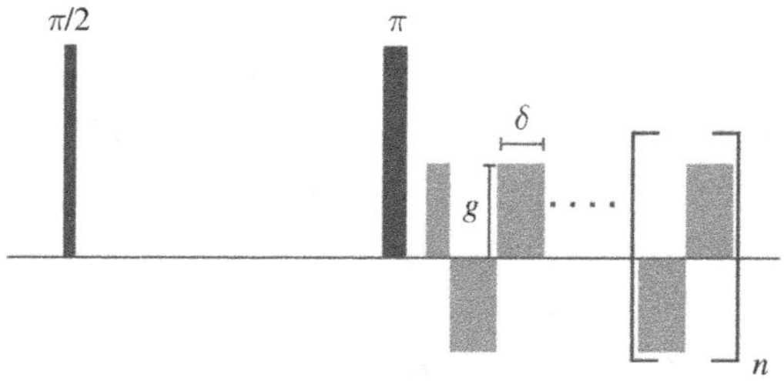

Figure 8.5 The single-shot diffusion sequence of Van Gelderen *et al.*[59]

$$S_n = A_n \exp\left(-\frac{\gamma^2 D}{4}\left(g^2 n\delta^3/3 - \mathbf{g}_0 \cdot \mathbf{g}\left(2\tau\delta^2 - n\delta^3\right)\right)\right) \text{ for } n \text{ odd} \tag{8.6}$$

and

$$S_n = A_n \exp\left(-\frac{\gamma^2 D}{4}\left(g^2 n\delta^3/3 - \mathbf{g}_0 \cdot \mathbf{g} n\delta^3\right)\right) \text{ for } n \text{ even,} \tag{8.7}$$

where A_n is the amplitude without diffusional attenuation and τ is the echo time. The difference in dependence of the even and odd echoes on the $\mathbf{g}_0 \cdot \mathbf{g}$ terms may be used to estimate static local gradients. Also, due to gradient nulling effects, so long as any bulk motion can be considered as constant over a time interval of 2δ, the even echoes will be motion-compensated. Diffusion can be determined from the experiment in two ways. In the first method, the experiment is performed twice, changing only the gradient strength. The ratios of the corresponding echoes can then be compared to extract the diffusion constant. In the second approach, just one experimental data set is acquired and the ratio of the peaks symmetric in time to the echo maximum are recorded. However the second method does not account for T_2 effects.[59] However, the ability of this experiment to retain chemical shift information is limited by the digitisation rate of the spectrometer and the maximum gradient strength available. Doran and Decorps[60] used multiple applications of small angle excitation in the presence of a read gradient followed by a π pulse and then acquired the multiple echoes in the presence of a read gradient as shown in Figure 8.6. Their method is related to the Burst imaging technique.[61] After some assumptions, it is found that the attenuation of the lth echo (corresponding to the n-l-1th rf pulse) is given by

$$E_l = \exp\left(-\gamma^2 g^2 \delta_l^2 (\Delta_l - \delta_l/3)\right) \tag{8.8}$$

where $\delta_l = l\tau + d_1$, $\Delta_l = l\tau + d_2$ and the τ, d_1 and d_2 are defined in the pulse sequence diagram. Clearly, the applicability of the sequence is limited since all of the signals are acquired in the presence of a read gradient. The method also requires a high S/N.

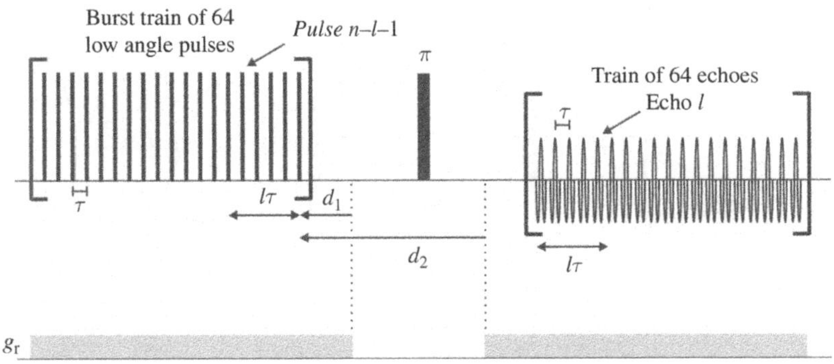

Figure 8.6 The spin-echo Burst pulse sequence.[60] The sample is excited in the presence of a read gradient (g_r) by a train of very low amplitude pulses ($\lesssim 1°$). The π pulse and second read gradient result in a series of echoes corresponding to each rf pulse. The delays are arranged so that the spin and gradient echoes coincide.

Peled *et al.*[62] presented a similar technique based on stimulated echoes and applied it to the study of laser polarised ^{3}He and ^{129}Xe. Song and Tang[63,64] have extended this approach, in which unequal time spacings are used, to provide more efficient use of the available signal and removed the limitation of small tipping angles, relying on multiple echoes, each following a different coherence pathway and thus each echo has associated with it a different b value. Their method, termed 'multiple modulation multiple-echo' (MMME), can also be used with static gradients. Tang *et al.*[64,65] extended the MMME method to measure diffusion along multiple non-collinear directions simultaneously and thus obtained a two-dimensional diffusion tensor.

The method of Sendhil Velan and Chandrakumar[66] is a two-scan procedure in which a modified CPMG sequence (see Figure 8.7) is run with and without gradient pulses. A large number of echo tops (~ 200) are sampled midway between the π pulses and upon Fourier transformation Lorentzian linewidths at the echo modulation frequency are analysed to determine the diffusion coefficient. Specifically, the linewidth of the Lorentzian resulting from the sequence run with gradients is given by

$$\Delta\nu_{\frac{1}{2}} = \frac{1}{\pi}\left[\frac{1}{T_2} + \gamma^2 g^2 \delta^2 D\left(\frac{\Delta}{\tau} - \frac{\delta}{3\tau}\right)\right], \tag{8.9}$$

and the relaxation term in the brackets can be accounted for from the linewidth measured from the second scan run in the absence of gradient pulses. Clearly, as this technique depends on linewidth measurements, the systems to which it is applicable are limited unless deconvolution or chemical shift selective excitation is employed.

In the Difftrain approach,[67,68] which is an extension of the alternating pulsed gradient stimulated-echo sequence of Cotts *et al.*,[69] there is a single encoding step but then many echoes are acquired with multiple applications of a small angle rf pulse

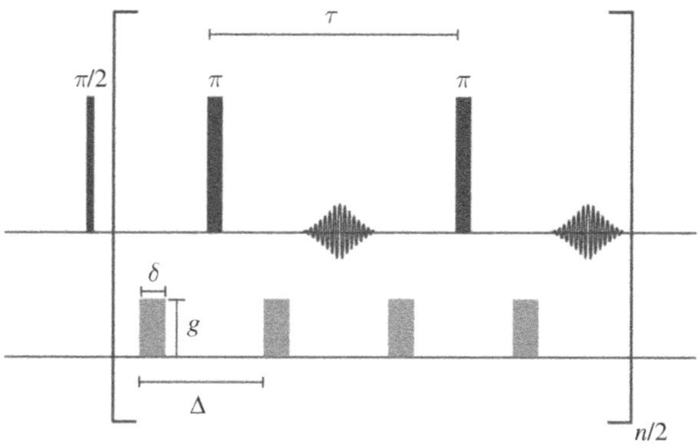

Figure 8.7 The fast multi-spin-echo diffusion sequence of Sendhil Velan and Chandrakumar.[66] This is based on a CPMG sequence with gradient pulses symmetrically placed around the π pulses. The echo tops are sampled midway between the π pulses. Note their definition of τ is twice as long as that traditionally used (see Figure 2.9). Purge gradient pulses (not shown) are used at the end of each scan to destroy residual transverse magnetisation.

(i.e., θ) gradient subsequence (Figure 8.8). Only $\cos(\theta)$ of the signal remains longitudinal after each θ rf pulse and consequently after n echoes the remaining signal is dropped to $\cos^n(\theta)$ of M_0 – this is in addition to relaxation and diffusion losses. There is some loss of chemical shift resolution because of the need to acquire numerous FIDs before the spatially encoded magnetisation is lost to spin–lattice relaxation. A further limitation is that the observed signal attenuation results from both diffusion and relaxation which necessitates the use of a control experiment in addition to a two-step phase cycle. Due to the need for the correction for relaxation the sequence is unsuitable for polydisperse samples.

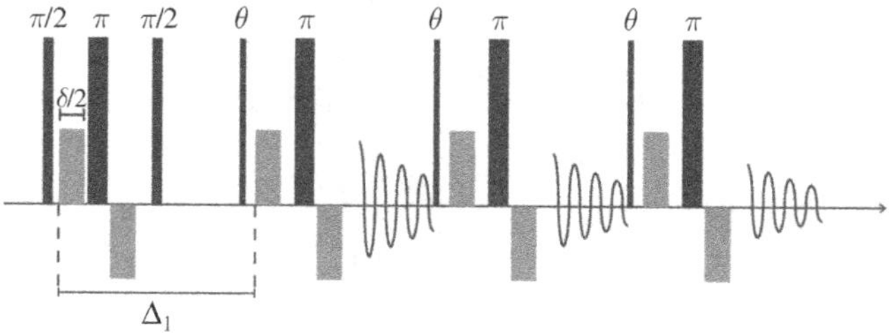

Figure 8.8 The Difftrain pulse sequence.[67] In this sequence, the spins are encoded once in the transverse plane and then stored longitudinally. Subsequently, small aliquots of this magnetisation are sampled using the small angle θ rf pulse and decoding gradient. After each such acquisition a homospoil gradient pulse (not shown) is applied to disperse residual transverse magnetisation. This sampling is typically repeated 16 times. Each echo delay adds to the diffusion time for every next echo (only the first Δ is depicted).

8.4.2 One-shot sequences (reduced phase cycling)

Millet and Pons[70] presented a variant of accordion spectroscopy[71] for measuring diffusion using a single 2D experiment (GAUDI: gradient accordion used for diffusion) in which the diffusion information is encoded into the linewidth. This is achieved by incrementing the strength of the gradients proportionally with the evolution time (i.e., $g = \kappa t_1$; where t_1 is the evolution time and κ is a constant). Neglecting relaxation, the signal decay during t_1 is related to the diffusion coefficient by

$$S(t_1) = S(0)\exp\left[-(\gamma\delta\kappa t_1)^2 D(\Delta - \delta/3)\right]. \tag{8.10}$$

After 2D Fourier transformation, both the diagonal- and cross-peaks have a Gaussian shape in F_1 from which the diffusion coefficients can be determined by regression of the Fourier transform of Eq. (8.10)

$$S(\omega) = (S(0)/2)\left(\frac{\pi}{(\gamma\delta\kappa)^2 D(\Delta - \delta/3)}\right)\exp\left[-\frac{\omega^2}{4(\gamma\delta\kappa)^2 D(\Delta - \delta/3)}\right], \tag{8.11}$$

(where ω is the frequency difference with respect to the centre of the peak) onto the peaks. This technique can be implemented into any 2D experiment with a diffusion filter and an example of a constant time GAUDI–NOESY sequence is given in Figure 8.9. This technique suffers from some significant limitations including the need to correct for intrinsic linewidth, diffusion-broadened lines resulting in lower sensitivity and resolution, and interference from scalar coupling unless a constant time version of the experiment is used.

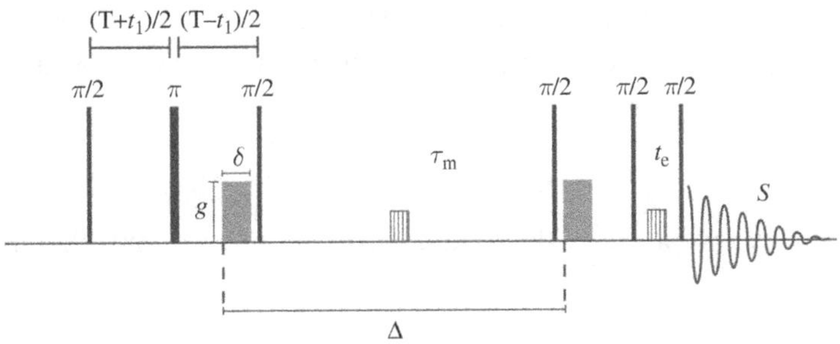

Figure 8.9 The constant-time GAUDI–NOESY sequence.[70] The vertical striped boxes represent gradient pulses used for coherence selection. T is a constant delay and τ_m is the mixing delay. The t_e delay acts as a z filter.

The 1D DOSY experiment of Loening *et al.*[72] uses a non-constant (i.e., z^2) gradient to encode the diffusion information into the lineshape. In a later modification of the 1D DOSY approach, Thrippleton *et al.*[73] used chirp rf pulses[74] in the presence of the diffusion gradients (see Figure 8.10) to vary the effect of the effective gradient at different points in the sample instead of using hard rf pulses

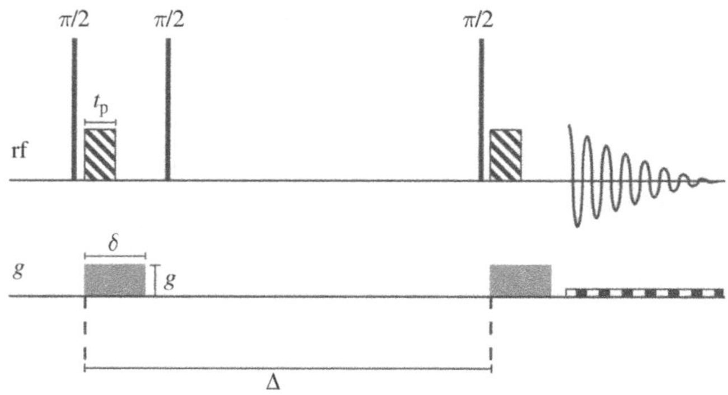

Figure 8.10 A simplistic representation of the 1D DOSY sequence which permits the measurement of diffusion in a single scan. In the real sequence, additional gradient pulses are added to effect coherence selection and reduce eddy-current effects.[73] The rectangles with slanted stripes represent adiabatic π chirp pulses of length t_p. Acquisition is performed in the presence of a weak read gradient (checked rectangle).

and a constant gradient as in the earlier study.[72] Specifically, during the gradient pulses, spins in different parts of the sample have different offsets and as the frequency-swept π pulse works by sweeping through a range of offsets, different parts of the sample are affected at different times during t_p. Thus, spins at one part of the sample immediately experience a π pulse and are then spatially encoded by the gradient. Spins at the centre of the sample are, in contrast, essentially unaffected by the gradient during t_p since they are gradient encoded for a period $t_p/2$, then experience a π pulse and are then refocussed unencoded for a period $t_p/2$. Simplistically assuming that a spin at position z experiences a π pulse the moment it is on resonance at a time $\alpha(z)t_p$ where $0 \leq \alpha(z) \leq 1$, then the net phase acquired by a spin at position z at the end of the frequency-swept pulse is

$$\phi(z) = \gamma g z t_p[1 - 2\alpha(z)]. \tag{8.12}$$

Thus, during the swept pulse the spin evolves as if under an effective gradient strength

$$g^{\text{eff}}(z) = g[1 - 2\alpha(z)]. \tag{8.13}$$

A more detailed treatment accounting for the finite duration of the π pulse is given elsewhere.[73]

The gradient pulse extends past t_p so that all spins experience some degree of spatial encoding but on account of the frequency swept pulse in combination with the usual constant (along z) gradient there is a z-dependence of the pitch of the magnetisation helix. Thus, this is equivalent to the spatial encoding having been performed by a (in the mathematically correct sense) linear gradient along z. Thus, the degree of attenuation in the resulting stimulated-echo sequence will vary along the sample due to the effectively spatially changing gradient. Acquiring the signal in the presence of a weak read gradient results in all of the resonances becoming one-dimensional diffusion-weighted images of the sample. Thus, the method is restricted to relatively strong samples. Although the read gradient is weak, there is nevertheless considerable loss of chemical shift resolution which effectively limits the sequence to relatively simple samples giving uncrowded spectra. The diffusion coefficient is then determined by lineshape analysis. The method is complicated by the effects of J-evolution and is only applicable to Fickian diffusion.

The use of asymmetric bipolar gradient pulses allow a great reduction in the required phase cycle for stimulated echo sequences.[75] And at present, this provides probably the most generally applicable fast means to acquire PGSE data (also see the related sequence in Figure 6.2B).

8.5 Distributions of relaxation times

Under some circumstances, such as in heterogeneous systems, the resonance of interest may be superimposed on other resonances but with different relaxation times. This situation commonly occurs where water is in an emulsion and is thus of particular relevance to the food industry. An approach to overcoming this is to apply relaxation-weighting to the diffusion measuring systems in order to winnow out those resonances not of interest. If the resonance of interest has a longer relaxation time than the others, the PGSE sequence can just be increased in length so that the resonance of interest is the only one remaining. Or alternatively, explicit relaxation-weighting can be added to the sequence (Figure 8.11). Heink *et al.*[78] have suggested replacing the initial $\pi/2$ pulse in the Hahn or STE sequences with either a primary or STE sequence. The durations in the preparatory sequence may be varied while leaving the diffusion part of the sequence unchanged. If the signal attenuation due to the gradients is independent of the durations in the preparatory sequence, artefacts due to a distribution of relaxation times can be excluded. Instead of simple relaxation filtering another approach is to explicitly use correlations between spin–spin relaxation time and diffusion coefficients by combining a PGSE sequence with a relaxation measurement.[79–81] And this type of approach is extended further in Section 8.7.

8.6 Multi-dimensional mixture separation and diffusion editing

Diffusion encoding gradients can be appended, prefixed or incorporated into essentially any 1D or higher experiment and this can be advantageous when a sample contains a mixture of compounds as it may facilitate the analysis of resonances that

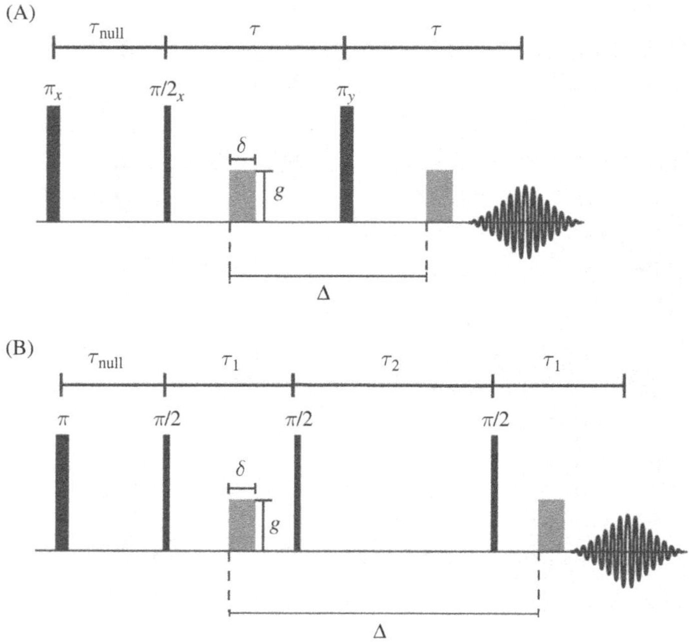

Figure 8.11 (A) A Hahn spin-echo sequence prefixed with a π pulse and a delay.[76] (B) A stimulated-echo sequence prefixed with a π pulse and a delay for nulling resonances not of interest as introduced by Van Den Enden *et al.*[77]

overlap in 1D PGSE measurements. In the simplest case, a single value of q can be used to edit out the most rapidly diffusing (and typically lowest molecular weight) species (e.g., solvent suppression as in Section 6.2.2). But it can also be used to add an extra dimension to an experiment when a full diffusion measurement is performed (see Section 2.5). However, adding extra spectral dimensions can result in dramatic increases in experimental times (e.g., 10–1000 times longer) making the experiment impractical to perform and it may also preclude the study of systems that undergo changes on timescales on the order (or less than) of the experimental time.

Diffusion editing can also be contained in more complicated sequences which also consider differences in relaxation, and differences in coupling constants.[82] For example, by placing a spin-echo before the PGSE sequence, the resonances of coupled spins can be nulled thereby retaining only the singlets in the ^{1}H NMR spectrum (Figure 8.12).[83]

Other examples include inclusion into sequences for generation of multiple quantum coherences and 1D off-resonance ROESY experiments,[84] a modification of the STE sequence for selectively measuring the self-diffusion of molecules with ^{13}C–^{1}H bonds using ^{1}H detection,[85] and an STE sequence incorporating WATERGATE and isotope filtering.[86]

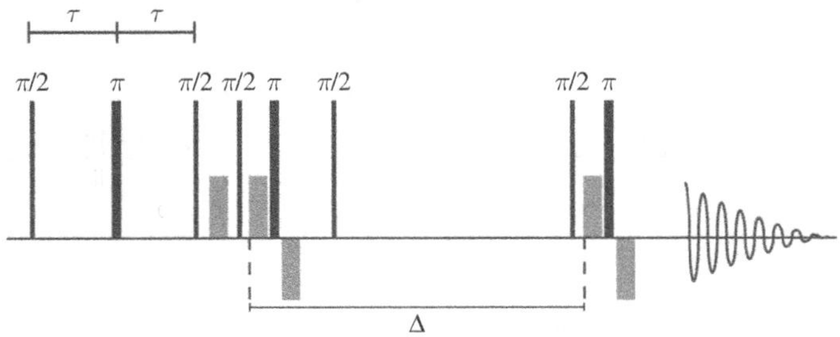

Figure 8.12 The GOSE BPP–STE sequence.[83] τ is set to $1/4J$ to null the coupled spins. The $\pi/2$ pulse after the echo is used to tip the refocussed singlet magnetisation to the z-axis. The first gradient pulse is used to remove any residual magnetisation from the coupled resonances. The next $\pi/2$ pulse returns the singlet magnetisation for the beginning of the bipolar PGSTE sequence.

The literature contains many examples of when diffusion is incorporated into 2D experiments and the ideal case is where the original 2D sequence contains an echo, where gradient pulses can be incorporated or a gradient pair separated by a mixing period. Incorporating the gradient pulses into the sequence is the first choice since the resulting sequence is shorter leading to less signal loss and perhaps also allowing a reduction in signal coherence pathways.[84,87] Nilsson *et al.* have suggested the terminology DOSY-X to indicate a 3D experiment where the diffusion encoding step precedes the 2D pulse sequence (X), X-DOSY for when the diffusion encoding succeeds the 2D sequence, or X-IDOSY for when the diffusion encoding is incorporated into the sequence.[87]

A large number of 3D sequences now exist including DOSY–COSY and COSY–IDOSY (see Figure 8.13),[88,89] DOSY–DEPT and DOSY–INEPT,[41,90] DOSY–HMQC and HMQC–IDOSY,[41,90,91] HSQC–IDOSY,[92] DOSY–NOESY,[93] DOSY–TOCSY and TOCSY–IDOSY (see Figure 8.14),[84,94–97] 2D-*J*-DOSY and 2D-*J*-IDOSY,[87,98] and convection-compensated versions of some of these sequences have also been presented.[99] Given the time required to conduct multi-dimensional experiments, we mention that Hadamard-encoded $^{1}H\{^{13}C\}$-resolved

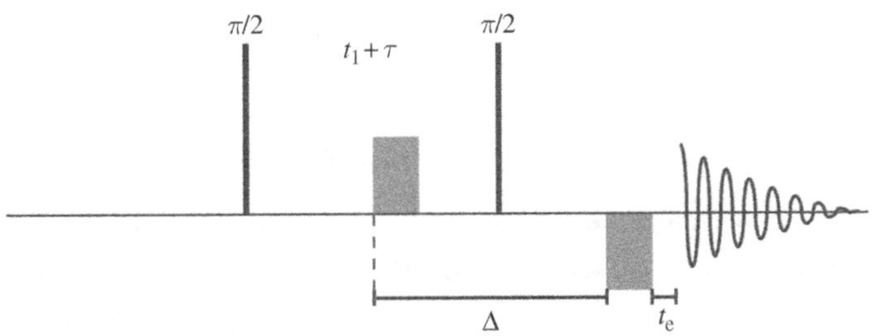

Figure 8.13 The COSY–IDOSY sequence with p-type coherence selection. The short delay t_e is to allow for field stabilisation.[89]

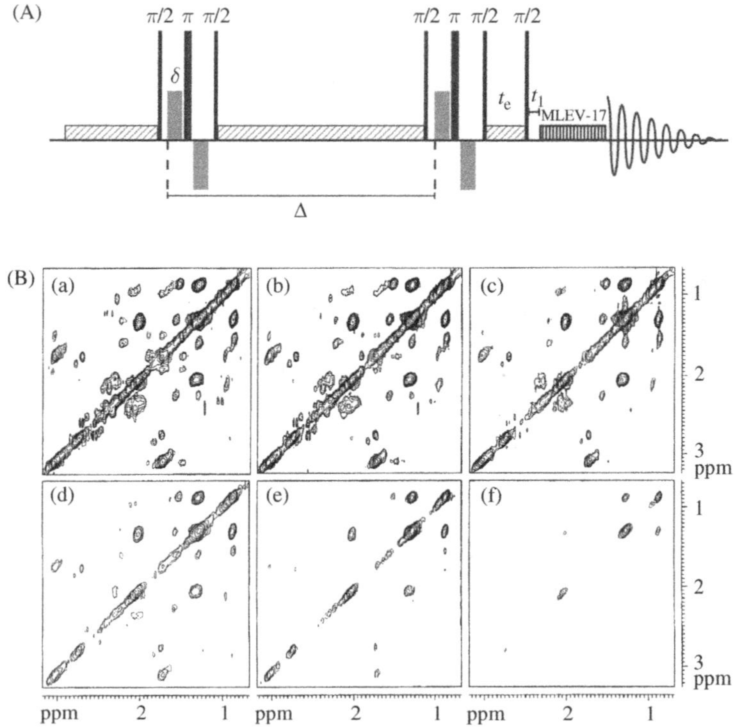

Figure 8.14 (A) Diffusion-edited TOCSY pulse sequence using the MLEV-17 spin-lock method.[94,96] The slanted striped areas indicate areas of water presaturation and t_1 is the two-dimensional increment time.[96] (B) Spectra obtained of human blood plasma using the sequence at 600 MHz with $\Delta = 300$ ms, $\delta = 2$ ms and $g =$ (a) 21.3, (b) 95.8, (c) 170.3, (d) 244.8, (e) 319.3 and (f) 393.8 mT m^{-1} showing increasing attenuation of the signal intensities. From Liu *et al.*[96]

diffusion measurement has been presented that offers something like a factor of 10 reduction in measuring time.[100]

Moonen *et al.*[101] developed gradient enhanced exchange spectroscopy (GEXSY) that allows the separation of the effects of exchange from cross-relaxation in order to better characterise the dynamic matrix in EXSY (or NOESY) experiments and used it to study amide exchange in the *N*-acetylaspartate and water system.

8.7 Double PGSE and multi-dimensional correlations

In Section 8.6, we were concerned with multi-dimensional sequences in which a traditional Fourier domain-based multi-dimensional sequence involving an oscillatory spin phase evolution (e.g., COSY) was combined with a diffusion measurement (i.e., the sequences probed joint probabilities of the form $P(\omega, D)$; where ω is frequency) and in the diffusion domain the signal merely decays (and is thus amenable to Laplace inversion; see Sections 2.5 and 6.5.2) and thus the analysis involves mixed Fourier–Laplace dimensions. Note, flow is a special case as it is oscillatory and thus Fourier analysis is appropriate. We illustrate a key point of the subsequent analysis using the SGP solution for free isotopic diffusion (Eq. (2.44)), namely

$$E(q,\Delta) = \exp\left(-\gamma^2 g^2 D \delta^2 \Delta\right) = \exp\left(-(2\pi q)^2 D\Delta\right). \tag{8.14}$$

Inverse Fourier transformation with respect to q (see Eq. (2.35)) yields the diffusion propagator (see Section 3.2, a Gaussian probability density in this case – see Eq. (1.61)), whereas (ideally) inverse Laplace transformation would lead to (see Section 2.5) delta functions in the diffusion space. Thus, the choice of analysis depends on the information sought. As noted in Sections 2.5 and 6.5.2, inverse Laplace transformation, especially of noisy data, is fraught with difficulty. And this difficulty is augmented when multi-dimensional inverse Laplace transforms are involved.[102–107]

Here we consider other types of correlations which probe different sorts of joint probabilities such as between various combinations of molecular displacements (i.e., diffusion or advection) measured with applied gradients, diffusion in internal gradients (g_{int}) and spin relaxation; however, we will only consider those combinations that involve translational motion. Such measurements are performed with some form of 'multi-wavevector' extension of a PGSE sequence – typically a double PGSE sequence or by combining a PGSE sequence with a sequence for measuring relaxation (e.g., a CPMG sequence). As noted by Callaghan *et al.*,[108] the various types of measurement discussed below and those in Section 8.5, together, span the three categories of multi-dimensional NMR experiments proposed by Ernst *et al.*,[109] namely correlation, exchange and separation.

In an oblique fashion, we have already come across the idea of using more than one pair of gradient pulses in an experiment when we considered convection compensation in Section 7.4. In that case, the two pairs of gradient pulses were designed to remove the effects of constant motion. However, by having more than one pair of gradient pulses, additional information on the translational dynamics can be obtained.[110,111] An early attempt by Li and Sotak[112] involved a pulse sequence that combined both a stimulated echo and spin-echo component with two pairs of gradient pulses with the amplitude of each echo type being dependent on different pairs of gradient pulses. They noted that this sequence has potential for measuring anisotropic and restricted diffusion since the two echoes can be used to observe diffusion in different directions and over different timescales. In a later work, it was realised that the echo attenuation observed in a sequence containing two independent pairs of gradient pulses is more sensitive to the effects of restriction at low q

values in contrast to a PGSE experiment performed with a single gradient pulse pair. And that the 'multiple wave-vector' extensions of PGSE sequence have the possibility of unambiguously distinguishing between multicompartment diffusion and diffraction-like behaviour due to restricted diffusion.[113] A recent study has indicated that different forms of diffraction patterns are observed when the PGSE sequence is extended to a larger number of gradient pulse pairs and that it may allow the measurement of smaller pore sizes.[114]

Many experimental variations exist even with just two pairs of gradient pulses and short contemplation of Figure 8.15 reveals that the gradient pairs could be arranged with the pairs having different directions either simultaneously or sequentially – with or without a delay between the pairs, the amplitudes of the pairs could be identical, correlated or completely independent and the settings of Δ_1 and Δ_2 present additional degrees of freedom. By encoding the same magnetisation again with a second pair of gradient pulses it becomes possible to correlate the motion at two different time intervals or to reveal changes in motion that have occurred between the pairs of gradient pulses.[115] Importantly, if the two pairs of gradient pulses are changed in unison then the result is a one-dimensional experiment (e.g., $E(q, \Delta)$, where q pertains to both q_1 and q_2); if they are changed independently, then the result is a two-dimensional experiment (e.g., $E(q_1, q_2, \Delta)$).

Below, we give a brief coverage of the following correlations in detail: diffusion–diffusion, diffusion–relaxation, diffusion–internal magnetic field gradient and dispersive flow. But this list is not exhaustive and other correlations are possible including spin–lattice relaxation,[119] flow and relaxation[120] and the non-local dispersion tensor, which describes the correlation of the flow field at different points in space and time.[121]

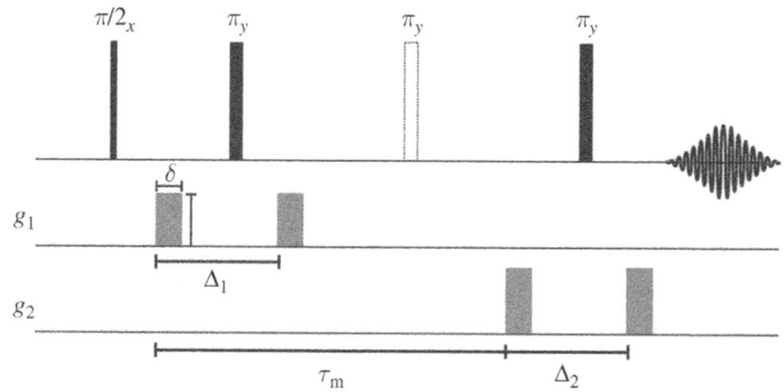

Figure 8.15 A generic double PGSE sequence based on Hahn echoes. Double PGSE sequences are also possible based on stimulated echoes. τ_m denotes an exchange or mixing time. In the absence of the dotted π rf pulse, the sequence is compensated for constant flow.[116,117] If the two pairs of gradient pulses are applied independently this sequence generates a two dimensional velocity map and is termed a VEXSY experiment.[118]

8.7.1 Diffusion – diffusion correlation (DDCOSY)

By applying two sequential encoding periods along arbitrary directions it has been shown that it is possible to probe anisotropy in a pore matrix and local diffusional anisotropy.[110,122,123] In a later work,[124] Chin *et al.* proposed that by simultaneously applying gradient pulses in two separate directions (but stepped in unison) as in Figure 8.16 it is possible to determine 2D q-space attenuation maps and transform the data using 2D Fourier transformation. Using a finite difference model of restricted diffusion, they simulated maps for water diffusing within pores with various degrees of asperity. They report that the observed ring patterns (of diffraction minima) reveal the boundary profiles of the pores and may have application to the study of the boundary morphology of structured materials and biological cells. By having the gradient pairs directed in orthogonal directions in the double PGSE sequence, it is also possible to measure (and construct using double Fourier transformation) 2D propagators (e.g., $P(x, y, \Delta)$) and higher propagators,[125–127] which are useful for characterising locally anisotropic motion.

The DDCOSY sequence[105,128] consists of two successive PGSE sequences with independent pairs of gradient pulses, g_i and g_j (Figure 8.17). The directions of the gradient pairs may be applied collinearly or orthogonally to correlate successive diffusion along the corresponding axis of the laboratory frame of reference. The signal attenuation is given by a discrete representation of a 2D Laplace integral

$$E(q_i, q_j) = \sum P(D_{ii}, D_{jj}) \exp(-q_i^2 D_{ii}\Delta) \exp\left(-q_j^2 D_{jj}\Delta\right), \tag{8.15}$$

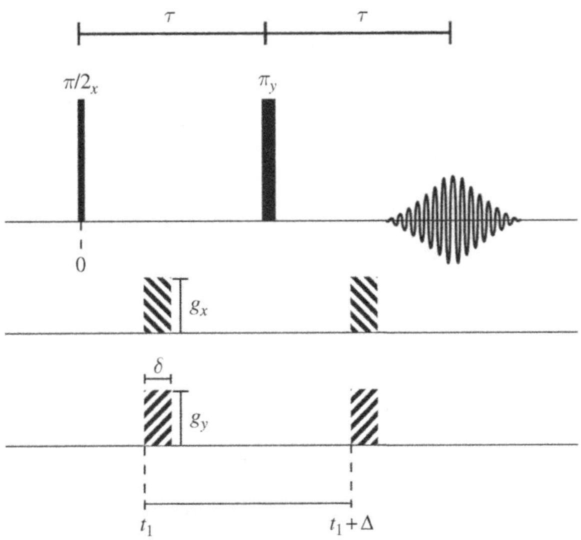

Figure 8.16 A PGSE pulse sequence that could be used for 2D q-space imaging.[124]

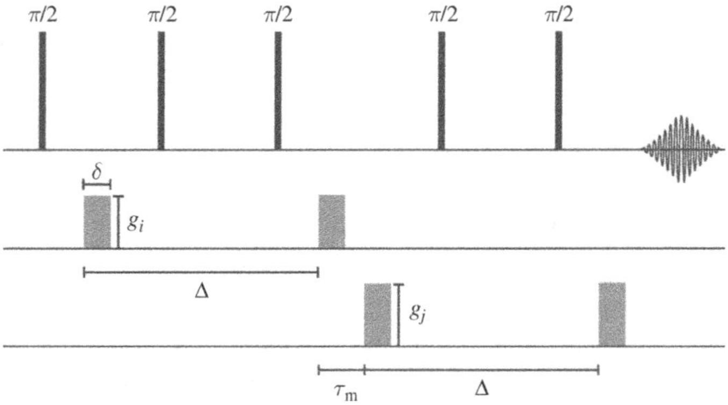

Figure 8.17 The DDCOSY sequence[105,128] consists of two successive PGSE sequences such that the mixing time τ_m is set as short as possible.

where $P(D_{ii}, D_{jj})$ is the joint probability of contribution to the signal from the diffusion tensor elements D_{ii} and D_{jj}. Inverse Laplace transform of Eq. (8.15) provides a 2D map of diffusion tensor elements in the corresponding spatial directions. Diagonal peaks correspond to isotropic diffusion behaviour, whereas off-diagonal peaks appear when the diffusion is described by a tensor. Thus, the DDCOSY experiment can be used to investigate the anisotropy of liquid diffusion in porous systems.

The DDCOSY and DRCOSY (covered below) methods have been used to investigate anisotropic water diffusion in chive leaves.[129] The DRCOSY measurement revealed six different peaks corresponding to six different states of water in the sample. The method proved useful in extracting the diffusion tensor at different points in the sample. A DDCOSY (see Figure 8.18) was then used to relate the water peaks to the different cell shapes and orientations.

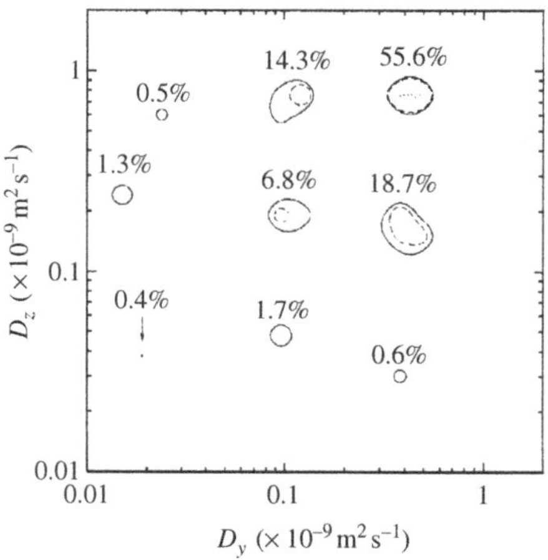

Figure 8.18 DDCOSY map of water diffusion in chopped chive leaves reveals local diffusional anisotropy by showing the correlations of diffusion in two different directions. From Qiao *et al.*[129]

The DEXSY pulse sequence (Figure 8.19),[105,128] similar to the DDCOSY sequence, consists of two successive PGSE sequences, but in contrast to the DDCOSY pulse sequence, the gradients in the DEXSY sequence are collinear and separated by a mixing time τ_m. The signal attenuation is given by a discrete representation of a 2D Laplace integral

$$E(q_1, q_2) = \sum P(D_1, D_2) \exp\left(-q_1^2 D_1 \Delta\right) \exp\left(-q_2^2 D_2 \Delta\right), \tag{8.16}$$

where $P(D_1, D_2)$ is the joint probability of contribution to the signal from the diffusion coefficients D_1 and D_2. Off-diagonal peaks in the 2D map resulting from the inverse Laplace transform of E represent molecules that have changed their diffusion coefficient during τ_m, whereas diagonal peaks in the map reflect those molecules that have not changed their diffusion coefficient during τ_m.

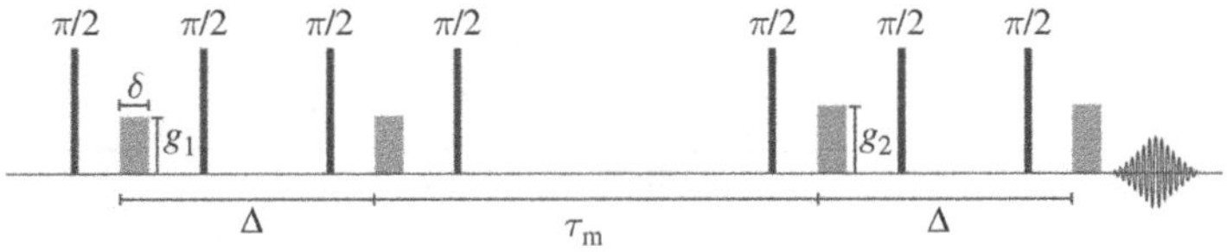

Figure 8.19 The DEXSY sequence[105,128] consists of two successive PGSE sequences with collinear gradient pairs g_1 and g_2 which can be varied independently.

8.7.2 Diffusion–relaxation correlation (DRCOSY)

The DRCOSY sequence is some combination of a PGSE sequence with a CPMG sequence[80,130] and an example is given in Figure 8.20. The signal attenuation is given by a discrete representation of a 2D Laplace integral

$$E(q, t) = \sum P(D, T_2) \exp\left(-q^2 D \Delta\right) \exp\left(-t/T_2\right), \tag{8.17}$$

where $P(D, T_2)$ is the joint probability of contribution to the signal from D and T_2. Inverse Laplace transformation yields a 2D map in which cross peaks correlate diffusion coefficients with the corresponding T_2.

DRCOSY experiments have been performed in the fringe field of superconducting magnets[18,131] using, for example, the sequence in Figure 8.20, but omitting the gradient pulses. However, since these experiments are effectively performed 'off resonance', complicated phase cycling is required to select the correct coherences. An example of a DRCOSY map for limestone saturated with water is given in Figure 8.21 and the results indicate a strong correlation between restricted diffusion and faster spin–spin relaxation.

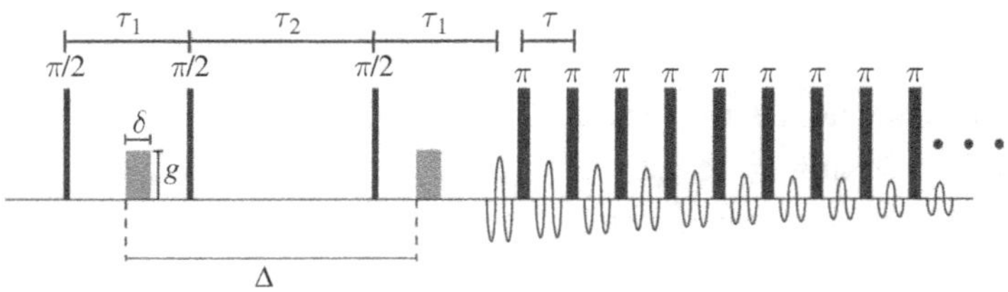

Figure 8.20 Pulse sequence for measuring diffusion–relaxation correlations.[80] This pulse sequence consists of an initial subsequence (i.e., a stimulated echo) of length $t_d = \Delta + \tau_1$ that probes diffusion diffraction followed by a second (CPMG) subsequence that measures relaxation effects. The same sequence with the omission of the gradient pulses can be used in steady gradient experiments, but in such case the effective diffusion time is now $\Delta_{eff} = 2\tau_1 + \tau_2$.[131]

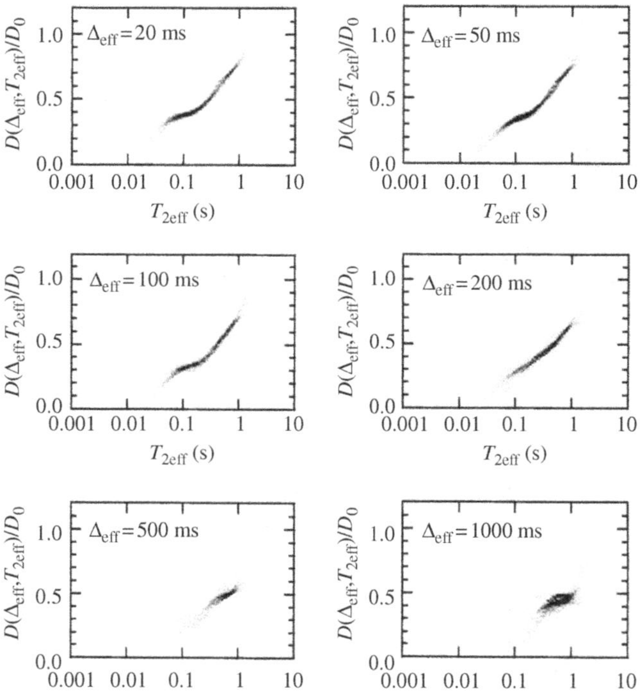

Figure 8.21 Correlation maps of the time-dependent diffusion coefficient, $D(t)$, versus the effective relaxation time (T_{2eff}) and diffusion time (Δ_{eff}) of water in Indiana Limestone. From Hürlimann and Venkataramanan.[131]

Marinelli *et al.*[132] have presented a modal analysis of q-space relaxation correlation experiments. And Callaghan *et al.*[130] have presented analytic expressions for the 2D echo amplitudes for diffusion in planar and spherical pores with relaxation boundary conditions and presented the relaxation diffusion maps which result for a range of Ma/D^0 and $D^0\Delta/a^2$ values (this work can be viewed as an extension to the theory discussed in Chapter 4). It was found that even for these simple geometries the behaviour was quite complex. Their analysis was also instructive in separating the Brownstein–Tarr modes associated with wall relaxation.[133] They remarked that for inverse Laplace transformation to be meaningful, Δ must be sufficiently small so that diffraction effects are not apparent.

8.7.3 Correlations with internal gradients

Sequences have also been presented for probing correlations with internal gradients such as the DICOSY, which probes $P(D, g_{\text{int}}^2 D)$; RICOSY, which probes $P(T_2, g_{\text{int}}^2 D)$; and DRICOSY (see Figure 8.22), which probes $P(D, T_2, g_{\text{int}}^2 D)$.[107,108,134,135]

8.7.4 Dispersive flow

As noted above, a double PGSE sequence can be engineered to remove the effects of constant motions such as flow, and so conversely, such a double PGSE sequence in which the amplitudes of the gradient pulse pairs are stepped simultaneously will be suitable for examining velocity fluctuations.[115,116,118,136,137]

We illustrate the power of the double PGSE sequence for probing the velocity auto-correlation by considering dispersive flow in a porous medium such as packed beads as described in Section 1.9 where the flow around the characteristic length scale is characterised by a correlation time τ_v.[116,138] This is related to the OGSE method outlined in Section 2.4 except that here we will study the velocity correlation function in the time domain instead of the frequency domain. Recalling Eq. (1.1), namely

$$\mathbf{v}(t) = \bar{\mathbf{V}} + \mathbf{u}(t), \tag{8.18}$$

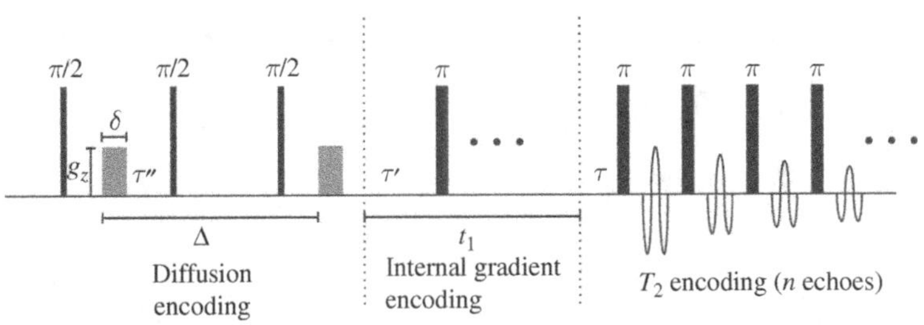

Figure 8.22 Pulse sequence for the 3D DRICOSY experiment.[107,108] The sequence is comprised of a PGSE segment and two CPMG segments. The internal encoding segment consists of a CPMG subsequence of constant length but with a variable number of π pulses, whilst the T_2 encoding segment consists of a train of echoes separated by τ.

and considering only displacement parallel to the gradient (taken to be along z), we can define the associated displacements, $\langle Z \rangle$ and $Z_u(t)$, by

$$\langle Z \rangle + Z_{\mathrm{u}}(t) = \int_0^t \{\bar{V}_z + u_z(t')\}dt'. \tag{8.19}$$

Thus, the echo amplitude is given by

$$\begin{aligned} E &= \langle \exp(i2\pi q Z) \rangle \\ &= \exp(i2\pi q \langle Z \rangle) \langle \exp(i2\pi q Z_{\mathrm{u}}) \rangle, \end{aligned} \tag{8.20}$$

and by taking the modulus

$$|E| = \langle \exp(i2\pi q Z_{\mathrm{u}}) \rangle, \tag{8.21}$$

which provides access to the fluctuating component alone. As before (e.g., see Eq. (3.8)), in the limit of low q it is possible to determine the positional variance, $\langle Z_{\mathrm{u}}^2 \rangle$, and thus the effective time-dependent dispersion coefficient

$$D^*_{\mathrm{eff}} = \frac{\langle Z_{\mathrm{u}}^2 \rangle}{2\Delta}. \tag{8.22}$$

When $\Delta \ll \tau_{\mathrm{v}}$, $D^*_{\mathrm{eff}} = D^0$ and when $\Delta \gg \tau_{\mathrm{v}}$, $D^*_{\mathrm{eff}} = D^*$ which is then equal to the asymptotic dispersion coefficient $D^*(\Delta \to \infty)$.

We now return to the sequence in Figure 8.15 and examine how if the two pulse pairs are applied with identical amplitude (i.e., as a one-dimensional experiment) in a compensated fashion it is possible to probe velocity fluctuations

$$E(q) = \langle \exp(i2\pi q[v_z(0) - v_z(\tau_{\mathrm{m}})]\Delta) \rangle, \tag{8.23}$$

which due to the cancellation of the $\langle \mathrm{V} \rangle$ term becomes[115,139]

$$E(q) = \langle \exp(i2\pi q[u_z(0) - u_z(\tau_{\mathrm{m}})]\Delta) \rangle, \tag{8.24}$$

which in the limit of small q gives the velocity auto-correlation function

$$E(q \to 0) \approx 1 - 4\pi^2 q^2 \Delta^2 \left[\langle u_z^2 \rangle - \langle u_z(0) u_z(\tau_{\mathrm{m}}) \rangle \right], \tag{8.25}$$

and thus

$$D^*_{\mathrm{eff}}(\tau_{\mathrm{m}}) \approx \frac{\Delta}{2} \left[\langle u_z^2 \rangle - \langle u_z(0) u_z(\tau_{\mathrm{m}}) \rangle \right]. \tag{8.26}$$

In the case of isotropic diffusion $\langle u_z^2 \rangle = \frac{1}{3} \langle u^2 \rangle$ and in the limit $\tau_{\mathrm{m}} \gg \tau_{\mathrm{v}}$, $\langle u(0) u(\tau) \rangle = 0$ and thus[115]

$$D^*_{\rm eff}(\Delta \to \infty) \approx \frac{\Delta}{6}\langle u^2 \rangle. \tag{8.27}$$

Thus, the auto-correlation function can be extracted from the ratio

$$\frac{D^*_{\rm eff}(\tau_{\rm m})}{D^*_{\rm eff}(\Delta \to \infty)} \approx 1 - \frac{\langle u_z(0)u_z(\tau_{\rm m})\rangle}{\langle u^2 \rangle}. \tag{8.28}$$

As noted in Section 1.3.2, the velocity auto-correlation function is often described by an exponential decay. Examples of such exponential decays are presented in the paper by Callaghan and Khrapitchev.[138]

In the two-dimensional velocity exchange experiment[116,118,136,137,140,141] for examining velocity distributions (VEXSY; see Figure 8.15), the displacements Z_1 and Z_2 are made over two time intervals separated by a mixing time $\tau_{\rm m}$. Z_1 and Z_2 correspond to the classical spectral domains F_1 and F_2. Thus, the experiment probes the joint probability $P(Z_1, Z_2)$. The echo attenuation is given by

$$E(q) = \langle \exp(i2\pi[q_1 v_z(0) - q_2 v_z(\tau_{\rm m})]\Delta)\rangle. \tag{8.29}$$

The resulting two-dimensional spectrum is reconstructed using Fourier Transformation in both dimensions. A species travelling at a constant velocity has identical displacements and consequently appears on the diagonal. Conversely, off-diagonal peaks arise from species that have migrated from one region of the displacement spectrum to another during $\tau_{\rm m}$. By varying $\tau_{\rm m}$ the molecular velocities can be correlated.

8.8 Flow measurements

Flow measurements have been reviewed by numerous authors including refs. 142–145 and consequently we present only a brief coverage of developments in flow measurements. Sequences which characterise the effects of flow in as short a timescale as possible are desirable – especially if non-steady state phenomena are to be observed. As noted in Chapter 2, PGSE sequences have differing dependencies on flow depending on the gradient moment. Indeed, in the (steady gradient) CPMG sequence only the odd-numbered echoes are susceptible to flow and this has led to the development of multiple pulse sequences with applied gradients for studying flow[54,146,147] (see also Section 8.4.1).

Many developments in measuring flow – especially at low velocities – have come about as a result of developments in measuring ionic drift (see also Section 8.9).[148]

Frydman *et al.*[149] developed a two-dimensional method for studying flow (see Figure 8.23) and thus has some relationship to the discussion in Section 8.7. In this method, a slice-selected spatial distribution of fluid displacements is measured. The

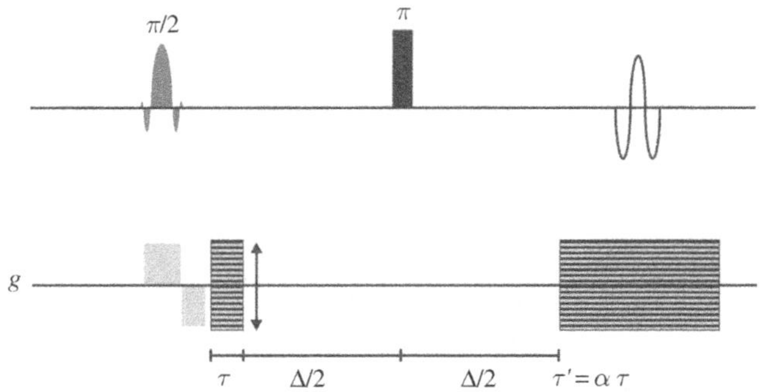

Figure 8.23 The sequence of Frydman *et al.*[149] for obtaining position correlation spectra.

PGSE echo-shape is analysed noting that the phase encoding of the echo evolves from a position dependence at the start of this pulse to a displacement dependence at the echo peak. Thus, the shape of the echo is determined by the joint probability distribution correlating initial particle positions with displacements occurring between the first and the second gradient pulses. Fourier analysis as a function of the gradient is used to extract this distribution. They were able to access Lagrangian flow statistics in Taylor–Couette flow.

We note that the MMME approach for fast diffusion measurements (see Section 8.4.1) can be adapted to perform single-shot flow measurements of flow up to 50 cm s^{-1}.[64,150,151]

8.9 Electrophoretic NMR

The possibility of measuring ionic velocities in the presence of electric fields was recognised early on in the history of PGSE by Packer in 1969[54] and followed by an unsuccessful experimental attempt in 1972.[152] The first successful attempt was by Holz *et al.* in 1982[153,154] and then further improvements were made by Johnson and co-workers.[155,156] This area of research, which has become known as electrophoretic NMR (ENMR), but has also been referred to as DCNMR and magnetic resonance electrophoresis, opens up the possibility of resolving NMR spectra according to the individual electrophoretic mobilities (or drift velocities; μ) of the ionic species in the sample and has been widely reviewed.[157–161] As noted in Section 1.2.4, there are also theoretical relationships between electrophoretic mobilities of the individual species and their self-diffusion coefficients.

The key difference between a PGSE sequence and an ENMR sequence is the addition of a pulsed electric field (E_{dc}). μ is related to the measured velocity (i.e., v) by

$$\mu_{\pm} = \frac{v_{\pm}}{E_{\mathrm{dc}}}, \tag{8.30}$$

where the subscripts '+' and '−' are related to the cationic and anionic species, respectively. It should be noted that upon application of the electric field the constant drift velocity is very quickly – but not instantly – obtained and is likely to be less than an ns.[162] Although one recent study on a concentrated battery electrolyte system indicated that a considerably longer time to reach constant drift velocity is required.[163] However, it is likely that some additional electrokinetic phenomena may be operative in such systems. Early sequences were based on the Stejskal and Tanner sequence[154,157] and a double-echo convection-compensated ENMR pulse sequence is depicted in Figure 8.24 and multi-dimensional sequences such as electrophoretic COSY sequences have also been proposed.[164] In order to deliver the electric field pulses to the sample, the NMR sample tube has to be constructed in the form of an electrophoretic cell (Figure 8.25). The design of the cell is critical since it must: (i) avoid any bubbles generated by electrolysis at the electrode surface from passing through the NMR-sensitive region of the sample (i.e., the volume enclosed by the rf coils) and (ii) prevent the formation of macroscopic flow induced by the bubbles. This explains the prevalence of the U-tube design of electrophoretic NMR cells. Importantly, the design also minimises possibility of the electrodes generating (detectable) rf interference. The development of better electrophoretic NMR cells and better measurement procedures remains an area of considerable interest.[165,166] Detailed considerations of ENMR instrumentation have been given elsewhere[158,159] and we note that at the time of writing the additional instrumentation required for ENMR was unavailable commercially.

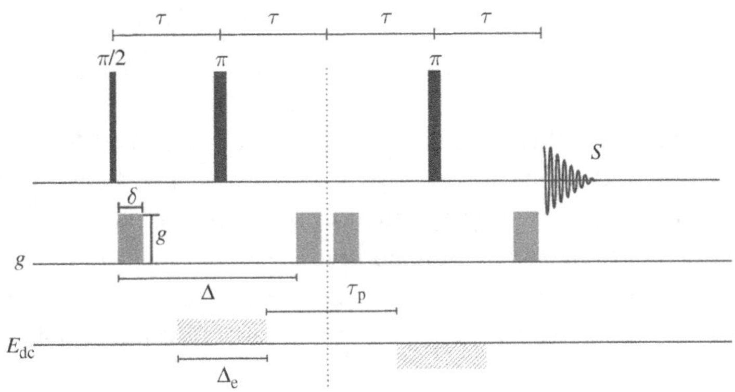

Figure 8.24 ENMR pulse sequence with convection compensation.[167] The line marked with an E_{dc} denotes the electric field pulses.

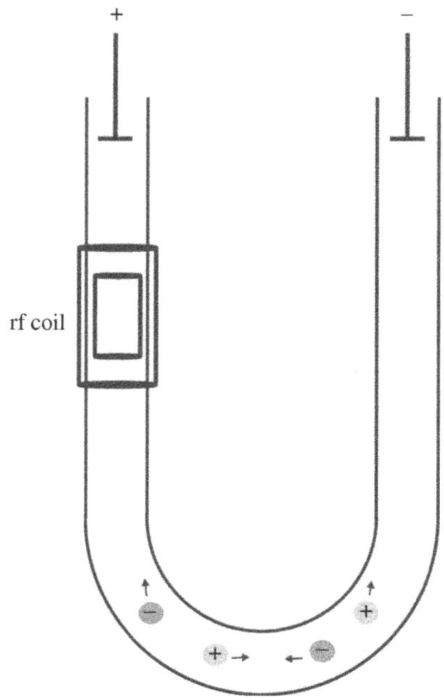

Figure 8.25 A U-tube electrophoretic NMR cell with the anode and cathode electrodes at the top of each arm of the U. The rf coil (in this case a saddle coil) surrounds part of one arm of the U.

The applied current (I) generates an electric field E_{dc} given by

$$E_{dc} = \frac{I}{\Lambda A}, \tag{8.31}$$

where Λ is the conductivity and A is the cross-sectional area of the sample tube. In the ENMR sequence in Figure 8.24, the polarity of E_{dc} was reversed in each half of the sequence. The attenuation of the echo signal for a particular species is given by[167]

$$E(E_{dc}) = \exp\left(-8\pi^2 q^2 D(\Delta - \delta/3)\right) \exp(i4\pi q E_{dc} \Delta\mu). \tag{8.32}$$

The electrophoretic mobility (and direction) is then determined from the complex phase modulation of the echo signal (Figure 8.26). Were the rf coil to encircle both arms of the U tube, only a cosine modulation would be observable since the phase modulation accompanying the magnetic gradient pulses will be opposite on each arm of the U tube. However, the ability to determine the direction of mobility comes at the cost of only being able to observe a small fraction of the sample (i.e., a small filling factor).

Given that only charged species are affected by the pulsed electric field, it is possible to use ENMR as a mobility filter such that in a complex NMR spectrum of a

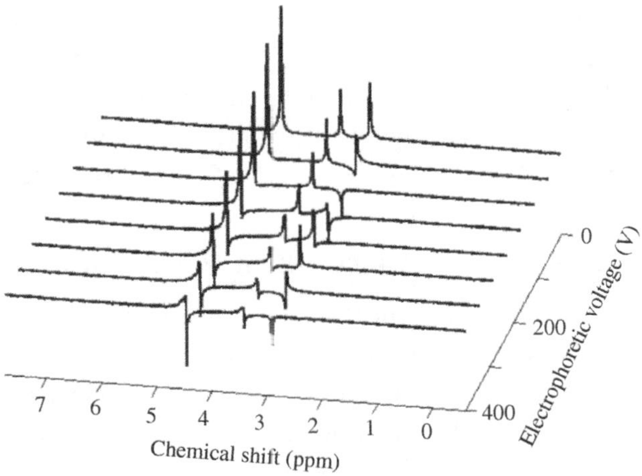

Figure 8.26 ENMR spectra of a solution consisting of 8.6 mM tetramethyl ammonium bromide (TMA) and 0.0071 mM PEO (M_w 100,000) in ^2H_2O at 298 K. The signals at 2.9, 3.4 and 4.5 ppm correspond to protons in TMA^+, PEO and HDO, respectively. From Hallberg *et al.*[166]

multicomponent liquid mixture, the resonances of the electrically charged species can be selectively filtered.[168] The method works by subtracting spectra acquired using an ENMR sequence with and without the pulsed electric field. The signals from the uncharged species should theoretically be equal in both spectra whilst those from the charged species should have a phase shift.

Similar to PGSE measurements being displayed as DOSY plots, it is possible to Fourier transform the ENMR signal to give a 2D plot (i.e., one dimension is the usual chemical shift and the second dimension is intensity versus Δ or E_{dc}).[156,159,169–173] Such displays have become known by the acronym MOSY (for mobility-ordered spectroscopy). In practice, one of the two quantities can be incremented to give the second dimension (i.e., Δ or v – by incrementing E_{dc}). If Δ is incremented it leads also to greater diffusive attenuation and thus a line broadening in the second dimension. Hence, it is preferable to increment E_{dc} as this will result only in a phase modulation (i.e., the second exponential in Eq. (8.32)). Nevertheless, as it is normally impossible to acquire many oscillations, the undamped signal that needs to be Fourier transformed is highly truncated and techniques such as linear prediction[156] or the maximum entropy method,[174] may need to be employed.

Like diffusion measurements, ENMR is susceptible to a number of phenomena which can lead to artificial dampening of the echo intensity and thus result in incorrect interpretation of mobilities.[158–160,165] These deleterious phenomena include electroosmosis (i.e., the flow of liquid outside a charged surface due to the application of an electric field that has components parallel to the surface), convection induced by resistive heating, and increasing diffusion with increase of temperature or flow due to formation and movement of bubbles in the sample capillary tube. In a recent paper, Hallberg *et al.*[166] have suggested measurement procedures that compensate for most of these complications. It requires no electroosmotic coating, and can be readily performed in standard field gradient-equipped probes.

8.10 Multiple spin-echoes and miscellaneous

Diffusion measurements and theoretical advances involving the demagnetising field and multiple spin-echoes have been presented. However, they are outside the scope of the present work and the interested reader should consult the relevant literature.[175–185]

Measurements involving MAS (including high-resolution MAS; HRMAS) are becoming increasingly commonplace and it has been shown that accurate self-diffusion coefficients of liquids can be obtained using PGSE of samples undergoing MAS on a probe equipped with a magic angle gradient coil, providing that the rotor volume was kept to a minimum.[186]

Due to the much more rapid diffusion of gases, it is necessary to study diffusion on much shorter timescales than those used for liquids in order to study the short-time PGSE attenuation behaviour. To this end, pulse sequences related to OGSE sequences but designed to extract time-domain not frequency-domain information at very short timescales have been developed for studying the diffusion of hyperpolarised gases.[187] It has recently been realized that fast translational motion can lead to significant signal intensity prior to the expected time of echo formation.[188] This has been termed a pseudo spin-echo.

Readers interested in diffusion measurements with single-sided NMR techniques are encouraged to consult the relevant literature.[189]

References

1. R. Kimmich, W. Unrath, G. Schnur, and E. Rommel, NMR Measurement of Small Self-Diffusion Coefficients in the Fringe Field of Superconducting Magnets. *J. Magn. Reson.* **91** (1991), 136–40.
2. G. Farrher, I. Ardelean, and R. Kimmich, Probing Four Orders of Magnitude of the Diffusion Time in Porous Silica Glass with Unconventional NMR Techniques. *J. Magn. Reson.* **182** (2006), 215–20.
3. T. J. Norwood and R. A. Quilter, A Robust NMR Method for Studying Diffusion. *J. Magn. Reson.* **97** (1992), 99–110.
4. T. J. Norwood, New NMR Methods for Measuring Diffusion. *J. Magn. Reson. A* **103** (1993), 258–67.
5. P. J. McDonald, Stray Field Magnetic Resonance Imaging. *Prog. NMR Spectrosc.* **30** (1997), 69–99.
6. P. J. McDonald and B. Newling, Stray Field Magnetic Resonance Imaging. *Rep. Prog. Phys.* **61** (1998), 1441–93.

7. B. Geil, Measurement of Translational Molecular Diffusion Using Ultrahigh Magnetic Field Gradient NMR. *Concepts Magn. Reson.* **10** (1998), 299–321.
8. D. E. Demco, A. Johansson, and J. Tegenfeldt, Constant-Relaxation Methods for Diffusion Measurements in the Fringe Field of Superconducting Magnets. *J. Magn. Reson. A* **110** (1994), 183–93.
9. E. Fischer and R. Kimmich, Constant Time Steady Gradient NMR Diffusometry Using the Secondary Stimulated Echo. *J. Magn. Reson.* **166** (2004), 273–9.
10. R. Kimmich and E. Fischer, One- and Two-Dimensional Pulse Sequences for Diffusion Experiments in the Fringe Field of Superconducting Magnets. *J. Magn. Reson. A* **106** (1994), 229–35.
11. T. Feiweier, B. Geil, O. Isfort, and F. Fujara, Demonstrating Spatial Resolution of Field Gradient NMR. *J. Magn. Reson.* **131** (1998), 203–8.
12. E. E. Sigmund and W. P. Halperin, Hole-Burning Diffusion Measurements in High Magnetic Field Gradients. *J. Magn. Reson.* **163** (2003), 99–104.
13. M. D. Hürlimann and D. D. Griffin, Spin Dynamics of Carr–Purcell–Meiboom–Gill-like Sequences in Grossly Inhomogeneous B_0 and B_1 Fields and Application to NMR Well Logging. *J. Magn. Reson.* **143** (2000), 120–35.
14. D. E. Freed, U. M. Scheven, L. J. Zielinski, P. N. Sen, and M. D. Hürlimann, Steady-State Free Precession Experiments and Exact Treatment of Diffusion in a Uniform Gradient. *J. Chem. Phys.* **115** (2001), 4249–58.
15. M. D. Hürlimann, Carr–Purcell Sequences with Composite Pulses. *J. Magn. Reson.* **152** (2001), 109–23.
16. M. D. Hürlimann, Optimization of Timing in the Carr–Purcell–Meiboom–Gill Sequence. *Magn. Reson. Imaging* **19** (2001), 375–8.
17. M. D. Hürlimann, Diffusion and Relaxation Effects in General Stray Field NMR Experiments. *J. Magn. Reson.* **148** (2001), 367–78.
18. G. Leu, E. J. Fordham, M. D. Hürlimann, and P. Frulla, Fixed and Pulsed Gradient Diffusion Methods in Low-Field Core Analysis. *Magn. Reson. Imaging* **23** (2005), 305–9.
19. R. C. Wilson and M. D. Hürlimann, Relationship Between Susceptibility Induced Field Inhomogeneities, Restricted Diffusion, and Relaxation in Sedimentary Rocks. *J. Magn. Reson.* **183** (2006), 1–12.
20. M. D. Hürlimann, Encoding of Diffusion and T_1 in the CPMG Echo Shape: Single-Shot D and T_1 Measurements in Grossly Inhomogeneous Fields. *J. Magn. Reson.* **184** (2007), 114–29.
21. A. R. Preston, P. Kinchesh, and E. W. Randall, Calibration of the Stray Field Gradient by a Heteronuclear Method and by Field Profiling. *J. Magn. Reson.* **146** (2000), 359–62.
22. P. T. Callaghan, C. D. Eccles, and J. D. Seymour, An Earth's Field Nuclear Magnetic Resonance Apparatus Suitable for Pulsed Gradient Spin Echo Measurements of Self-Diffusion under Arctic Conditions. *Rev. Sci. Instrum.* **68** (1997), 4263–70.
23. P. T. Callaghan, A. Coy, R. Dykstra, C. D. Eccles, M. E. Halse, M. W. Hunter, O. R. Mercier, and J. N. Robinson, New Zealand Developments in Earth's Field NMR. *Appl. Magn. Reson.* **32** (2007), 63–74.
24. J. Stepišnik, Violation of the Gradient Approximation in NMR Self-Diffusion Measurements. *Z. Phys. Chem.* **190** (1995), 51–62.
25. A. Mohorič, J. Stepišnik, M. Kos, and G. Planinšič, Self-Diffusion Imaging by Spin Echo in Earth's Magnetic Field. *J. Magn. Reson.* **136** (1999), 22–6.
26. P. T. Callaghan and J. Stepišnik, Spatially-Distributed Pulsed Field Gradient Spin Echo NMR using Single-Wire Proximity. *Phys. Rev. Lett.* **75** (1995), 4532–5.

27. J. F. Martin, L. S. Selwyn, R. R. Vold, and R. L. Vold, The Determination of Translational Diffusion Constants in Liquid Crystals from Pulsed Field Gradient Double Quantum Spin Echo Decays. *J. Chem. Phys.* **76** (1982), 2632–4.
28. D. Zax and A. Pines, Study of Anisotropic Diffusion of Oriented Molecules by Multiple Quantum Spin Echoes. *J. Chem. Phys.* **78** (1983), 6333–4.
29. L. E. Kay and J. H. Prestegard, An Application of Pulse-Gradient Double Quantum Spin Echoes to Diffusional Measurements on Molecules with Scalar-Coupled Spins. *J. Magn. Reson.* **67** (1986), 103–13.
30. B. E. Chapman and P. W. Kuchel, Sensitivity in Heteronuclear Multiple-Quantum Diffusion Experiments. *J. Magn. Reson. A* **102** (1993), 105–9.
31. P. W. Kuchel and B. E. Chapman, Heteronuclear Double-Quantum-Coherence Selection with Magnetic-Field Gradients in Diffusion Experiments. *J. Magn. Reson. A* **101** (1993), 53–9.
32. L. van Dam, Bo. Andreasson, and L. Nordenskiöld, Multiple-Quantum Pulsed Gradient NMR Diffusion Experiments on Quadrupolar ($I > 1/2$) Spins. *Chem. Phys. Lett.* **262** (1996), 737–43.
33. C. H. Sotak, A Method for Measuring the Apparent Self-Diffusion Coefficient of in Vivo Lactic Acid Using Double-Quantum Coherence-Transfer Spectroscopy. *J. Magn. Reson.* **90** (1990), 198–204.
34. C. Dalvit and J. M. Böhlen, Analysis of Biofluids and Chemical Mixtures in Non-deuterated Solvents with ^{1}H Diffusion-Weighted PFG Phase-Sensitive Double Quantum NMR Spectroscopy. *NMR Biomed.* **10** (1997), 285–91.
35. K. I. Momot and P. W. Kuchel, Convection-Compensating Diffusion Experiments with Phase-Sensitive Double-Quantum Filtering. *J. Magn. Reson.* **174** (2005), 229–36.
36. T. J. Norwood, An Eddy-Current-Independent Multiple-Quantum Method for Measuring the Diffusion of Coupled Spins. *J. Magn. Reson.* **99** (1992), 208–13.
37. C. Dalvit and J. M. Böhlen, Multiple-Solvent Suppression in Double-Quantum NMR Experiments with Magic Angle Pulsed Field Gradients. *Magn. Reson. Chem.* **34** (1996), 829–33.
38. P. T. Callaghan, M. A. Le Gros, and D. N. Pinder, The Measurement of Diffusion Using Deuterium Pulsed Gradient Nuclear Magnetic Resonance. *J. Chem. Phys.* **79** (1983), 6372–81.
39. I. Furó and B. Halle, 2D Quadrupolar-Echo Spectroscopy with Coherence Selection and Optimized Pulse Angle. *J. Magn. Reson.* **98** (1992), 388–407.
40. I. Furó and H. Jóhannesson, Accurate Anisotropic Water-Diffusion Measurements in Liquid Crystals. *J. Magn. Reson. A* **119** (1996), 15–21.
41. D. Wu, A. Chen, and C. S. Johnson, Jr., Heteronuclear-Detected Diffusion-Ordered NMR Spectroscopy through Coherence Transfer. *J. Magn. Reson. A* **123** (1996), 215–18.
42. A. J. Dingley, J. P. Mackay, G. L. Shaw, B. D. Hambly, and G. F. King, Measuring Macromolecular Diffusion Using Heteronuclear Multiple-Quantum Pulsed-Field-Gradient NMR. *J. Biomol. NMR* **10** (1997), 1–8.
43. C. Dalvit, P. Ramage, and U. Hommel, Heteronuclear X-Filter ^{1}H PFG Double-Quantum Experiments for the Proton Resonance Assignment of a Ligand Bound to a Protein. *J. Magn. Reson.* **131** (1998), 148–53.
44. M. L. Tillett, M. A. Horsfield, L.-Y. Lian, and T. J. Norwood, Protein-Ligand Interactions Measured by ^{15}N-Filtered Diffusion Experiments. *J. Biomol. NMR* **13** (1999), 223–32.
45. M. Liu, X.-A. Mao, C. Ye, J. K. Nicholson, and J. C. Lindon, Enhanced Effect of Magnetic Field Gradients Using Multiple Quantum NMR Spectroscopy Applied to Self-Diffusion Coefficient Measurement. *Mol. Phys.* **93** (1998), 913–20.

46. R.-S. Luo, M. Liu, and X.-A. Mao, Eliminating Systematic Error in Multiple Quantum Diffusion Measurements by Bipolar Gradient Pulses. *Meas. Sci. Technol.* **9** (1998), 1347–50.
47. F. Ferrage, M. Zoonens, D. E. Warschawski, J.-L. Popot, and G. Bodenhausen, Slow Diffusion of Macromolecular Assemblies by a New Pulsed Field Gradient NMR Method. *J. Am. Chem. Soc.* **125** (2003), 2541–5.
48. F. Ferrage, M. Zoonens, D. E. Warschawski, J.-L. Popot, and G. Bodenhausen, Slow Diffusion of Macromolecular Assemblies by a New Pulsed Field Gradient NMR Method. *J. Am. Chem. Soc.* **126** (2004), 5654.
49. T. J. Norwood, Magnetic Field Gradients in NMR: Friend or Foe? *Chem. Soc. Rev.* **23** (1994), 59–66.
50. L. D. Hall and T. J. Norwood, A New Method for Studying Diffusion Using a Static Magnetic Field Gradient. *J. Magn. Reson.* **88** (1990), 192–8.
51. S. Cavadini, J. Dittmer, S. Antonijevic, and G. Bodenhausen, Slow Diffusion by Singlet State NMR Spectroscopy. *J. Am. Chem. Soc.* **127** (2005), 15744–48.
52. S. Cavadini and P. R. Vasos, Singlet States Open the Way to Longer Time-Scales in the Measurement of Diffusion by NMR Spectroscopy. *Concepts Magn. Reson.* **32A** (2008), 68–78.
53. P. L. de Sousa, D. Abergel, and J.-Y. Lallemand, Experimental Time Saving in NMR Measurement of Time Dependent Diffusion Coefficients. *Chem. Phys. Lett.* **342** (2001), 45–50.
54. K. J. Packer, The Study of Slow Coherent Molecular Motion by Pulsed Nuclear Magnetic Resonance. *Mol. Phys.* **17** (1969), 355–68.
55. Y.-Q. Song, M. D. Hürlimann, and C. Flaum, A Method for Rapid Characterization of Diffusion. *J. Magn. Reson.* **161** (2003), 222–33.
56. L. Li and C. H. Sotak, Diffusion Measurements by Pulsed Field-Gradient Multiple Spin Echoes. *J. Magn. Reson.* **92** (1991), 411–20.
57. L. Li and C. H. Sotak, Self-Diffusion Measurements by Pulsed-Gradient Multiple Spin-Echo Imaging. *J. Magn. Reson. B* **101** (1993), 8–16.
58. R. W. Mair, D. G. Cory, S. Peled, C.-H. Tseng, S. Patz, and R. L. Walsworth, Pulsed-Field-Gradient Measurements of Time-Dependent Gas Diffusion. *J. Magn. Reson.* **135** (1998), 478–86.
59. P. van Gelderen, A. Olson, and C. T. W. Moonen, A Single-Shot Diffusion Experiment. *J. Magn. Reson. A* **103** (1993), 105–8.
60. S. J. Doran and M. Décorps, A Robust, Single-Shot Method for Measuring Diffusion Coefficients Using the 'Burst' Sequence. *J. Magn. Reson. A* **117** (1995), 311–16.
61. S. J. Doran, M. E. Bourgeois, and M. O. Leach, Burst Imaging – Can It Ever be Useful in the Clinic? *Concepts Magn. Reson.* **26A** (2005), 11–34.
62. S. Peled, C.-H. Tseng, A. Sodickson, R. W. Mair, R. L. Walsworth, and D. G. Cory, Single-Shot Diffusion Measurement in Laser-Polarized Gas. *J. Magn. Reson.* **140** (1999), 320–4.
63. Y.-Q. Song and X. Tang, A One-Shot Method for Measurement of Diffusion. *J. Magn. Reson.* **170** (2004), 136–48.
64. E. E. Sigmund, H. Cho, and Y.-Q. Song, Multiple-Modulation-Multiple-Echo Magnetic Resonance. *Concepts Magn. Reson.* **30A** (2007), 358–77.
65. X.-P. Tang, E. E. Sigmund, and Y.-Q. Song, Simultaneous Measurement of Diffusion Along Multiple Directions. *J. Am. Chem. Soc.* **126** (2004), 16336–7.
66. S. Sendhil Velan and N. Chandrakumar, High-Resolution NMR Measurement of Molecular Self-Diffusion by Fast Multi-Spin-Echo Diffusion Studies. *J. Magn. Reson. A* **123** (1996), 122–5.

67. J. P. Stamps, B. Ottink, J. M. Visser, J. P. M. van Duynhoven, and R. Hulst, Difftrain: A Novel Approach to a True Spectroscopic Single-Scan Diffusion Measurement. *J. Magn. Reson.* **151** (2001), 28–31.
68. C. Buckley, C. A. Hollingsworth, A. J. Sederman, D. J. Holland, M. L. Johns, and L. F. Gladden, Applications of Fast Diffusion Measurement Using Difftrain. *J. Magn. Reson.* **161** (2003), 112–17.
69. R. M. Cotts, M. J. R. Hoch, T. Sun, and J. T. Markert, Pulsed Field Gradient Stimulated Echo Methods for Improved NMR Diffusion Measurements in Heterogeneous Systems. *J. Magn. Reson.* **83** (1989), 252–66.
70. O. Millet and M. Pons, A New Method for Measuring Diffusion Coefficients by 2D NMR Using Accordion Spectroscopy. *J. Magn. Reson.* **131** (1998), 166–9.
71. G. Bodenhausen and R. R. Ernst, Direct Determination of Rate Constants of Slow Dynamic Processes by Two-Dimensional 'Accordion' Spectroscopy in Nuclear Magnetic Resonance. *J. Am. Chem. Soc.* **104** (1982), 1304–9.
72. N. M. Loening, J. Keeler, and G. A. Morris, One-Dimensional DOSY. *J. Magn. Reson.* **153** (2001), 103–12.
73. M. J. Thrippleton, N. M. Loening, and J. Keeler, A Fast Method for the Measurement of Diffusion Coefficients: One-Dimensional DOSY. *Magn. Reson. Chem.* **41** (2003), 441–7.
74. J. M. Böhlen, I. Burghardt, M. Rey, and G. Bodenhausen, Frequency-Modulated 'Chirp' Pulses for Broadband Inversion Recovery in Magnetic Resonance. *J. Magn. Reson.* **90** (1990), 183–91.
75. M. D. Pelta, G. A. Morris, M. J. Stchedroff, and S. J. Hammond, A One-Shot Sequence for High Resolution Diffusion Ordered Spectroscopy. *Magn. Reson. Chem.* **40** (2002), S147–52.
76. A. Métais and F. Mariette, Determination of Water Self-Diffusion Coefficient in Complex Food Products by Low Field ^{1}H PFG-NMR: Comparison Between the Standard Spin-Echo Sequence and the T_1-Weighted Spin-Echo Sequence. *J. Magn. Reson.* **165** (2003), 265–75.
77. J. C. Van Den Enden, D. Waddington, H. Van Aalst, C. G. Van Kralingen, and K. J. Packer, Rapid Determination of Water Droplet Size Distributions by PFG-NMR. *J. Colloid Interface Sci.* **140** (1990), 105–13.
78. W. Heink, J. Kärger, and H. Pfeifer, A Simple Pulse Sequence to Exclude Artifacts in Self-Diffusion Measurements by Means of the NMR Pulsed Field Gradient Technique. *Z. Phys. Chem.* **170** (1991), 199–206.
79. D. van Dusschoten, P. A. De Jager, and H. Van As, Extracting Diffusion Constants from Echo-Time-Dependent PFG NMR Data using Relaxation-Time Information. *J. Magn. Reson. A* **116** (1995), 22–8.
80. D. van Dusschoten, C. T. W. Moonen, P. A. De Jager, and H. Van As, Unraveling Diffusion Constants in Biological Tissue by Combining Carr–Purcell–Meiboom–Gill Imaging and Pulsed Field Gradient NMR. *Magn. Reson. Med.* **36** (1996), 907–13.
81. G. J. Stanisz, J. G. Li, G. A. Wright, and R. M. Henkelman, Water Dynamics in Human Blood via Combined Measurements of T_2 Relaxation and Diffusion in the Presence of Gadolinium. *Magn. Reson. Med.* **39** (1998), 223–33.
82. A. M. Dixon and C. K. Larive, Modified Pulsed-Field Gradient NMR Experiments for Improved Selectivity in the Measurement of Diffusion Coefficients in Complex Mixtures: Application to the Analysis of the Suwannee River Fulvic Acid. *Anal. Chem.* **69** (1997), 2122–8.
83. W. H. Otto and C. K. Larive, Improved Spin-Echo-Edited NMR Diffusion Measurements. *J. Magn. Reson.* **153** (2001), 273–6.

84. N. Birlirakis and E. Guittet, A New Approach in the Use of Gradients for Size-Resolved 2D-NMR Experiments. *J. Am. Chem. Soc.* **118** (1996), 13083–4.
85. C. Casieri, C. Testa, M. Paci, and F. De Luca, Molecular Self-Diffusion Measurement by Stimulated Echo of Selected ^{13}C–^{1}H Bonds: Case of the Glucose Metabolites. *Chem. Phys. Lett.* **387** (2004), 295–300.
86. I. V. Nesmelova, D. Idiyatullin, and K. H. Mayo, Measuring Protein Self-Diffusion in Protein–Protein Mixtures Using a Pulsed Gradient Spin-Echo Technique with WATERGATE and Isotope Filtering. *J. Magn. Reson.* **166** (2004), 129–33.
87. M. Nilsson, A. M. Gil, I. Delgadillo, and G. A. Morris, Improving Pulse Sequences for 3D Diffusion-Ordered NMR Spectroscopy: 2DJ-IDOSY. *Anal. Chem.* **76** (2004), 5418–22.
88. D. Wu, A. Chen, and C. S. Johnson, Jr., Three-Dimensional Diffusion-Ordered NMR Spectroscopy: The Homonuclear COSY-DOSY Experiment. *J. Magn. Reson. A* **121** (1996), 88–91.
89. M. Nilsson, A. M. Gil, I. Delgadillo, and G. A. Morris, Improving Pulse Sequences for 3D DOSY: COSY-IDOSY. *Chem. Commun.* (2005), 1737–9.
90. M. J. Stchedroff, A. M. Kenwright, G. A. Morris, M. Nilsson, and R. K. Harris, 2D and 3D DOSY Methods for Studying Mixtures of Oligomeric Dimethylsiloxanes. *Phys. Chem. Chem. Phys.* **6** (2004), 3221–7.
91. H. Barjat, G. A. Morris, and A. G. Swanson, A Three-Dimensional DOSY-HMQC Experiment for the High-Resolution Analysis of Complex Mixtures. *J. Magn. Reson.* **131** (1998), 131–8.
92. B. Vitorge and D. Jeanneat, NMR Diffusion Measurements in Complex Mixtures Using Constant-Time-HSQC-IDOSY and Computer-Optimized Spectral Aliasing for High Resolution in the Carbon Dimension. *Anal. Chem.* **78** (2006), 5601–6.
93. E. K. Gozansky and D. G. Gorenstein, DOSY-NOESY: Diffusion-Ordered NOESY. *J. Magn. Reson. B* **111** (1996), 94–6.
94. M. Liu, J. K. Nicholson, and J. C. Lindon, High-Resolution Diffusion and Relaxation Edited One and Two-Dimensional ^{1}H NMR Spectroscopy of Biological Fluids. *Anal. Chem.* **68** (1996), 3370–6.
95. A. Jerschow and N. Müller, 3D Diffusion-Ordered TOCSY for Slowly Diffusing Molecules. *J. Magn. Reson. A* **123** (1996), 222–5.
96. M. Liu, J. K. Nicholson, J. A. Parkinson, and J. C. Lindon, Measurement of Biomolecular Diffusion Coefficients in Blood Plasma Using Two-Dimensional ^{1}H-^{1}H Diffusion-Edited Total-Correlation NMR Spectroscopy. *Anal. Chem.* **69** (1997), 1504–9.
97. S. A. Bradley, K. Krishnamurthy, and H. Hu, Simplifying DOSY Spectra with Selective TOCSY Edited Preparation. *J. Magn. Reson.* **172** (2005), 110–17.
98. L. H. Lucas, W. H. Otto, and C. K. Larive, The 2D-*J*-DOSY Experiment: Resolving Diffusion Coefficients in Mixtures. *J. Magn. Reson.* **156** (2002), 138–45.
99. M. Nilsson and G. A. Morris, Improving Pulse Sequences for 3D DOSY: Convection Compensation. *J. Magn. Reson.* **177** (2005), 203–11.
100. C. A. Steinbeck and B. F. Chmelka, Rapid $^{1}H\{^{13}C\}$ – Resolved Diffusion and Spin-Relaxation Measurements by NMR Spectroscopy. *J. Am. Chem. Soc.* **127** (2005), 11624–35.
101. C. T. W. Moonen, P. van Gelderen, G. W. Vuister, and P. C. M. Van Zijl, Gradient-Enhanced Exchange Spectroscopy. *J. Magn. Reson.* **97** (1992), 419–25.
102. J. H. Lee, C. Labadie, C. S. Springer, Jr., and G. S. Harbison, Two-Dimensional Inverse Laplace Transform NMR: Altered Relaxation Times Allow Detection of Exchange Correlation. *J. Am. Chem. Soc.* **115** (1993), 7761–4.

103. Y.-Q. Song, L. Venkataramanan, M. D. Hürlimann, M. Flaum, P. Frulla, and C. Straley, T_1–T_2 Correlation Spectra Obtained Using a Fast Two-Dimensional Laplace Inversion. *J. Magn. Reson.* **154** (2002), 261–8.
104. L. Venkataramanan, Y.-Q. Song, and M. D. Hürlimann, Solving Fredholm Integrals of the First Kind with Tensor Product Structure in 2 and 2.5 Dimensions. *IEEE Trans. Signal Process.* **50** (2002), 1017–26.
105. P. T. Callaghan, S. Godefroy, and B. N. Ryland, Use of the Second Dimension in PGSE NMR Studies of Porous Media. *Magn. Reson. Imaging* **21** (2003), 243–8.
106. Y.-Q. Song, L. Venkataramanan, and L. Burcaw, Determining the Resolution of Laplace Inversion Spectrum. *J. Chem. Phys.* **122** (2005), 104104-1–104104-10.
107. C. H. Arns, K. E. Washburn, and P. T. Callaghan, Multidimensional NMR Inverse Laplace Spectroscopy in Petrophysics. *Petrophysics.* **48** (2007), 380–92.
108. P. T. Callaghan, C. H. Arns, P. Galvosas, M. J. Hunter, Y. Qiao, and K. E. Washburn, Recent Fourier and Laplace Perspectives for Multidimensional NMR in Porous Media. *Magn. Reson. Imaging* **25** (2007), 441–4.
109. R. R. Ernst, G. Bodenhausen, and A. Wokaun, *Principles of Magnetic Resonance in One and Two Dimensions*. (London: Clarendon Press, 1987).
110. D. G. Cory, A. N. Garroway, and J. B. Miller, Applications of Spin Transport as a Probe of Local Geometry. *Polym. Prepr.* **31** (1990), 149–52.
111. P. T. Callaghan, How Two Pairs of Gradient Pulses Give Access to New Information about Molecular Dynamics. In *Diffusion Fundamentals*, ed. J. Kärger, F. Grinberg, and P. Heitjans. (Leipzig: University of Leipzig, 2005), pp. 321–38.
112. L. M. Li and C. H. Sotak, A Method for Evaluating Anisotropic and Restricted Diffusion by Simultaneous Use of Spin and Stimulated Echoes. *J. Magn. Reson.* **96** (1992), 501–13.
113. P. P. Mitra, Multiple Wave-Vector Extensions of the NMR Pulsed-Field-Gradient Spin-Echo Diffusion Measurement. *Phys. Rev. B* **51** (1995), 15074–8.
114. E. Özarslan and P. J. Basser, MR Diffusion – 'Diffraction' Phenomenon in Multi-Pulse-Field-Gradient Experiments. *J. Magn. Reson.* **188** (2007), 285–94.
115. P. T. Callaghan, *Principles of Nuclear Magnetic Resonance Microscopy.* (Oxford: Clarendon Press, 1991).
116. P. T. Callaghan, S. L. Codd, and J. D. Seymour, Spatial Coherence Phenomena Arising from Translational Spin Motion in Gradient Spin Echo Experiments. *Concepts Magn. Reson.* **11** (1999), 181–202.
117. A. Caprihan and J. D. Seymour, Correlation Time and Diffusion Coefficient Imaging: Application to a Granular Flow System. *J. Magn. Reson.* **144** (2000), 96–107.
118. P. T. Callaghan and B. Manz, Velocity Exchange Spectroscopy. *J. Magn. Reson. A* **106** (1994), 260–5.
119. M. D. Silva, K. G. Helmer, J.-H. Lee, S. S. Han, and C. S. Springer, Jr., Deconvolution of Compartmental Water Diffusion Coefficients in Yeast-Cell Suspensions Using Combined T_1 and Diffusion Measurements. *J. Magn. Reson.* **156** (2002), 52–63.
120. M. M. Britton, R. G. Graham, and K. J. Packer, Relationships Between Flow and NMR Relaxation of Fluids in Porous Solids. *Magn. Reson. Imaging* **19** (2001), 325–31.
121. M. W. Hunter and P. T. Callaghan, NMR Measurement of Nonlocal Dispersion in Complex Flows. *Phys. Rev. Lett.* **99** (2007), 210602-1–210602-4.
122. Y. Cheng and D. G. Cory, Multiple Scattering by NMR. *J. Am. Chem. Soc.* **121** (1999), 7935–6.
123. P. T. Callaghan and M. E. Komlosh, Locally Anisotropic Motion in a Macroscopically Isotropic System: Displacement Correlations Measured Using Double Pulsed Gradient Spin-Echo NMR. *Magn. Reson. Chem.* **40** (2002), S15–19.

124. C.-L. Chin, F. W. Wehrli, S. N. Hwang, D. L. Jaggard, D. B. Hackney, and S. W. Wehrli, Feasibility of Probing Boundary Morphology of Structured Materials by 2D NMR *q*-Space Imaging. *J. Magn. Reson.* **160** (2003), 20–5.
125. K. J. Packer, S. Stapf, J. J. Tessier, and R. A. Damion, The Characterisation of Fluid Transport in Porous Solids by Means of Pulsed Magnetic Field Gradient NMR. *Magn. Reson. Imaging* **16** (1998), 463–9.
126. S. Stapf, K. J. Packer, R. G. Graham, J.-F. Thovert, and P. M. Adler, Spatial Correlations and Dispersion for Fluid Transport Through Packed Glass Beads Studied by Pulsed Field-Gradient NMR. *Phys. Rev. E* **58** (1998), 6206–21.
127. R. G. Graham, W. M. Holmes, C. De Panfilis, and K. J. Packer, Characterisation of Locally Anisotropic Structures Within Isotropic Porous Solids Using 2-D Pulsed Field Gradient NMR. *Chem. Phys. Lett.* **332** (2000), 319–23.
128. P. T. Callaghan and I. Furó, Diffusion-Diffusion Correlation and Exchange as a Signature for Local Order and Dynamics. *J. Chem. Phys.* **120** (2004), 4032–8.
129. Y. Qiao, P. Galvosas, and P. T. Callaghan, Diffusion Correlation NMR Spectroscopic Study of Anisotropic Diffusion of Water in Plant Tissues. *Biophys. J.* **89** (2005), 2899–905.
130. P. T. Callaghan, S. Godefroy, and B. N. Ryland, Diffusion–Relaxation Correlation in Simple Pore Structures. *J. Magn. Reson.* **162** (2003), 320–7.
131. M. D. Hürlimann and L. Venkataramanan, Quantitative Measurement of Two-Dimensional Distribution Functions of Diffusion and Relaxation in Grossly Inhomogeneous Fields. *J. Magn. Reson.* **157** (2002), 31–42.
132. L. Marinelli, M. D. Hürlimann, and P. N. Sen, Modal Analysis of *q*-Space – Relaxation Correlation Experiments. *J. Chem. Phys.* **118** (2003), 8927–40.
133. K. R. Brownstein and C. E. Tarr, Importance of Classical Diffusion in NMR Studies of Water in Biological Cells. *Phys. Rev. A* **19** (1979), 2446–53.
134. B. Sun and K.-J. Dunn, Probing the Internal Field Gradients of Porous Media. *Phys. Rev. E* **65** (2002), 051309-1–051309-7.
135. J. G. Seland, K. E. Washburn, H. W. Anthonsen, and J. Krane, Correlations Between Diffusion, Internal Magnetic Field Gradients, and Transverse Relaxation in Porous Systems Containing Oil and Water. *Phys. Rev. E* **70** (2004), 051305-1–051305-10.
136. A. A. Khrapitchev and P. T. Callaghan, Double PGSE NMR with Stimulated Echoes: Phase Cycles for the Selection of Desired Encoding. *J. Magn. Reson.* **152** (2001), 259–68.
137. P. T. Callaghan, Some Perspectives on Dispersion and the Use of Ensemble-Averaged PGSE NMR. *Magn. Reson. Imaging* **23** (2005), 133–7.
138. P. T. Callaghan and A. A. Khrapitchev, Time-Dependent Velocities in Porous Media Dispersive Flow. *Magn. Reson. Imaging* **19** (2001), 301–5.
139. S. L. Codd, B. Manz, J. D. Seymour, and P. T. Callaghan, Taylor Dispersion and Molecular Displacements in Poiseuille Flow. *Phys. Rev. E* **60** (1999), R3491–4.
140. P. T. Callaghan and J. Stepišnik, Generalized Analysis of Motion Using Magnetic Field Gradients. *Adv. Magn. Opt. Reson.* **19** (1996), 325–88.
141. A. A. Khrapitchev, S. Stapf, and P. T. Callaghan, NMR Visualization of Displacement Correlations for Flow in Porous Media. *Phys. Rev. E* **66** (2002), 051203-1–051203-16.
142. A. Caprihan and E. Fukushima, Flow Measurements by NMR. *Phys. Rep.* **198** (1990), 195–235.
143. K. J. Packer, Diffusion & Flow in Liquids. In *Encyclopedia of Nuclear Magnetic Resonance*, ed. D. M. Grant and R. K. Harris. vol. 3. (New York: Wiley, 1996), pp. 1615–26.
144. E. Fukushima, Nuclear Magnetic Resonance as a Tool to Study Flow. *Annu. Rev. Fluid Mech.* **31** (1999), 95–123.
145. B. Newling, Gas Flow Measurements by NMR. *Prog. NMR Spectrosc.* **52** (2008), 31–48.

146. A. N. Garroway, Velocity Measurements in Flowing Fluids by NMR. *J. Phys. D. Appl. Phys.* **7** (1974), L159–63.
147. D. N. Guilfoyle, P. Mansfield, and K. J. Packer, Fluid Flow Measurement in Porous Media by Echo-Planar Imaging. *J. Magn. Reson.* **97** (1992), 342–58.
148. M. Holz, C. Müller, and A. M. Wachter, Modification of the Pulsed Magnetic Field Gradient Method for the Determination of Low Velocities by NMR. *J. Magn. Reson.* **69** (1986), 108–15.
149. L. Frydman, J. S. Harwood, D. N. Garnier, and G. C. Chingas, Position-Displacement Correlations in Fluids from Magnetic Resonance Gradient-Echo Shapes. *J. Magn. Reson. A* **101** (1993), 240–8.
150. H. Cho, X.-H. Ren, E. E. Sigmund, and Y.-Q. Song, A Single-Scan Method for Measuring Flow Along an Arbitrary Direction. *J. Magn. Reson.* **186** (2007), 11–16.
151. L. G. Raguin and L. Ciobanu, Multiple Echo NMR Velocimetry: Fast and Localized Measurements of Steady and Pulsatile Flows in Small Channels. *J. Magn. Reson.* **184** (2007), 337–43.
152. K. J. Packer, C. Rees, and D. J. Tomlinson, Studies of Diffusion and Flow by Pulsed NMR Techniques. *Adv. Mol. Relax. Process.* **3** (1972), 119–31.
153. M. Holz and C. Müller, Direct Measurement of Single Ionic Drift Velocities in Electrolyte Solutions. An NMR Method. *Ber. Bunsenges. Phys. Chem.* **86** (1982), 141–7.
154. M. Holz, NMR in the Presence of an Electric Current. Simultaneous Measurements of Ionic Mobilities, Transference Numbers, and Self-Diffusion Coefficients using an NMR Pulsed-Gradient Experiment. *J. Magn. Reson.* **58** (1984), 294–305.
155. T. R. Saarinen and C. S. Johnson, Jr., High-Resolution Electrophoretic NMR. *J. Am. Chem. Soc.* **110** (1988), 3332–3.
156. K. F. Morris and C. S. Johnson, Jr., Mobility-Ordered Two-Dimensional Nuclear Magnetic Resonance Spectroscopy. *J. Am. Chem. Soc.* **114** (1992), 776–7.
157. C. S. Johnson, Jr. and Q. He, Electrophoretic Nuclear Magnetic Resonance. *Adv. Magn. Reson.* **13** (1989), 131–59.
158. M. Holz, Electrophoretic NMR. *Chem. Soc. Rev.* **23** (1994), 165–74.
159. C. S. Johnson, Jr., Electrophoretic NMR. In *Encyclopedia of Nuclear Magnetic Resonance*, ed. D. M. Grant and R. K. Harris. vol. 3. (New York: Wiley, 1996), pp. 1886–95.
160. M. Holz, Field-Assisted Diffusion Studied by Electrophoretic NMR. In *Diffusion in Condensed Matter*, ed. J. Kärger and P. Heitjans. (Berlin: Springer, 2005), pp. 717–42.
161. P. C. Griffiths, A. Paul, and N. Hirst, Electrophoretic NMR Studies of Polymer and Surfactant Systems. *Chem. Soc. Rev.* **35** (2006), 134–45.
162. C. R. Cantor and P. R. Schimmel, *Biophysical Chemistry, Part II: Techniques for the Study of Biological Structure and Function*. (New York: W.H. Freeman, 1980), pp. 555–6.
163. K. Hayamizu, S. Seki, H. Miyashiro, and Yo. Kobayashi, Direct in Situ Observation of Dynamic Transport for Electrolyte Components by NMR Combined with Electrochemical Measurements. *J. Phys. Chem. B* **110** (2006), 22302–5.
164. Q. He, W. Lin, Y. Liu, and E. Li, Three-Dimensional Electrophoretic NMR Correlation Spectroscopy. *J. Magn. Reson.* **147** (2000), 361–5.
165. E. Pettersson, I. Furó, and P. Stilbs, On Experimental Aspects of Electrophoretic NMR. *Concepts Magn. Reson.* **22A** (2004), 61–8.
166. F. Hallberg, I. Furó, P. V. Yushmanov, and P. Stilbs, Sensitive and Robust Electrophoretic NMR. Instrumentation and Experiments, *J. Magn. Reson.* **192** (2008), 69–77.
167. Q. He and Z. Wei, Convection Compensated Electrophoretic NMR. *J. Magn. Reson.* **150** (2001), 126–31.

168. S. R. Heil and M. Holz, A Mobility Filter for the Detection and Identification of Charged Species in Complex Liquid Mixtures by ENMR Phase Difference Spectroscopy. *Angew. Chem. (Int. Ed. Engl.)* **35** (1996), 1717–20.
169. Q. He and C. S. Johnson, Jr., Two-Dimensional Electrophoretic NMR for the Measurement of Mobilities and Diffusion in Mixtures. *J. Magn. Reson.* **81** (1989), 435–9.
170. Q. He and C. S. Johnson, Jr., Stimulated Echo Electrophoretic NMR. *J. Magn. Reson.* **85** (1989), 181–5.
171. Q. He, D. P. Hinton, and C. S. Johnson, Jr., Measurement of Mobility Distributions for Vesicles by Electrophoretic NMR. *J. Magn. Reson.* **91** (1991), 654–8.
172. K. F. Morris and C. S. Johnson, Jr., Mobility-Ordered 2D NMR Spectroscopy for the Analysis of Ionic Mixtures. *J. Magn. Reson. A* **101** (1993), 67–73.
173. E. Li and Q. He, Constant-Time Multidimensional Electrophoretic NMR. *J. Magn. Reson.* **156** (2002), 181–6.
174. S. B. Thakur and Q. He, High Flow-Resolution for Mobility Estimation in 2D-ENMR of Proteins Using Maximum Entropy Method (MEM-ENMR). *J. Magn. Reson.* **183** (2006), 32–40.
175. H. Körber, E. Dormann, and G. Eska, Multiple Spin Echoes for Protons in Water. *J. Magn. Reson.* **93** (1991), 589–95.
176. R. Bowtell and P. Robyr, Structural Investigations with the Dipolar Demagnetizing Field in Solution NMR. *Phys. Rev. Lett.* **76** (1996), 4971–4.
177. P. Robyr and R. Bowtell, Measuring Diffusion in Liquids with a Single Gradient Pulse. *J. Magn. Reson. A* **121** (1996), 206–8.
178. R. Ruf and E. Dormann, How to Measure Diffusion Constant and Absolute Susceptibility of Charge Carriers at the Same Time. *Z. Phys. B* **102** (1997), 157–61.
179. A. Bifone, G. S. Payne, and M. O. Leach, In Vivo Multiple Spin Echoes. *J. Magn. Reson.* **135** (1998), 30–6.
180. I. Ardelean and R. Kimmich, Diffusion Measurements Using Nonlinear Stimulated Echo. *J. Magn. Reson.* **143** (2000), 101–5.
181. J. Jeener, Macroscopic Molecular Diffusion in Liquid NMR, Revisited. *Concepts Magn. Reson.* **14** (2002), 79–88.
182. I. Ardelean and R. Kimmich, Principles and Unconventional Aspects of NMR Diffusometry. In *Annual Reports on NMR Spectroscopy.* ed. G. A. Webb. vol. 49 (London: Elsevier, 2003), pp. 43–115.
183. C. Cai, Z. Chen, S. Cai, and J. Zhong, Propagator Formalism and Computer Simulation of Restricted Diffusion Behaviors of Inter-Molecular Multiple-Quantum Coherences. *Physica B* **366** (2005), 127–37.
184. W. Barros, Jr. and D. F. Gochberg, Fast Single-Gradient Simultaneous Measurement of D and T_2 in Liquids via the Distant Dipolar Field. *Chem. Phys. Lett.* **431** (2006), 174–8.
185. W. Barros, Jr., J. C. Gore, and D. F. Gochberg, Simultaneous Measurement of D and T_2 Using the Distant Dipolar Field. *J. Magn. Reson.* **178** (2006), 166–9.
186. S. Viel, F. Ziarelli, G. Pagès, C. Carrara, and S. Caldarelli, Pulsed Field Gradient Magic Angle Spinning NMR Self-Diffusion Measurements in Liquids. *J. Magn. Reson.* **190** (2008), 113–23.
187. M. Carl, G. W. Miller, J. P. Mugler III, S. Rohrbaugh, W. A. Tobias, and G. D. Cates, Jr., Measurement of Hyperpolarized Gas Diffusion at Very Short Time Scales. *J. Magn. Reson.* **189** (2007), 228–40.
188. P. Zänker, J. Schmidt, J. Schmiedeskamp, R. H. Acosta and H. W. Spiess, Spin Echo Formation in the Presence of Stochastic Dynamics. *Phys. Rev. Lett.* **99** (2007), 263001-1–263001-4.
189. C. Casieri, S. Bubici, and F. De Luca, Self-Diffusion Coefficient by Single-Sided NMR. *J. Magn. Reson.* **162** (2003), 348–55.

9

NMR imaging studies of translational motion

9.1 Introduction

Most simplistically, mutual diffusion can be probed by imaging diffusion profiles (e.g., the ingress of a solvent into a material). However, the integration of MRI techniques with the gradient-based measurements of translational motion that we have discussed in previous chapters allows for potentially more information to be obtained – especially from spatially inhomogeneous samples. It also provides additional techniques for measuring such motions. Diffusion is extremely important in MRI,[1,2] and, amongst other effects, at very high resolutions it determines the ultimate resolution limit when the distance moved by a molecule is comparable to voxel dimensions.[3] Further, since motion is more restricted near a boundary, the spins near the boundary are less dephased (attenuated) during the application of imaging gradients in high resolution imaging, consequently a stronger signal is obtained near the boundary and this has become known as diffusive edge enhancement.[4–10] Relatedly, since the length scales that can be probed with NMR diffusion measurements encompass those that restrict diffusion in cellular systems, the combination of PGSE with imaging techniques can result in MRI contrasts. Whilst there can be diffusion anisotropy at the microscopic level (e.g., diffusion in a biological cell), the MRI sampling is coarse and thus if there is too much inhomogeneity of the ordering of the microscopic anisotropic systems, the information obtained from the voxel will appear isotropic. Thus, for the measured diffusion to appear anisotropic in an MRI-based measurement of an anisotropic system also requires that there be coherent macroscopic arrangement of the anisotropic microscopic architecture.

We have considered almost all of the relevant background theory for understanding the combination of PGSE techniques with MRI measurements in previous chapters and as the area of translational motion measurements in MRI is already well-covered in the literature,[1,11–15] here we only briefly consider this area. We note that measurements of translational motion using stray-field imaging were considered in Section 8.2. Imaging-based techniques for measuring translational motion are also possible using B_1 gradients (see Chapter 10).

9.2 Imaging – basic concepts

9.2.1 Spatial and temporal averaging

As noted with Eq. (2.2) and in Section 2.2.3, the ability to obtain localised information comes at a high cost of S/N in k-space imaging and sample permitting, the current highest available resolutions involve voxels of the order of $(10\,\mu\text{m})^3$. As noted by Fukushima,[12] a compromise would be to do a coarse imaging-based experiment to probe a parameter of the statistical distribution, such as the parameter width, for each voxel, and thereby obtain knowledge at the imaging scale of the heterogeneities. In principle, the spatial scales over which such measurements could take place range from 10^{-5} m ($\sim 10^{-4}$ m for macroscopic flow measurements) to the effective volume of the rf coil.

MRI techniques are particularly powerful for studying multiphase flows (i.e., those consisting of more than one phase or component) since it is possible to vary both spatial and temporal averaging. Just as in a PGSE experiment the measurement is on the timescale of Δ, in an MRI experiment the dynamics of the system is sampled over the total time taken for image acquisition, T – typically on the order of minutes or longer, although fast imaging sequences such as echo-planar imaging (EPI) can be used to narrow temporal resolution. Thus, in a flow measurement, the velocity measured is[12,16]

$$\langle \bar{v}(\mathbf{r}) \rangle = \frac{1}{T}\int_0^T \langle v(\mathbf{r}, t)\rangle \, dt, \tag{9.1}$$

where the brackets indicate spatial averaging over the voxel dimension and the overbar indicates the average over T. The distribution of the velocity, $P(v)$, can be measured by forfeiting spatial resolution.[17] These velocities are averaged over the sample as well as over Δ, namely[12]

$$V = \int \rho(\mathbf{r}) \left[\frac{1}{\Delta}\int_0^\Delta V(\mathbf{r}, t) dt\right] d\mathbf{r}. \tag{9.2}$$

9.3 Combining PGSE measurements with imaging

9.3.1 Dynamic NMR microscopy

‘Dynamic NMR Microscopy’ describes the incorporation of a PGSE pulse sequence into an imaging experiment (see, for example, Figures 9.1 and 9.2) and the acquisition

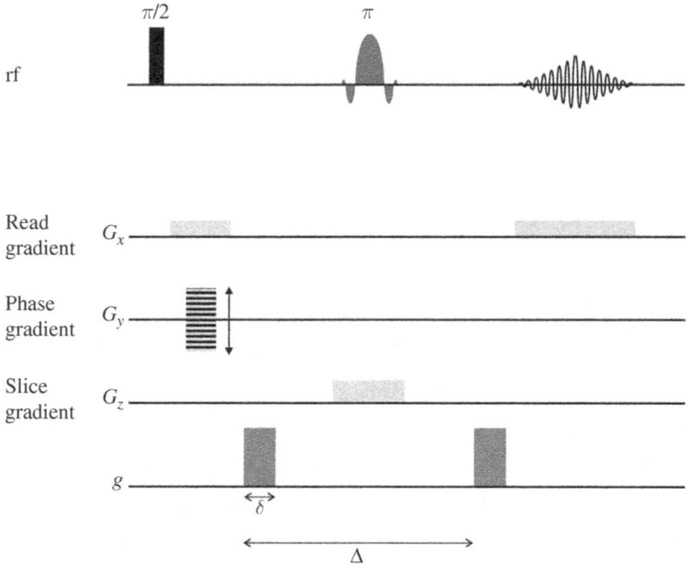

Figure 9.1 An example of a dynamic NMR microscopy sequence.[16]

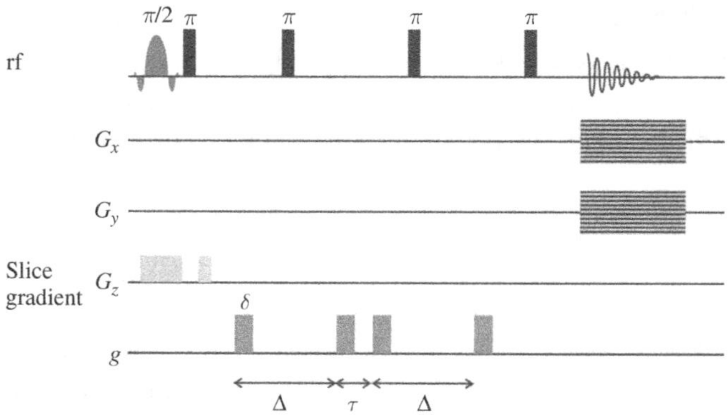

Figure 9.2 An example of an imaging sequence for velocity-compensated dynamic NMR microscopy[19] which incorporates the Hahn PGSE sequence. The sequence contains two PGSE trains resulting in flow compensation. The precursor read gradient is not shown.

of a sequence of high resolution images such that each corresponds to slices in q-space.[1,18–20] As a result diffusion and flow can be used as sources of contrast in the image. The spectrum of displacements at each pixel is then calculated using Fourier methods and maps of the velocity and diffusion can be obtained simultaneously with high precision.

Since the dynamic NMR microscopy sequences have elements of both PGSE and conventional imaging, the signal is modulated in both k- and q-space,[18]

$$S(\mathbf{k},\mathbf{q}) = \int \rho(\mathbf{r}_0)\exp(i2\pi\mathbf{k}\cdot\mathbf{r}_0)\int P(\mathbf{r}_0,\mathbf{r}_1,\Delta)\exp(i2\pi\mathbf{q}\cdot(\mathbf{r}_1-\mathbf{r}_0))d\mathbf{r}_1 d\mathbf{r}_0$$
$$= \int \rho(\mathbf{r}_0)E(\mathbf{q},\mathbf{r}_0,\Delta)\exp(i2\pi\mathbf{k}\cdot\mathbf{r}_0)d\mathbf{r}_0. \quad (9.3)$$

In Eq. (9.3) an infinitesimally small volume $d\mathbf{r}$ is assumed and we can consider $E(\mathbf{q}, \mathbf{r}_0, \Delta)$ as the Fourier transform of the local dynamic profile (cf. Eq. (2.35))

$$E(\mathbf{q},\mathbf{r}_0,\Delta) = \int \bar{P}(\mathbf{R},\Delta)\exp(i2\pi\mathbf{q}\cdot\mathbf{R})d\mathbf{R}, \quad (9.4)$$

where $\bar{P}(\mathbf{R},\Delta)$ is now a function of the pixel coordinate and consequently $E(\mathbf{q}, \mathbf{r}_0, \Delta)$ refers to the normalised echo contribution at the pixel coordinate $\mathbf{r}$. $\mathbf{q}$ is applied in only a single dimension (e.g., along z) and from a series of images in which $\mathbf{q}$ is incremented, $E(\mathbf{q}, \Delta)$ can be calculated for each pixel and thus $\bar{P}(Z,\Delta)$, v and D can be determined for each pixel in the image. As is evident from Eq. (9.3), dynamic NMR microscopy contains elements of both a Lagrangian (via motion encoding by the PGSE sequence) and Eulerian (via the spatial localisation) information encoding.[16]

The calculation of the so called b-values in imaging sequences can be rather more complicated than that outlined in Section 2.3.2 due to the influence of the imaging gradients.[21,22]

9.3.2 *Diffusion-weighted imaging (DWI) and diffusion tensor imaging (DTI)*

Generally the voxel size used in imaging is sufficiently large that there is some heterogeneity in the sample and thus variation in $\mathbf{D}$. Thus, MRI measurements of $\mathbf{D}$ return a volume average and this is normally referred to in the literature as an apparent diffusion coefficient (ADC) or apparent diffusion tensor. There are countless variations of imaging approaches to measuring diffusion[21,23,24] and we have only presented two simple sequences above. The interested reader is asked to consult one of the many review papers. We remark that the measurement of diffusion in MRI has become extremely important as a source of image contrast and has opened up new diagnostic possibilities and for characterising microscopic tissue architecture. DWI and DTI have been the subject of many books and reviews including refs. 15, 25–34. Kingsley has presented detailed reviews of the mathematics related to diffusion tensor imaging.[35–37] Somewhat expectedly the presence of noise can have deleterious effects in diffusion tensor imaging.[38]

Other scalar imaging parameters can be derived from **D** including the mean diffusivity (= Tr(**D**)), and many metrics of anisotropy such as the relative anisotropy and fractional anisotropy have been proposed. Importantly, these features, which may be used as contrast factors, are rotationally and translationally invariant and are thus free of orientational artefacts.[39–42] We note that higher-order generalisations of the diffusion tensor models have been developed and the different forms of analysis have recently been reviewed by Minati and Węglarz.[34] Also, better models for diffusion including tortuosity in brain tissue and solutions linking PGSE attenuation with anomalous diffusion (see Section 4.5) as a means of describing diffusion in complex tissue architecture (e.g., brain) are becoming increasingly important from a clinical perspective in MRI diffusion measurements.[43–53]

9.3.3 Fibre tracking

An area of increasing importance in MRI-based diffusion studies is in fibre tracking. In fibre tracking it is assumed that the white matter fibres are oriented along the direction of greatest diffusion. Hence, a fibre is tracked by following this direction from a diffusion ellipsoid representation of the diffusion tensor for each voxel.[15,54–57]

A particular problem with simply using the diffusion tensor approach is the inability to resolve intravoxel orientational heterogeneity since the tensor approach only gives a single fibre direction in each voxel due to the implicit assumption that there is only a single Gaussian diffusion compartment in each voxel. Thus, intra-voxel fibre crossings cannot be resolved. Since diffusion in tissue is in general non-Gaussian due to the effects of restriction, the diffusion cannot be fully described by the diffusion tensor. Whilst the diffusion tensor is evaluated in low q measurements and hence contains information about the width or second moment of the propagator, attempts to generalise the diffusion tensor by the inclusion of higher moments have been found to be mathematically inconsistent for non-Gaussian diffusion.[58] This has prompted the development of new methods and theory including propagator and q-space approaches which do not make the assumption of (multi)-Gaussian diffusion.[59–62] It has also been noted that diffusion–diffusion correlation experiments (see Section 8.7) and related experiments may help to remove such ambiguities as they can prove the existence of local anisotropy and yield information on the domain size and orientation.[63,64]

9.4 Imaging-based flow measurements

There are essentially three categories of flow-imaging techniques: inflow/outflow, tagging and time-of-flight and phase-sensitive methods. As noted by Callaghan, the range of average velocities, v, that are measurable is limited by $v_{max} \approx l/t_{echo}$ (where l

is the length of the receiver coil and t_{echo} is the echo time) from above and by the molecular self-diffusion from below. For water at room temperature the minimum velocity measurable is about 20 μm s^{-1}.[1] However, rapid-imaging techniques might extend the applicable range.

9.4.1 Inflow/outflow

Flow-sensitive (or steady-state) techniques rely on the build-up of longitudinal magnetisation being limited by the transit time.[65,66] The rf and gradient pulses are rapidly repeated such that the saturation recovery or the steady-state free precession signal amplitude depends on the molecular motion since moving spins will not be permanently in the excitation slice.

9.4.2 Tagging and time-of-flight

Time-of-flight methods involve the selective destruction (or excitation) of a particular group of spins (a 'tag') and then following their displacement during the pulse sequence.[12,67] As tagging is based on selective saturation the effects only last on the order of T_1. In its simplest form a planar tag is applied in exactly the same way as slice selection of a 3D object and an example of a time-of-flight pulse sequence is given in Figure 9.3. This method differs from the flow-sensitive method in that the slice that is eventually imaged is different from that which was initially excited. The excited slice is chosen to be normal to the principal flow direction, whereas the detected slice can be parallel or perpendicular to the flow (e.g., to measure the velocity field in steady pipe flow). More elaborate tagging is also possible

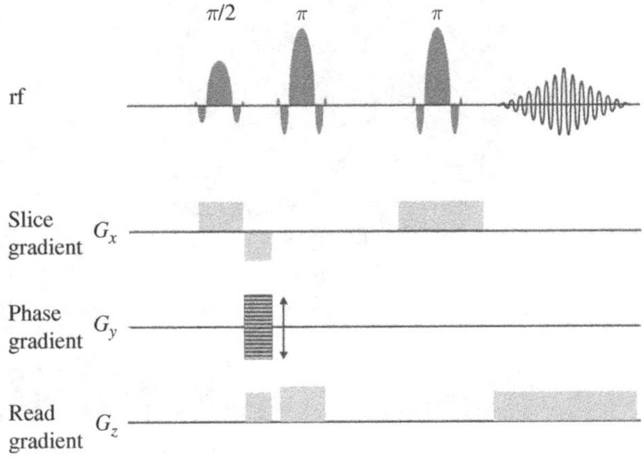

Figure 9.3 An example of a time-of-flight imaging pulse sequence for measuring spin displacement.

where multiple tags or even grids are used to visualise more complex flows (see Figure 9.4).[12,67–70] A limitation with tagging methods is that the grid resolution determines the limit at which flow can be measured quantitatively and thus the phase evolution methods, as discussed below, afford lower limits of detection of flow velocities. Nevertheless, tagging techniques have enormous potential for studying fluid dynamics and are able to provide information such as shear velocity.

9.4.3 Phase-sensitive methods

The phase-sensitive techniques are based on the phase shift of the transverse magnetisation that occurs in the presence of a magnetic field gradient in the direction of flow (see Section 2.2.2) and are the most precise of the three methods for measuring flow.[11,12,71] By nulling all moments up to a required order it is possible to obtain images which detect, for example, acceleration. A typical flow-selective excitation sequence is $\pi/2_x - \tau - \pi/2_{-x}$ with a bipolar gradient pulse pair between the rf pulses. Many examples of such gradient waveforms exist.[11,72] An imaging sequence incorporating such a flow selection sequence is shown in Figure 9.5.

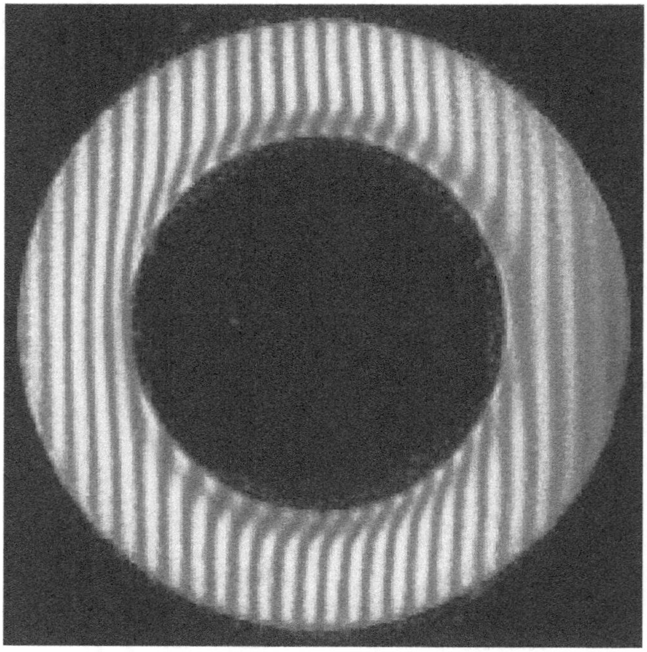

Figure 9.4 Spin-tagged MRI image of granular Couette flow between two rotating concentric cylinders. The shear band shows up as the narrow region of deformed stripes near the inner rotating cylinder. From Mueth *et al.*[70]

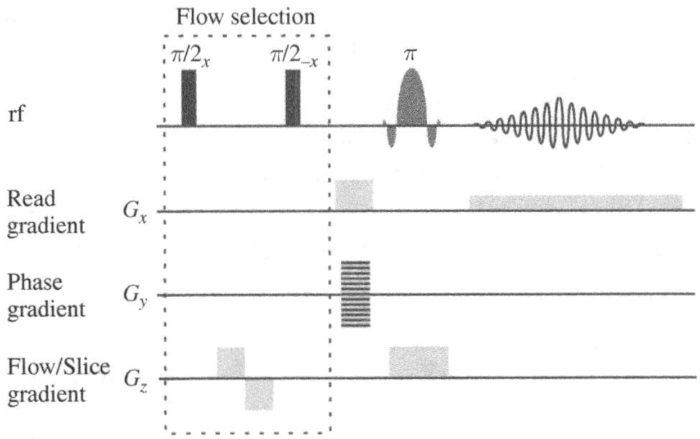

Figure 9.5 An example of a single slice imaging sequence with a flow-selective excitation sequence (i.e., $\pi/2_x - G - (\text{-}G) - \pi/2_{-x}$).

Madio *et al.*[73] have developed a sequence combining bipolar gradients to encode the velocity with an ultrafast rotating imaging sequence that enables the velocity to be encoded in 3 ms and the imaging in 8 ms. Using this sequence they were able to image turbulent jets in a 15 mm i.d. glass tube with a 75% stenosis with Reynolds numbers from 560 to 3750. In their measurements they were able to determine a maximal velocity of 59.4 cm s^{-1}.

Buhai *et al.* have presented a phase-encoding-based imaging technique that permits direct mapping of local accelerations and demonstrated its use on percolation model objects.[74]

References

1. P. T. Callaghan, *Principles of Nuclear Magnetic Resonance Microscopy.* (Oxford: Clarendon Press, 1991).
2. P. T. Callaghan, Susceptibility & Diffusion Effects in NMR Microscopy. In *Encyclopedia of Nuclear Magnetic Resonance*, ed. D. M. Grant and R. K. Harris. vol. 7. (New York: Wiley, 1996), pp. 4665–71.
3. P. T. Callaghan and C. D. Eccles, Diffusion-Limited Resolution in Nuclear Magnetic Resonance Microscopy. *J. Magn. Reson.* **78** (1988), 1–8.
4. W. B. Hyslop and P. C. Lauterbur, Effects of Restricted Diffusion on Microscopic NMR Imaging. *J. Magn. Reson.* **94** (1991), 501–10.
5. D. Barsky, B. Pütz, K. Schulten, J. Schoeniger, E. W. Hsu, and S. Blackband, Diffusional Edge Enhancement Observed by NMR in Thin Glass Capillaries. *Chem. Phys. Lett.* **200** (1992), 88–96.
6. B. Pütz, D. Barsky, and K. Schulten, Edge Enhancement by Diffusion in Microscopic Magnetic Resonance Imaging. *J. Magn. Reson.* **97** (1992), 27–53.
7. P. T. Callaghan, A. Coy, L. C. Forde, and C. J. Rofe, Diffusive Relaxation and Edge Enhancement in NMR Microscopy. *J. Magn. Reson. A* **101** (1993), 347–50.
8. T. M. De Swiet, Diffusive Edge Enhancement in Imaging. *J. Magn. Reson. B* **109** (1995), 12–18.

9. Y.-Q. Song, B. M. Goodson, B. Sheridan, T. M. De Swiet, and A. Pines, Effects of Diffusion on Magnetic Resonance Imaging of Laser-Polarized Xenon Gas. *J. Chem. Phys.* **108** (1998), 6233–9.
10. J. Stepišnik, A. Duh, A. Mohorič, and I. Serša, MRI Edge Enhancement as a Diffusive Discord of Spin Phase Structure. *J. Magn. Reson.* **137** (1999), 154–60.
11. J. M. Pope and S. Yao, Quantitative NMR Imaging of Flow. *Concepts Magn. Reson.* **5** (1993), 281–302.
12. E. Fukushima, Nuclear Magnetic Resonance as a Tool to Study Flow. *Annu. Rev. Fluid Mech.* **31** (1999), 95–123.
13. Y. Cohen and Y. Assaf, High *b*-Value q-Space Analyzed Diffusion-Weighted MRS and MRI in Neuronal Tissues – A Technical Review. *NMR Biomed.* **15** (2002), 516–42.
14. M. A. Bernstein, K. F. King, and X. J. Zhou, *Handbook of MRI Pulse Sequences*. (London: Elsevier Academic Press, 2004).
15. S. Mori, *Introduction to Diffusion Tensor Imaging*. (Oxford: Elsevier, 2007).
16. J. D. Seymour and P. T. Callaghan, Generalized Approach to NMR Analysis of Flow and Dispersion in Porous Media. *AIChE J.* **43** (1997), 2096–111.
17. A. N. Garroway, Velocity Measurements in Flowing Fluids by NMR. *J. Phys. D. Appl. Phys.* **7** (1974), L159–63.
18. P. T. Callaghan, C. D. Eccles, and Y. Xia, NMR Microscopy of Dynamic Displacements: ***k***-Space and ***q***-Space Imaging. *J. Phys. E: Sci. Instrum.* **21** (1988), 820–2.
19. P. T. Callaghan and Y. Xia, Velocity and Diffusion Imaging in Dynamic NMR Microscopy. *J. Magn. Reson.* **91** (1991), 326–52.
20. P. T. Callaghan, K. R. Jeffrey, and Y. Xia, Translational Motion Imaging with Pulsed Gradient Spin Echo Methods. In *Magnetic Resonance Microscopy*, ed. B. Blümich and W. Kuhn. (Weinheim: VCH, 1992), pp. 328–47.
21. J. Mattiello, P. J. Basser, and D. Le Bihan, Analytical Expressions for the *b* Matrix in NMR Diffusion Imaging and Spectroscopy. *J. Magn. Reson. A* **108** (1994), 131–41.
22. D. Güllmar, J. Haueisen, and J. R. Reichenbach, Analysis of *b*-Value Calculations in Diffusion Weighted and Diffusion Tensor Imaging. *Concepts Magn. Reson.* **25A** (2005), 53–66.
23. P. J. Basser, J. Mattiello, and D. Le Bihan, Estimation of the Effective Self-Diffusion Tensor from the NMR Spin Echo. *J. Magn. Reson. B* **103** (1994), 247–54.
24. J. Mattiello, P. J. Basser, and D. Le Bihan, The b Matrix in Diffusion Tensor Echo-Planar Imaging. *Magn. Reson. Med.* **37** (1997), 292–300.
25. D. Le Bihan, Molecular Diffusion Nuclear Magnetic Resonance Imaging. *Magn. Reson. Q.* **7** (1991), 1–30.
26. P. J. Basser, Inferring Microstructural Features and the Physiological State of Tissues from Diffusion-Weighted Images. *NMR Biomed.* **8** (1995), 333–44.
27. D. Le Bihan, Methods and Applications of Diffusion MRI. In *Methods in Biomedical Magnetic Resonance Imaging and Spectroscopy*, ed. I. M. Young. vol. 1. (New York: Wiley, 2000).
28. D. Le Bihan, Diffusion & Perfusion in MRI. In *Encyclopedia of Nuclear Magnetic Resonance*, ed. D. M. Grant and R. K. Harris. vol. 3. (New York: Wiley, 1996), pp. 1645–56.
29. D. G. Norris, The Effects of Microscopic Tissue Parameters on the Diffusion Weighted Magnetic Resonance Imaging Experiment. *NMR Biomed.* **14** (2001), 77–93.
30. P. J. Basser and D. K. Jones, Diffusion-Tensor MRI: Theory, Experimental Design and Data Analysis – A Technical Review. *NMR Biomed.* **15** (2002), 456–67.
31. R. Bammer, Basic Principles of Diffusion-Weighted Imaging. *Eur. J. Radiol.* **45** (2003), 169–84.

32. K. Nicolay, K. P. Braun, R. A. De Graaf, R. M. Dijkhuizen, and M. J. Kruiskamp, Diffusion NMR Spectroscopy. *NMR Biomed.* **14** (2001), 94–111.
33. G. J. Strijkers, M. R. Drost, A. M. Heemskerk, M. J. Kruiskamp, and K. Nicolay, Diffusion MRI and MRS of Skeletal Muscle. *Isr. J. Chem.* **43** (2003), 71–80.
34. L. Minati and W. P. Węglarz, Physical Foundations, Models, and Methods of Diffusion Magnetic Resonance Imaging of the Brain: A Review. *Concepts Magn. Reson.* **30A** (2007), 278–307.
35. P. B. Kingsley, Introduction to Diffusion Tensor Imaging Mathematics: Part I. Tensors, Rotations, and Eigenvectors. *Concepts Magn. Reson.* **28A** (2006), 101–22.
36. P. B. Kingsley, Introduction to Diffusion Tensor Imaging Mathematics: Part II. Anisotropy, Diffusion-Weighting Factors, and Gradient Encoding Schemes. *Concepts Magn. Reson.* **28A** (2006), 123–54.
37. P. B. Kingsley, Introduction to Diffusion Tensor Imaging Mathematics: Part III. Tensor Calculation, Noise, Simulations, and Optimization. *Concepts Magn. Reson.* **28A** (2006), 155–79.
38. P. J. Basser and S. Pajivec, Statistical Artifacts in Diffusion Tensor MRI (DT-MRI) Caused by Background Noise. *Magn. Reson. Med.* **44** (2000), 41–50.
39. C. Pierpaoli and P. J. Basser, Toward a Quantitative Assessment of Diffusion Anisotropy. *Magn. Reson. Med.* **36** (1996), 893–906.
40. A. M. Uluğ and P. C. M. Van Zijl, Orientation-Independent Diffusion Imaging Without Tensor Diagonalization: Anisotropy Definitions Based on Physical Attributes of the Diffusion Ellipsoid. *J. Mag. Res. Imaging* **9** (1999), 804–13.
41. P. J. Basser, J. Mattiello, and D. Le Bihan, MR Diffusion Tensor Spectroscopy and Imaging. *Biophys. J.* **66** (1994), 259–67.
42. P. J. Basser and C. Pierpaoli, Microstructural and Physiological Features of Tissues Elucidated by Quantitative-Diffusion-Tensor MRI. *J. Magn. Reson. B* **111** (1996), 209–19.
43. C.-L. Chin, F. W. Wehrli, S. N. Hwang, D. L. Jaggard, D. B. Hackney, and S. W. Wehrli, Feasibility of Probing Boundary Morphology of Structured Materials by 2D NMR *q*-Space Imaging. *J. Magn. Reson.* **160** (2003), 20–5.
44. C.-L. Chin, F. W. Wehrli, Y. Fan, S. N. Hwang, E. D. Schwartz, J. Nissanov, and D. B. Hackney, Assessment of Axonal Fiber Tract Architecture in Excised Rat Spinal Cord by Localized NMR *q*-Space Imaging: Simulations and Experimental Studies. *Magn. Reson. Med.* **52** (2004), 733–40.
45. S. N. Hwang, C.-L. Chin, F. W. Wehrli, and D. B. Hackney, An Image-Based Finite Difference Model for Simulating Restricted Diffusion. *Magn. Reson. Med.* **50** (2003), 373–82.
46. C. H. Meier, W. Dreher, and D. Leibfritz, Diffusion in Compartmental Systems. I. A Comparison of An Analytical Model with Simulations. *Magn. Reson. Med.* **50** (2003), 500–7.
47. C. H. Meier, W. Dreher, and D. Leibfritz, Diffusion in Compartmental Systems. II. Diffusion-Weighted Measurements of Rat Brain Tissue In Vivo and Postmortem at Very Large *b*-Values. *Magn. Reson. Med.* **50** (2003), 510–14.
48. Y. Assaf, R. Z. Freidlin, G. K. Rohde, and P. J. Basser, New Modeling and Experimental Framework to Characterize Hindered and Restricted Water Diffusion in Brain White Matter. *Magn. Reson. Med.* **52** (2004), 965–78.
49. J. Hrabe, S. Hrabtová, and K. Segeth, A Model of Effective Diffusion and Tortuosity in the Extracellular Space of the Brain. *Biophys. J.* **87** (2004), 1606–17.
50. E. Özarslan, P. J. Basser, T. M. Shepherd, P. E. Thelwall, B. C. Vemuri, and S. J. Blackband, Observation of Anomalous Diffusion in Excised Tissue by Characterizing

the Diffusion-Time Dependence of the MR Signal. *J. Magn. Reson.* **183** (2006), 315–23.

51. P. N. Sen and P. J. Basser, A Model for Diffusion in White Matter in the Brain. *Biophys. J.* **89** (2005), 2927–38.
52. P. N. Sen and P. J. Basser, Modeling Diffusion in White Matter in the Brain: A Composite Porous Medium. *Magn. Reson. Imaging* **23** (2005), 215–20.
53. R. L. Magin, O. Abdullah, D. Baleanu, and X. J. Zhou, Anomalous Diffusion Expressed Through Fractional Order Differential Operators in the Bloch–Torrey Equation. *J. Magn. Reson.* **190** (2008), 255–70.
54. T. E. Conturo, N. F. Lori, T. S. Cull, E. Akbudak, A. Z. Snyder, J. S. Shimony, R. C. McKinstry, H. Burton, and M. E. Raichle, Tracking Neuronal Fiber Pathways in the Living Human Brain. *Proc. Natl. Acad. Sci. U.S.A.* **96** (1999), 10422–27.
55. N. F. Lori, E. Akbudak, J. S. Shimony, T. S. Cull, A. Z. Snyder, R. K. Guillory, and T. E. Conturo, Diffusion Tensor Fiber Tracking of Human Brain Connectivity: Aquisition Methods, Reliability Analysis and Biological Results. *NMR Biomed.* **15** (2002), 493–515.
56. S. Mori and P. C. M. Van Zijl, Fiber Tracking: Principles and Strategies – A Technical Review. *NMR Biomed.* **15** (2002), 468–80.
57. R. Watts, C. Liston, S. Niogi, and A. M. Uluğ, Fiber Tracking Using Magnetic Resonance Diffusion Tensor Imaging and Its Applications to Human Brain Development. *Ment. Retard. Dev. Disabil. Res. Rev.* **9** (2003), 168–77.
58. C. Liu, R. Bammer, and M. E. Moseley, Limitations of Apparent Diffusion Coefficient-Based Models in Characterizing Non-Gaussian Diffusion. *Magn. Reson. Med.* **54** (2005), 419–28.
59. Y. Assaf, D. Ben-Bashat, J. Chapman, S. Peled, I. E. Biton, M. Kafri, Y. Segev, T. Hendler, A. D. Korczyn, M. Graif, and Y. Cohen, High *b*-Value *q*-Space Analyzed Diffusion-Weighted MRI: Application to Multiple Sclerosis. *Magn. Reson. Med.* **47** (2002), 115–26.
60. D. S. Tuch, Q-Ball Imaging. *Magn. Reson. Med.* **52** (2004), 1358–72.
61. E. Özarslan, B. C. Vemuri, and T. H. Mareci, Generalized Scalar Measures for Diffusion MRI Using Trace, Variance, and Entropy. *Magn. Reson. Med.* **53** (2005), 866–76.
62. E. Özarslan, T. M. Shepherd, B. C. Vemuri, S. J. Blackband, and T. H. Mareci, Resolution of Complex Tissue Microarchitecture Using the Diffusion Orientation Transform (DOT). *Neuroimage* **31** (2006), 1086–106.
63. D. Topgaard, Probing Biological Tissue Microstructure with Magnetic Resonance Diffusion Techniques. *Curr. Opin. Colloid Interface Sci.* **11** (2006), 7–12.
64. M. E. Komlosh, F. Horkay, R. Z. Freidlin, U. Nevo, Y. Assaf, and P. J. Basser, Detection of Microscopic Anisotropy in Gray Matter and in a Novel Tissue Phantom Using Double Pulsed Gradient Spin Echo MR. *J. Magn. Reson.* **189** (2007), 38–45.
65. S. Patz, Steady State Free Precession: An Overview of Basic Concepts and Applications. In *Advances in Magnetic Resonance Imaging*, ed. E. Feig. (Norwood, NJ: Ablex Publishing Corporation, 1989), pp. 73–102.
66. H. Gudbjartsson and S. Patz, Simultaneous Calculation of Flow and Diffusion Sensitivity in Steady-State Free Precession Imaging. *Magn. Reson. Med.* **34** (1995), 567–79.
67. R. Kimmich, *NMR: Tomography, Diffusometry, Relaxometry*. (Berlin: Springer Verlag, 1997).
68. T. J. Mosher and M. B. Smith, A DANTE Tagging Sequence for the Evaluation of Translational Sample Motion. *Magn. Reson. Med.* **15** (1990), 334–9.
69. S. Chandra and Y. Yang, Simulations and Demonstrations of Localized Tagging Experiments. *J. Magn. Reson. B* **111** (1996), 285–8.

70. D. M. Mueth, G. M. Debregeas, G. F. Karczmar, P. J. Eng, S. R. Nagel, and H. M. Jaeger, Signatures of Granular Microstructure in Dense Shear Flows. *Nature* **406** (2000), 385–9.
71. A. Caprihan and E. Fukushima, Flow Measurements by NMR. *Phys. Rep.* **198** (1990), 195–235.
72. J. L. Duerk and O. P. Simonetti, Review of MRI Gradient Waveform Design Methods with Application in the Study of Motion. *Concepts Magn. Reson.* **5** (1993), 105–22.
73. D. P. Madio, H. N. Gach, and I. J. Lowe, Ultra-Fast Velocity Imaging in Stenotically Produced Turbulent Jets Using RUFIS. *Magn. Reson. Med.* **39** (1998), 574–80.
74. B. Buhai, A. Hakimov, I. Ardelean, and R. Kimmich, NMR Acceleration Mapping in Percolation Model Objects. *J. Magn. Reson.* **168** (2004), 175–85.

10

B_1 gradient methods

10.1 Introduction

Although having some distinct limitations (e.g., relatively weak gradients and poor directionality), B_1-based measurements have some particular advantages over B_0 gradient-based methods. However, B_1-based techniques have so far received only limited usage and consequently in this chapter we provide only a cursory coverage of these techniques and the interested reader is referred to the pertinent reviews on the subject.[1–5]

10.2 B_1 gradients

B_1 gradients are more complex than B_0 gradients. Apart from purely technical considerations, there are three main differences between B_0 and B_1 gradients:[6] (i) A B_0 field couples only into the spin system along the z-axis, thus the effective gradient tensor is always truncated into an effective vector (see Section 2.2.2). Radio frequency fields, however, couple into the spin system from any orientation within the transverse plane. As a result the B_1 gradient generally retains its tensor form when it couples into the spin system. (ii) When the same rf coil is used for both excitation and detection, any phase variation is cancelled during the measurement. But when an experiment involves two rf fields at the same frequency this cancellation no longer occurs and phase variations need to be considered. This spatial dependence of the phase difference between the two rf fields presents an additional complication (or opportunity). (iii) The third difference is that B_1 fields are non-secular and so do not commute with internal Hamiltonians. Thus, unlike a B_0 gradient, a B_1 gradient cannot be treated additively with respect to internal Hamiltonians.

A formalism in which the steps of coherence transformation and gradient evolution are clearly separated has recently been introduced for describing the spin dynamics of B_1 gradient experiments.[7] In B_0-based experiments only the gradient strength is varied spatially. In B_1 experiments there are two mechanisms that lead to spatially varying spin dynamics: the amplitude variation of the gradient rf field and the phase difference between the gradient and homogeneous rf fields. The amplitude variation is most directly analogous to B_0 experiments. The phase variation arises from the symmetry of the rf field and current flow through the coils.

Two types of B_1 gradients can be employed depending on the coil geometry: planar or radial (i.e., quadrupolar). The amplitude of a planar field (i.e., dB_x/dx) increases along only one axis in the laboratory frame, while the amplitude of a radial field (i.e., dB_x/dx, $-dB_y/dy$) increases along two orthogonal axes. The dephasing due to a planar gradient occurs in a plane perpendicular to the rotating frame axis along which the rf gradient field is applied. For a radial gradient, however, the dephasing, being the result of both an amplitude and a phase variation has the effect of scattering the magnetisation over the surface of a sphere. Hence, when trying to dephase longitudinal magnetisation, a radial gradient is more efficient than a planar gradient. Also, because of the radial phase distribution, the gradient phase does not have to be adjusted to the phase of the homogeneous rf coil.

At a technical level, B_1 gradients have some significant advantages over their B_0 counterparts. The main advantages are:[2,8–10] (i) the switching times are much shorter, (ii) the lock channel is unaffected, (iii) there is no need for pre-emphasis, (iv) the lineshape is not distorted, (v) no eddy currents are induced and so there is no need for shield coils, etc., (vi) they do not suffer from the effects of background gradients (and thus cross-terms), at least to first order, as do B_0 gradients, and (vii) the gradient is frequency selective. However, at present, the absolute strength of B_1 gradients ($\sim 0.75\,\mathrm{T m^{-1}}$) is far less than that available with B_0 gradients.[11] And because B_1 gradients generally preserve their tensor form when they couple into a spin system, the design of truly planar rf gradient fields is difficult.

10.3 B_1 gradient diffusion measurements

The use of B_1 gradients for flow and diffusion measurements was first proposed and demonstrated for flow in 1988 by Karczmar[12] (and later demonstrated for diffusion by Dupeyre *et al.*[13]) with a method related to the concept of a rotary spin-echo[14] (see Figure 10.1A) and then demonstrated by Canet *et al.*[15,16] using a B_1 analogue of the B_0-based Stejskal and Tanner sequence (see Figure 10.1B). In a later work Humbert *et al.* considered the possible artefacts in B_1 diffusion measurements and presented an improved sequence.[17] A one-shot diffusion sequence has also been presented.[18] Also a B_1 gradient-based sequence for measuring diffusion using longitudinal spin orders ($2I_zS_z$ and $4I_z\,I_z\,I_z$) of weakly coupled AX and AX_2 spin systems that allow the determination of comparatively low diffusion coefficients has been presented.[19]

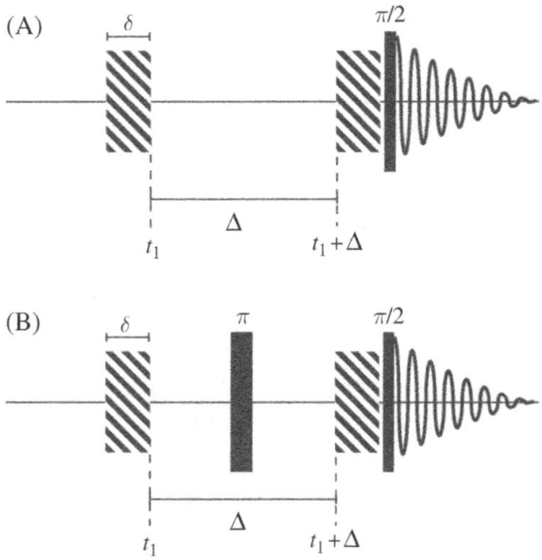

Figure 10.1. B_1 Gradient pulse sequences for measuring diffusion: (A) The B_1 analogue of the B_0 STE sequence and (B) The B_1 analogue of the B_0 spin-echo sequence. The B_1 gradient pulses are represented as striped rectangles.

The theory used to interpret B_1 gradient experiments is essentially the same as that for B_0 gradients.[16] And the attenuation function for the sequences in Figure 10.1 is

$$E(g_1) = \exp\left(-\gamma^2 g_1^2 D\delta^2(\Delta - 2\delta/3)\right), \tag{10.1}$$

where g_1 is the amplitude of the B_1 gradient pulse. Note the correction for the effects of the finite gradient pulse is now $2\delta/3$ due to the slightly different convention with the delay Δ.

Local self-diffusion coefficients can be mapped using B_1 gradient-based sequences.[20–22] And it is noted that diffusive diffraction phenomena have recently been observed in a porous material (72% open porosity polystyrene) using B_1 gradients.[23]

10.4 Miscellaneous

If B_0 inhomogeneities exist in addition to B_1 inhomogeneities in one or two of the rf pulses, one or two nutation echoes can result.[5,24–27] Scharfenecker *et al.* have suggested ways of using these effects to measure diffusion coefficients.[28]

Another recently introduced technique for measuring diffusion based on rf gradients is the MAGROFI (magnetisation rotating frame imaging) technique in which B_1 gradients are combined with rotating frame imaging.[5,21,29,30]

References

1. D. Canet, Radiofrequency Gradient Pulses. In *Encyclopedia of Nuclear Magnetic Resonance*, ed. D. M. Grant and R. K. Harris. vol. 6. (New York: Wiley, 1996), pp. 3938–44.
2. D. Canet, Radiofrequency Field Gradient Experiments. *Prog. NMR Spectrosc.* **30** (1997), 101–35.
3. D. Canet, Radiofrequency Field Gradients in NMR, Theory. In *Encyclopedia of Spectroscopy and Spectrometry*, ed. J. C. Lindon, G. E. Tranter, and J. L. Holmes. vol. 3. (London: Academic Press, 2000), pp. 1937–44.
4. F. Humbert, Potentials of Radio-Frequency Field Gradient NMR Microscopy in Environmental Science. *J. Ind. Microbiol. Biotechnol.* **26** (2001), 53–61.
5. I. Ardelean and R. Kimmich, Principles and Unconventional Aspects of NMR Diffusometry. In *Annual Reports on NMR Spectroscopy*. ed. G. A. Webb. vol. 49. (London: Academic Press, 2003), pp. 43–115.
6. D. G. Cory, F. H. Laukien, and W. E. Maas, Double-Quantum-Filtered-COSY B_1-Gradient Experiments. *J. Magn. Reson. A* **105** (1993), 223–9.
7. Y. Zhang, W. E. Maas, and D. G. Cory, Analysis of Homonuclear RF Gradient NMR Spectroscopy. *Mol. Phys.* **86** (1995), 347–58.
8. W. E. Maas, F. Laukien, and D. G. Cory, Coherence Selection by Radiofrequency Gradients. *J. Magn. Reson. A* **103** (1993), 115–17.
9. R. Raulet, J.-M. Escanyé, F. Humbert, and D. Canet, Quasi-Immunity of B_1 Gradient NMR Microscopy to Magnetic Susceptibility Distortions. *J. Magn. Reson. A* **119** (1996), 111–14.
10. J. M. Dereppe, C. Moreaux, and F. Humbert, NMR Diffusion Measurements in Heterogeneous Media Using Pulsed Radio-Frequency Field Gradients. *Microporous Mesoporous Mater.* **21** (1998), 645–50.
11. F. Humbert, B. Diter, and D. Canet, NMR Microscopy by Strong Radiofrequency-Field Gradients with Spatial Resolution Better Than Five Micrometers. *J. Magn. Reson. A* **123** (1996), 242–5.
12. G. S. Karczmar, D. B. Twieg, T. J. Lawry, G. B. Matson, and M. W. Weiner, Detection of Motion Using B_1 Gradients. *Magn. Reson. Med.* **7** (1988), 111–16.
13. R. Dupeyre, Ph. Devoulon, D. Bourgeois, and M. Decorps, Diffusion Measurements Using Stimulated Rotary Echoes. *J. Magn. Reson.* **95** (1991), 589–96.
14. I. Solomon, Rotary Spin Echoes. *Phys. Rev. Lett.* **2** (1959), 301–2.
15. A. Belmajdoub, D. Boudot, C. Tondre, and D. Canet, NMR Self-Diffusion Measurements Using a Radio-Frequency Field Gradient Combined with Water Signal Suppression. Application to the pH-Dependent Solubilization of Hydroxyquinoline in SDS Micellar Solutions. *Chem. Phys. Lett.* **150** (1988), 194–8.
16. D. Canet, B. Diter, A. Belmajdoub, J. Brondeau, J.-C. Boubel, and K. Elbayed, Self-Diffusion Measurements Using a Radio-Frequency Field Gradient. *J. Magn. Reson.* **81** (1989), 1–12.
17. F. Humbert, M. Valtier, A. Retournard, and D. Canet, Diffusion Measurements Using Radiofrequency Field Gradient: Artifacts, Remedies, Practical Hints. *J. Magn. Reson.* **134** (1998), 245–54.
18. E. Mischler, F. Humbert, and D. Canet, A One-Shot Diffusion Sequence Using a B_1 Gradient. *J. Magn. Reson. B* **109** (1995), 121–5.
19. P. Mutzenhardt and D. Canet, Behavior of Longitudinal Spin Orders in NMR Measurements of Self-Diffusion Coefficients Using Radiofrequency Field Gradients. *J. Chem. Phys.* **105** (1996), 4405–11.

20. E. Mischler, F. Humbert, B. Diter, and D. Canet, Measurement of One-Dimensional Spatially Resolved Self-Diffusion Coefficients and Longitudinal Relaxation Times with a Single B_1 Gradient. *J. Magn. Reson. B* **106** (1995), 32–9.
21. R. Kimmich, B. Simon, and H. Köstler, Magnetization-Grid Rotating-Frame Imaging Technique for Diffusion and Flow Measurements. *J. Magn. Reson. A* **112** (1995), 7–12.
22. M. Valtier, F. Humbert, and D. Canet, Maps of Self-Diffusion Coefficients by Radiofrequency Field Gradient NMR Microscopy. *J. Magn. Reson.* **141** (1999), 7–17.
23. J.-F. Kuntz, G. Trausch, P. Palmas, P. Mutzenhardt, and D. Canet, Diffusive Diffraction Phenomenon in a Porous Polymer Material Observed by NMR Using Radio-Frequency Field Gradients. *J. Chem. Phys.* **126** (2007), 134904-1–134904-6.
24. R. Kimmich and I. Ardelean, Intermolecular Multiple-Quantum Coherence Transfer Echoes and Multiple Echoes in Nuclear Magnetic Resonance. *J. Chem. Phys.* **110** (1999), 3708–13.
25. A. Jerschow, Multiple Echoes Initiated by a Single Radio Frequency Pulse in NMR. *Chem. Phys. Lett.* **296** (1998), 466–70.
26. I. Ardelean, A. Scharfenecker, and R. Kimmich, Two-Pulse Nutation Echoes Generated by Gradients of the Radiofrequency Amplitude and of the Main Magnetic Field. *J. Magn. Reson.* **144** (2000), 45–52.
27. I. Ardelean, R. Kimmich, and A. Klemm, The Nutation Spin Echo and Its Use for Localized NMR. *J. Magn. Reson.* **146** (2000), 43–8.
28. A. Scharfenecker, I. Ardelean, and R. Kimmich, Diffusion Measurements with the Aid of Nutation Spin Echoes Appearing after Two Inhomogeneous Radiofrequency Pulses in Inhomogeneous Magnetic Fields. *J. Magn. Reson.* **148** (2001), 363–6.
29. B. Simon, R. Kimmich, and H. Köstler, Rotating-Frame-Imaging Technique for Spatially Resolved Diffusion and Flow Studies in the Fringe Field of RF Probe Coils. *J. Magn. Reson. A* **118** (1996), 78–83.
30. G. Farrher, I. Ardelean, and R. Kimmich, Probing Four Orders of Magnitude of the Diffusion Time in Porous Silica Glass with Unconventional NMR Techniques. *J. Magn. Reson.* **182** (2006), 215–20.

11

Applications

11.1 Introduction and reviews

The applications of NMR techniques to the study of translational motion is enormous and it is impossible to give anything approaching a comprehensive review. Consequently, only a smattering of papers from the different areas of application is presented and, in general, instead of citing the first paper with respect to each application, more recent papers have been chosen and the interested reader should consult the references listed therein. The classification of different studies is complicated since many studies have significance in more than one area. Numerous reviews on PGSE NMR have already appeared in the literature including ones of a general nature.[1–13] Similarly, there are many books and review articles devoted entirely or in part to the use and applications of MRI techniques to study translational motion and mass transfer including clinical applications and rheological studies.[4,14–25]

There are also a large number of more specialised reviews (or reviews on specialised areas including sections on gradient-based NMR techniques) dealing with NMR measurements of translational motion on diffusion-weighted spectroscopy for studying intact mammalian tissues,[26,27] drug binding, exchange and combinatorial chemistry,[28–34] flow,[6,35,36] heterogeneous systems,[37] liquid crystals, membranes and surfactants,[38–45]organometallics,[46] polymers,[47,48] porous systems including zeolites,[49–52] and solids.[53]

Reviews have also been presented on the complementarity of the structural information that can be obtained from NMR diffusion measurements with that obtained from NOE experiments,[54] the use of PGSE NMR in the studies of physicochemical processes in molecular systems,[55] applications to environmental science,[56] ENMR,[57–59] the spectral editing of complex mixtures with particular emphasis on techniques involving diffusion,[60] and B_1 gradient-based measurements.[61,62]

11.2 Applications to high-resolution NMR

It is of great interest to edit and/or separate complex mixtures on a spectroscopic basis alone thereby avoiding extensive sample preparation and perhaps changes to the properties of the species being analysed. Of the NMR observable parameters, differences in relaxation and diffusion are the most likely means of resolving such mixtures since both parameters are sensitive to molecular weight. Examples of studies involving diffusion filters (i.e., where fast diffusing species are edited out) and mixture separation on the basis of diffusion (where all species are retained in the spectrum but separated on the basis of diffusion; e.g., see Figure 2.18) are given in the following two sections.

11.2.1 Diffusion filters

As shown in Figure 2.13, it is possible to set PGSE parameters to screen a mixture such that only the larger molecules or small molecules with apparently smaller diffusion coefficients resulting from spending time bound to macromolecules are recorded. Apart from requiring a sufficient difference in diffusion coefficient between the solvent and solute, the applicability of PGSE as a means of suppression is limited by the T_2's of the solute resonances. Nevertheless, diffusion provides a very powerful means of suppression since, with the exception of resonances originating from labile spins, all of the resonances of a species are equally attenuated. When used in this way PGSE NMR is sometimes referred to as affinity NMR.[63]

A number of sequences for the selective editing of spectra based on diffusion coefficient differences and on the exchange modulation of diffusion coefficients which contain PGSE subsequences have been proposed. For example, HMQC type sequences that contain a PGSE subsequence have been presented.[64] The STE-based WEX filter only allows through those signals that result from exchanged water.[65] The non-exchanged water can then be selectively suppressed leaving only those resonances that result from the exchange of the water. Wider *et al.*[66] presented a 1D NMR difference experiment, HYDRA, in which a diffusion filter is used to separate intermolecular water-protein NOEs from intramolecular NOEs. A diffusion-edited NOESY experiment was used to suppress the signals of small organic compounds to facilitate the observation of intermolecular NOEs between *N,N*-dimethylformamide (DMF) and lysozyme.[67] Gonnella *et al.*[68] have developed a double-editing pulse sequence which allows direct observation of protein binding ligands. The sequence edits both by diffusion coefficient and by ^{13}C isotope filtering. The sequence was demonstrated using ^{13}C-/^{15}N-labelled stromelysin catalytic domain.

PGSE sequences have been combined with TOCSY sequences,[69] which have been used to study the binding of a series of D-amino acid tetrapeptides.[70] In a later work the PGSE–TOCSY sequence was improved and a heteronuclear diffusion-edited HSQC presented.[71] Diffusion-editing has been used in analysing biofluids including human plasma.[72] An example is given in Figure 8.14. Diffusion-edited ^{1}H HRMAS NMR spectra and COSY spectra have been used to analyse solid-phase resin resulting in compounds that were not covalently attached to the resin, including solvent, to be edited out.[73]

A diffusion-assisted NOE-pumping experiment has been proposed in which a diffusion experiment is prefixed to an NOE experiment for the unambiguous detection of ligands that bind to macromolecules.[74] The diffusion sequence suppresses all of the ligand signals whilst preserving signals from macromolecules. Thus, any ligand signals observed at the end of the sequence must arise from polarisation transferred from the macromolecule (i.e., the polarisation reservoir).

11.2.2 Mixture separation

Due in part to now being part of the standard software on many commercially available spectrometers, DOSY type analysis has been used in a diverse range of studies including the analysis of polydisperse systems,[75] studying equilibria involving binding, absorption and partitioning in surfactant systems,[76] perchloric acid extracts of gerbil brain,[77] non-Newtonian to Newtonian transition in a hexadecyltrimethylammonium bromide (CTAB)/sodium salicylate/water viscoelastic micellar system induced by the addition of a soluble yet slightly hydrophobic polymer.[78] By using sucrose trapped inside vesicles as a marker it was possible to measure the diffusion coefficients (and thence the size) of vesicles.[79] The method also allowed the determination of the trapped and free fractions of sucrose. ^{1}H and ^{13}C-detected DOSY spectra have also been used to analyse and assign the spectra of mixtures of hydrocarbons.[80]

DOSY type analysis has also been applied to spectral assignment of liquid foods such as fruit juices and beer,[81] and to study a range of Port wine samples of different ages,[82] providing approximate MW estimates for polyphenolic compounds in young Ports. Thus, such analysis may be able to characterise the evolution of polyphenol size with age. Nevertheless, limitations of the DOSY technique became obvious in areas of strong spectral overlap. DOSY has also been performed on natural products such as mushroom extracts,[83] and extracts of organic matter from the surface horizon of an oak forest soil.[84] ^{13}C HSQC–DOSY measurements have been used to study the interactions of charged porphyrins with non-ionic triblock copolymer hosts in aqueous solution.[85] The DOSY–NOESY sequence has been demonstrated on a sample of a dinucleotide, d(pAG) and a 14-mer duplex d(ACAATATATATTGT)$_2$.[86] DOSY is not restricted to proton detection and Si^{29} DOSY has been used to study the speciation of aqueous silicates.[87]

It has been shown[88] that by adding a typical chromatographic stationary phase (e.g., silica gel) used in HPLC, HRMAS measurements incorporating PGSE techniques showed significantly enhanced ability to separate the NMR spectra of a mixture.

11.2.3 Electrophoresis, electroosmosis and electroconvection

ENMR has been used to study the tetramethylammonium ion, *N,N,N′,N′*-tetramethylenediamine and tetrahexylammonium ion in polyacrylamide gels[89] and surfactant systems including vesicles and micelles,[90,91] colloidal particles and biological macromolecules (e.g., proteins such as albumin, and ubiquitin),[92,93] and polyelectrolytes.[94] He and Wei used ENMR to study a mixture of L-aspartic acid and 4,9-dioxa-1,12-dodecanediamine.[95] Pettersson *et al.* investigated the interactions between polyethylene oxide and various surfactants using ^{1}H PGSE measurements and ^{1}H ENMR measurements.[96] ENMR has also been used to measure transference numbers in lithium salt–polymer electrolytes.[97] ENMR studies of porous systems have also been performed.[98]

Electroosmotic flows have been imaged using NMR techniques,[99] and observed in natural porous media (i.e., fluid saturated sand beds),[100] and electroosmotic drag coefficients have been measured in polymer electrolyte membranes as would be used in fuel cells.[101] Wu *et al.*[102] using one-dimensional flow imaging have determined the distribution of velocities (actually the distribution of mobilities, μ) and constructed a flow profile of the electric field–driven flow for an oil-in-water microemulsion sample containing a non-ionic surfactant and a trace amount of an ionic surfactant to give a solution of oil-swollen micelles with charged interfaces.

11.3 Biological and pharmaceutical studies

11.3.1 Protein studies

11.3.1.1 Diffusion and hydrodynamics

NMR diffusion measurements have now become one of the basic techniques in the protein chemist's arsenal as they have numerous advantages over traditional techniques including being able to work over large concentration ranges and in non-standard solvents such as liquid CO_2[103] and under high pressure.[104] A very large number of protein diffusion studies are now to be found in the literature including BPTI,[103] c-Jun leucine zipper,[105] cyclosporin A,[103] haemoglobin,[106] lysozyme,[107] myoglobin,[108] myosin light chain 2,[109] ovalbumin,[110] p53 DNA-binding domain,[111] parvalbumin,[112] ubiquitin,[105] antifreeze glycoproteins,[113] and others.[114] The diffusion of the *cis* and *trans* isomers of some proline containing peptides was studied and it was found that the *cis* isomers diffused faster on account of their more compact shape.[115] The protein products of the N-terminal SRCR domain of human CD5 (a type-I transmembrane glycoprotein) have been studied using PGSE NMR.[116]

Cross-linked dextran hydrogels form model systems for protein and peptide drugs. Aso *et al.*[117] have used PGSE measurements to study the diffusion coefficients of lysozyme and insulin in such gels. In such systems the diffusion coefficient of the protein is a measure of the drug release from the gel.

11.3.1.2 Obstruction and hydration

Obstruction has been observed for small molecules moving in protein solutions such as H_2 and H_2O in aqueous bovine serum albumin,[118] H_2O in collagen and cartilage.[119] Kimmich *et al.*[120] have comprehensively studied the effects of geometrical restrictions on water diffusion in aqueous protein solutions of bovine serum albumin, gelatin and myoglobin.

Protein hydration has been studied both as a bulk property by measuring the obstruction effect on water diffusion as proposed in the model by Wang (see Eq. (1.149)) and by using diffusion filters to edit out bulk water, thereby allowing identification of individual hydration sites on a range of proteins such as BPTI, lysozyme, albumin and fibrinogen.[121,122] Studies of the water diffusion coefficient in solutions containing two different concentrations of human serum albumin (HSA) (2.8 and 4.3 w/w%) from 273 to 313 K revealed a linear trend for D versus T/η (see the Einstein–Sutherland equation, Eq. (1.13)), but with η being the bulk solution viscosity measured via Ostwald viscometry.[123] However, it was noted the Einstein–Sutherland equation seems to be violated in the protein concentration dependence of the effective hydrodynamic radius of water and that the deviation of the measured water diffusion coefficient and viscosity data from the Einstein–Sutherland equation was consistent with an enhancement of the solvent structure around the protein surface. However, the significance of obstruction factors in this study were not fully investigated and the question arises as to what is the correct viscosity to consider in the Einstein–Sutherland equation, the bulk viscosity or a microscopic ('local') viscosity. The change in water self-diffusion in casein solutions as a function of casein concentration was analysed using a cell model for obstruction (see Eq. (1.157)) and it was found that the self-diffusion of the water was insensitive to the structure of the casein in solution or in a gelled state.[124]

OGSE has been used to study the diffusion of water in the hydration shells of myoglobin single crystals and bovine serum albumin.[125] It was found that high mobility is retained even below the freezing point of water and is consistent with the water diffusing along the surface of the protein. Percolation transitions were observed with respect to the hydration shells and the free water phase.

11.3.1.3 Folding and aggregation

Protein folding and protein aggregation are inextricably linked since part of the aggregation process generally involves conformational changes. These are in general thermodynamically delicate processes that generally do not involve covalent bonds. PGSE NMR diffusion measurements are particularly suited to studying such processes on account of their non-invasive nature and ability to measure diffusion over a wide range of protein concentrations.

The last decade has seen an enormous increase in the number of papers dealing with diffusion measurements of protein association.[126,127] Associating protein solutions are crowded systems in that the average spacing between protein molecules is much less than the RMSD (see Eq. (1.104)) of the particles over the timescale of the PGSE diffusion experiment (i.e., Δ). For example, the average spacing between lysozyme molecules in a 0.5 mM solution is of the order of 9 nm. Yet, taking the monomer diffusion coefficient at 298 K to be of the order of $1 \times 10^{-10}\,\mathrm{m^2 s^{-1}}$, $\Delta = 30$ ms, the RMSD is about 4 μm. Thus, during Δ there is a high probability for the protein molecules to collide numerous times. This has two consequences, firstly even in the absence of aggregation the measured diffusion coefficient of any oligomeric species will decrease due to self-obstruction (see Section 1.8.6) and second, as noted in Section 2.5, there is evidence of an ensemble averaging of the diffusion coefficients of the different oligomers on the microscopic scale which results in a narrower distribution of diffusion coefficients than would be expected for an isolated ensemble of molecules of the same mass distribution.

In a pioneering work in 1968, Moll[128] studied the helix to random coil transition of poly-L-glutamic acid as a function of pH at 298 K. Since then the folding and unfolding of many proteins have been investigated using PGSE NMR, including a range of peptide 33-mers with some derived from β-sheet domains of interleukin-8 and Gro-α,[129] the SH3 domain of the p85α subunit of bovine phosphatidylinositol 3-kinase which is involved in amyloid fibril formation,[130] BPTI,[131] size distributions of molecules within an unfolded state of the N-terminal SH3 domain of the Drosophila signal transduction protein, drk,[132] and HSQC–DOSY has been used to characterise the folding of the collagen triple helix motif.[133] The urea-induced unfolding of lysozyme resulted in a 28% reduction in the diffusion coefficient which corresponded to a 38% increase in hydrodynamic radius.[134] Similar results were observed for the unfolding of the β-subunit of the heterodimeric salt-mediated killer toxin (SMKT) from *Pichia farinosa*.[135] Analysis of PGSE data has been used to determine empirical relationships between the measured hydrodynamic radius and the number of residues in the polypeptide chain (N).[136] For native folded proteins $r_S = 4.75N^{0.29}$ Å and for highly denatured states $r_S = 2.21N^{0.57}$ Å. Thus, PGSE diffusion measurements can be used to analyse the conformational properties of a range of non-native states of proteins such as partially structured molten globule states. Wilkins *et al.*[136] noted the effective dimensions of the polypeptide chain depend significantly on the level of persistence of regions of secondary structure or features such as hydrophobic clusters within a conformational ensemble.

11.3.1.4 Ligand-protein binding

As noted previously, NMR diffusion studies can be used just to screen a mixture to see if a ligand (or ligands) bind or to characterise the dissociation constant. Some examples of binding were given in Section 11.2. Diffusion NMR studies of binding are myriad and here we mention only a few: the binding of (trimethylsilyl)propionic acid (TSP) to a 17-residue peptide derived from the Alzheimer's associated Aβ peptide,[137] the association of leucine and methionine enkephalin peptides to SDS micelles,[138] the interaction between the cell-penetrating peptide, penetratin, in neutral and negatively charged bicelles, in SDS micelles and in aqueous solution,[139] the binding of protein kinase C substrate, $NG_{(28\text{–}43)}$ to SDS,[140] the binding of 4-trifluoromethylbenzoic acid, ibuprofen, and flurbiprofen to HSA,[141] and the self-association (monomer-dimer) of bathocuproine and its subsequent binding to Alzheimer's associated Aβ peptide.[142]

From diffusion measurements Ma *et al.*[143] showed that racemic, *R*− and *S*+ ibuprofen bound to the low affinity and high capacity binding sites on HSA in a similar manner. In a later study,[144] by measuring the diffusion of ibuprofen as a function of concentration in the presence of 0.1 mM HSA and using the fast exchange model outlined in Section 4.2, it was determined that the association constant K_d was about 0.017 M and that the number of binding sites was about 50. The binding of salicylate to albumin has also been studied using diffusion measurements[145] as shown in Figures 4.6 and 6.3. The interaction between ibuprofen and lipoproteins in blood plasma has also been studied.[146]

^{31}P PGSE NMR has been used to measure the diffusion coefficient of 2,3-bisphosphoglycerate (DPG) in haemoglobin solutions in both free solution and in intact erythrocytes.[147] The dependence of the measured diffusion coefficients on the amount of DPG bound to haemoglobin was used to estimate the dissociation constants for DPG complexed to carbonmonoxygenated, oxygenated and deoxygenated haemoglobin.

11.3.2 DNA and RNA

Similar to the case of proteins, the last decade has seen NMR diffusion measurements become a major tool for characterising the solution behaviour of DNA and RNA. For example, diffusion coefficient measurements of nucleic acids have allowed duplex RNAs to be distinguished from RNA hairpins.[148] Measurements on three single-stranded DNA dodecanucleotides and the related duplexes allowed a direct assessment of polymeric and conformational states such as hairpin formation.[149] The diffusion of polyamine spermidine and the polyamine analog *N*-methylated spermidine in solutions of calf thymus DNA in the presence of different salts has been measured to determine the importance of electrostatic interactions on polyamine–DNA association.[150]

PGSE measurements have been used to study the anisotropic diffusion of $^{7}Li^{+}$ and $^{133}Cs^{+}$ in hydrated oriented DNA fibres.[151] Kaucher *et al.*[152] studied the influence of cation, solvent and anion on the cation-templated self-assembly of lipophilic guanosines in organic solvents. In a later work diffusion measurements were used to clarify the influence of the cation and anion in the supramolecular structure of self-assembled ionophores from isoguanosine.[153]

A water-selective one-dimensional diffusion experiment was used to measure the exchange rates of the rapidly exchanging protons of a 16 base-pair DNA sequence,[154] and the binding of ethidium bromide to a DNA hairpin (dU_5-hairpin) has been studied using diffusion-weighted COSY experiments.[155] Interestingly, the diffusion of DNA in agarose gels has been observed to display reptational behaviour[156] and obstruction effects have been observed in PGSE diffusion measurements of polyammonium cations in aqueous solutions of DNA.[157]

11.3.3 Drugs, biomolecules and biomaterials

The solution properties of many drugs and biomolecules have been studied using NMR diffusion measurements including the hydration of *N*-acetylaspartic acid[158] and the self-association of the shellfish toxin okadaic acid upon complexation with potassium ions.[159] The results showed that the K^{+} okadaic acid complex is a dimer. The diffusion of monensin in chloroform solution has been studied as a function of concentration and the results explained in terms of free volume theory.[160] Parkinson *et al.* have used HSQC–DOSY experiments to detect the cluster complexes formed by a bismuth(III) anti-ulcer complex in aqueous solution.[161] Diffusion measurements have revealed that at low concentrations the antibiotic Ramoplanin in $C^{2}H_{3}OD$ is in an equilibrium between monomeric and dimeric states.[162] The diffusion of psychosine (β-galactosyl-sphingosine) was studied in $^{2}H_{2}O$ and was found to be pH-dependent and indicative of aggregation.[163] The aggregation of steroid compounds in solution has been probed using PGSE NMR diffusion measurements.[164]

PGSE NMR diffusion measurements have been used to investigate a diverse range of biomaterials including the diffusion of water in fibrin gels, plasma and blood clots,[165] of metabolites in blood and seminal plasma,[166] and of amino acids in gels.[167]

Of relevance to drug release and cell immobilisation and to more generally provide insight into the diffusion of micelles into hydrogels, the diffusion of small organic solutes and large micelles has been studied in dextran solutions and gels revealing drastic reductions in diffusion in the case of Triton X-100 micelles.[168] Mistry *et al.*[169] have used ^{19}F NMR diffusion measurements in impurity profiling in pharmaceutical batches. In particular, the diffusion measurements allowed discrimination between monomeric and dimeric impurities.

There have been numerous reports on the use of diffusion measurements, and combined with relaxation editing, to study and simplify the spectra of the International Humic Substances Society Suwannee River fulvic acid standard.[170] Lead *et al.*[171] measured the diffusion coefficient and polydispersity of Suwannee River fulvic acid as a function of pH and ionic strength using PGSE NMR and compared the results with those obtained with fluorescence correlation spectroscopy, and flow-field fractionation. The diffusion coefficients ranged between 1.9 and $3.5 \times 10^{-10}\, m^2s^{-1}$.

11.3.4 Liquid crystals, lipid membranes and membrane proteins

PGSE NMR has been used to measure lipid lateral diffusion in aligned lamellar membranes,[172] including the effects of microdomains such as cholesterol,[173] sphingomyelin[174] and raft mixtures.[175] Orädd and Lindblom also studied the effects of obstruction by the transmembrane protein gramicidin D and gel patches on the lateral diffusion of dimyristoylphosphatidylcholine in oriented bilayers.[176] Diffusion measurements were also used to probe lateral phase separation in dioleoylphosphatidylcholine–sphingomyelin bilayers.[177] Diffusion has been studied in oriented phospholipid bilayers using fringe field measurements.[178]

By using MAS NMR in conjunction with magic angle gradients it is possible to measure lipid translational diffusion without aligning the lipid bilayers[179] including the diffusion of ibuprofen in membranes[180] and measurements in bilayers supported in nanopores.[181] However, in an unoriented sample the analysis of the diffusion measurement must account for the random orientations.[180] MAS PGSE has also been used to study water, ubiquinone and lipid mixtures in cylindrical aluminium oxide nanotubes.[182]

11.3.5 Restricted diffusion and obstruction

11.3.5.1 Food, plants and natural products

Yeast cells have been a popular model system from the earliest days of PGSE NMR with Stejskal and Tanner observing an intracellular water diffusion coefficient of $6.8 \times 10^{-10}\, m^2s^{-1}$ and a mean cell size of 4.1 μm.[183] In later studies diffusive diffraction effects have been observed in yeast giving a characteristic distance of 5 μm.[184] Most food can be viewed as being a porous material. PGSE NMR of oil and water in peanuts reveals regions of complete restriction for the oil and the cell structure of the peanuts was identified.[185] Diffusion measurements of water in native potato, maize and pea starch granules have shown that the water is confined to different types of restricted environments.[186] The diffusion of oil and water in rape seeds has been investigated.[187] For $\Delta > 30$ ms the diffusion of the oil was observed to be completely restricted. The results were modelled using a Gaussian mass distribution of spherical droplets with a mean radius of 0.7 μm. At maximum moisture content the water diffusion coefficient was $4.2 \times 10^{-10}\,\mathrm{m^2 s^{-1}}$. The diffusion coefficients of CO_2 in various carbonated beverages (e.g., champagne, beer) have been measured using ^{13}C PGSE NMR and was found to deviate from that predicted with the Stokes–Einstein equation.[188]

Wood, cellulose and starch systems have been widely studied using NMR diffusion techniques. In wood cellulose is in the form of rod like microfibrils about 10 nm wide and μm length scales. Starch occurs in two major forms: linear amylase and branched amylopectin. In plants starch is found as rounded granules with radii of tenths of μm. In an early study Peemoeller *et al.*[189] used PGSE measurements to study cell diameters and volume fractions in samples of Douglas-fir. Wycoff *et al.* used PGSE to study the tangential dimensions of cells in a number of wood samples including eastern red cedar, eastern white pine, Redwood and sugar pine.[190] PGSE NMR has been used to measure water diffusion and pore volume in cellulose fibres.[191] Two components were observed: one with a self-diffusion coefficient independent of time and the other with a time-dependent 'apparent' diffusion coefficient. The components were attributed to bulk water between the cellulose fibres and water in anisotropic pores with lengths from several μm to 20 μm within the fibres. The average pore volume of the pulp fibres was determined to be $1.6 \pm 0.3\,\mathrm{cm^3\,g^{-1}}$. The bulk water diffusion coefficient was about 30% less than that of pure water and was thought to be due to the effects of (mainly) obstruction and hydration. In a later study,[192] Li *et al.* used q-space imaging to investigate molecular displacement profiles of the water in cellulose fibres for Δ ranging between 5–1200 ms. Ek *et al.* have estimated the pore size and studied the swelling and tortuosity of highly porous cellulose beads used as pharmaceutical excipients.[193] Water diffusion in starch has also been studied.[194] Newling and Batchelor have studied the diffusion of water and polyethylene glycol (PEG) in the supramolecular structure of wet cotton.[195] The data was analysed in terms of the modified Kärger model in which one site is considered to be restricted (see Section 4.3.1). Topgaard and Södermann have studied water diffusion in cellulose and starch fibres, biological porous materials consisting of randomly oriented domains with anisotropic supermolecular organisation.[196]

11.3.5.2 Biomedical

Diffusion and MRI studies performed in biological tissues generally exhibit non-exponential signal decays. This could result from multicomponent Gaussian (i.e., free) diffusion or restricted diffusion or a combination of both. Being able to discriminate the two cases is extremely important for the development of diagnostic applications of diffusion MRI. In simple cases, examining the dependence of the measured diffusion coefficient as a function of Δ will allow discrimination between free and restricted diffusion; however, membrane permeability complicates this approach. Attempts have been made to separate intracellular from extracellular components by curve stripping[197] and biexponential fitting[198] of PGSE data. However, exchange between compartments during the experiment complicates the separation of the signals. Malmborg *et al.*[199] have shown that restricted components in multiexponential echo decay curves can be identified by varying δ.

In an early study Cooper *et al.* studied restricted diffusion in various biophysical systems including blood, heart and liver using PGSE NMR.[200] Later studies include the intracellular diffusion coefficient of water in frog (*Rana pipiens*) muscle which was measured and found to be $1.6 \times 10^{-9}\,\mathrm{m^2 s^{-1}}$, while analysis of the diffusion data gave a membrane permeability coefficient of $0.01\,\mathrm{cm\,s^{-1}}$,[201] and ^{13}C PGSE NMR diffusion measurements of glycine inside human red blood cells.[202] The time-dependent water diffusion measurements has been studied in packed erythrocytes and it was observed that the long-time diffusion coefficient, D^{∞}, was very sensitive to the extracellular volume fraction.[203] Also from the short-time behaviour of the diffusion coefficient, the surface-to-volume ratio of the cells was estimated to be approximately $(0.72\,\mu\mathrm{m})^{-1}$. The diffusion of phosphocreatine in the cylindrically shaped fibres of rabbit skeletal muscle have been studied using diffusion measurements in three orthogonal directions enabling the determination of the trace of the diffusion tensor.[204] From this data the radius of the cells was found to be in the range of 8–9 μm and the phosphocreatine diffusion coefficient to be in the range of $7\text{–}9 \times 10^{-10}\,\mathrm{m^2 s^{-1}}$. Diffusional anisotropy of phosphocreatine has also been observed in fish skeletal muscle.[205] De Graaf *et al.*[206] studied the diffusion of ATP and phosphocreatine in intact rat skeletal muscle using ^{31}P NMR. The apparent diffusion was measured as a function of Δ. Orientation effects were eliminated by measuring the trace of the diffusion tensor. The bulk diffusion coefficients and radial dimensions of the restriction were determined by fitting to the cylindrical restriction model (Eq. (4.60)). Their estimates of the bulk diffusion values were ~90% of the in vitro values and the estimates for the diameters of the restriction were ~16 and 22 μm for the ATP and phosphocreatine, respectively – which are both considerably smaller than the known diameters of rat skeletal muscle fibres (60–80 μm). By analysing apparent diffusion coefficients obtained in HRMAS measurements of testicular tissue Griffin *et al.*[207] were able to show that a large proportion of the creatine observed was extracellular. The relationship between molecular crowding, viscosity and the apparent diffusion coefficients of metabolites in subcellular organelles has also been studied.[208]

q-Space coherence features have been observed in suspensions of red blood cells. The features have been assigned to water undergoing restricted diffusion both inside cells and in the cavities in the extracellular medium.[209] These coherence features are sensitive to factors such as haematocrit, cell size, and the transmembrane exchange rate.

Stray-field diffusion measurements conducted on a 9.4 T magnet have been used to characterise biological tissues and their pathological alterations.[210] In particular, water diffusion was studied in carcinomas, fibrous mastopathies, adipose and liver tissues. The results provided evidence for the fractal dimension of the underlying cellular structure of the tissues.

Due to the impeding effects of cell membranes, the effective diffusion tensor of water in brain tissue is, in general, anisotropic and 2–10 times slower than bulk water. Stroke is accompanied by the swelling of cells and consequent reduction in extracellular space which has been suggested to lead to a reduced effective diffusion coefficient. Latour *et al.*[203] have argued that the effective diffusion coefficient is dominated by the tortuosity of the extracellular space and that the tortuosity increases on cell swelling. As would be expected for restricting geometries, strong diffusional anisotropy effects have been noted in studies of biological cells.[211] In fact, the observation of the anisotropy has been noted as being a useful clinical probe of demyelinating disorders, white matter infarcts, neoplasms and of neonatal brain and spinal cord development.[212] Schoeniger *et al.*[213] studied water diffusion in neurons. They noted that water in the nucleus has different diffusion properties than that in the cytoplasm. Similarly, Le Bihan and co-workers[214] have proposed that measurements of diffusion coefficients have clinical applications including functional assessment, tissue characterisation and treatment monitoring. ^{3}He diffusion in the lungs has been investigated using in vivo PGSE measurements and a 1D imaging technique and finite difference simulations.[215]

Carlton *et al.* have performed diffusion-weighted imaging of bacteria colonies in the STRAFI plane and were able to detect bacteria in concentrations down to 4×10^9 cells ml^{-1}.[216]

11.3.6 Multi-dimensional correlations

Numerous studies have attempted to correlate relaxation time with diffusion in biological samples.[217,218] For example, Peled *et al.*[218] measured water diffusion as a function of T_2 in frog sciatic nerve.

DDCOSY and DEXSY experiments have been used to study anisotropic water self-diffusion and defects in the mesophases of lyotropic liquid crystals.[219] Komlosh *et al.* have used a double PGSE sequence to measure and assess the degree of local anisotropy in brain grey matter using nine different combinations of gradient direction for the two gradient pulse pairs.[220] DRCOSY experiments have been used to distinguish between oil and water in a mixture,[221] and to detect restricted diffusion in porous media such as sandstones and dolomites.[222] DEXSY and DRCOSY have been used to investigate the dynamics of water and oil in food and microemulsion systems,[223] and to investigate the diffusion exchange of dextran with molecular weights 4.4 and 77 kDa through polyelectrolyte multilayer hollow capsules consisting of four bilayers of polystyrene sulfonate/polydiallyldimethylammonium chloride.[224] The DRCOSY experiments revealed that the diffusion process of dextran 77 kDa exhibits an observation time-dependence indicative of restricted diffusion. Further, the 77 kDa dextran molecules are in diffusive exchange through the capsules with an exchange time of around 1 s whereas the diffusion process of the 4.4 kDa dextran is unaffected by the capsules.

11.3.7 Transport, exchange and binding

Intracellular species tend to diffuse more slowly due to being in a more viscous environment (the intracellular milieu typically has a high protein concentration) and also have greater motional restriction due to being enclosed within a pore with either limited or no permeability. In an early study, Andrasko used differences in diffusional restriction between the inside of red blood cells and the extracellular medium to observe intracellular Li^+ and thereby study the uptake of (very slowly permeating) Li ions.[225] Jiang *et al.* have studied water exchange through red blood cells using PGSE NMR measurements and determined the water residence time to be 10 ms.[226] Differences in PGSE attenuation have also been used to discriminate between the intra- and extracellular milieux to monitor the intracellular metabolism of human breast cancer cells,[197] to separate intra- and extracellular sodium signals in red blood cells,[227] and to measure the cell density of *Pseudomonas cepacia*.[228] The ionophore mediated transmembrane exchange of Li^+ in liposomes has been studied using PGSE NMR and modelled using the Kärger equations.[229]

The partition of adrenocorticotropin peptides in water and SDS and dodecylphosphatidylcholine micelles has been studied using diffusion measurements.[230] Using a simple two-site model and assuming fast exchange of the peptides (see Eq. (4.18)), the diffusion of the bound form was taken to be equal to that of the micellar diffusion coefficient and the true diffusion of the free form was extracted accounting for obstruction by the spherical micelles using Eq. (1.142).

Proton exchange, including amide protons, have been studied in a number of systems using diffusion NMR, including the amide protons of the acyl carrier protein,[231] the amide protons of viomycine,[232] the hydroxyl protons of sucrose,[233] and the two amido protons of uracil for which the on rates were determined to be 8 and 18 s^{-1} for H_1 and H_3, respectively.[234] The selective complexation of the *cis*-isomer of phenylalanylproline over the *trans*-isomer with β-cyclodextrin has been studied using diffusion measurements.[235] Diffusion measurements have been used to study the binding of the zwitterionic opioid peptide D-penicillamine^{2,5}-enkephalin (DPDPE) to a zwitterionic phospholipid micellar bilayer and the bound populations were determined using Eq. (4.18).[236] Orfi *et al.* have studied the association of two simple tripeptides (glycyl-histidyl-glycine and phenylalanyl-histidyl-phenylalanine) with SDS micelles using ^{1}H diffusion measurements.[237]

11.4 Chemical and material studies

11.4.1 Diffusion-based studies

11.4.1.1 Diffusion measurements

Supercooled water has been measured down to 238 K in the case of ^1H_2O[238] and down to 244 K for ^2H_2O.[239] Using a specially constructed probe, Yoshida *et al.* have measured the diffusion coefficients of ^1H_2O and ^2H_2O in the sub-critical temperature range 303–623 K on the liquid vapour coexisting curve and at the supercritical temperature of 673 K as a function of water density between 0.071–0.251 g cm^{-3}.[240] ^{2}H PGSE NMR has been used to study the solution dynamics of the D_4-methane-water system.[241] Klenø *et al.* have studied the diffusion of dimethyl sulfoxide, ^{15}N-labelled DNO_3 and ^{17}O-labelled ^2H_2O in solutions of 0.6–0.8 weight fraction D_2SO_4.[242] Dippel and Kreuer[243] measured the diffusion coefficients of ^{1}H and ^{17}O in different concentrations of aqueous hydrochloric acid in the temperature range 5–90 °C. The self-diffusion of CCl_4 and CS_4 has been measured using B_1 gradient-based ^{13}C PGSE measurements.[244]

The diffusion of liquid lithium has been of considerable interest from the very early days of PGSE NMR.[245] In almost isotopically pure samples Murday and Cotts[246] determined $D(Li^6)$ and $D(Li^7)$ to be $6.8 \pm 0.7 \times 10^{-9}$ m^2s^{-1} and $5.8 \pm 0.6 \times 10^{-9}$ m^2s^{-1} at the melting point of 453 K, giving $D(Li^6)/D(Li^7) = 1.18 \pm 0.07$, which was slightly larger than the ratio of 1.09 ± 0.06 obtained by Krüger *et al.* at 463 K.[247] This is interesting from the perspective of the Einstein–Sutherland equation since in both cases the ratio was less than the viscosity ratio $\eta^7/\eta^6 = 1.44$.

Using the steady gradient spin-echo McCall and Douglass studied self-diffusion in binary solutions (i.e., benzene–cyclohexane, acetone–chloroform, acetone–benzene and acetone–water) in which one component at a time was deuterated to allow the measurement of the other. The results were discussed in connection with mutual diffusion.[248] Self-diffusion in binary mixtures of the system benzene-cyclohexane has been studied at pressures up to 200 MPa,[249] and in four binary systems consisting of ammonia with benzene, acetonitrile, 1-pentanol and methanol.[250] PGSE has been used to investigate the 1,4-cyclohexanedione-bromate-acid oscillating system.[251]

STRAFI measurements were combined with incoherent neutron scattering experiments to study glycerol for microscopic and mesoscopic wave numbers and it was possible to discern signs of a crossover between relaxation-like and diffusion-like motion at $q \lesssim 0.3\,\text{Å}^{-1}$.[252] The diffusion coefficients of water and polyethylene glycol were measured in aqueous mixtures of polydisperse polyethylene glycol with various average molecular weights ranging from 200–10,000 Da.[253] The aim of the work was to characterise the formation of water-PEG networks purported to occur under the conditions used for protein crystallisation. Despite the polydispersity, the diffusion measurements of each sample was observed to be single exponential. At high concentrations the diffusion of the PEG molecules approached a constant value indicative of the formation of a dynamic network between the PEG and the water. Above this concentration the system can be microscopically described as a double domain: bulk water and a water-PEG network.

The diffusion of neat CO_2 has been measured between 223 and 450 K at pressures up to 200 MPa with the diffusion data showed significant deviation from that expected for a hard sphere fluid.[254] The diffusion of five binary systems consisting of CO_2 with benzene, hydrogen, palmitic acid, methyl ester, acetic acid or methanol were measured at low temperature and pressures up to 200 MPa,[255] and the diffusion coefficients of both $^{13}CO_2$ and $C_{16}{}^1H_{34}$ were measured in supercritical CO_2-*n*-hexadecane mixtures.[256] The diffusion of the noble gases have also been measured in water.[257]

The self-diffusion and mutual diffusion of the cation and anion of sodium monofluorophosphate and water in silica pastes have been studied using PGSE and 1D MRI, respectively.[258] In trying to model the solidification of metals, Lutz and Mendenhall[259] measured the self-diffusion coefficient of ^{15}N labelled ammonium ions in aqueous ammonium chloride, and the diffusion coefficients of water and succinonitrile in aqueous succinonitrile and for succinonitrile and acetone in a mixture of acetone–succinonitrile. The diffusion of the cation and anion of tetrabutylammonium tetrahydridoborate ($(C_4H_9)_4N^+BH_4^-$) have been studied in C^2HCl_3.[260] It was found that the BH_4^- ion diffused only slightly faster than the

$((C_4H_9)_4N^+$ ion. Interestingly, both ions diffused significantly slower than the uncharged tetrabutylsilane ion which was of similar size to the tetrabutylammonium ion, indicating that the tight ion pair is the primary diffusive species.

Ionic liquids (molten salts) have been increasingly studied using NMR diffusion measurements. ^1H diffusion measurements have been conducted on chloroaluminate (e.g., 2:1 $AlCl_3$:1-ethyl-3-methylimidazolium chloride) melts.[261] Measurements have also been performed on phosphoric acids.[262]

11.4.1.2 Hydrodynamics

The diffusion of the *n*-alkanes – *n*-pentane, *n*-hexane, cyclohexane and *o*-xylene – at saturation have been measured in ^2H_2O at 298 K and the data analysed to give scaling relations for the molecular weight dependence of the self-diffusion coefficients.[263] The diffusion of 15 liquid *n*-paraffins (C_8–C_{60}) in the temperature range 30 to 170 °C has been studied.[264] The data was consistent with a scaling relation $D \sim M^{\beta}$ with β changing approximately linearly from −2.72 to − 1.85 with increasing temperature.

The diffusion of *N*-methylformamide has been studied at pressures up to 200 MPa between the melting pressure curves up to 420 K.[265] Interestingly, the self-diffusion coefficient of the *cis*-conformer was 17% lower than that of the *trans*-conformer at the same temperature and pressure. The self-diffusion of C^2H_4 and N^2H_3 over considerable temperature and pressure ranges up to 450 K and 200 Mpa were studied to investigate dynamic isotope effect ($D_r = D_{X-H}/D_{X-{}^2H}$) in the liquids.[266] D_r was determined to be 1.3 and 1.4 for C^2H_4 (at 150 K) and N^2H_3 (at 200 K), respectively. The D_r values and temperature-dependence could not be explained by theories for single particle motions nor could the results be explained on the basis of collective phenomena due to hydrogen bonding since the hydrogen bonding properties of C^2H_4 and N^2H_3 are quite distinct. Theoretical models for obstruction have been compared with PGSE data of toluene, aniline and phenol in polymer solutions in order to separate the effects of binding and obstruction.[267]

Haselmeier *et al.*[268] have measured the diffusion of C_{60} and C_{70} and of the solvent molecules C_6H_6 and CS_2 in liquid fullerene solutions. In CS_2 solution at 298 K they found $D(C_{60}) = 1.85 \pm 0.18 \times 10^{-9}\ m^2s^{-1}$ and $D(C_{70}) = 1.56 \pm 0.16 \times 10^{-9}\ m^2s^{-1}$. They remarked that C_{70} may be regarded as an ideal ellipsoidal particle (short semi-axis = 5.165 Å and long semi-axis = 5.775 Å) with which to test hydrodynamic theories. Indeed, their data showed that the long semi-axis of the ellipsoid determined the diffusional behaviour. Measurements of C_{60} in Benzene-D_6 have also been reported.[269]

Feiweier *et al.*[270] used a static gradient STE sequence in the presence of ultrahigh magnetic gradients generated by superconducting anti-Helmholtz coils (gradients up to 184 T m^{-1}) to measure the self-diffusion of ^7Li and ^1H in the glass forming electrolyte $LiCl:7H_2O$ (glass transition temperature $T_g = 139$ K). The measurements were conducted over the temperature range 313 K down to 173 K over which D_{Li} and D_H changed by five orders of magnitude and the self-diffusion exhibited strong non-Arrhenius behaviour typical of fragile glass formers. Whilst the lithium diffusion was closely coupled to other transport modes such as viscosity, electrical conductivity and water reorientation, as the temperature decreased the ratio D_H/D_{Li} increased indicating decoupling of these diffusion modes with D_H decoupling at $T < 1.5\,T_g$.

11.4.1.3 Carbohydrates

PGSE diffusion measurements have potentially a lot to offer in the field of carbohydrate chemistry. Examples of studies include measurements of a series of carbohydrates that differ in size and branching pattern in dilute aqueous solutions,[271] diffusion in aqueous solutions of the linear polysaccharide hyaluronan,[272] the diffusion of ^{13}C-labelled glucose in molasses.[273] And in a perhaps more unusual study it was observed that the sweetness of sugar solutions (sucrose, fructose and glucose) did not correlate with the translational diffusion coefficients.[274]

The diffusion of aqueous trehalose solutions has been studied to determine if strong attractive interactions effect the prevision of the Einstein–Sutherland theory.[275] It was found that the strong trehalose–water attractive interactions result in significantly reduced friction and that the temperature-dependence of the trehalose solution was described by a Vogel–Fulcher–Tamman relationship. The results confirm that trehalose has a strong destructuring effect on the tetrahedral network structure of water. Diffusion measurements on sucrose and trehalose in aqueous solution revealed that although their diffusion coefficients were comparable at low disaccharide concentrations, they differ from each other with increasing concentration.[276]

11.4.1.4 Hydrides and organometallics

PGSE NMR has been used to study hydrogen diffusion in Laves-phase hydrides over a wide range of temperature and hydrogen ion concentrations.[277]

PGSE is now widely used in studying organometallic compounds and in conjunction with other NMR techniques, such as NOE data, has proven to be an important tool for elucidating problems related to molecular volume, hydrogen bonding, mass distributions, solvent-dependence and ion pairing which thus includes problems related to catalysis and coordination chemistry.[278] ^{195}Pt PGSE NMR measurements were used in studies on the solvent-dependence of aggregation of the hexachloroplatinate dianion in Na_2PtCl_6 and H_2PtCl_6.[279] PGSE NMR studies of the aggregation of *n*-butyllithium in tetrahydrofuran enabled dimeric and tetrameric forms to be distinguished.[280] PGSE NMR has also been used to study the distribution of species in solutions of metallomacrocycles.[281] Zuccacia *et al.*[282] used ^1H PGSE NMR to study the diffusion of the complexes *trans*-[$Ru(PMe_3)_2(CO)(COMe)(pz_2$-$CH_2$)]$BPh_4$ and *trans*-[$Ru(PMe_3)_2(CO)(COMe)(pz_2$-$BH_2$)] in nitromethane, chloroform and methylene chloride as a function of concentration. They found indications of ions (in nitromethane), ion pairs (in chloroform at low concentration) and ion quadruples (in chloroform at high concentration). Diffusion measurements of the aggregates of dimethyl- and bis[(trimethylsilyl)methyl]cuprates in dimethyl ether revealed that aggregates larger than dimers were formed.[283]

11.4.1.5 Surfactants, emulsions and liquid crystals

PGSE diffusion measurements have found enormous application in the estimation of droplet sizes and distributions in emulsions (including concentrated emulsions) as outlined in Section 3.8 and has extensive applications to the food industry,[284] cosmetics[285] and others. PGSE measurements have also been used to study the effects of temperature on droplet size and also the effects of freezing.[286] Freezing causes coalescence of the droplets. In a study of water-in-oil emulsions, diffraction-like effects have been observed in a highly concentrated water/oil emulsion.[287] From the experiment the mean droplet size of the emulsion was determined. PGSE measurements have also been applied to multiple emulsions. Lönnqvist *et al.* studied a water-in-oil-in-water (W/O/W) emulsion in which they were able to obtain the size distribution in the starting emulsion and the double emulsion.[288] Intercompartmental exchange and droplet size in W/O/W emulsions have been studied using PGSE NMR and modelled with the modified Kärger relations.[289] Topgaard *et al.* were able to probe micrometre-sized compartments in water-in-oil emulsions using OGSE experiments.[290] PGSE was combined with imaging techniques to study the spatial dependence of droplet size in the cream layer of water-in-oil emulsions.[291] Using a stimulated echo 2D imaging sequence and the inverse Abel transform, Hollingsworth and John[292] have demonstrated that it is possible to measure spatially resolved emulsion droplet sizes in cylinders in which the emulsion is undergoing laminar flow.

In an early study, Stilbs and Lindman[293] studied the micellisation of a cationic surfactant (decylammonium) with organic counterions ($CHCl_2COO^-$) in ^2H_2O. The diffusion of sodium alkylsulfonates in ^2H_2O has been studied using PGSE NMR to determine the critical micelle concentration (CMC).[294] Solubilised TMS was used to study the diffusion in the micellar composition concentration range and it was observed that the diffusion of both the surfactant and the micelles sharply decreased above the CMC. This decrease was interpreted in terms of obstruction by the micelles which was thought to be enhanced by the electrostatic repulsion between the particles. The interaction between non-ionic micelles and a non-ionic polymer have been studied using PGSE.[295] It was found that the micelle diffusion is influenced by the polymer in two ways: obstruction by the polymer network and association of the micelles to the polymer chain.

Xu *et al.*[296] have studied water-in-CO_2 (W/C) microemulsion systems formed with phosphorus fluorosurfactants of bis[2-(F-hexyl)ethyl] phosphate (DiF_8) salts with different counterions at high pressure. The diffusion measurements revealed that the counterions and the surfactant molecules move together with a diffusion coefficient associated with that of the microemulsion droplets. The average hydrodynamic radius of the ammonium counterion-DiF_8 microemulsions droplets was 2 nm at 298 K, 206 bar at a water–surfactant mole ratio of 5. Self-diffusion measurements were used to investigate the micellar behaviour of malonamides in the presence of dodecane.[297] Malmborg *et al.* have studied diffusion in highly concentrated emulsions (96% aqueous solutions of salt or other additives with heptane as the oil phase). They observed that acetic acid/acetate ions could diffuse between the droplets and had exchange times that were functions of pH. And diffraction peaks were observed at the same values of q for both water and the acetic acid/acetate ions (see Figure 11.1).[298]

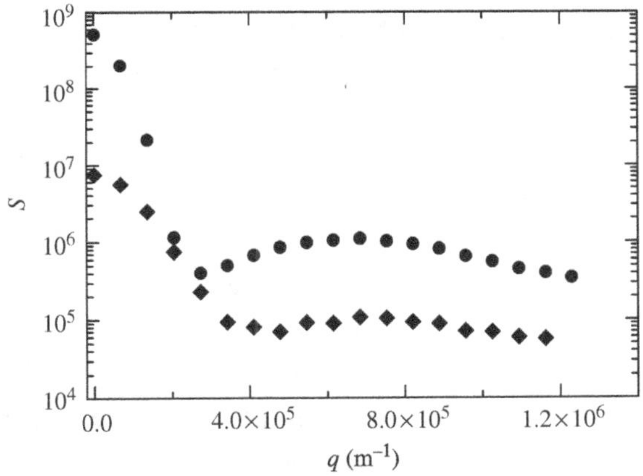

Figure 11.1 Water (●) and acetic acid/acetate ions (◆) in a concentrated (gel) w/o emulsion of 96 wt% water. The ratio of acetic acid/acetate ions is 5/95. From Malmborg *et al.*[298]

The concentration dependence of the self-diffusion coefficient has been investigated for a range of micelle and mixed micelle systems.[299] Angelico *et al.* have observed curvilinear diffusion of surfactants in giant wormlike micelles.[300] The MSD of the surfactant was found to scale with $t^{1/2}$. B_1 gradients have been used to measure the self-diffusion coefficient of hydroxyquinoline in aqueous solutions of micellised SDS, and also to determine its partition coefficient as a function of pH.[301]

PGSE NMR is now a major tool in the study of liquid crystal systems with studies including water self-diffusion in polycrystalline lamellar systems,[302] hexagonal mesophases[303] and smectic liquid crystals.[304] Coppola *et al.*[305] studied water self-diffusion in water lithium/perfluorononanoate (LiPFN) and cetyltrimethylammonium bromide (CTABr) micellar systems. The diffusion of the water was rationalised by including the effects of hydration and obstruction. Cifelli *et al.*[306] studied the diffusion of ^{129}Xe in the chiral liquid crystal 1-methylheptyl 4′-(4″-*n*-decyloxybenzoyloxy) biphenyl-4-carboxylate (10B1M7) over a range of temperatures covering the isotropic, smectic, and ferroelectric, ferrielectric and antiferroelectric smectic phases. They observed that the anisotropy ($D_\perp/D_\parallel$) of the diffusion tensor increases with decreasing temperature. In a later study[307] they used fringe field NMR diffusometry measurements to probe from the isotropic to the anticlinic phase of the ferroelectric liquid crystal 1-methylheptyl 4′(4″-*n*-decyloxybenzoyloxy) biphenyl-4-carboxylate.

^{129}Xe diffusion has been studied as a function of temperature in different phases of the ferroelectric liquid crystal FELIX-R&D.[308] In the smectic mesophases significantly faster diffusion was observed in the perpendicular direction than in the direction parallel to $\mathbf{B}_0$. ^{129}Xe diffusion has also been studied in the so-called critical mixture of two nematic liquid crystals with opposite diamagnetic anisotropy as a function of temperature.[309] In this mixture the liquid crystal director rotates 90° from the parallel direction to the perpendicular direction with respect to $\mathbf{B}_0$ as the critical point is approached from the high temperature side.

Solutions of block copolymers are interesting systems for the study of self-organisation since due to hydrophobic and hydrophilic moieties, the polymer molecules may form micelles over a certain temperature interval. As a result, the diffusion of some polymers can have strange temperature dependencies, such as have been observed with the triblock copolymer PEO-PPO-PEO in ^2H_2O.[310] Interestingly, it was observed that even for the case of $\Delta = 3$ ms the attenuation of the echo signal was single exponential. Thus, the exchange time of the polymer molecules between the micelles and monomers must be much shorter than the observation time. The influence of polydispersity on the micellisation of triblock copolymers has been investigated using PGSE NMR.[311] Ambrosone *et al.*[312] have studied diffusion of a surfactant confined in a branched cylindrical micellar network formed from lecithin and small amounts of water in the solvent isooctane. At certain concentrations and at shorter diffusion times they observed that the MSD of the surfactant scaled as $t^{1/2}$ consistent with curvilinear diffusion.

11.4.1.6 Colloids

The long-time self-diffusion (D^∞) of spherical polybutadiene latex particles, which were highly charged due to the adsorption of anionic surfactant, were measured in water using PGSE NMR.[313] Roberts *et al.*[314] studied the self-diffusion of $^7Li^+$ in a suspension of sulfonated polystyrene latex particles over a wide range of electrolyte concentrations. The diffusion of the lithium was considerably less than that in a solution of $LiClO_4$ of the same concentration indicating that the Li^+ counterions are electrostatically bound to the large latex particles. Further, the ion-exchange behaviour of the latex suspension was able to be characterised by observing the change in Li^+ diffusion with addition of $LiClO_4$.

Schipper and Leyte[315] have studied the self-diffusion of the species in a ^2H_2O solution of polymethacrylic acid neutralised with tetramethyl ammonium hydroxide. At concentrations below 0.01 mol monomer kg^{-1} the polyion self-diffusion coefficient approaches that of a freely diffusing rod; above this concentration the diffusion of the polyions and counterions are independent of concentration but at concentrations greater than 0.1 mol monomer kg^{-1} the diffusion of the solvent counterions decrease with concentration due to topological constraints. They found that the relative counterion diffusion coefficient (i.e., D/D^0; the obstruction factor) was predicted almost quantitatively by the Poisson–Boltzmann–Smoluchowski model (see Section 1.8.6).

Böhme and Scheler[316] used PGSE to study the fractal dimension in the polyelectrolyte poly(styrenesulfonate). From the variation of the hydrodynamic radius with molecular weight for each ionic strength the fractal dimensions were determined.

11.4.1.7 Membranes

PGSE measurements of water and glycerol in cellulose acetate membranes have been performed and the results compared with simulations of the anomalous (and anisotropic) diffusion on the membrane using a two-dimensional lattice model.[317] Measurements of ethanol–water mixtures in polyacrylic acid-polysulfone composite membranes[318] and in cellulose derivative membranes[319] have also been performed. In a more recent work the diffusion of water and fluorine ions in anion-exchange membranes and resins was investigated.[320]

11.4.1.8 Polymer and macromolecules

PGSE can provide an enormous amount of information about polymers both in solution and in polymer melts. For example, PGSE NMR has recently been used to measure the critical overlap concentration for polystyrene in tetrachloromethane.[321] Callaghan and Coy,[322] using very large values of q ($2\pi q \sim (130\ \text{Å})^{-1}$), have obtained evidence for reptational motion of high molar mass polystyrene in semidilute solution using ^{1}H PGSE measurements. In perhaps the first spin-echo study on the diffusion of a polymer melt, in 1959, McCall *et al.*[323] measured the self-diffusion of low molecular weight polystyrenes. The molecular 'reptation' model in polymers as described in Section 1.8.7 has been confirmed with PGSE NMR for both polymer melts[324] and solutions.[322] Cheng *et al.*[325] studied the temperature-dependence of the diffusion of poly(ethylene oxide) in the melt. Their results were at variance with those predicted by de Gennes (Eq. (1.160)).

Cosgrove *et al.*[326] observed Rouse-like behaviour (Eq. (1.159)) in blends of linear and cyclic polydimethylsiloxane melts. Using stray-field-based measurements Fischer *et al.* studied the diffusometry of segment displacements in melts of entangled polymers.[327] Appel *et al.*[324] used the fringe field of a superconducting magnet to measure the self-diffusion of poly(dimethylsiloxane), polybutadiene and polyisoprene for times smaller than the reptation time. A time-dependent self-diffusion coefficient was observed.

PGSE NMR has been able to probe internal diffusion modes in semi-dilute solutions of high molecular weight polystyrene in toluene revealing reptational scaling laws at length scales as small as 20 nm.[328] And the diffusion of block copolymers has been studied in liquid CO_2.[329] The structures of PEG-*rotaxa*-(α-cyclodextrin)s and (PEG-*rotaxa*-αCD)s in dilute DMSO-D_6 have been studied using NMR diffusion measurements. It was found that the diffusion of the (PEG-*rotaxa*-αCD)s and PEGs scaled with their respective molecular weights as $D_{(\text{PEG-}rotaxa\text{-}\alpha\text{CD})} \sim Mn^{-0.60 \pm 0.05}$ and $D_{\text{PEG}} \sim Mn^{-0.55 \pm 0.03}$.[330]

Polymeric microgels are quite distinct from polymers. They are semi-rigid and are unable to penetrate each other. While, similar to polymers, their diffusivity is determined by mutual interactions and their transport properties are quite different from polymer chains. In a study with cross-linked polystyrene beads, it was noted that with increasing concentration, a large decrease in mobility is observed and the deviation from ordinary diffusion increased.[331] This indicated that the individual microgels are in a cage formed by their neighbours. However, as the cages are not perfectly confining at large Δ, the motion of the microgels appears to occur via normal diffusion. Rosén *et al.*[332] observed that the diffusion of dodecyl sulphate in cross-linked gels of ethyl(hydroxyethyl cellulose) above the critical association concentration was anomalous. However, below the critical association constant it was Gaussian. They ascribed the anomalous diffusion to inhomogeneities within the gel.

^{1}H PGSE measurements have been used to study the morphology and structure of polyvinyl alcohol cryogels,[333] and to study contact lens hydrogels.[334] Anomalous diffusion and evidence of fractal structure in the dynamics of water in a hydrogel was observed using ^{1}H fringe field diffusion measurements.[335] Skirda *et al.*[336] studied the diffusion of PEG in weakly cross-linked poly(methacrylic acid) hydrogel. They found that a fraction of the PEG inside the collapsed gel has diffusion characteristics similar to the network chains, suggesting the formation of an interpolymer complex, while another fraction of the PEG diffused freely. However, in the swollen gel (PEG concentrations above 5 wt%) the diffusion of all of the PEG molecules was independent of Δ indicating either no interpolymer complex or very rapid exchange.

PGSE has been used in combination with other physico-chemical techniques to evaluate the nanomorphology and molecular accessibility of gel-type resins used for supporting palladium nanoclusters.[337] The dynamics of water and cations (Li and Na) in gel-type cation exchange resins has also been examined using PGSE NMR.[338] It was found that the diffusion of the water and the cations decreased in the higher cross-linked resins and evidence of restricted diffusion was observed suggesting a three-dimensional network structure in the gels. Anomalous diffusion has been observed in an aqueous system of a PEO-PEO-PEO block copolymer during gelation.[339]

Meresi *et al.* have used PGSE to study the diffusion of pentane in a random copolymer of tetrafluoroethylene (TFE) and 2,2-bis(trifluoromethyl)-4,5-difluoro-1,3-dioxole (PDD) as a function of the time allowed for diffusion to occur (i.e., Δ). Using the change in echo amplitude at low values of q they found that the apparent diffusion at 12.4 ms of $2.2 \times 10^{-11}\ \mathrm{m^2 s^{-1}}$ decreased to a constant value of $5.87 \times 10^{-12}\ \mathrm{m^2 s^{-1}}$ at 1 s.[340] This was interpreted in terms of morphological structure of this completely amorphous glassy polymer on the micron length scale. It is thought that regions of high free volume are interspersed in lower free volume regions and that this leads to the observed changes in the measured diffusion constant as a function of Δ.

11.4.1.9 Diffusion in polymer electrolytes and ionic conductivity

^{23}Na PGSE NMR has been used in conjunction with T_1 measurements to study the relationship between the diffusion coefficient, ionic conductivity and τ_c of sodium ions in water–glycerol solutions.[341] ^{7}Li and other heteronuclei are commonly used to study diffusion in lithium salt–polymer (esp. poly(ethylene oxide)) electrolyte systems and there are now an enormous number of studies.[342] Golodnitsky *et al.*[343] have used PGSE NMR to investigate the poly(ethylene oxide)–LiI system. Their results showed that the diffusivity of the Li^+ is anisotropic and enhanced in the direction of the stretched polymer. Li–salt polymer electrolytes have complicated molecular dynamics and appear to undergo structural rearrangement with age.[344]

11.4.1.10 Diffusion in dendrimers, nanoparticles and supramolecular polymers

Gorman *et al.*[345] have used PGSE NMR to study two series of redox-active, iron–sulfur core dendrimers dissolved in DMF and THF and to determine the corresponding Stokes radii. Newkome *et al.*[346] have investigated the pH dependence of the molecular size distribution of a cascade series of polyacids dissolved in ^2H_2O with formula weights ranging from 1341 to 111373 using diffusion measurements. For each cascade generation they found that the hydrodynamic radius was largest at neutral pH and smallest at acidic pH. In a later study,[347] the diffusion of three series of cascade polymers possessing identical internal hierarchical structures but with either acidic, basic or neutral functionality were compared. The results were consistent with the flexible internal structure and coulombic repulsion between the terminal groups being responsible for the pH sensitive ('smart') behaviour of these dendrimers. The diffusion of nanoparticles[348] and of three different poly(propyleneimine) dendrimers with hydrophilic triethylenoxy methyl ether terminal groups (generations 2, 4 and 5) in poly(vinyl alcohol) aqueous solutions and gels,[349] and the association of homochiral cyclic dimers of 2-ureido-4[1*H*]-pyrimidinone (UPy),[350] have also been studied using NMR diffusion measurements.

11.4.1.11 Host–guest chemistry

There have been numerous NMR diffusion studies involving crown ethers,[351] complexes of γ-cyclodextrin with macrocycles such as 12-crown-4 in the presence and absence of inorganic and organic salts,[352] encapsulation of guests in self-assembled tetraurea calix[4]arene dimers,[353] complexes of cyclohexylacetic acid and cholic acid with β-cyclodextrin.[354] The host–guest complexes of resorcin[4]arenes with glutaric acid and β-methyl D-glucopyranoside in chloroform were investigated with a range of NMR methods including PGSE.[355] From the results it was inferred that the complex is a self-assembled capsule of six resorcinarenes that surround three guest molecules of β-methyl D-glucopyranoside or six of glutaric acid.

11.4.1.12 Porous media

Although our coverage is essentially limited to studies using purposely applied gradient pulses, it is noted that a large literature exists on studying porous media based on the presence of the internal gradients. We stress that emulsions and biological tissues (see above) can also constitute porous media.

Mesoporous and microporous materials In the mesoporous and microporous domain (also sometimes referred to as nanopores), PGSE is only able to provide transport information on length scales larger than the characteristic distances of the pores. Consequently, for nanoporous media the information obtained from PGSE studies is related to averaged fluid–porous matrix interactions.

PGSE diffusion measurements reveal that water diffused anisotropically in (the porous silicate) MCM-41 and that the data was best represented by an axisymmetrical diffusion tensor.[356] Stallmach *et al.*[357] have studied the diffusion of benzene, *n*-hexadecane, propylenecarbonate, ethylbenzene and diethyl ether in ordered mesoporous materials of type MCM-41. The fast diffusivities observed with benzene, ethylbenzene and diethyl ether in MCM-41 suggest that the transport processes for these molecules are governed by gas or vapour phase diffusion, which is reduced by interactions with the silica walls. Whilst the results for *n*-hexadecane revealed that its diffusion in the hexagonal channels of MCM-41 is anisotropic. In a later study, Valiullin *et al.*[358] have studied the diffusion of cyclohexane in MCM-41 at different external gas pressures from zero to saturated vapour pressure.

Single file diffusion has been verified in $AlPO_4-5$ zeolites and the attenuation data was well-described by Eq. (4.57),[359] and explored in the ~500 pm nanochannels in tris(*o*-phenylenedioxi)cyclophosphazene that persisted for 10s of seconds using laser polarised ^{129}Xe.[360] By analysing short-time diffusion data and extrapolating to zero observation time, Gjerdåker *et al.*[361] were able to determine the unrestricted intracrystalline diffusion coefficient for ethane confined in H-ZSM-5 crystallites. Pampel *et al.*[362] have used MAS PGSE to study the diffusion of *n*-butane adsorbed in silicalite-1. They noted that the combination of the techniques brought two benefits from the additional line narrowing: (i) a prolongation of the intervals during which the magnetic field gradients may be applied and a corresponding enhancement in the sensitivity towards small molecular displacements, and (ii) an enhanced chemical shift resolution. Geil *et al.* have used static field ^{1}H diffusion measurements to study benzene translational dynamics on zeolite NaY.[363] Analysis of the data suggests that the elementary motional process is related to jumps along well-defined adsorption sites in the guest.

Interconnected pores Many nanoporous materials are produced as particles with dimensions in the realms of micrometres, although they may be packaged in larger structures (e.g., chromatographic media). Thus, such systems have both nanopores and macropores with the transport properties being dominated by the interconnected macropores. Importantly, as the pore size increases, the fluid–matrix interactions decrease which results in longer transverse relaxation times. Much can be deduced about pore structure from the time-dependence of the measured diffusion coefficient, and when surface relaxation is not too great, measurement of D^{∞} and thus the tortuosity. The situation is made much easier when gases are used as probes and the use of hyperpolarised gases obviates the loss in sensitivity when moving to gas-based diffusion studies. The physical parameters determined from PGSE measurements have important practical implications such as catalytic performance in fluid cracking catalysts.[364] Indeed, PGSE measurements and MRI experiments have been used to assess the spatial distribution of voidage, and simulations of diffusion based on the images to examine the diffusion and tortuosity of water in Al_2O_3 and SiO_2 catalyst support pellets.[365]

The pore connectivity in rock samples has been studied using the gases methane and ethane by measuring the diffusion as a function of relaxation time using a diffusion-editing sequence in which diffusion editing was performed prior to collection of relaxation data. The increased diffusivities and long relaxation times allowed greater length scales to be probed.[366]

Callaghan *et al.* have used PGSE measurements in the Earth's magnetic field to study Antarctic sea ice. The resulting data revealed restricted diffusion of brine water in sea ice with pore size of the order of 40 μm^{-1} and that the water can tortuously move between pores connected by narrow throats.[367] Vargas-Florencia *et al.*[368] have studied the size distribution and permeability of pores in nanostructured TiO_2 films using PGSE measurements in combination with other advanced techniques.

Sørland[369] has investigated the use of the short-time diffusion equation (Eq. (1.122)) in a model system consisting of soda-lime glass spheres and observed that a better fit to the data was obtained when the effects of restricted diffusion during the gradient pulses were accounted for.[370] Latour *et al.*[371] and Hürlimann *et al.*[372] have used measurements of $D(t)$ and RTOP to study artificial porous media consisting of polystyrene or glass spheres as well as sandstones.

Unlike model porous materials like packed spheres, carbonate rocks and sandstones have smooth and featureless deviations from Gaussian behaviour and do not generally show clear diffraction behaviour.[373] Interestingly, the curves giving the non-Gaussian behaviour of $E(q, \Delta)$ can be collapsed onto each other where the momentum is scaled by the diffusion length.[373] This has been interpreted that on the timescale of Δ the diffusion is locally Gaussian but anisotropic. Indeed, from the behaviour of RTOP it has been proposed that diffusion is locally one-dimensional in some rock samples on the NMR timescale.[373]

Minerals, construction materials and sediments In general, mineral-based construction materials and sediments pose special experimental difficulties for performing PGSE experiments due to the inclusion of paramagnetic species/large susceptibility differences. Thus, specialised PGSE sequences that reduce the effects of background gradients as discussed in Section 7.6 are normally required. However, Vasenkov *et al.* have demonstrated using measurements with hexadecane in sand and *n*-octanol in doped MgO pastes that true diffusion coefficients can be determined using a standard PGSTE sequence by extrapolation.[374] Nestle *et al.*[375] studied the diffusion of water in white cement pastes at various degrees of hydration. In a later study, the diffusion of water in porous building materials including fired clay brick was studied using a constant gradient STE technique, and the results were used to evaluate the tortuosity of the pore space.[376]

By measuring the diffusion coefficient of water in samples of glacial sand over a range of observation times (Δ) from 2.5 to 250 ms, Stallmach *et al.*[377] were able to determine the surface-to-volume ratios and specific surface areas of the grains. In both cases the quantities exhibited non-integer power-law dependencies as a function of grain diameter. However, a later paper questioned the analysis and presented an alternative method for analysing the PGSE data.[378] Mutina and Skirda[379] have used DDIF experiments to probe porous geometry and the localisation of tridecane in quartz sand. Specialised imaging equipment for performing one-dimensional images has been used to study water and salt transport in materials such as brick during drying.[380]

Almost all natural sediments studied necessarily have a well-connected pore space allowing the introduction of the fluids being probed in the NMR measurements. However, these systems tend to be complex and even using background suppression sequences, the PGSE amplitudes in porous sandstones and carbonate rocks do not decay single exponentially due to the effects of randomly orientated and partially ordered restricted diffusion.[381] The diffusion of water and polypropylene oxide in the pores of hardened gypsum and their data were analysed to form a model of the pore network inside the gypsum.[382] Stray-field measurements have been used to study flow displacement distributions in Bentheimer sandstone.[383] Numerous studies have involved measuring the time-dependent diffusion coefficient to investigate surface-to-volume ratios in water-saturated sedimentary rocks.[372]

Takahashi *et al.*[384] observed diffusional anisotropy for water in suspensions of the colloidal clay mineral Na-montmorillonite using PGSE measurements. They interpreted their data to indicate that in strong magnetic fields the Na-montmorillonite particles aligned with their basal planes parallel to the static field and their nanocrystalline *c*-axes perpendicular to it.

Chromatography PGSE diffusion and flow measurements have been shown to be useful in characterising the diffusion and flow behaviour in chromatographic media and thereby relating macroscopic chromatography column performance with the underlying physico-chemical fundamentals[385] including information on stagnant mobile phase transfer kinetics,[386] axial and transverse dispersion characteristics,[387,388] scaling behaviour,[389] structure-flow and dispersion correlations,[390] and to discriminate between pressure and electrokinetically driven flows through open and packed columns.[391] The evolution of viscous fingering instability has been investigated in packed chromatography beds using MRI.[392] Finally, we note that there have been a number of PGSE NMR studies of the diffusion of proteins such as conalbumin, lysozyme and ovalbumin in porous chromatographic media.[393]

11.4.1.13 Gas diffusion

Low S/N is a particular problem of gas samples due to their extremely low density. The earliest gas diffusion measurements were performed using thermally polarised gases.[394] The use of optical pumping techniques to produce large nuclear spin polarisations has led to a surge in applications of gas-phase NMR including diffusion measurements.[395] In the simplest case a one-dimensional imaging technique is used to measure diffusion by observing the signal return to an area that has been bleached or the evolution of an excited slice.[396] Gradient-echo-based methods have also been used.[397] NMR measurements can be complicated by the very long relaxation times with T_1s greater than 50 min being often observed[397] and thus either specialised sequences are used or the gas is mixed with O_2 to reduce the relaxation time. A further complication of experiments involving hyperpolarised samples is that there is no recovery of magnetisation after an rf pulse. Consequently, specialised variants of the standard diffusion sequences are required. Mair *et al.*[395] have presented modified PGSE sequences in attempts to obviate these problems including a multiple spin-echo single-shot sequence.

Measurements of the time-dependent diffusion of gases in porous media should be a good probe of multi-pore length scales due to the rapid diffusion of gas and the weak interaction of gases with surfaces and of surface-to-volume ratios.[398] However, such samples are extremely magnetically heterogeneous due to the presence of gas–solid interfaces and the diffusion sequences that account for such effects need to be used (see Section 7.6).[395] Mair *et al.*[395] have observed time-dependent gas diffusion in a mixture of thermally polarised xenon and O_2 in a pack of glass beads. From their measurements they determined that the diffusion coefficient of pure Xe at 1 atm was $5.71 \times 10^{-6}\,m^2s^{-1}$. Codd and Altobelli[399] studied the probability density for molecular displacement of propane gas flowing through three types of porous bead packs (PVC particles, glass spheres and polystyrene spheres) and compared the results to that for water flowing in the polystyrene pack. They found that the exchange time between the moving and stagnant portions of the flow is strongly related to the diffusion coefficient of the fluid.

11.5 Spin-diffusion

In extremely slowly diffusing systems, for example, high molecular weight polymers with sufficiently slow reptational motion, spin-diffusion may play a role in determining the RMSDs of magnetisation quasi-particles as predicted by Fatkullin.[400] Using PGSE NMR spin-diffusion has been measured in a single crystal of CaF_2,[401] in polymer melts,[402] and in high molecular mass polystyrene dissolved in deuterated toluene.[403]

11.6 Imaging-based studies

As whole body MRI imagers become available at increasingly greater field strengths the applicability of many of the methods for measuring translational motion will increase. Interestingly, recent studies have revealed that diffusion MRI could be used to look at brain activity leading to improved spatial and temporal resolution compared to previous functional MRI approaches based on observation of the BOLD effect.[404]

11.6.1 Diffusion measurement by imaging profiles

Blackband and Mansfield used 1D imaging to study the ingress of water into blocks of nylon at 100 °C and thereby determine the mutual diffusion coefficient.[405] Similarly one-dimensional proton concentration profiles in Laponite clay gels prepared from $^1H_2O/^2H_2O$ mixtures were measured using MRI to determine macroscopic interdiffusion (i.e., mutual diffusion) coefficients and these were compared with PGSE NMR measurements of the self-diffusion.[406] Song *et al.*[407] have used 1D MRI-based diffusion measurements to study the diffusion of laser-polarised Xe gas between planes. Nestle and Kimmich[408] used NMR images to probe the progress of isovalent competitive ion exchange processes in alginate gels.

11.6.2 Diffusion-weighted imaging

Diffusion-weighted MRI is now widely used to detect and characterise ischaemic, malignant and neurodegenerative diseases. In a study of myelin deficient rat spinal cords, Biton *et al.*[409] observed that lack of myelin has a pronounced effect on the diffusion properties of water in white matter in high b-value q-space diffusion weighted MRI. A later study found that high b-value q-space diffusion weighted MRI that emphasises the water component which exhibits restricted diffusion has potential for diagnosing Alzheimer's disease and vascular dementia whereas conventional diffusion tensor imaging did not show significant changes between either of the groups and controls.[410] Does *et al.* combined OGSE with the echo planar imaging sequence to study water diffusion in normal and globally ischaemic rat brain.[411]

Boujraf *et al.*[412] studied anisotropic diffusion in asparagus stems. And Ellegood *et al.*[413] have measured the mean apparent diffusion coefficient and fractional anisotropy of metabolites including *N*-acetylaspartate, creatinine and phosphocreatinine in peripheral nerve using diffusion-weighted magnetic resonance.

11.6.3 Diffusion tensor imaging

Diffusion tensor MRI has been used to investigate the white matter of patients with focal temporal lobe epilepsy.[414] The results suggested that diffusion anisotropy may reveal abnormalities in such patients. DTI has been used to detect and quantify the infiltration of tumours into white matter since the destruction of white matter alters the local diffusion properties.[415] Fibre tracking using diffusion tensor MRI has also been used to gauge brain development since it provides a means of quantifying changes in neural tissue microstructure (see Figure 11.2).[416]

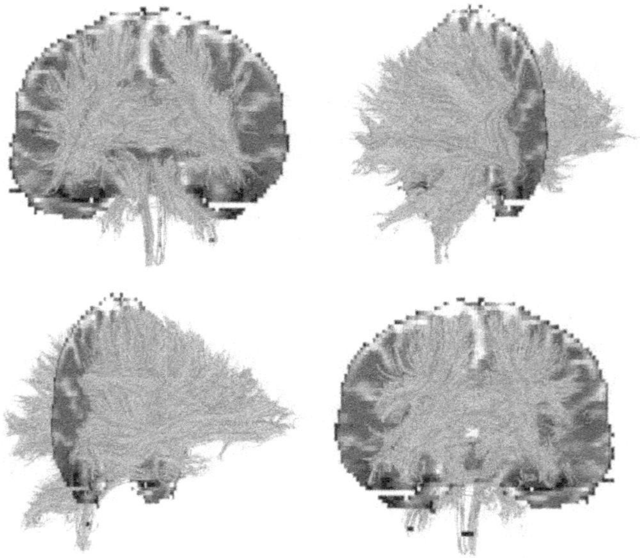

Figure 11.2 Four different views of fibre tracking of an entire human adult brain. Coronal sections of T_2-weighted images are overlaid. From Watts *et al.*[416]

11.6.4 Imaging-based flow studies

MRI-based flow techniques enable measurements to be performed that are simply not possible with traditional techniques,[37] and also provides the possibility of visualising flow in real time.[417] Couette flow has been widely studied.[418] Rheometry by NMR has become an area of increasing interest with applications such as measuring velocity profiles of a Newtonian liquid, a shear thinning liquid and a suspension undergoing Poiseuille flow in a capillary.[419] Britton and Callaghan have studied non-linear viscosities and observed properties such as apparent slip, shear thinning, shear thickening, shear banding and yield-stress behaviour in a range of systems including food substances.[420] Using a cone-and-plate rheometer they verified the uniform shear rate assumption in Newtonian and simple non-Newtonian fluids but observed anomalous behaviour such as apparent slip, shear banding and fracture rates in systems including worm-like surfactants, semi-dilute solutions of 18 MDa polyacrylamide and dispersed silica in silicone grease.[421] Xia and Callaghan have imaged velocity profiles in a tubeless siphon with a visco-elastic liquid of high extensional mobility.[422]

Multiphase flows in porous media typically involve two liquid phases such as oil and water. As the two phases generally have quite different NMR relaxation times, they can be easily distinguished. Alternatively, differences in chemical shifts can be used to study separate phases in multiphase flows. For example, chemical shift imaging was used to quantitatively investigate the oil concentration polarisation layers which formed during hollow fibre membrane filtration.[423]

In porous media flow imaging studies it is the liquid phase that is imaged as the solid matrix produces no signal. PGSE and MRI-based measurements of diffusion and flow have proven to be very powerful methods for characterising chromatographic media. Tallarek *et al.* used dynamic MRI to study a fluid percolating through a packed bed of a dynamic radial compression column for liquid chromatography.[424] They were able to determine the distribution of the fluid dynamic displacement probability at each point of the image, the local axial dispersion coefficient, the fluid velocity and external porosity. Koptyug *et al.* have studied liquid and gas flow in monolithic alumina catalysts using thermally polarised ^{1}H NMR imaging.[425] Klemm *et al.* have used NMR microscopy and velocity mapping to visualise flow patterns of percolating liquids in porous materials.[426] Their analysis revealed anomalous diffusion.

Magnetic resonance flow imaging has been used to study thermal convection[427] and was found to be capable of resolving details at a level comparable to optical techniques. Weis *et al.*[428] demonstrated that the combined application of Fourier encoding velocity imaging and multistripe/multiplane tagging imaging allows the quantitative examination of thermal convection for arbitrary boundary conditions. In a later imaging study it was shown that convection in a vertical cylindrical tube lead to axially antisymmetric flow, multiple vertical scrolls and twisted node planes. Morhorič and Stepišnik have studied the effect of convection in horizontally oriented cylinder on the distribution of diffusivity using imaging measurements conducted in the earth's magnetic field.[429] Weber *et al.* have measured convection as a function of porosity in random-site percolation model objects.[430] They found that the maximum velocity as a function of the porosity indicated a combined percolation/Rayleigh–Bénard transition. Convection of laser-polarised ^{129}Xe gas has also been imaged.[431]

Diffusion-weighted MRI has proven to be very sensitive to microstructural changes in brain tissue – especially those resulting from stroke or ischaemic injury.[432] Since therapeutically induced changes in tumour water diffusion are detectable with diffusion MRI, the measured ADCs can be used as an early surrogate marker of therapeutic efficiency in brain tumours.[433] Helmer *et al.*[434] have compared the use of RTOP and apparent diffusion coefficients in NMR imaging experiments as indicators of necrosis in tumours. DTI has been used in characterising the nature of white matter lesions in the brains of multiple sclerosis sufferers.[435] Tagging techniques have application to clinical medicine for studying heart wall motion[436] and even to studying coherent and incoherent flows in fertilised bird eggs.[437]

Noting that due to differences in relaxation rates and quadrupolar splittings the signals of water in the three different compartments (endoneurium, epineurium and axon) of rat sciatic nerve are resolved in ^{2}H double-quantum-filtered NMR spectrum, Seo *et al.*[438] were able to measure the water diffusion coefficients in each compartment. And the diffusion was found to be anisotropic in all three compartments. Parallel to the nerve fibre the average intra-axonal water diffusion coefficient was $1.11 \times 10^{-9}\,\mathrm{m^2 s^{-1}}$, while perpendicular to the nerve fibre the diffusion was heavily restricted. The average perpendicular diffusion coefficient ranged from $0.29 \times 10^{-9}\,\mathrm{m^2 s^{-1}}$ to $0.05 \times 10^{-9}\,\mathrm{m^2 s^{-1}}$ for $\Delta = 7\,\mathrm{ms}$ and $50\,\mathrm{ms}$, respectively. Assuming restricted diffusion in non-permeable cylinders, intra-axonal mean diameters of 6.0, 7.4 and 9.0 μm were obtained for nerves taken from three different rats.

The transport of osteoid water across the mineralised matrix of bone has been studied using ^{1}H NMR imaging to gauge the exchange of tissue water whilst the bone was immersed in ^2H_2O.[439]

Velocity imaging has been used to probe pipe flow of blood.[440] And Loureiro de Sousa *et al.*[441] have noted that the effect of flow on the distant dipolar field could be used to obtain information on blood perfusion at an intravoxel scale in MRI experiments.

References

1. P. Stilbs, Fourier Transform Pulsed-Gradient Spin-Echo Studies of Molecular Diffusion. *Prog. NMR Spectrosc.* **19** (1987), 1–45.
2. J. Kärger, H. Pfeifer, and W. Heink, Principles and Applications of Self-Diffusion Measurements by Nuclear Magnetic Resonance. *Adv. Magn. Reson.* **12** (1988), 1–89.
3. R. L. Haner and T. Schleich, Measurement of Translational Motion by Pulse-Gradient Spin-Echo Nuclear Magnetic Resonance. *Methods Enzymol.* **176** (1989), 418–46.
4. P. T. Callaghan, *Principles of Nuclear Magnetic Resonance Microscopy.* (Oxford: Clarendon Press, 1991).
5. D. Canet and M. Décorps, Applications of Field Gradients in NMR. In *Dynamics of Solutions and Fluid Mixtures*, ed. J.-J. Delpuech. (New York: Wiley, 1995), pp. 309–43.
6. K. J. Packer, Diffusion & Flow in Liquids. In *Encyclopedia of Nuclear Magnetic Resonance*, ed. D. M. Grant and R. K. Harris. (New York: Wiley, 1996), pp. 1615–26.
7. W. S. Price, Gradient NMR. In *Annual Reports on NMR Spectroscopy*, ed. G. A. Webb. vol. 32. (London: Academic Press, 1996), pp. 51–142.
8. R. Kimmich, *NMR: Tomography, Diffusometry, Relaxometry.* (Berlin: Springer Verlag, 1997).
9. C. S. Johnson, Jr., Diffusion Ordered Nuclear Magnetic Resonance Spectroscopy: Principles and Applications. *Prog. NMR Spectrosc.* **34** (1999), 203–56.
10. W. S. Price, Probing Molecular Dynamics in Biochemical and Chemical Systems Using Pulsed Field Gradient NMR Diffusion Measurements. In *New Advances in Analytical Chemistry*, ed. Atta-Ur-Rahman. vol. 1. (Amsterdam: Harwood Academic Publishers, 2000), pp. 31–72.
11. P. Stilbs, Diffusion Studied Using NMR Spectroscopy. In *Encyclopedia of Spectroscopy and Spectrometry*, ed. J. C. Lindon, G. E. Tranter, and J. L. Holmes. vol. 1. (London: Academic Press, 2000), pp. 369–75.

12. H. Weingärtner and M. Holz, NMR Studies of Self-Diffusion in Liquids. In *Annu. Rep. Prog. Chem., Sect. C*, ed. G. A. Webb. vol. 98. (Cambridge: Royal Society of Chemistry, 2002), pp. 121–55.
13. I. Ardelean and R. Kimmich, Principles and Unconventional Aspects of NMR Diffusometry. In *Annual Reports on NMR Spectroscopy*, ed. G. A. Webb. vol. 49. (London: Academic Press, 2003), pp. 43–115.
14. B. Blümich and W. Kuhn, (eds) *Magnetic Resonance Microscopy*. (Berlin: VCH, 1992).
15. J. V. Hajnal and I. R. Young, Perfusion and Diffusion Imaging. *Magn. Reson. Mater. Phys. Bio. Med.* **2** (1994), 225–31.
16. P. Blümich, B. Blümich, R. Botto, and E. Fukushima, (eds) *Spatially Resolved Magnetic Resonance*. (Weinheim: VCH, 1998).
17. E. M. Haacke, R. W. Brown, M. R. Thompson, and R. Venkatesan, *Magnetic Resonance Imaging–Physical Principles and Sequence Design*. (New York: Wiley, 1999).
18. I. V. Koptyug and R. Z. Sagdeev, Modern Applications of NMR Tomography in Physical Chemistry. The Characteristic Features of the Technique and Its Applications to Studies of Liquid-Containing Objects. *Russ. Chem. Rev.* **71** (2002), 593–617.
19. D. Le Bihan, Looking into the Functional Architecture of the Brain with Diffusion MRI. *Nature Reviews: Neuroscience* **4** (2003), 469–80.
20. M. D. Mantle and A. J. Sederman, Dynamic MRI in Chemical Process and Reaction Engineering. *Prog. NMR Spectrosc.* **43** (2003), 3–60.
21. P. T. Narasimhan and R. E. Jacobs, Microscopy in Magnetic Resonance Imaging. In *Annual Reports on NMR Spectroscopy*, ed. G. A. Webb. vol. 55. (London: Academic Press, 2005), pp. 259–97.
22. P. T. Callaghan, Rheo-NMR and Velocity Imaging. *Curr. Opin. Colloid Interface Sci.* **11** (2006), 13–18.
23. S. Stapf and S.-I. Han, (eds) *NMR Imaging in Chemical Engineering*. (New York: Wiley, 2006).
24. P. T. Callaghan, *Rheo-NMR: A New Window on the Rheology of Complex Fluids*. (New York: Wiley, 2007).
25. S. Mori, *Introduction to Diffusion Tensor Imaging*. (Oxford: Elsevier, 2007).
26. K. Nicolay, K. P. Braun, R. A. De Graaf, R. M. Dijkhuizen, and M. J. Kruiskamp, Diffusion NMR Spectroscopy. *NMR Biomed.* **14** (2001), 94–111.
27. G. J. Strijkers, M. R. Drost, A. M. Heemskerk, M. J. Kruiskamp, and K. Nicolay, Diffusion MRI and MRS of Skeletal Muscle. *Isr. J. Chem.* **43** (2003), 71–80.
28. A. R. Waldeck, P. W. Kuchel, A. J. Lennon, and B. E. Chapman, NMR Diffusion Measurements to Characterise Membrane Transport and Solute Binding. *Prog. NMR Spectrosc.* **30** (1997), 39–68.
29. M. J. Shapiro and J. S. Gounarides, NMR Methods Utilized in Combinatorial Chemistry Research. *Prog. NMR Spectrosc.* **35** (1999), 153–200.
30. B. J. Stockman and C. Dalvit, NMR Screening Techniques in Drug Discovery and Drug Design. *Prog. NMR Spectrosc.* **41** (2002), 187–231.
31. B. Meyer and T. Peters, NMR Spectroscopy Techniques for Screening and Identifying Ligand Binding to Protein Receptors. *Angew. Chem. (Int. Ed.)* **42** (2003), 864–90.
32. L. H. Lucas and C. K. Larive, Measuring Ligand-Protein Binding Using NMR Diffusion Experiments. *Concepts Magn. Reson.* **20A** (2004), 24–41.
33. Y. Cohen, L. Avram, and L. Frish, Diffusion NMR Spectroscopy in Supramolecular and Combinatorial Chemistry: An Old Parameter – New Insights. *Angew. Chem. (Int. Ed.)* **44** (2005), 520–54.
34. L. Fielding, NMR Methods for the Determination of Protein–Ligand Dissociation Constants. *Prog. NMR Spectrosc.* **51** (2007), 219–42.

35. E. Fukushima, Nuclear Magnetic Resonance as a Tool to Study Flow. *Ann. Rev. Fluid Mech.* **31** (1999), 95–123.
36. B. Newling, Gas Flow Measurements by NMR. *Prog. NMR Spectrosc.* **52** (2008), 31–48.
37. J. Kärger and G. Fleischer, NMR Diffusion Studies in Heterogeneous Systems. *Trends Anal. Chem.* **13** (1994), 145–57.
38. G. Lindblom and G. Orädd, NMR Studies of Translational Diffusion in Lyotropic Liquid Crystals and Lipid Membranes. *Prog. NMR Spectrosc.* **26** (1994), 483–515.
39. G. Lindblom and G. Orädd, Liquid Crystalline Samples: Diffusion. In *Encyclopedia of Nuclear Magnetic Resonance*, ed. D. M. Grant and R. K. Harris. vol. 4 (New York: Wiley, 1996), pp. 2760–8.
40. G. Celebre, G. Chidichimo, L. Coppola, C. La Mesa, R. Muzzalupo, L. Pogliani, G. A. Ranieri, and M. Terenzi, Water Self-Diffusion in Lyotropic Liquid Crystals: Pulsed Gradient Spin-Echo NMR and Simulation Techniques. *Gazz. Chim. Ital.* **126** (1996), 489–503.
41. O. Söderman, P. Stilbs, and W. S. Price, NMR Studies of Surfactants. *Concepts Magn. Reson.* **23A** (2004), 121–35.
42. I. Furó, NMR Spectroscopy of Micelles and Related Systems. *J. Mol. Liquids* **117** (2005), 117–37.
43. S. V. Dvinskikh and I. Furó, Nuclear Magnetic Resonance Studies of Translational Diffusion in Thermotropic Liquid Crystals. *Russ. Chem. Rev.* **75** (2006), 497–506.
44. G. Lindblom and G. Gröbner, NMR on Lipid Membranes and Their Proteins. *Curr. Opin. Colloid Interface Sci.* **11** (2006), 24–9.
45. M. L. Johns and K. G. Hollingsworth, Characterisation of Emulsion Systems Using NMR and MRI. *Prog. NMR Spectrosc.* **50** (2007), 51–70.
46. P. S. Pregosin, Ion Pairing Using PGSE Diffusion Studies. *Prog. NMR Spectrosc.* **49** (2006), 261–88.
47. T. Nose, Pulsed-Field-Gradient NMR Studies of the Diffusion of Chain Molecules in Polymer Matrices. In *Annual Reports on NMR Spectroscopy*, ed. G. A. Webb. vol. 27. (London: Academic Press, 1993), pp. 217–53.
48. Y. Yamane, S. Kanesaka, S. Kim, K. Kamiguchi, M. Matsui, S. Kuroki, and I. Ando, Diffusion in Soft Polymer Systems as Approached by Field-Gradient NMR. In *Annual Reports on NMR Spectroscopy*, ed. G. A. Webb. vol. 58. (London: Elsevier, 2006), pp. 51–154.
49. J. Kärger, Diffusion in Porous Media. In *Encyclopedia of Nuclear Magnetic Resonance*, ed. D. M. Grant and R. K. Harris. vol. 3. (New York: Wiley, 1996), pp. 1656–63.
50. A. T. Watson and C. T. P. Chang, Characterizing Porous Media with NMR Methods. *Prog. NMR Spectrosc.* **31** (1997), 343–86.
51. R. Kimmich, Strange Kinetics, Porous Media, and NMR. *Chem. Phys.* **284** (2002), 253–85.
52. F. Stallmach and P. Galvosas, Spin Echo NMR Diffusion Studies. In *Annual Reports on NMR Spectroscopy*, ed. G. A. Webb. vol. 61. (New York: Elsevier, 2007), pp. 51–131.
53. R. M. Cotts, Diffusion in Solids. In *Encyclopedia of Nuclear Magnetic Resonance*, ed. D. M. Grant and R. K. Harris. vol. 3. (New York: Wiley, 1996), pp. 1670–85.
54. T. Brand, E. J. Cabrita, and S. Berger, Intermolecular Interaction as Investigated by NOE and Diffusion Studies. *Prog. NMR Spectrosc.* **46** (2005), 159–96.
55. V. D. Skirda and V. I. Volkov, NMR with Pulsed Magnetic Field Gradient in Studies of Physicochemical Processes in Molecular Systems. *Zh. Fiz. Khim. A* **73** (1999), 298–309.
56. L. A. Cardoza, A. K. Korir, W. H. Otto, C. J. Wurrey, and C. K. Larive, Applications of NMR Spectroscopy in Environmental Science. *Prog. NMR Spectrosc.* **45** (2005), 209–38.
57. C. S. Johnson, Jr., Electrophoretic NMR. In *Encyclopedia of Nuclear Magnetic Resonance*, ed. D. M. Grant and R. K. Harris. vol. 3. (New York: Wiley, 1996), pp. 1886–95.

58. M. Holz, Field-Assisted Diffusion Studied by Electrophoretic NMR. In *Diffusion in Condensed Matter*, ed. J. Kärger and P. Heitjans. (Berlin: Springer, 2005), pp. 717–42.
59. P. C. Griffiths, A. Paul, and N. Hirst, Electrophoretic NMR Studies of Polymer and Surfactant Systems. *Chem. Soc. Rev.* **35** (2006), 134–45.
60. A. M. Dixon and C. K. Larive, NMR Spectroscopy with Spectral Editing for the Analysis of Complex Mixtures. *Appl. Spectrosc.* **53** (1999), 426A–40A.
61. D. Canet, Radiofrequency Field Gradient Experiments. *Prog. NMR Spectrosc.* **30** (1997), 101–35.
62. D. Canet, Radiofrequency Field Gradients in NMR, Theory. In *Encyclopedia of Spectroscopy and Spectrometry*, ed. J. C. Lindon, G. E. Tranter, and J. L. Holmes. vol .3. (London: Academic Press, 2000), pp. 1937–44.
63. M. Lin, M. J. Shapiro, and J. R. Wareing, Diffusion-Edited NMR-Affinity NMR for Direct Observation of Molecular Interactions. *J. Am. Chem. Soc.* **119** (1997), 5249–50.
64. R. W. Kriwacki, R. B. Hill, J. M. Flanagan, J. P. Caradonna, and J. H. Prestegard, New NMR Methods for the Characterization of Bound Waters in Macromolecules. *J. Am. Chem. Soc.* **115** (1993), 8907–11.
65. S. Mori, M. O. Johnson, J. M. Berg, and P. C. M. Van Zijl, Water Exchange Filter (WEX Filter) for Nuclear Magnetic Resonance Studies of Macromolecules. *J. Am. Chem. Soc.* **116** (1994), 11982–4.
66. G. Wider, R. Riek, and K. Wüthrich, Diffusion Filters for Separation of Solvent–Protein and Protein–Protein Nuclear Overhauser Effects (HYDRA). *J. Am. Chem. Soc.* **118** (1996), 11629–34.
67. H. Ponstingl and G. Otting, Detection of Protein – Ligand NOEs with Small, Weakly Binding Ligands by Combined Relaxation and Diffusion Filtering. *J. Biomol. NMR* **9** (1997), 441–4.
68. N. Gonnella, M. Lin, M. J. Shapiro, J. R. Wareing, and X. Zhang, Isotope-Filtered Affinity NMR. *J. Magn. Reson.* **131** (1998), 336–8.
69. M. Lin and M. J. Shapiro, Mixture Analysis in Combinatorial Chemistry. Application of Diffusion-Resolved NMR Spectroscopy. *J. Org. Chem.* **61** (1996), 7617–19.
70. K. Bleicher, M. Lin, M. J. Shapiro, and J. R. Wareing, Diffusion Edited NMR: Screening Compound Mixtures by Affinity NMR to Detect Binding Ligands to Vancomycin. *J. Org. Chem.* **63** (1998), 8486–90.
71. R. T. Williamson, E. L. Chapin, A. W. Carr, J. R. Gilbert, P. R. Graupner, P. Lewer, P. McKamey, J. R. Carney, and W. H. Gerwick, New Diffusion-Edited NMR Experiments to Expedite the Dereplication of Known Compounds from Natural Product Mixtures. *Org. Lett.* **2** (2000), 289–92.
72. B. M. Beckwith-Hall, N. A. Thompson, J. K. Nicholson, J. C. Lindon, and E. Holmes, A Metabonomic Investigation of Hepatotoxicity Using Diffusion-Edited ^{1}H NMR Spectroscopy of Blood Serum. *Analyst* **128** (2003), 814–18.
73. J. Chin, A. Chen, and M. J. Shapiro, Improved High-Resolution Diffusion Filtered ^{1}H MAS NMR. *Magn. Reson. Chem.* **38** (2000), 782–4.
74. A. Chen and M. J. Shapiro, NOE Pumping: A Novel NMR Technique for Identification of Compounds with Binding Affinity to Macromolecules. *J. Am. Chem. Soc.* **120** (1998), 10258–9.
75. K. F. Morris and C. S. Johnson, Jr., Resolution of Discrete and Continuous Molecular Size Distributions by Means of Diffusion-Ordered 2D NMR Spectroscopy. *J. Am. Chem. Soc.* **115** (1993), 4291–9.
76. K. F. Morris, P. Stilbs, and C. S. Johnson, Jr., Analysis of Mixtures Based on Molecular Size and Hydrophobicity by Means of Diffusion-Ordered 2D NMR. *Anal. Chem.* **66** (1994), 211–15.

77. H. Barjat, G. A. Morris, S. Smart, A. G. Swanson, and S. C. R. Williams, High-Resolution Diffusion-Ordered 2D Spectroscopy (HR-DOSY) – A New Tool for the Analysis of Complex Mixtures. *J. Magn. Reson. B* **108** (1995), 170–2.
78. K. F. Morris, C. S. Johnson, Jr., and T. C. Wong, Polymer-Induced Non-Newtonian to Newtonian Transition in Viscoelastic CTAB/Sodium Salicylate/Water as Studied by Diffusion-Ordered 2D NMR. *J. Phys. Chem.* **98** (1994), 603–8.
79. D. P. Hinton and C. S. Johnson, Jr., Simultaneous Measurements of Vesicle Diffusion Coefficients and Trapping Efficiencies by Means of Diffusion Ordered 2D NMR Spectroscopy. *Chem. Phys. Lipids* **69** (1994), 175–8.
80. G. S. Kapur, M. Findeisen, and S. Berger, Analysis of Hydrocarbon Mixtures by Diffusion-Ordered NMR Spectroscopy. *Fuel* **79** (2000), 1347–51.
81. A. M. Gil, I. Duarte, E. Cabrita, B. J. Goodfellow, M. Spraul, and R. Kerssebaum, Exploratory Applications of Diffusion Ordered Spectroscopy to Liquid Foods: An Aid Towards Spectral Assignment. *Anal. Chim. Acta* **506** (2004), 215–23.
82. M. Nilsson, I. F. Duarte, C. Almeida, I. Delgadillo, B. J. Goodfellow, A. M. Gil, and G. A. Morris, High-Resolution NMR and Diffusion-Ordered Spectroscopy of Port Wine. *J. Agric. Food Chem.* **52** (2004), 3736–43.
83. M. Politi, P. Groves, M. I. Chávez, F. J. Cañada, and J. Jiménez-Barbero, Useful Applications of DOSY Experiments for the Study of Mushroom Polysaccharides. *Carbohydr. Res.* **341** (2006), 84–9.
84. A. J. Simpson, W. L. Kingery, M. Spraul, E. Humpfer, P. Dvortsak, and R. Kerssebaum, Separation of Structural Components in Soil Organic Matter by Diffusion Ordered Spectroscopy. *Environ. Sci. Technol.* **35** (2001), 4421–25.
85. C. A. Steinbeck, N. Hedin, and B. F. Chmelka, Interactions of Charged Porphyrins with Nonionic Triblock Copolymer Hosts in Aqueous Solutions. *Langmuir* **20** (2004), 10399–412.
86. E. K. Gozansky and D. G. Gorenstein, DOSY-NOESY: Diffusion-Ordered NOESY. *J. Magn. Reson. B* **111** (1996), 94–6.
87. R. K. Harris, K. A. Kinnear, G. A. Morris, M. J. Stchedroff, A. Samadi-Maybodi, and N. Azizi, Silicon-29 Diffusion-Ordered NMR Spectroscopy (DOSY) as a Tool for Studying Aqueous Silicates. *Chem. Commun.* (2001), 2422–3.
88. G. Pages, C. Delaurent, and S. Caldarelli, Simplified Analysis of Mixtures of Small Molecules by Chromatographic NMR Spectroscopy. *Angew. Chem. (Int. Ed.)* **49** (2006), 5950–3.
89. S. J. Gibbs and C. S. Johnson, Jr., Pulsed Field Gradient NMR Study of Probe Motion in Polyacrylamide Gels. *Macromolecules* **24** (1991), 6110–13.
90. D. P. Hinton and C. S. Johnson, Jr., Diffusion Coefficients, Electrophoretic Mobilities, and Morphologies of Charged Phospholid Vesicles by Pulsed Field Gradient NMR and Electron Microscopy. *J. Colloid Interface Sci.* **173** (1995), 364–71.
91. P. C. Griffiths, A. Y. F. Cheung, C. Farley, A. Paul, R. K. Heenan, S. M. King, E. Pettersson, P. Stilbs, and R. Ranganathan, Small-Angle Neutron Scattering, Electron Paramagnetic Resonance, Electrophoretic NMR, and Time-Resolved Fluorescence Quenching Studies of Sodium Dodecyl Sulfate and Tetra(ethylene oxide) Dodecyl Ether Mixed Surfactant Micelles. *J. Phys. Chem. B* **108** (2004), 1351–6.
92. P. C. Griffiths, E. Pettersson, P. Stilbs, A. Y. F. Cheung, A. M. Howe, and A. R. Pitt, Electrophoretic Nuclear Magnetic Resonance Studies of Mixed Anionic-Nonionic Surfactant Micelles. *Langmuir* **17** (2001), 7178–81.
93. U. Böhme and U. Scheler, Effective Charge of Bovine Serum Albumin Determined by Electrophoresis NMR. *Chem. Phys. Lett.* **435** (2007), 342–5.

94. U. Böhme and U. Scheler, Counterion Mobility and Effective Charge of Polyelectrolytes in Solution. *Macromol. Symp.* **211** (2004), 87–92.
95. Q. He and Z. Wei, Convection Compensated Electrophoretic NMR. *J. Magn. Reson.* **150** (2001), 126–31.
96. E. Pettersson, D. Topgaard, P. Stilbs, and O. Söderman, Surfactant/Nonionic Polymer Interaction. A NMR Investigation and NMR Electrophoretic Investigation. *Langmuir* **20** (2004), 1138–43.
97. H. J. Walls and T. A. Zawodzinski, Jr., Anion and Cation Transference Numbers Determined by Electrophoretic NMR of Polymer Electrolytes Sum to Unity. *Electrochem. Solid State Lett.* **3** (2000), 321–4.
98. M. Holz, S. R. Heil, and I. A. Schwab, Electrophoretic NMR Studies of Electrical Transport in Fluid-Filled Porous Systems. *Magn. Reson. Imaging* **19** (2001), 457–63.
99. B. Manz, P. Stilbs, B. Jönsson, O. Söderman, and P. T. Callaghan, NMR Imaging of the Time Evolution of Electroosmotic Flow in a Capillary. *J. Phys. Chem.* **99** (1995), 11297–301.
100. P. Bendel, M. Bernado, J. H. Dunsmuir, and H. Thomann, Electric Field Driven Flow in Natural Porous Media. *Magn. Reson. Imaging* **21** (2003), 321–7.
101. M. Ise, K. D. Kreuer, and J. Maier, Electroosmotic Drag in Polymer Electrolyte Membranes: An Electrophoretic NMR Study. *Solid State Ionics* **125** (1999), 213–23.
102. D. Wu, A. Chen, and C. S. Johnson, Jr., Flow Imaging by Means of 1D Pulsed-Field-Gradient NMR with Application to Electroosmotic Flow. *J. Magn. Reson. A* **115** (1995), 123–6.
103. S. Gaemers, C. J. Elsevier, and Ad. Bax, NMR of Biomolecules in Low Viscosity, Liquid CO_2. *Chem. Phys. Lett.* **301** (1999), 138–44.
104. Y. O. Kamatari, H. Yamada, K. Akasaka, J. A. Jones, C. M. Dobson, and L. J. Smith, Response of Native and Denatured Hen Lysozyme to High Pressure Studied by $^{15}N/^{1}H$ NMR Spectroscopy. *Eur. J. Biochem.* **268** (2001), 1782–93.
105. J. P. Mackay, G. L. Shaw, and G. F. King, Backbone Dynamics of the c-Jun Leucine Zipper: ^{15}N NMR Relaxation Studies. *Biochemistry* **35** (1996), 4867–77.
106. C. H. Everhart and C. S. Johnson, Jr., The Determination of Tracer Diffusion Coefficients for Proteins by Means of Pulsed Field Gradient NMR with Applications to Hemoglobin. *J. Magn. Reson.* **48** (1982), 466–74.
107. W. S. Price, F. Tsuchiya, and Y. Arata, Lysozyme Aggregation and Solution Properties Studied Using PGSE NMR Diffusion Measurements. *J. Am. Chem. Soc.* **121** (1999), 11503–12.
108. I. V. Nesmelova and V. D. Fedotov, Self-Diffusion of Myoglobin and Water Molecules in Solutions. *Polym. Sci. A* **39** (1997), 361–5.
109. A. J. Dingley, J. P. Mackay, B. E. Chapman, M. B. Morris, P. W. Kuchel, B. D. Hambly, and G. F. King, Measuring Protein Self-Association Using Pulsed-Field-Gradient NMR Spectroscopy: Application to Myosin Light Chain 2. *J. Biomol. NMR* **6** (1995), 321–8.
110. S. J. Gibbs, A. S. Chu, E. N. Lightfoot, and T. W. Root, Ovalbumin Diffusion at Low Ionic Strength. *J. Phys. Chem.* **95** (1991), 467–71.
111. A. Dehner and H. Kessler, Diffusion NMR Spectroscopy: Folding and Aggregation of Domains in P53. *ChemBioChem* **6** (2005), 1550–65.
112. W. S. Price, M. Nara, and Y. Arata, A Pulsed Field Gradient NMR Study of the Aggregation and Hydration of Parvalbumin. *Biophys. Chem.* **65** (1997), 179–87.
113. S. R. Inglis, M. J. McGann, W. S. Price, and M. M. Harding, Diffusion NMR Studies on Fish Antifreeze Proteins and Synthetic Analogues. *FEBS Lett.* **580** (2006), 3911–15.

114. V. V. Krishnan, Determination of Oligomeric State of Proteins in Solution from Pulsed-Field-Gradient Self-Diffusion Coefficient Measurements. A Comparison of Experimental, Theoretical, and Hard-Sphere Approximated Values. *J. Magn. Reson.* **124** (1997), 468–73.
115. T. S. Derrick and C. K. Larive, Use of PFG-NMR for Mixture Analysis: Measurement of Diffusion Coefficients of *Cis* and *Trans* Isomers of Proline-Containing Peptides. *Appl. Spectrosc.* **53** (1999), 1595–600.
116. M. S. B. McAlister, B. Davis, M. Pfuhl, and P. C. Driscoll, NMR Analysis of the N-Terminal SRCR Domain of Human CD5: Engineering of a Glycoprotein for Superior Characteristics in NMR Experiments. *Protein Eng.* **11** (1998), 847–53.
117. Y. Aso, S. Yoshioka, and S. Kojima, Determination of the Diffusion Coefficient of Insulin and Lysozyme in Crosslinked Dextran Hydrogels by Pulsed-Field-Gradient NMR. *Chem. Pharm. Bull.* **46** (1998), 1836–9.
118. P. W. Kuchel, B. E. Chapman, and A. J. Lennon, Diffusion of Hydrogen in Aqueous Solutions Containing Protein. Pulsed Field Gradient NMR Measurements. *J. Magn. Reson. A* **103** (1993), 329–31.
119. R. Knauss, J. Schiller, G. Fleischer, J. Kärger, and K. Arnold, Self-Diffusion of Water in Cartilage and Cartilage Components as Studied by Pulsed Field Gradient NMR. *Magn. Reson. Med.* **41** (1999), 285–92.
120. R. Kimmich, F. Klammler, V. D. Skirda, I. A. Serebrennikova, A. I. Maklakov, and N. Fatkullin, Geometrical Restrictions of Water Diffusion in Aqueous Protein Systems. A Study Using NMR Field-Gradient Techniques. *Appl. Magn. Reson.* **4** (1993), 425–40.
121. V. Dötsch and G. Wider, Exchange Rates of Internal Water Molecules in Proteins Measured Using Pulsed Field Gradients. *J. Am. Chem. Soc.* **117** (1995), 6064–70.
122. H. M. Baranowska and K. J. Olszewski, The Hydration of Proteins in Solutions by Self-Diffusion Coefficients NMR Study. *Biochim. Biophys. Acta* **1289** (1996), 312–14.
123. R. Lamanna, M. Delmelle, and S. Cannistraro, Solvent Stokes–Einstein Violation in Aqueous Protein Solutions. *Phys. Rev. E* **49** (1994), 5878–80.
124. F. Mariette, D. Topgaard, B. Jönsson, and O. Söderman, ^{1}H NMR Diffusometry Study of Water in Casein Dispersions and Gels. *J. Agric. Food Chem.* **50** (2002), 4295–302.
125. K. Kotitschke, R. Kimmich, E. Rommel, and F. Parak, NMR Study of Diffusion in Protein Hydration Shells. *Progr. Colloid Polym. Sci.* **83** (1990), 211–15.
126. W. S. Price, NMR Gradient Methods in the Study of Proteins. In *Annual Reports on the Progress in Chemistry Section C*, ed. G. A. Webb. vol. 96. (Cambridge: Royal Society of Chemistry, 2000), pp. 3–53.
127. W. S. Price, F. Tsuchiya, and Y. Arata, Time-Dependence of Aggregation in Crystallizing Lysozyme Solutions Probed Using NMR Self-Diffusion Measurements. *Biophys. J.* **80** (2001), 1585–90.
128. R. E. Moll, Observation of a Helix-Coil Transition by Pulsed-Field-Gradient Spin-Echo Nuclear Magnetic Resonance. *J. Am. Chem. Soc.* **90** (1968), 4739.
129. K. H. Mayo, E. Ilyina, and H. Park, A Recipe for Designing Water-Soluble, β-Sheet-Forming Peptides. *Protein Eng.* **5** (1996), 1301–15.
130. J. I. Guijarro, M. Sunde, J. A. Jones, I. D. Campbell, and C. M. Dobson, Amyloid Fibril Formation by an SH3 Domain. *Proc. Natl. Acad. Sci. U.S.A.* **95** (1998), 4224–8.
131. H. Pan, G. Barany, and C. Woodward, Reduced BPTI is Collapsed. A Pulsed Field Gradient NMR Study of Unfolded and Partially Folded Bovine Pancreatic Trypsin Inhibitor. *Protein Sci.* **6** (1997), 1985–92.
132. W. Y. Choy, F. A. A. Mulder, K. A. Crowhurst, D. R. Muhandiram, I. S. Millett, S. Doniach, J. D. Forman-Kay, and L. E. Kay, Distribution of Molecular Size within

An Unfolded State Ensemble Using Small-Angle X-Ray Scattering and Pulse Field Gradient NMR Techniques. *J. Mol. Biol.* **316** (2002), 101–12.

133. A. V. Buevich and J. Baum, Residue-Specific Real-Time NMR Diffusion Experiments Define the Association States of Proteins during Folding. *J. Am. Chem. Soc.* **124** (2002), 7156–62.

134. J. A. Jones, D. K. Wilkins, L. J. Smith, and C. M. Dobson, Characterization of Protein Unfolding by NMR Diffusion Measurements. *J. Biomol. NMR* **10** (1997), 199–203.

135. W. S. Price, F. Tsuchiya, C. Suzuki, and Y. Arata, Characterization of the Solution Properties of *Pichia farinosa* Killer Toxin Using PGSE NMR Diffusion Measurements. *J. Biomol. NMR* **13** (1999), 113–17.

136. D. K. Wilkins, S. B. Grimshaw, V. Receveur, C. M. Dobson, J. A. Jones, and L. J. Smith, Hydrodynamic Radii of Native and Denatured Proteins Measured by Pulse Field Gradient NMR Techniques. *Biochemistry* **38** (1999), 16424–31.

137. D. A. Jayawickrama and C. K. Larive, Analysis of the (Trimethylsilyl)Propionic Acid-β(12–28) Peptide Binding Equilibrium with NMR Spectroscopy. *Anal. Chem.* **71** (1999), 2117–22.

138. B. A. Begotka, J. L. Hunsader, C. Oparaeche, J. K. Vincent, and K. F. Morris, A Pulsed Field Gradient NMR Diffusion Investigation of Enkephalin Peptide-Sodium Dodecyl Sulfate Micelle Association. *Magn. Reson. Chem.* **44** (2006), 586–93.

139. A. Andersson, J. Almqvist, F. Hagn, and L. Mäler, Diffusion and Dynamics of Penetratin in Different Membrane Mimicking Media. *Biochim. Biophys. Acta* **1661** (2007), 18–25.

140. W. J. Chien, S. F. Cheng, and D. K. Chang, Determination of the Binding Constant of a Protein Kinase C Substrate, $NG_{(28-43)}$, to Sodium Dodecyl Sulfate via the Diffusion Coefficient Measured by Pulsed Field Gradient Nuclear Magnetic Resonance. *Anal. Biochem.* **264** (1998), 211–15.

141. M. Liu, J. K. Nicholson, and J. C. Lindon, Analysis of Drug-Protein Binding Using Nuclear Magnetic Resonance Based Molecular Diffusion Measurements. *Anal. Commun.* **34** (1997), 225–8.

142. S. Yao, R. A. Cherny, A. I. Bush, C. L. Masters, and K. J. Barnham, Characterizing Bathocuproine Self-Association and Subsequent Binding to Alzheimer's Disease Amyloid β-Peptide by NMR. *J. Peptide Sci.* **10** (2004), 210–17.

143. Y. Ma, M. Liu, X.-A. Mao, J. K. Nicholson, and J. C. Lindon, NMR Spectroscopic Diffusion, Chemical Shift and Linewidth Measurements of Low-Affinity Binding of Ibuprofen Enantiomers to Human Serum Albumin. *Magn. Reson. Chem.* **37** (1999), 269–73.

144. R.-S. Luo, M.-L. Liu, and X.-A. Mao, NMR Diffusion and Relaxation Study on Ibuprofen-HSA Interaction. *Appl. Spectrosc.* **53** (1999), 776–9.

145. W. S. Price, F. Elwinger, C. Vigouroux, and P. Stilbs, PGSE-WATERGATE, a New Tool for NMR Diffusion-Based Studies of Ligand-Macromolecule Binding. *Magn. Reson. Chem.* **40** (2002), 391–5.

146. Y. Yang, G. Bai, Xu. Zhang, C. Ye, and M. Liu, ^{1}H Spectroscopic Evidence of Interaction Between Ibuprofen and Lipoproteins in Human Blood Plasma. *Anal. Biochem.* **324** (2004), 292–7.

147. A. J. Lennon, N. R. Scott, B. E. Chapman, and P. W. Kuchel, Hemoglobin Affinity for 2,3-Bisphosphoglycerate in Solutions and Intact Erythrocytes: Studies Using Pulsed-Field Gradient Nuclear Magnetic Resonance and Monte Carlo Simulations. *Biophys. J.* **67** (1994), 2096–109.

148. J. Lapham, J. P. Rife, P. B. Moore, and D. M. Crothers, Measurement of Diffusion Constants for Nucleic Acids by NMR. *J. Biomol. NMR* **10** (1997), 255–62.

149. X. Yang, Y. S. Sanghvi, and X. Gao, Conformational Studies of Antisense DNA by PFG NMR. *J. Biomol. NMR* **10** (1997), 383–8.
150. B. Andreasson, L. Nordenskiöld, and J. Schultz, Interactions of Spermidine and Methylspermidine with DNA Studied by Nuclear Magnetic Resonance Self-Diffusion Measurements. *Biophys. J.* **70** (1996), 2847–56.
151. L. van Dam, A. P. Lyubartsev, A. Laaksonen, and L. Nordenskiöld, Self-Diffusion and Association of Li^+, Cs^+, and H_2O in Oriented DNA Fibres. An NMR and MD Simulation Study. *J. Phys. Chem. B* **102** (1998), 10636–42.
152. M. S. Kaucher, Y.-F. Lam, S. Pieraccini, G. Gotarelli, and J. T. Davis, Using Diffusion NMR to Characterize Guanosine Self-Association: Insights into Structure and Mechanism. *Chemistry: A European Journal* **11** (2005), 164–73.
153. T. Evan-Salem, L. Frish, F. W. B. Van Leeuwen, D. N. Reinhoudt, W. Verboom, M. S. Kaucher, J. T. Davis, and Y. Cohen, Self-Assembled Ionophores from Isoguanosine: Diffusion NMR Spectroscopy Clarifies Cation's and Anion's Influence on Supramolecular Structure. *Chemistry: A European Journal* **13** (2007), 1969–77.
154. A. Böckmann and E. Guittet, Determination of Fast Proton Exchange Rates of Biomolecules by NMR Using Water Selective Diffusion Experiments. *FEBS Lett.* **418** (1997), 127–30.
155. W. H. Gmeiner, C. J. Hudalla, A. M. Soto, and L. Marky, Binding of Ethidium to DNA Measured Using a 2D Diffusion-Modulated Gradient COSY NMR Experiment. *FEBS Lett.* **465** (2000), 148–52.
156. A. Pluen, P. A. Netti, R. K. Jain, and D. A. Berk, Diffusion of Macromolecules in Agarose Gels: Comparison of Linear and Globular Configurations. *Biophys. J.* **77** (1999), 542–52.
157. S. J. Gibbs and C. S. Johnson, Jr., Polyammonium Cation Diffusion in Aqueous Solutions of DNA as Studied by Pulsed Field Gradient NMR. *Macromolecules* **24** (1991), 5224–5.
158. M. H. Baslow and D. N. Guilfoyle, Effect of *N*-Acetylaspartic Acid on the Diffusion Coefficient of Water: A Proton Magnetic Resonance Phantom Method for Measurement of Osmolyte-Obligated Water. *Anal. Biochem.* **311** (2002), 133–8.
159. A. H. Daranas, J. H. Fernández, E. Q. Morales, M. Norte, and J. A. Gavín, Self-Association of Okadaic Acid upon Complexation with Potassium Ion. *J. Med. Chem.* **47** (2004), 10–13.
160. K. Adachi, M. Natsuisaka, and A. Tanioka, Measurements of Self-Diffusion Coefficients of Monensin in Chloroform Solution by PFG-NMR. *J. Chem. Soc., Faraday Trans.* **93** (1997), 3347–50.
161. J. A. Parkinson, H. Sun, and P. J. Sadler, New Approach to the Solution Chemistry of Bismuth Citrate Antiulcer Complexes. *Chem. Commun.* (1998), 881–2.
162. M.-C. Lo, J. S. Helm, G. Sarngadharan, I. Pelczer, and S. Walker, A New Structure for the Substrate-Binding Antibiotic Ramoplanin. *J. Am. Chem. Soc.* **123** (2001), 8640–1.
163. L. Orfi, C. K. Larive, and S. M. LeVine, Physicochemical Characterization of Psychosine by ^1H Nuclear Magnetic Resonance and Electron Microscopy. *Lipids* **32** (1997), 1035–40.
164. K. Shikii, S. Sakamoto, H. Seki, H. Utsumi, and K. Yamaguchi, Narcissistic Aggregation of Steroid Compounds in Diluted Solution Elucidated by CSI-MS, PFG NMR and X-Ray Analysis. *Tetrahedron* **60** (2004), 3487–92.
165. A. Blinc, G. Lahajnar, R. Blinc, A. Zidanšek, and A. Sepe, Proton NMR Study of the State of Water in Fibrin Gels, Plasma, and Blood Clots. *Magn. Reson. Med.* **14** (1990), 105–22.
166. Xu. Zhang, C.-G. Li, C.-H. Ye, and M.-L. Liu, Determination of Molecular Self-Diffusion Coefficient Using Multiple Spin-Echo NMR Spectroscopy with Removal of Convection and Background Gradient Artifacts. *Anal. Chem.* **73** (2001), 3528–34.

167. Y. Yamane, M. Kobayashi, H. Kimura, S. Kuroki, and I. Ando, Diffusional Behavior of Amino Acids in Solid-Phase Reaction Field as Studied by ^{1}H Pulsed-Field-Gradient Spin-Echo NMR Method. *Polymer* **43** (2002), 1767–72.
168. S. Kwak and M. Lafleur, NMR Self-Diffusion of Molecular and Macromolecular Species in Dextran Solutions and Gels. *Macromolecules* **36** (2003), 3189–95.
169. N. Mistry, M. I. Ismail, R. D. Farrant, M. Liu, J. K. Nicholson, and J. C. Lindon, Impurity Profiling in Bulk Pharmaceutical Batches Using ^{19}F NMR Spectroscopy and Distinction Between Monomeric and Dimeric Impurities by NMR-Based Diffusion Measurements. *J. Pharm. Biomed. Anal.* **19** (1999), 511–17.
170. K. F. Morris, B. J. Cutak, A. M. Dixon, and C. K. Larive, Analysis of Diffusion Coefficient Distributions in Humic and Fulvic Acids by Means of Diffusion Ordered NMR Spectroscopy. *Anal. Chem.* **71** (1999), 5315–21.
171. J. R. Lead, K. J. Wilkinson, E. Balnois, B. J. Cutak, C. K. Larive, S. Assemi, and R. Beckett, Diffusion Coefficients and Polydispersities of the Suwannee River Fulvic Acid: Comparison of Fluorescence, Pulsed-Field Gradient Nuclear Magnetic Resonance, and Flow Field-Flow Fractionation. *Environ. Sci. Technol.* **34** (2000), 3508–13.
172. G. Orädd and G. Lindblom, Lateral Diffusion Studied by Pulsed Field Gradient NMR on Oriented Lipid Membranes. *Magn. Reson. Chem.* **42** (2004), 123–31.
173. A. Filippov, G. Orädd, and G. Lindblom, The Effect of Cholesterol on the Lateral Diffusion of Phospholipids in Oriented Bilayers. *Biophys. J.* **84** (2003), 3079–86.
174. A. Filippov, G. Orädd, and G. Lindblom, Sphingomyelin Structure Influences the Lateral Diffusion and Raft Formation in Lipid Bilayers. *Biophys. J.* **90** (2006), 2086–92.
175. G. Orädd, P. W. Westerman, and G. Lindblom, Lateral Diffusion Coefficients of Separate Lipid Species in a Ternary Raft-Forming Bilayer: A Pfg-NMR Multinuclear Study. *Biophys. J.* **89** (2005), 315–20.
176. G. Orädd and G. Lindblom, NMR Studies of Lipid Lateral Diffusion in the DMPC/Gramicidin D/Water System: Peptide Aggregation and Obstruction Effects. *Biophys. J.* **87** (2004), 980–7.
177. V. Shahedi, G. Orädd, and G. Lindblom, Domain-Formation in DOPC/SM Bilayers Studied by pfg-NMR: Effect of Sterol Structure. *Biophys. J.* **91** (2006), 2501–7.
178. P. Karakatsanis and T. M. Bayerl, Diffusion Measurements in Oriented Phospholipid Bilayers by ^{1}H- NMR in a Static Fringe Field Gradient. *Phys. Rev. E* **54** (1996), 1785–90.
179. A. Pampel, K. Zick, H. Glauner, and F. Engelke, Studying Lateral Diffusion in Lipid Bilayers by Combining a Magic Angle Spinning NMR Probe with a Microimaging Gradient System. *J. Am. Chem. Soc.* **126** (2004), 9534–5.
180. H. C. Gaede and K. Gawrisch, Multi-Dimensional Pulsed Field Gradient Magic Angle Spinning NMR Experiments on Membranes. *Magn. Reson. Chem.* **42** (2004), 115–22.
181. H. C. Gaede, K. C. Luckett, I. V. Polozov, and K. Gawrisch, Multinuclear NMR Studies of Single Lipid Bilayers Supported in Cylindrical Aluminum Oxide Nanopores. *Langmuir* **20** (2004), 7711–19.
182. O. Wattraint and C. Sarazin, Diffusion Measurements of Water, Ubiquinone and Lipid Bilayer Inside a Cylindrical Nanoporous Support: A Stimulated Echo Pulsed-Field Gradient MAS-NMR Investigation. *Biochim. Biophys. Acta* **1713** (2005), 65–72.
183. J. E. Tanner and E. O. Stejskal, Restricted Self-Diffusion of Protons in Colloidal Systems by the Pulsed-Gradient, Spin-Echo Method. *J. Chem. Phys.* **49** (1968), 1768–77.
184. D. G. Cory and A. N. Garroway, Measurement of Translational Displacement Probabilities by NMR: An Indicator of Compartmentation. *Magn. Reson. Med.* **14** (1990), 435–44.

185. N. L. Zakhartchenko, V. D. Skirda, and R. R. Valiullin, Self-Diffusion of Water and Oil in Peanuts Investigated by PFG NMR. *Magn. Reson. Imaging* **16** (1998), 583–6.
186. H.-R. Tang, J. Godward, and B. Hills, The Distribution of Water in Native Starch Granules – A Multinuclear NMR Study. *Carbohydr. Polym.* **43** (2000), 375–87.
187. G. Fleischer, V. Skirda, and A. Werner, NMR-Investigation of Restricted Self-Diffusion of Oil in Rape Seeds. *Eur. Biophys. J.* **19** (1990), 25–30.
188. G. Liger-Belair, É. Prost, M. Parmentier, P. Jeandet, and J.-M. Nuzillard, Diffusion Coefficient of CO_2 Molecules as Determined by ^{13}C NMR in Various Carbonated Beverages. *J. Agric. Food Chem.* **51** (2003), 7560–3.
189. H. Peemoeller, M. E. Hale, M. H. Schneider, A. R. Sharp, and D. W. Kydon, Study of Restricted Diffusion in Wood. *Wood Fiber Sci* **17** (1985), 110–16.
190. W. Wycoff, S. Pickup, B. Cutter, W. Miller, and T. C. Wong, The Determination of the Cell Size in Wood by Nuclear Magnetic Resonance Diffusion Techniques. *Wood Fiber Sci* **32** (2000), 72–80.
191. T.-Q. Li, U. Henriksson, T. Klason, and L. Ödberg, Water Diffusion in Wood Pulp Cellulose Fibers Studied by Means of the Pulsed Gradient Spin-Echo Method. *J. Colloid Interface Sci.* **154** (1992), 305–15.
192. T.-Q. Li, M. Häggkvist, and L. Ödberg, Porous Structure of Cellulose Fibers Studied by Q-Space NMR Imaging. *Langmuir* **13** (1997), 3570–4.
193. R. Ek, T. Gren, U. Henriksson, H. Nyqvist, C. Nyström, and L. Ödberg, Prediction of Drug Release by Characterisation of the Tortuosity in Porous Cellulose Beads Using a Spin Echo NMR Technique. *Int. J. Pharm.* **124** (1995), 9–18.
194. D. Topgaard and O. Söderman, A NMR Self-Diffusion Study of the Porous Structure of Starch Granules. *Progr. Colloid Polym. Sci.* **120** (2002), 47–51.
195. B. Newling and S. N. Batchelor, Pulsed Field Gradient NMR Study of the Diffusion of H_2O and Polyethylene Glycol Polymers in the Supramolecular Structure of Wet Cotton. *J. Phys. Chem. B* **107** (2003), 12391–7.
196. D. Topgaard and O. Söderman, Self-Diffusion in Two- and Three-Dimensional Powders of Anisotropic Domains: An NMR Study of the Diffusion of Water in Cellulose and Starch. *J. Phys. Chem. B* **106** (2002), 11887–92.
197. P. C. M. Van Zijl, C. T. W. Moonen, P. Faustino, J. Pekar, O. Kaplan, and J. S. Cohen, Complete Separation of Intracellular and Extracellular Information in NMR Spectra of Perfused Cells by Diffusion-Weighted Spectroscopy. *Proc. Natl. Acad. Sci. U.S.A.* **88** (1991), 3228–32.
198. Y. Assaf and Y. Cohen, Non-Mono-Exponential Attenuation of Water and *N*-Acetyl Aspartate Signals Due to Diffusion in Brain Tissue. *J. Magn. Reson.* **131** (1998), 69–85.
199. C. Malmborg, M. Sjöbeck, S. Brockstedt, E. Englund, O. Söderman, and D. Topgaard, Mapping the Intracellular Fraction of Water by Varying the Gradient Pulse Length in *q*-Space Diffusion MRI. *J. Magn. Reson.* **180** (2006), 280–5.
200. R. L. Cooper, D. B. Chang, A. C. Young, C. J. Martin, and B. Ancker-Johnson, Restricted Diffusion in Biophysical Systems: Experiment. *Biophys. J.* **14** (1974), 161–77.
201. J. E. Tanner, Self-Diffusion of Water in Frog Muscle. *Biophys. J.* **28** (1979), 107–16.
202. W. S. Price, B. E. Chapman, B. A. Cornell, and P. W. Kuchel, Translational Diffusion of Glycine in Erythrocytes Measured at High Resolution with Pulsed Field Gradients. *J. Magn. Reson.* **83** (1989), 160–6.
203. L. L. Latour, K. Svoboda, P. P. Mitra, and C. H. Sotak, Time-Dependent Diffusion of Water in a Biological Model System. *Proc. Natl. Acad. Sci. U.S.A.* **91** (1994), 1229–33.

204. P. van Gelderen, D. DesPres, P. C. M. Van Zijl, and C. T. W. Moonen, Evaluation of Restricted Diffusion in Cylinders. Phosphocreatine in Rabbit Leg Muscle. *J. Magn. Reson. B* **103** (1994), 255–60.
205. S. T. Kinsey, B. R. Locke, B. Penke, and T. S. Moerland, Diffusional Anisotropy is Induced by Subcellular Barriers in Skeletal Muscle. *NMR Biomed.* **12** (1999), 1–7.
206. R. A. De Graaf, A. van Kranenburg, and K. Nicolay, In Vivo ^{31}P-NMR Diffusion Spectroscopy of ATP and Phosphocreatine in Rat Skeletal Muscle. *Biophys J.* **78** (2000), 1657–64.
207. J. L. Griffin, J. Troke, L. A. Walker, R. F. Shore, J. C. Lindon, and J. K. Nicholson, The Biochemical Profile of Rat Testicular Tissue as Measured by Magic Angle Spinning ^{1}H Spectroscopy. *FEBS Lett.* **486** (2000), 225–9.
208. A. I. García-Pérez, E. A. López-Beltrán, P. Klüner, J. Luque, P. Ballesteros, and S. Cerdan, Molecular Crowding and Viscosity as Determinants of Translational Diffusion of Metabolites in Subcellular Organelles. *Arch. Biochem. Biophys.* **362** (1999), 329–38.
209. D. G. Regan and P. W. Kuchel, NMR Studies of Diffusion-Coherence Phenomena in Red Cell Suspensions: Current Status. *Isr. J. Chem.* **43** (2003), 45–54.
210. M. Köpf, C. Corinth, O. Haferkamp, and T. F. Nonnenmacher, Anomalous Diffusion of Water in Biological Tissues. *Biophys. J.* **70** (1996), 2950–8.
211. F. A. Howe, A. G. Filler, B. A. Bell, and J. R. Griffiths, Magnetic Resonance Neurography. *Magn. Reson. Med.* **28** (1992), 328–38.
212. M. E. Moseley, J. Kucharczyk, H. S. Asgari, and D. Norman, Anisotropy in Diffusion-Weighted MRI. *Magn. Reson. Med.* **19** (1991), 321–6.
213. J. S. Schoeniger, N. Aiken, E. Hsu, and S. J. Blackband, Relaxation-Time and Diffusion NMR Microscopy of Single Neurons. *J. Magn. Reson. B* **103** (1994), 261–73.
214. D. Le Bihan, Molecular Diffusion Nuclear Magnetic Resonance Imaging. *Magn. Reson. Q.* **7** (1991), 1–30.
215. S. Fichele, M. N. J. Paley, N. Woodhouse, P. D. Griffiths, E. J. R. van Beek, and J. M. Wild, Measurements and Modeling of Long Range ^{3}He Diffusion in the Lung Using a 'Slice-Washout' Method. *J. Magn. Reson.* **174** (2005), 28–33.
216. K. J. Carlton, M. R. Halse, and J. H. Strange, Diffusion-Weighted Imaging of Bacteria Colonies in the STRAFI Plane. *J. Magn. Reson.* **143** (2004), 24–9.
217. G. J. Stanisz and R. M. Henkelman, Diffusional Anistropy of T_2 Components in Bovine Optic Nerve. *Magn. Reson. Med.* **40** (1998), 405–10.
218. S. Peled, D. G. Cory, S. A. Raymond, D. A. Kirschner, and F. A. Jolesz, Water Diffusion, T_2, and Compartmentation in Frog Sciatic Nerve. *Magn. Reson. Med.* **42** (1999), 911–18.
219. P. L. Hubbard, K. M. McGrath, and P. T. Callaghan, A Study of Anisotropic Water Self-Diffusion and Defects in the Lamellar Mesophase. *Langmuir* **21** (2005), 4340–6.
220. M. E. Komlosh, F. Horkay, R. Z. Freidlin, U. Nevo, Y. Assaf, and P. J. Basser, Detection of Microscopic Anisotropy in Gray Matter and in a Novel Tissue Phantom Using Double Pulsed Gradient Spin Echo MR. *J. Magn. Reson.* **189** (2007), 38–45.
221. M. D. Hürlimann, L. Venkataramanan, and C. Flaum, The Diffusion–Spin Relaxation Time Distribution Function as An Experimental Probe to Characterize Fluid Mixtures in Porous Media. *J. Chem. Phys.* **117** (2002), 10223–32.
222. M. D. Hürlimann, M. Flaum, L. Venkataramanan, C. Flaum, R. Freedman, and G. J. Hirasaki, Diffusion-Relaxation Distribution Functions of Sedimentary Rocks in Different Saturation States. *Magn. Reson. Imaging* **21** (2003), 305–10.
223. S. Godefroy and P. T. Callaghan, 2D Relaxation/Diffusion Correlations in Porous Media. *Magn. Reson. Imaging* **21** (2003), 381–3.

224. Y. Qiao, P. Galvosas, T. Adalsteinsson, M. Schönhoff, and P. T. Callaghan, Diffusion Exchange NMR Spectroscopic Study of Dextran Exchange Through Polyelectrolyte Multilayer Capsules. *J. Chem. Phys.* **122** (2005), 214912-1–214912-9.
225. J. Andrasko, Measurement of Membrane Permeability to Slowly Penetrating Molecules by a Pulse Gradient NMR Method. *J. Magn. Reson.* **21** (1976), 479–84.
226. P.-C. Jiang, T.-Y. Yu, W.-C. Perng, and L.-P. Hwang, Pore-to-Pore Hopping Model for the Interpretation of the Pulsed Gradient Spin Echo Attenuation of Water Diffusion in Cell Suspension Systems. *Biophys. J.* **80** (2001), 2493–504.
227. J. W. C. van der Veen, P. van Gelderen, J. H. N. Creyghton, and W. M. M. J. Bovée, Diffusion in Red Blood Cell Suspensions: Separation of the Intracellular and Extracellular NMR Sodium Signal. *Magn. Reson. Med.* **29** (1995), 571–4.
228. K. Potter, R. L. Kleinberg, F. J. Brockman, and E. W. McFarland, Assay for Bacteria in Porous Media by Diffusion-Weighted NMR. *J. Magn. Reson. B* **113** (1996), 9–15.
229. A. R. Waldeck, A. J. Lennon, B. E. Chapman, and P. W. Kuchel, ^{7}Li and ^{23}Na Nuclear Magnetic Resonance Studies of Transport and Diffusion in Liposomes. Comparison of Transport Rate Constants Estimated Using Pulsed Field Gradient and Magnetization-Transfer Procedures. *J. Chem. Soc., Faraday Trans.* **89** (1993), 2807–14.
230. X. Gao and T. C. Wong, Studies of the Binding and Structure of Adrenocorticotoprin Peptides in Membrane Mimics by NMR Spectroscopy and Pulsed-Field Gradient Diffusion. *Biophys. J.* **74** (1998), 1871–88.
231. M. Andrec and J. H. Prestegard, Quantitation of Chemical Exchange Rates Using Pulsed-Field-Gradient Diffusion Measurements. *J. Biomol. NMR* **9** (1997), 136–50.
232. M. Liu, H. C. Toms, G. E. Hawkes, J. K. Nicholson, and J. C. Lindon, Determination of the Relative NH Proton Lifetimes of the Peptide Analogue Viomycin in Aqueous Solution by NMR-Based Diffusion Measurement. *J. Biomol. NMR* **13** (1999), 25–30.
233. E. J. Cabrita and S. Berger, HR-DOSY as a New Tool for the Study of Chemical Exchange Phenomena. *Magn. Reson. Chem.* **40** (2002), S122–7.
234. P. Thureau, B. Ancian, S. Viel, and A. Thévand, Determining Chemical Exchange Rates of the Uracil Labile Protons by NMR Diffusion Experiments. *Chem. Commun.* (2006), 200–2.
235. M. Lin, D. A. Jayawickrama, R. A. Rose, J. A. DelViscio, and C. K. Larive, Nuclear Magnetic Resonance Spectroscopic Analysis of the Selective Complexation of the *cis* and *trans* Isomers of Phenylalanylproline by β-Cyclodextrin. *Anal. Chim. Acta* **307** (1995), 449–57.
236. F. Rinaldi, M. Lin, M. J. Shapiro, and M. Petersheim, δ-Opiate DPDE in Magnetically Oriented Phospholipid Micelles: Binding and Arrangement of Aromatic Pharmacophores. *Biophys. J.* **73** (1997), 3337–48.
237. L. Orfi, M. Lin, and C. K. Larive, Measurement of SDS Micelle-Peptide Association Using ^{1}H NMR Chemical Shift Analysis and Pulsed-Field Gradient NMR Spectroscopy. *Anal. Chem.* **70** (1998), 1339–45.
238. W. S. Price, H. Ide, and Y. Arata, Self-Diffusion of Supercooled Water to 238 K Using PGSE NMR Diffusion Measurements. *J. Phys. Chem. A* **103** (1999), 448–50.
239. W. S. Price, H. Ide, Y. Arata, and O. Söderman, Temperature Dependence of the Self-Diffusion of Supercooled Heavy Water to 244 *K. J. Phys. Chem. B* **104** (2000), 5874–6.
240. K. Yoshida, C. Wakai, N. Matubayasi, and M. Nakahara, A New High-Temperature Multinuclear-Magnetic-Resonance Probe and the Self-Diffusion of Light and Heavy Water in Sub- and Supercritical Conditions. *J. Chem. Phys.* **123** (2005), 164506-1–164506-10.
241. A. Laaksonen and P. Stilbs, Molecular-Dynamics and NMR Study of Methane-Water Systems. *Molec. Phys.* **74** (1991), 747–64.

242. J. G. Klenø, M. W. Kristiansen, C. J. Nielsen, E. J. Pedersen, L. R. William, and T. Pedersen, Diffusion Coefficients in Cold Sulfuric Acid. *J. Phys. Chem. A* **105** (2001), 8440–4.
243. Th. Dippel and K. D. Kreuer, Proton Transport Mechanism in Concentrated Aqueous Solutions and Solid Hydrates of Acids. *Solid State Ionics* **46** (1991), 3–9.
244. C. Malveau, B. Diter, F. Humbert, and D. Canet, Self-Diffusion Measurements by Carbon-13 NMR Using Radiofrequency Field Gradients. *J. Magn. Reson.* **130** (1998), 131–4.
245. J. S. Murday and R. M. Cotts, Self-Diffusion Coefficient of Liquid Lithium. *J. Chem. Phys.* **48** (1968), 4938–45.
246. J. S. Murday and R. M. Cotts, Self-Diffusion in Molten Lithium. *Z. Naturforsch.* **26a** (1971), 85–93.
247. G. J. Krüger, W. Müller-Warmuth, and A. Klemm, Self-Diffusion of Liquid ^{6}Li and ^{7}Li as Measured by Nuclear Magnetic Resonance. *Z. Naturforsch.* **26a** (1971), 94–8.
248. D. W. McCall and D. C. Douglass, Diffusion in Binary Systems. *J. Phys. Chem.* **71** (1967), 987–97.
249. A. C. Jacob and M. D. Zeidler, Self-Diffusion in Binary Mixtures: The System Benzene/Cyclohexane at High Pressures. *Phys. Chem. Chem. Phys.* **5** (2003), 538–42.
250. T. Groβ, L. Chen, J. Buchhauser, and H.-D. Lüdemann, *T,p* Dependence of Intradiffusion in Binary Fluid Mixtures with Ammonia as One Component. *Phys. Chem. Chem. Phys* **3** (2001), 3701–6.
251. M. M. Britton, Nuclear Magnetic Resonance Studies of the 1,4-Cyclohexanedione-Bromate-Acid Oscillatory System. *J. Phys. Chem. A* **107** (2003), 5033–41.
252. J. Wuttke, I. Chang, O. G. Randl, F. Fujara, and W. Petry, Tagged-Particle Motion in Viscous Glycerol: Diffusion–Relaxation Crossover. *Phys. Rev. E* **54** (1996), 5364–9.
253. A. Vergara, L. Paduano, G. D'Errico, and R. Sartorio, Network Formation in Polyethyleneglycol Solutions. An Intradiffusion Study. *Phys. Chem. Chem. Phys.* **1** (1999), 4875–9.
254. T. Groβ, J. Buchhauser, and H.-D. Lüdemann, Self-Diffusion in Fluid Carbon Dioxide at High Pressures. *J. Chem. Phys.* **109** (1998), 4518–22.
255. T. Groβ, L. Chen, J. Buchhauser, and H.-D. Lüdemann, *T,p* Dependence of Intradiffusion in Binary Fluid Mixtures with Carbon Dioxide as One Component. *Phys. Chem. Chem. Phys.* **3** (2001), 2845–51.
256. P. Etesse, W. G. Chapman, and R. Kobayashi, Nuclear Magnetic Resonance Measurement of Spin-Lattice Relaxation and Self-Diffusion in Supercritical CO_2-*n*-hexadecane Mixtures. *Mol. Phys.* **80** (1993), 1145–64.
257. M. Holz, R. Haselmeier, R. K. Mazitov, and H. Weingärtner, Self-Diffusion of Neon in Water by ^{21}Ne NMR. *J. Am. Chem. Soc.* **116** (1994), 801–2.
258. E. G. Smith, J. W. Rockliffe, and P. I. Riley, Fourier Transform Pulsed Field Gradient and 1D NMR Imaging Studies of Molecular and Ionic Diffusion in Aqueous Dispersions. *J. Colloid Interface Sci.* **131** (1989), 29–37.
259. J. L. Lutz and G. D. Mendenhall, Diffusion Coefficients by NMR-Spin Echo Methods for the Systems Water-Ammonium Chloride, Water-Succinonitrile, and Acetone-Succinonitrile. *J. Cryst. Growth* **217** (2000), 183–8.
260. H. Mo and T. C. Pochapsky, Self-Diffusion Coefficients of Paired Ions. *J. Phys. Chem. B* **101** (1997), 4485–6.
261. C. K. Larive, M. Lin, B. S. Kinnear, B. J. Piersma, C. E. Keller, and W. R. Carper, ^{13}C and ^{27}Al NMR Relaxation, Viscosity, and ^{1}H Diffusion Studies of an Ethylaluminum Dichloride Melt. *J. Phys. Chem. B* **102** (1998), 1717–23.

262. Y. Aihara, A. Sonai, M. Hattori, and K. Hayamizu, Ion Conduction Mechanisms and Thermal Properties of Hydrated and Anhydrous Phosphoric Acids Studied with ^{1}H, ^{2}H, and ^{31}P NMR. *J. Phys. Chem. B* **110** (2006), 24999–5006.
263. W. S. Price and O. Söderman, Self-Diffusion Coefficients of Some Hydrocarbons in Water: Measurements and Scaling Relations. *J. Phys. Chem. A* **104** (2000), 5892–4.
264. E. von Meerwall, S. Beckman, J. Jang, and W. L. Mattice, Diffusion of Liquid *n*-Alkanes: Free-Volume and Density Effects. *J. Chem. Phys.* **108** (1998), 4299–304.
265. L. Chen, T. Groβ, H.-D. Lüdemann, H. Krienke, and R. Fischer, First Observation of Different Diffusion Coefficients for Two Conformers in a Neat Liquid. *Naturwissenschaften* **87** (2000), 225–8.
266. J. Buchhauser, T. Groβ, N. Karger, and H.-D. Lüdemann, Self-Diffusion in CD_4 and ND_3: With Notes on the Dynamic Isotope Effect in Liquids. *J. Chem. Phys.* **110** (1999), 3037–42.
267. J. A. Davies and P. C. Griffiths, A Phenomenological Approach to Separating the Effects of Obstruction and Binding for the Diffusion of Small Molecules in Polymer Solutions. *Macromolecules* **36** (2003), 950–2.
268. R. Haselmeier, M. Holz, M. M. Kappes, R. H. Michel, and D. Fuchs, Translational Diffusion in C_{60} and C_{70} Fullerene Solutions. *Ber. Bunsenges. Phys. Chem.* **98** (1994), 878–81.
269. T. Kato, K. Kikuchi, and Y. Achiba, Measurement of the Self-Diffusion Coefficient of C_{60} in Benzene-d_6 Using ^{13}C Pulsed-Gradient Spin-Echo. *J. Phys. Chem.* **97** (1993), 10251–3.
270. T. Feiweier, O. Isfort, B. Geil, F. Fujara, and H. Weingärtner, Decoupling of Lithium and Proton Self-Diffusion in Supercooled LiCl:$7H_2O$: A Nuclear Magnetic Resonance Study in Ultrahigh Magnetic Field Gradients. *J. Chem. Phys.* **105** (1996), 5737–44.
271. C. Monteiro and C. H. du Penhoat, Translational Diffusion of Dilute Aqueous Solutions of Sugars as Probed by NMR and Hydrodynamic Theory. *J. Phys. Chem. A* **105** (2001), 9827–33.
272. A. Masuda, K. Ushida, H. Koshida, K. Yamashita, and T. Kluge, Novel Distance Dependence of Diffusion Constants in Hyaluronan Aqueous Solution Resulting from Its Characteristic Nano-Microstructure. *J. Am. Chem. Soc.* **123** (2001), 11468–71.
273. M. Rampp, C. Buttersack, and H.-D. Lüdemann, Self-Diffusion of Sucrose in Molasses. *Ind. Eng. Chem. Res.* **39** (2000), 4400–07.
274. T. Mahawanich and S. J. Schmidt, Molecular Mobility and the Perceived Sweetness of Sucrose, Fructose, and Glucose Solutions. *Food Chem.* **84** (2003), 169–79.
275. C. Branca, S. Magazù, G. Maisano, P. Migliardo, and E. Tettamanti, Anomalous Translational Diffusive Processes in Hydrogen-Bonded Systems Investigated by Ultrasonic Technique, Raman Scattering and NMR. *Physica B* **291** (2000), 180–9.
276. N. Ekdawi-Sever, J. J. de Pablo, E. J. Feick, and E. von Meerwall, Diffusion of Sucrose and α,α-Trehalose in Aqueous Solutions. *J. Phys. Chem. A* **107** (2003), 936–43.
277. G. Majer, W. Renz, A. Seeger, R. G. Barnes, J. Shinar, and A. V. Skripov, Pulsed-Field-Gradient Nuclear Magnetic Resonance Studies of Hydrogen Diffusion in Laves-Phase Hydrides. *J. Alloys Compd.* **231** (1995), 220–5.
278. P. S. Pregosin, E. Martínez-Viviente, and P. G. A. Kumar, Diffusion and NOE NMR Spectroscopy. Applications to Problems Related to Coordination Chemistry and Homogeneous catalysis. *J. Chem. Soc., Dalton Trans.* (2003), 4007–14.
279. D. Nama, A. P. G. Kumar, and P. S. Pregosin, ^{195}Pt, ^{1}H and ^{31}P PGSE Diffusion Studies on Platinum Complexes. *Magn. Reson. Chem.* **43** (2005), 246–50.
280. I. Keresztes and P. G. Williard, Diffusion-Ordered NMR Spectroscopy (DOSY) of THF Solvated *n*-Butyllithium Aggregates. *J. Am. Chem. Soc.* **122** (2000), 10228–9.

281. J. W. Beves, E. C. Constable, C. E. Housecroft, M. Neuberger, S. Schaffner, and E. J. Shardlow, [$n + n$]-Heterometallomacrocyclic Complexes ($n \geqslant 2$) Prepared From Platinum(II)-Centred Ditopic 2,2′:6′,2″-Terpyridine Ligands: Dimensional Cataloguing by Pulsed-Field Gradient Spin-Echo NMR Spectroscopy. *J. Chem. Soc., Dalton Trans.* (2007), 1593–1602.
282. C. Zuccaccia, G. Bellachioma, G. Cardaci, and A. Macchioni, Self-Diffusion Coefficients of Transition-Metal Complex Ions, Ion Pairs, and Higher Aggregates by Pulsed Field Gradient Spin-Echo NMR Measurements. *Organometallics* **19** (2000), 4663–5.
283. X. Xie, C. Auel, W. Henze, and R. M. Gschwind, Dimethyl- and Bis[(trimethylsilyl) Methyl]cuprates Show Aggregates Higher than Dimers in Diethyl Ether: Molecular Diffusion Studies by PFG NMR and Aggregation-Reactivity Correlations. *J. Am. Chem. Soc.* **125** (2003), 1595–601.
284. P. Callaghan, K. W. Jolley, and R. S. Humphrey, Diffusion of Fat and Water in Cheese as Studied by Pulsed Field Gradient Nuclear Magnetic Resonance. *J. Colloid Interface Sci.* **93** (1983), 521–9.
285. T. Berg, P. Arlt, R. Brummer, D. Emeis, W.-M. Kulicke, S. Wiesner, and K.-P. Wittern, Insights into the Structure and Dynamics of Complex W/O-Emulsions by Combining NMR, Rheology and Electron Microscopy. *Colloids Surf. A* **238** (2004), 59–69.
286. J. P. Hindmarsh, K. G. Hollingsworth, D. I. Wilson, and M. L. Johns, An NMR Study of the Freezing of Emulsion-Containing Drops. *J. Colloid Interface Sci.* **275** (2004), 165–71.
287. B. Balinov, O. Söderman, and J. C. Ravey, Diffraction-like Effects Observed in the PGSE Experiment When Applied to a Highly Concentrated Water/Oil Emulsion. *J. Phys. Chem.* **98** (1994), 393–5.
288. I. Lönnqvist, B. Håkansson, B. Balinov, and O. Söderman, NMR Self-Diffusion Studies of the Water and the Oil Components in a W/O/W Emulsion. *J. Colloid Interface Sci.* **192** (1997), 66–73.
289. J. P. Hindmarsh, J. Su, J. Flanagan, and H. Singh, PFG-NMR Analysis of Intercompartment Exchange and Inner Droplet Size Distribution of W/O/W Emulsions. *Langmuir* **21** (2005), 9076–84.
290. D. Topgaard, C. Malmborg, and O. Söderman, Restricted Self-Diffusion of Water in a Highly Concentrated W/O Emulsion Studied Using Modulated Gradient Spin-Echo NMR. *J. Magn. Reson.* **156** (2002), 195–201.
291. P. J. McDonald, E. Ciampi, J. L. Keddie, M. Heidenreich, and R. Kimmich, Magnetic-Resonance Determination of the Spatial Dependence of the Droplet Size Distribution in the Cream Layer of Oil-in-Water Emulsions: Evidence for the Effects of Depletion Flocculation. *Phys. Rev. E* **59** (1999), 874–84.
292. K. G. Hollingsworth and M. L. Johns, Spatially Resolved Emulsion Droplet Sizing Using Inverse Abel Transforms. *J. Magn. Reson.* **176** (2005), 71–8.
293. P. Stilbs and B. Lindman, Determination of Organic Counterion Binding to Micelles through Fourier Transform NMR Self-Diffusion Measurements. *J. Phys. Chem.* **85** (1981), 2587–9.
294. O. Annunziata, L. Costantino, G. D'Errico, L. Paduano, and V. Vitagliano, Transport Properties for Aqueous Sodium Sulfonate Surfactants 2. Intradiffusion Measurements: Influence of the Obstruction Effect on the Monomer and Micelle Mobilities. *J. Colloid Interface Sci.* **216** (1999), 16–24.
295. K. Zhang, M. Jonströmer, and B. Lindman, Interaction Between Nonionic Micelles and a Nonionic Polymer Studied by Fourier Transform NMR Self-Diffusion. *J. Phys. Chem.* **98** (1994), 2459–63.

296. B. Xu, G. W. Lynn, Ji. Guo, Y. B. Melnichenko, G. D. Wignall, J. B. McClain, J. M. DeSimone, and C. S. Johnson, Jr., NMR and SANS Studies of Aggregation and Microemulsion Formation by Phosphorus Fluorosurfactants in Liquid and Supercritical Carbon Dioxide. *J. Phys. Chem. B* **109** (2005), 10261–9.
297. H. Dozol and C. Berthon, Characterisation of the Supramolecular Structure of Malonamides by Application of Pulsed Field Gradients in NMR Spectroscopy. *Phys. Chem. Chem. Phys.* **9** (2007), 5162–70.
298. C. Malmborg, D. Topgaard, and O. Söderman, Diffusion in An Inhomogeneous System: NMR Studies of Diffusion in Highly Concentrated Emulsions. *J. Colloid Interface Sci.* **263** (2003), 270–6.
299. J. J. Chou, J. L. Baber, and Ad. Bax, Characterization of Phospholipid Mixed Micelles by Translational Diffusion. *J. Biomol. NMR* **29** (2004), 299–303.
300. R. Angelico, U. Olsson, G. Palazzo, and A. Ceglie, Surfactant Curvilinear Diffusion in Giant Wormlike Micelles. *Phys. Rev. Lett.* **81** (1998), 2823–6.
301. A. Belmajdoub, D. Boudot, C. Tondre, and D. Canet, NMR Self-Diffusion Measurements Using a Radio-Frequency Field Gradient Combined with Water Signal Suppression. Application to the pH-Dependent Solubilization of Hydroxyquinoline in SDS Micellar Solutions. *Chem. Phys. Lett.* **150** (1988), 194–8.
302. L. Coppola, C. La Mesa, G. A. Ranieri, and M. Terenzi, Analysis of Water Self-Diffusion in Polycrystalline Lamellar Systems by Pulsed Field Gradient Nuclear Magnetic Resonance Experiments. *J. Chem. Phys.* **98** (1993), 5087–90.
303. G. Celebre, L. Coppola, G. A. Ranieri, and M. Terenzi, Analysis of the PFG-SE NMR Experiments in Lyotropic Mesophases: The Hexagonal Case. *Mol. Cryst. Liquid Cryst.* **238** (1994), 117–23.
304. V. Domenici, M. Geppi, and C. A. Veracini, NMR in Chiral and Achiral Smectic Phases: Structure, Orientational Order and Dynamics. *Prog. NMR Spectrosc.* **50** (2007), 1–50.
305. L. Coppola, C. La Mesa, G. A. Ranieri, and M. Terenzi, Water Self-Diffusion in Micellar Solutions: PFG-NMR Study. *Ann. Chim.* **80** (1990), 271–81.
306. M. Cifelli, J. Saunavaara, J. Jokisaari, and C. A. Veracini, ^{129}Xe Nuclear Shielding and Diffusion in the A and C* Phases of a Chiral Smectogen. *J. Phys. Chem. A* **108** (2006), 3973–9.
307. M. Cifelli, V. Domenici, and C. A. Veracini, From the Synclinic to the Anticlinic Smectic Phases: a Deuterium NMR and Diffusion NMR Study. *Mol. Cryst. Liquid Cryst.* **429** (2005), 167–79.
308. J. Ruohonen, M. Ylihautala, and J. Jokisaari, ^{129}Xe diffusion in a Ferroelectric Liquid Crystal. *Mol. Phys.* **99** (2001), 711–19.
309. J. Ruohonen and J. Jokisaari, ^{129}Xe NMR Shielding and Self-Diffusion in the Mixture of Two Thermotropic Nematogens with Opposite Diamagnetic Anisotropy. *Phys. Chem. Chem. Phys.* **3** (2001), 3208–12.
310. G. Fleischer, Micellization in Aqueous Solution of a Poly(ethylene oxide)/Poly (propylene oxide)/Poly(ethylene oxide) Triblock Copolymer Investigated with Pulsed Field Gradient NMR. *J. Phys. Chem.* **97** (1993), 517–21.
311. M. Nilsson, B. Håkansson, O. Söderman, and D. Topgaard, Influence of Polydispersity on the Micellization of Triblock Copolymers Investigated by Pulsed Field Gradient Nuclear Magnetic Resonance. *Macromolecules* **40** (2007), 8250–8.
312. L. Ambrosone, R. Angelico, A. Ceglie, U. Olsson, and G. Palazzo, Molecular Diffusion in a Living Network. *Langmuir* **17** (2001), 6822–30.
313. M. H. Blees, J. M. Geurts, and J. C. Leyte, Self-Diffusion of Charged Polybutadiene Latex Particles in Water Measured by Pulsed Field Gradient NMR. *Langmuir* **12** (1996), 1947–57.

314. J. M. Roberts, H. Sierzputowska-Gracz, E. O. Stejskal, and J. G. Osteryoung, Determination of Li^+ Self-Diffusion Coefficients in an Aqueous Suspension of Sulfonated Polystyrene Latex by Pulsed-Field-Gradient, Spin-Echo NMR. *J. Phys. Chem. B* **102** (1998), 7735–9.
315. F. J. M. Schipper and J. C. Leyte, Mass Transport in Polyelectrolyte Solutions. *J. Phys. Condens. Matter* **11** (1999), 1409–21.
316. U. Böhme and U. Scheler, Effective Size and Fractal Dimension of Polyelectrolytes Determined by Diffusion NMR. *Macromol. Symp.* **184** (2002), 349–56.
317. V. A. Daragan and E. E. Il'ina, Pulsed Field Gradient NMR for the Study of the Structure of Membrane Systems. *Chem. Phys.* **158** (1991), 105–11.
318. V. I. Volkov, S. A. Korotchkova, H. Ohya, and Q. Guo, Self-Diffusion of Water-Ethanol Mixtures in Polyacrylic Acid-Polysulfone Composite Membranes Obtained by Pulsed-Field Gradient Nuclear Magnetic Resonance Spectroscopy. *J. Memb. Sci.* **100** (1995), 273–86.
319. V. I. Volkov, S. A. Korotchkova, I. A. Nesterov, H. Ohya, Q. Guo, J. Huang, and J. Chen, The Self-Diffusion of Water and Ethanol in Cellulose Derivative Membranes and Particles with the Pulsed Field Gradient NMR Data. *J. Memb. Sci.* **110** (1996), 1–11.
320. V. I. Volkov, Yu. M. Popkov, S. F. Timashev, D. G. Bessarabov, R. D. Sanderson, and Z. Twardowski, Self-Diffusion of Water and Fluorine Ions in Anion-Exchange Polymeric Materials (Membranes and Resin) as Determined by Pulsed-Field Gradient Nuclear Magnetic Resonance Spectroscopy. *J. Memb. Sci.* **180** (2000), 1–13.
321. T. Cosgrove and P. C. Griffiths, The Critical Overlap Concentration Measured by Pulsed Field Gradient Nuclear Magnetic Resonance Techniques. *Polymer* **35** (1994), 509–13.
322. P. T. Callaghan and A. Coy, Evidence for Reptational Motion and the Entanglement Tube in Semidilute Polymer Solutions. *Phys. Rev. Lett.* **68** (1992), 3176–9.
323. D. W. McCall, D. C. Douglass, and E. W. Anderson, Diffusion in Ethylene Polymers. IV. *J. Chem. Phys.* **30** (1959), 771–3.
324. M. Appel, G. Fleischer, J. Kärger, F. Fujara, and I. Chang, Anomalous Segment Diffusion in Polymer Melts. *Macromolecules* **27** (1994), 4274–7.
325. S. Z. D. Cheng, J. S. Barley, and E. D. von Meerwall, Self-Diffusion of Poly(ethylene oxide) Fractions and Its Influence on the Crystalline Texture. *J. Polym. Sci. B* **29** (1991), 515–25.
326. T. Cosgrove, M. J. Turner, P. C. Griffiths, J. Hollingshurst, M. J. Shenton, and J. A. Semlyen, Self-Diffusion and Spin-Spin Relaxation in Blends of Linear and Cyclic Polydimethylsiloxane Melts. *Polymer* **37** (1996), 1535–40.
327. E. Fischer, R. Kimmich, and N. Fatkullin, NMR Field Gradient Diffusometry of Segment Displacements in Melts of Entangled Polymers. *J. Chem. Phys.* **104** (1996), 9174–8.
328. M. E. Komlosh and P. T. Callaghan, Segmental Motion of Entangled Random Coil Polymers Studied by Pulsed Gradient Spin Echo Nuclear Magnetic Resonance. *J. Chem. Phys.* **109** (1998), 10053–67.
329. J. B. Cain, K. Zhang, D. E. Betts, J. M. DeSimone, and C. S. Johnson, Jr., Diffusion of Block Copolymers in Liquid CO_2: Evidence of Self-Assembly from Pulsed Field Gradient NMR. *J. Am. Chem. Soc.* **120** (1998), 9390–1.
330. T. Zhao and H. W. Beckham, Direct Synthesis of Cyclodextrin-Rotaxanated Poly (ethylene glycol)s and Their Self-Diffusion Behavior in Dilute Solution. *Macromolecules* **36** (2003), 9859–65.
331. G. Fleischer, H. Sillescu, and V. D. Skirda, Molecular Motion in Concentrated Solutions of Spherical Polystyrene Microgels Studied with the Pulsed Field Gradient NMR. *Polymer* **35** (1994), 1936–41.

332. O. Rosén, M. Boström, M. Nydén, and L. Piculell, Anomalous Surfactant Diffusion in a Gel of Chemically Cross-Linked Ethyl(hydroxyethyl) Cellulose. *J. Phys. Chem. B* **107** (2003), 4074–9.
333. Y. E. Shapiro, Compartmentation in the Poly(vinyl alcohol) Cryogels. ^{1}H NMR Self-Diffusion Study. *Colloids Surf. A* **164** (2000), 71–83.
334. P. McConville and J. M. Pope, A Comparison of Water Binding and Mobility in Contact Lens Hydrogels from NMR Measurements of the Water Self-Diffusion Coefficient. *Polymer* **41** (2000), 9081–8.
335. H. F. Azurmendi and M. E. Ramia, Anomalous Diffusion of Water in a Hydrogel of Sucrose and Diepoxide Monomers. *J. Chem. Phys.* **114** (2001), 9657–62.
336. V. D. Skirda, I. Yu. Aslanyan, O. E. Philippova, N. S. Karybiants, and A. R. Khokhlov, Investigation of Translational Motion of Poly(ethylene glycol) Macromolecules in Poly(methacrylic acid) Hydrogels. *Macromol. Chem. Phys.* **200** (1999), 2152–9.
337. B. Corain and M. Kralik, Generating Palladium Nanoclusters Inside Functional Cross-Linked Polymer Frameworks. *J. Mol. Catal. A* **173** (2001), 99–115.
338. M. Ohuchi, P. Meadows, H. Horiuchi, Y. Sakai, and K. Furihata, Dynamics of Sodium and Lithium Counter-Ions and Water Molecules in Cation-Exchange Resins as Shown by NMR Spectroscopy. *Polym. J.* **32** (2000), 760–70.
339. H. Walderhaug and Bo. Nyström, Anomalous Diffusion in an Aqueous System of a Poly(ethylene oxide)-Poly(propylene oxide)-Poly(ethylene oxide) Triblock Copolymer during Gelation Studied by Pulsed Field Gradient NMR. *J. Phys. Chem. B* **101** (1997), 1524–8.
340. G. Meresi, Y. Wang, J. Cardoza, W. Y. Wen, A. A. Jones, J. Gosselin, D. Azar, and P. T. Inglefield, Pulse Field Gradient NMR Study of Diffusion of Pentane in Amorphous Glassy Perfluorodioxole. *Macromolecules* **34** (2001), 4852–6.
341. W. S. Price, B. E. Chapman, and P. W. Kuchel, Correlation of Viscosity and Conductance with $^{23}Na^+$ NMR T_1 Measurements. *Bull. Chem. Soc. Jpn.* **63** (1990), 2961–5.
342. K. Hayamizu, Y. Aihara, and W. S. Price, Correlating the NMR Self-Diffusion and Relaxation Measurements with Ionic Conductivity in Polymer Electrolytes Composed of Cross-Linked Poly(ethylene oxide-propylene oxide) Doped with $LiN(SO_2CF_3)_2$. *J. Chem. Phys.* **113** (2000), 4785–93.
343. D. Golodnitsky, E. Livshits, A. Ulus, Z. Barkay, I. Lapides, E. Peled, S. H. Chung, and S. Greenbaum, Fast Ion Transport in Oriented Semicrystalline LiI-P(EO)$_n$-Based Polymer Electrolytes. *J. Phys. Chem. A* **105** (2001), 10098–106.
344. W. S. Price, Y. Aihara, and K. Hayamizu, NMR Studies of Nanoscale Organization and Dynamics in Polymer Electrolytes. *Aust. J. Chem.* **57** (2004), 1185–90.
345. C. B. Gorman, J. C. Smith, M. W. Hager, B. L. Parkhurst, H. Sierzputowska-Gracz, and C. A. Haney, Molecular Structure-Property Relationships for Electron-Transfer Rate Attenuation in Redox-Active Core Dendrimers. *J. Am. Chem. Soc.* **121** (1999), 9958–66.
346. G. R. Newkome, J. K. Young, G. R. Baker, R. L. Potter, L. Audoly, D. Cooper, C. D. Weis, K. Morris, and C. S. Johnson, Jr., Cascade Polymers. pH Dependence of Hydrodynamic Radii of Acid-Terminated Dendrimers. *Macromolecules* **26** (1993), 2394–6.
347. J. K. Young, G. R. Baker, G. R. Newkome, K. F. Morris, and C. S. Johnson, Jr., 'Smart' Cascade Polymers. Modular Syntheses of Four-Directional Dendritic Macromolecules with Acidic, Neutral, or Basic Terminal Groups and the Effect of pH Changes on Their Hydrodynamic Radii. *Macromolecules* **27** (1994), 3464–71.
348. M. Valentini, A. Vaccaro, A. Rehor, A. Napoli, J. A. Hubbell, and N. Tirelli, Diffusion NMR Spectroscopy for the Characterization of the Size and Interactions of Colloidal

Matter: The Case of Vesicles and Nanoparticles. *J. Am. Chem. Soc.* **126** (2004), 2142–7.
349. W. E. Baille, C. Malveau, X. X. Zhu, Y. H. Kim, and W. T. Ford, Self-Diffusion of Hydrophilic Poly(propyleneimine) Dendrimers in Poly(vinyl alcohol) Solutions and Gels by Pulsed Field Gradient NMR Spectroscopy. *Macromolecules* **36** (2003), 839–47.
350. A. T. ten Cate, P. Y. W. Dankers, H. Kooijman, A. L. Spek, R. P. Sijbesma, and E. W. Meijer, Enantioselective Cyclization of Racemic Supramolecular Polymers. *J. Am. Chem. Soc.* **125** (2003), 6860–1.
351. W. S. Price, F. Hallberg, and P. Stilbs, A PGSE Diffusion and Electrophoretic NMR Study of Cs^+ and Na^+ Dynamics in Aqueous Crown Ether Systems. *Magn. Reson. Chem.* **45** (2007), 152–6.
352. A. Gafni and Y. Cohen, Complexes of Macrocycles with γ-Cyclodextrin as Deduced from NMR Diffusion Measurements. *J. Org. Chem.* **62** (1997), 120–5.
353. L. Frish, S. E. Matthews, V. Böhmer, and Y. Cohen, A Pulsed Gradient Spin Echo NMR Study of Guest Encapsulation by Hydrogen-Bonded Tetraurea Calix[4]arene dimers. *J. Chem. Soc., Perkin Trans. 2* **2** (1999), 669–71.
354. K. S. Cameron and L. Fielding, NMR Diffusion Spectroscopy as a Measure of Host–Guest Complex Association Constants and as a Probe of Complex Size. *J. Org. Chem.* **66** (2001), 6891–5.
355. T. Evan-Salem, I. Baruch, L. Avram, Y. Cohen, L. C. Palmer, and J. Rebek, Jr., Resorcinarenes are Hexameric Capsules in Solution. *Proc. Natl. Acad. Sci. U.S.A.* **103** (2006), 12296–300.
356. F. Stallmach, J. Kärger, C. Krause, M. Jeschke, and U. Oberhagemann, Evidence of Anisotropic Self-Diffusion of Guest Molecules in Nanoporous Materials of MCM-41 Type. *J. Am. Chem. Soc.* **122** (2000), 9237–42.
357. F. Stallmach, A. Gräser, J. Kärger, C. Krause, M. Jeschke, U. Oberhagemann, and S. Spange, Pulsed Field Gradient NMR Studies of Diffusion in MCM-41 Mesoporous Solids. *Microporous Mesoporous Mater.* **44–45** (2001), 745–53.
358. R. Valiullin, M. Dvoyashkin, P. Kortunov, C. Krause, and J. Kärger, Diffusion of Guest Molecules in MCM-41 Agglomerates. *J. Chem. Phys.* **126** (2007), 054705-1–054705-6.
359. V. Kukla, J. Kornatowski, D. Demuth, I. Girnus, H. Pfeifer, L. V. C. Rees, S. Schunk, K. K. Unger, and J. Kärger, NMR Studies of Single-File Diffusion in Unidimensional Channel Zeolites. *Science* **272** (1996), 702–4.
360. T. Meersmann, J. W. Logan, R. Simonutti, S. Caldarelli, A. Comotti, P. Sozzani, L. G. Kaiser, and A. Pines, Exploring Single-File Diffusion in One-Dimensional Nanochannels by Laser-Polarized ^{129}Xe NMR Spectroscopy. *J. Phys. Chem. A* **104** (2000), 11665–70.
361. L. Gjerdåker, G. H. Sørland, and D. W. Aksnes, Application of the Short Diffusion Time Model to Diffusion Measurements by NMR in Microporous Crystallites. *Microporous Mesoporous Mater.* **32** (1999), 305–10.
362. A. Pampel, M. Fernandez, D. Freude, and J. Kärger, New Options for Measuring Molecular Diffusion in Zeolites by MAS PFG NMR. *Chem. Phys.* **407** (2005), 53–7.
363. B. Geil, O. Isfort, B. Boddenberg, D. E. Favre, B. F. Chmelka, and F. Fujara, Reorientational and Translational Dynamics of Benzene in Zeolite NaY as Studied by One- and Two-Dimensional Exhange Spectroscopy and Static-Field-Gradient Nuclear Magnetic Resonance. *J. Chem. Phys.* **116** (2002), 2184–93.
364. P. Kortunov, S. Vasenkov, J. Kärger, M. Fé Elía, M. Perez, M. Stöcker, G. K. Papadopoulos, D. Theodorou, B. Dresecher, G. McElhiney, B. Bernauer,

V. Krystl, M. Kocirik, A. Zikanova, H. Jirglova, C. Berger, R. Gläser, J. Weitkamp, and E. W. Hansen, Pulsed-Field Gradient Nuclear Magnetic Resonance Study of Transport Properties of Fluid Catalytic Cracking Catalysts. *Magn. Reson. Imaging* **23** (2005), 233–7.
365. S. P. Rigby and L. F. Gladden, The Use of Magnetic Resonance Images in the Simulation of Diffusion in Porous Catalyst Support Pellets. *J. Catal.* **173** (1998), 484–9.
366. M. Flaum, G. J. Hirosaki, C. Flaum, and C. Straley, Measuring Pore Connectivity by Pulsed Field Gradient Diffusion Editing with Hydrocarbon. *Magn. Reson. Imaging* **23** (2005), 337–9.
367. P. T. Callaghan, C. D. Eccles, T. G. Haskell, P. J. Langhorne, and J. D. Seymour, Earth's Field NMR in Antarctica: A Pulsed Gradient Spin Echo NMR Study of Restricted Diffusion in Sea Ice. *J. Magn. Reson.* **133** (1998), 148–54.
368. D. Vargas-Florencia, T. Edvinsson, A. Hagfeldt, and I. Furó, Pores in Nanostructured TiO_2 Films. Size Distribution and Pore Permeability. *J. Phys. Chem. C* **111** (2007), 7605–11.
369. G. H. Sørland, Short-Time PFGSTE Diffusion Measurements. *J. Magn. Reson.* **126** (1997), 146–8.
370. E. J. Fordham, P. P. Mitra, and L. L. Latour, Effective Diffusion Times in Multiple-Pulse PFG Diffusion Measurements in Porous Media. *J. Magn. Reson. A* **121** (1996), 187–92.
371. L. L. Latour, P. P. Mitra, R. L. Kleinberg, and C. H. Sotak, Time-Dependent Diffusion Coefficient of Fluids in Porous Media as a Probe of Surface-to-Volume Ratio. *J. Magn. Reson. A* **101** (1993), 342–6.
372. M. D. Hürlimann, K. G. Helmer, L. L. Latour, and C. H. Sotak, Restricted Diffusion in Sedimentary Rocks. Determination of Surface-Area-to-Volume Ratio and Surface Relaxivity. *J. Magn. Reson. A* **111** (1994), 169–78.
373. P. P. Mitra, L. L. Latour, R. L. Kleinberg, and C. H. Sotak, Pulsed-Field-Gradient NMR Measurements of Restricted Diffusion and the Return-to-the-Origin Probability. *J. Magn. Reson. A* **114** (1995), 47–58.
374. S. Vasenkov, P. Galvosas, O. Geier, N. Nestle, F. Stallmach, and J. Kärger, Determination of Genuine Diffusivities in Heterogeneous Media Using Stimulated Echo Pulsed Field Gradient NMR. *J. Magn. Reson.* **149** (2001), 228–33.
375. N. Nestle, P. Galvosas, O. Geier, C. Zimmerman, M. Dakkouri, and J. Kärger, Nuclear Magnetic Resonance Study of Diffusion and Relaxation in Hydrating White Cement Pastes of Different Water Content. *J. Appl. Phys.* **89** (2001), 8061–5.
376. J. Petković, H. P. Huinink, L. Pel, and K. Kopinga, Diffusion in Porous Building Materials with High Internal Magnetic Field Gradients. *J. Magn. Reson.* **167** (2004), 97–106.
377. F. Stallmach, C. Vogt, J. Kärger, K. Helbig, and F. Jacobs, Fractal Geometry of Surface Areas of Sand Grains Probed by Pulsed Field Gradient NMR. *Phys. Rev. Lett.* **88** (2002), 105505-1–105505-4.
378. D. Candela and P.-Z. Wong, Using NMR to Measure Fractal Dimensions. *Phys. Rev. Lett.* **90** (2003), 039601-1.
379. A. R. Mutina and V. D. Skirda, Porous Media Characterization by PFG and IMFG NMR. *J. Magn. Reson.* **188** (2007), 122–8.
380. L. Pel, H. Huinink, and K. Kopinga, Salt Transport and Crystallization in Porous Building Materials. *Magn. Reson. Imaging* **21** (2003), 317–20.
381. E. J. Fordham, S. J. Gibbs, and L. D. Hall, Partially Restricted Diffusion in a Permeable Sandstone: Observations by Stimulated Echo PFG NMR. *Magn. Reson. Imaging* **12** (1994), 279–84.

382. A. V. Filippov, E. V. Khosina, and V. G. Khosin, Liquid Self-Diffusion in Pores of Hardened Gypsum: Pulsed Field Gradient NMR Study. *J. Mater. Sci.* **31** (1996), 1809–14.
383. U. M. Scheven, Stray Field Measurements of Flow Displacement Distributions Without Pulsed Field Gradients. *J. Magn. Reson.* **174** (2005), 338–42.
384. T. Takahashi, T. Ohkubo, and Y. Ikeda, Montmorillonite Alignment Induced by Magnetic Field: Evidence Based on the Diffusion Anisotropy of Water Molecules. *J. Colloid Interface Sci.* **299** (2006), 198–203.
385. U. Tallarek, T. W. J Scheenen, and H. Van As, Macroscopic Heterogeneities in Electroosmotic and Pressure-Driven Flow through Fixed Beds at Low Column-to-Particle Diameter Ratio. *J. Phys. Chem. B* **105** (2001), 8591–9.
386. U. Tallarek, F. J. Vergeldt, and H. Van As, Stagnant Mobile Phases Mass Transfer in Chromatographic Media: Intraparticle Diffusion and Exchange Kinetics. *J. Phys. Chem. B* **103** (1999), 7654–64.
387. U. Tallarek, E. Bayer, and G. Guiochon, Study of Dispersion in Packed Chromatographic Columns by Pulsed Field Gradient Nuclear Magnetic Resonance. *J. Am. Chem. Soc.* **120** (1998), 1494–505.
388. S. Stapf, K. J. Packer, S. Békri, and P. M. Adler, Two-Dimensional Nuclear Magnetic Resonance Measurements and Numerical Simulations of Fluid Transport in Porous Rocks. *Phys. Fluids* **12** (2000), 566–80.
389. B. Manz, L. F. Gladden, and P. B. Warren, Flow and Dispersion in Porous Media: Lattice-Boltzmann and NMR Studies. *AIChE J.* **45** (1999), 1845–54.
390. B. Manz, P. Alexander, and L. F. Gladden, Correlations Between Dispersion and Structure in Porous Media Probed by Nuclear Magnetic Resonance. *Phys. Fluids* **11** (1999), 259–67.
391. U. Tallarek, E. Rapp, T. Scheenen, E. Bayer, and H. Van As, Electroosmotic and Pressure-Driven Flow in Open and Packed Capillaries: Velocity Distributions and Fluid Dispersion. *Anal. Chem.* **72** (2000), 2292–301.
392. E. J. Fernandez, C. A. Grotegut, G. W. Braun, K. J. Kirshner, J. R. Staudaher, M. L. Dickson, and V. L. Fernandez, The Effects of Permeability Heterogeneity on Miscible Viscous Fingering: A Three-Dimensional Magnetic Resonance Imaging Analysis. *Phys. Fluids* **7** (1995), 468–77.
393. J. L. Coffman, E. N. Lightfoot, and T. W. Root, Protein Diffusion in Porous Chromatographic Media Studied by Proton and Fluorine PFG-NMR. *J. Phys. Chem. B* **101** (1997), 2218–23.
394. R. S. Ehrlich and H. Y. Carr, Xenon Self-Diffusion Near the Critical Point and on the Liquid Branch of the Coexistence Curve. *Phys. Rev. Lett.* **25** (1970), 341–4.
395. R. W. Mair, D. G. Cory, S. Peled, C.-H. Tseng, S. Patz, and R. L. Walsworth, Pulsed-Field-Gradient Measurements of Time-Dependent Gas Diffusion. *J. Magn. Reson.* **135** (1998), 478–86.
396. G. R. Davies and T. K. Halstead, High-Resolution NMR of Low Pressure Laser-Polarized ^{129}Xe gas. *Chem. Phys. Lett.* **230** (1991), 237–42.
397. B. R. Patyal, J.-H. Gao, R. F. Williams, J. Roby, B. Saam, B. A. Rockwell, R. J. Thomas, D. J. Stolarski, and P. T. Fox, Longitudinal Relaxation and Diffusion Measurements Using Magnetic Resonance Signals from Laser-Hyperpolarized ^{129}Xe Nuclei. *J. Magn. Reson.* **126** (1997), 58–65.
398. R. W. Mair, M. D. Hürlimann, P. N. Sen, L. M. Schwartz, S. Patz, and R. L. Walsworth, Tortuosity Measurement and the Effects of Finite Pulse Widths on Xenon Gas Diffusion NMR Studies of Porous Media. *Magn. Reson. Imaging* **19** (2001), 345–51.

399. S. L. Codd and S. A. Altobelli, A PGSE Study of Propane Gas Flow Through Model Porous Bead Packs. *J. Magn. Reson.* **163** (2003), 16–22.
400. N. Fatkullin, Theory of Stimulated Spin Echo in Polymer System. *Sov. Phys. JETP* **72** (1991), 563–9.
401. W. Zhang and D. G. Cory, First Direct Measurement of the Spin Diffusion Rate in a Homogeneous Solid. *Phys. Rev. Lett.* **80** (1998), 1324–7.
402. E. Fischer, R. Kimmich, and N. Fatkullin, Spin Diffusion in Melts of Entangled Polymers. *J. Chem. Phys.* **106** (1997), 9883–8.
403. M. E. Komlosh and P. T. Callaghan, Spin Diffusion in Semidilute Random Coil Polymers Studied by Pulsed Gradient Spin-Echo NMR. *Macromolecules* **33** (2000), 6824–7.
404. D. Le Bihan, The 'Wet Mind': Water and Functional Neuroimaging. *Phys. Med. Biol.* **52** (2007), R57–90.
405. S. Blackband and P. Mansfield, Diffusion in Liquid-Solid Systems by NMR Imaging. *J. Phys. C: Solid State Phys.* **19** (1986), L49–52.
406. F. P. Duval, P. Porion, and H. Van Damme, Microscale and Macroscale Diffusion of Water in Colloidal Gels. A Pulsed Field Gradient and NMR Imaging Investigation. *J. Phys. Chem. B.* **103** (1999), 5730–5.
407. Y.-Q. Song, B. M. Goodson, B. Sheridan, T. M. De Swiet, and A. Pines, Effects of Diffusion on Magnetic Resonance Imaging of Laser-Polarized Xenon Gas. *J. Chem. Phys.* **108** (1998), 6233–9.
408. N. F. E. I. Nestle and R. Kimmich, Concentration-Dependent Diffusion Coefficients and Sorption Isotherms. Application to Ion Exchange Processes as an Example. *J. Phys. Chem.* **100** (1996), 12569–73.
409. I. E. Biton, I. D. Duncan, and Y. Cohen, High *b*-Value *q*-Space Diffusion MRI in Myelin-Deficient Rat Spinal Cords. *Magn. Reson. Imaging* **24** (2006), 161–66.
410. O. Mayzel, Y. Assaf, A. Gigi, D. Ben-Bashat, R. Verchovsky, M. Mordohovitch, M. Graif, T. Hendler, A. Korczyn, and Y. Cohen, High *b*-Value Diffusion Imaging of Dementia: Application to Vascular Dementia and Alzheimer Disease. *J. Neuro. Sci.* **257** (2007), 105–13.
411. M. D. Does, E. C. Parsons, and J. C. Gore, Oscillating Gradient Measurements of Water Diffusion in Normal and Globally Ischemic Rat Brain. *Magn. Reson. Med.* **49** (2008), 206–15.
412. S. Boujraf, R. Luypaert, H. Eisendrath, and M. Osteaux, Echo Planar Magnetic Resonance Imaging of Anisotropic Diffusion in Asparagus Stems. *Magn. Reson. Mater. Phys. Bio. Med.* **13** (2001), 82–90.
413. J. Ellegood, R. T. McKay, C. C. Hanstock, and C. Beaulieu, Anisotropic Diffusion of Metabolites in Peripheral Nerve Using Diffusion Weighted Magnetic Resonance Spectroscopy at Ultra-High Field. *J. Magn. Reson.* **184** (2007), 20–8.
414. K. Arfanakis, B. P. Hermann, B. P. Rogers, J. D. Carew, M. Seidenberg, and M. E. Meyerand, Diffusion Tensor MRI in Temporal Lobe Epilepsy. *Magn. Reson. Imaging* **20** (2002), 511–19.
415. M. Schlüter, B. Stieltjes, H. K. Hahn, J. Rexilius, O. Konrad-Verse, and H. O. Peitgen, Detection of Tumour Infiltration in Axonal Fibre Bundles Using Diffusion Tensor Imaging. *Int. J. Med. Robot. Comput. Assist. Surg.* **1** (2006), 80–6.
416. R. Watts, C. Liston, S. Niogi, and A. M. Uluğ, Fiber Tracking Using Magnetic Resonance Diffusion Tensor Imaging and Its Applications to Human Brain Development. *Ment. Retard. Dev. Disabil. Res. Rev.* **9** (2003), 168–77.
417. K. Kose, T. Haishi, A. Caprihan, and E. Fukushima, Real-Time NMR Imaging Systems Using Personal Computers. *J. Magn. Reson.* **124** (1997), 35–41.

418. C. J. Rofe, R. K. Lambert, and P. T. Callaghan, Nuclear Magnetic Resonance Imaging of Flow for a Shear-Thinning Polymer in Cylindrical Couette Geometry. *J. Rheol.* **38** (1994), 875–87.
419. S. J. Gibbs, K. L. James, L. D. Hall, D. E. Haycock, W. J. Frith, and S. Ablett, Rheometry and Detection of Apparent Wall Slip for Poiseuille Flow of Polymer Solutions and Particulate Dispersions by Nuclear Magnetic Resonance Velocimetry. *J. Rheol.* **40** (1996), 425–40.
420. M. M. Britton and P. T. Callaghan, NMR Microscopy and the Non-Linear Rheology of Food Materials. *Magn. Reson. Chem.* **35** (1997), S37–46.
421. M. M. Britton and P. T. Callaghan, Nuclear Magnetic Resonance Visualization of Anomalous Flow in Cone-and-Plate Rheometry. *J. Rheol.* **41** (1997), 1365–86.
422. Y. Xia and P. T. Callaghan, Imaging the Velocity Profiles in Tubeless Siphon Flow by NMR Microscopy. *J. Magn. Reson.* **164** (2003), 365–8.
423. S. Yao, M. Costello, A. G. Fane, and J. M. Pope, Non-Invasive Observation of Flow Profiles and Polarization Layers in Hollow Fiber Membrane Filtration Modules Using NMR Micro-Imaging. *J. Membr. Sci.* **99** (1995), 207–16.
424. U. Tallarek, D. van Dusschoten, T. Scheenen, H. Van As, E. Bayer, G. Guiochon, and U. D. Neue, Dynamic NMR Microscopy of Chromatographic Columns. *AIChE J.* **44** (1998), 1962–75.
425. I. V. Koptyug, S. A. Altobelli, E. Fukushima, A. V. Matveev, and R. Z. Sagdeev, Thermally Polarized ^{1}H NMR Microimaging Studies of Liquid and Gas Flow in Monolithic Catalysts. *J. Magn. Reson.* **147** (2000), 36–42.
426. A. Klemm, H.-P. Müller, and R. Kimmich, Evaluation of Fractal Parameters of Percolation Model Objects and Natural Porous Media by Means of NMR Microscopy. *Physica A* **266** (1999), 242–6.
427. S. J. Gibbs, T. A. Carpenter, and L. D. Hall, Magnetic Resonance Imaging of Thermal Convection. *J. Magn. Reson. A* **105** (1993), 209–14.
428. J. Weis, R. Kimmich, and H.-P. Müller, NMR Imaging of Thermal Convection Patterns. *Magn. Reson. Imaging* **14** (1996), 319–27.
429. A. Mohorič and J. Stepišnik, Effect of Natural Convection in a Horizontally Oriented Cylinder on NMR Imaging of the Distribution of Diffusivity. *Phys. Rev. E* **62** (2000), 6628–35.
430. M. Weber, A. Klemm, and R. Kimmich, Rayleigh-Bénard Percolation Transition Study of Thermal Convection in Porous Media: Numerical Simulation and NMR Experiments. *Phys. Rev. Lett.* **86** (2001), 4302–5.
431. R. W. Mair, C.-H. Tseng, G. P. Wong, D. G. Cory, and R. L. Walsworth, Magnetic Resonance Imaging of Convection in Laser-Polarized Xenon. *Phys. Rev. E* **61** (2000), 2741–8.
432. M. E. Moseley, Y. Cohen, J. Mintorovich, L. Chileuitt, H. Shimizu, J. Kucharczyk, M. F. Wendland, and P. R. Weinstein, Early Detection of Regional Cerebral Ischemia in Cats: Comparison of Diffusion- and T2-Weighted MRI and Spectroscopy. *Magn. Reson. Med.* **14** (1990), 330–46.
433. T. L. Chenevert, L. D. Stegman, J. M. G. Taylor, P. L. Robertson, H. S. Greenberg, A. Rehemtulla, and B. D. Ross, Diffusion Magnetic Resonance Imaging: An Early Surrogate Marker of Therapeutic Efficacy in Brain Tumors. *J. Natl. Cancer Inst.* **92** (2000), 2029–36.
434. K. G. Helmer, M. R. Meiler, C. H. Sotak, and J. D. Petrucelli, Comparison of the Return-to-the-Origin Probability and the Apparent Diffusion Coefficient of Water as Indicators of Necrosis in RIF-1 Tumors. *Magn. Reson. Med.* **49** (2003), 468–78.

435. D. Goldberg-Zimring, A. U. J. Mewes, M. Maddah, and S. K. Warfield, Diffusion Tensor Magnetic Resonance Imaging in Multiple Sclerosis. *Journal of Neuroimaging* **15** (2005), S61–81.
436. C. J. Fonseca, H. C. Oxenham, B. R. Cowan, C. J. Occleshaw, and A. A. Young, Aging Alters Patterns of Regional Nonuniformity in LV Strain Relaxation: A 3-D MR Tissue Tagging Study. *Am. J. Physiol. Heart Circ. Physiol.* **285** (2003), H621–30.
437. U. Görke, R. Kimmich, and J. Weis, Detection of Anisotropic Pulsating Flow and Its Velocity-Fluctuation Rate in Fertilized Bird Eggs by NMR Microimaging. *J. Magn. Reson. B* **111** (1996), 236–42.
438. Y. Seo, H. Shinar, Y. Morita, and G. Navon, Anisotropic and Restricted Diffusion of Water in the Sciatic Nerve: A ^{2}H Double Quantum Filtered Study. *Magn. Reson. Med.* **42** (1999), 461–6.
439. M. A. Fernández-Seara, S. L. Wehrli, and F. W. Wehrli, Diffusion of Exchangeable Water in Cortical Bone Studied by Nuclear Magnetic Resonance. *Biophys. J.* **82** (2002), 522–9.
440. S.-I. Han, O. Marseille, C. Gehlen, and B. Blümich, Rheology of Blood by NMR. *J. Magn. Reson.* **152** (2001), 87–94.
441. P. Loureiro de Sousa, D. Gounot, and D. Grucker, Flow Effects in Long-Range Dipolar Field MRI. *J. Magn. Reson.* **162** (2003), 356–63.

Appendix

Cumulant expansion

A cumulant expansion is a logarithmic function expansion. Here we illustrate the concept using only a one-dimensional case (the general case is considered elsewhere[1,2]).

Let $P(x)$ be a distribution of a random variable x and $\phi(k)$ be its characteristic function

$$\phi(k) = \int_{-\infty}^{+\infty} P(x) e^{ikx} dx. \tag{A.1}$$

When expanded in a power series this function becomes

$$\phi(k) = \sum_{n=0}^{\infty} \frac{(ik)^n}{n!} \langle x^n \rangle, \tag{A.2}$$

where

$$\langle x^n \rangle = \int_{-\infty}^{+\infty} x^n P(x) dx \tag{A.3}$$

are the moments of the distribution $P(x)$. Thus $\phi(k)$ is the generating function for moments. By taking the logarithm of $\phi(k)$ we obtain another important function (termed the generating function for cumulants)

$$\Phi(k) = \ln \phi(k), \tag{A.4}$$

which when expanded gives

$$\Phi(k) = \sum_{n=0}^{\infty} \frac{(ik)^n}{n!} \langle\langle x^n \rangle\rangle. \tag{A.5}$$

The coefficients $\langle\langle x^n \rangle\rangle$ in Eq. (A.5) are termed cumulants of variable x.

Maple Worksheet for the Stejskal and Tanner equation

The definitions of the Maple code can be found in the appropriate User's Guide. Define the integral used in determining q(*t*) and *b*.

```
> q:=(geff,ti)->gamma*1/(2*pi)*int(geff,td=ti..t): # Eq. (2.14)
> b:=(q,ti)->(2*pi)^2*int(q^2,t=ti..td): # Eq. (2.67)
```

Define the time intervals and the relevant value of geff for each integral. Then use these to calculate the values of q(*t*) and *b* for each interval remembering that they contain contributions from all of the intervals from the start of the pulse sequence.

```
> l1:=0: # start of 1st subinterval
> geff:=g0:
> q1:=q(geff,l1):
> b1:=b(q1,l1):
> l2:=t1: # start of 2nd subinterval
> geff:=g+g0:
> q2:=subs(t=l2,q1)+q(geff,l2):
> b2:=b(q2,l2)+subs(td=l2,b1):
> l3:=t1+delta: # start of 3rd subinterval
> geff:=g0:
> q3:=simplify(subs(t=l3,q2)+q(geff,l3)):
> b3:=simplify(b(q3,l3)+subs(td=l3,b2)):
> l4:=tau: # start of 4th subinterval
> geff:=-g0:
> q4:=subs(t=l4,q3)+q(geff,l4):
> b4:=simplify(b(q4,l4)+subs(td=l4,b3)):
> l5:=t1+Delta: # start of 5th subinterval
> geff:=-g-g0:
> q5:=subs(t=l5,q4)+q(geff,l5):
> b5:=simplify(b(q5,l5)+subs(td=l5,b4)):
> l6:=t1+Delta+delta: # start of 6th subinterval
> geff:=-g0:
> q6:=simplify(subs(t=l6,q5)+q(geff,l6)):
> b6:=simplify(b(q6,l6)+subs(td=l6,b5)):
> l7:=2*tau: # signal acquisition (i.e., echo time)
> S:=S0*exp(-subs(td=l7,b6)*D); # Eq. (2.65)
```

$$S := S0\,\mathrm{e}^{-\frac{1}{3}\gamma^2\left(3g^2\delta^2\Delta + 2g0^2\tau^3 - 3g\,g0\,\Delta\delta^2 - 2g\delta^3 g0 - 3g\delta g0\Delta^2 + 6g\delta^2 g0\,\tau - 6g\delta g0\,t1^2 - 6g\delta^2 g0\,t1 - g^2\delta^3 - 6g\delta g0\,\tau^2 + 12g\delta g0\,\tau\,t1 + 12g\delta g0\,\tau\Delta - 6g\delta g0\,t1\,\Delta\right)\mathrm{D}}$$

```
> S:=subs(g0=0,S); # Eq. (2.75)
```

$$S := S0\,\mathrm{e}^{-\frac{1}{3}\gamma^2\left(3g^2\delta^2\Delta - g^2\delta^3\right)\mathrm{D}}$$

Derivation of echo shapes and 1D images for gradient calibration

Gradient directed across the cylinder

With the gradient directed across a cylindrical sample (i.e., $g = g_x$ in Fig. 6.5), the phase distribution of spins starting at x_0 is

$$P(\phi) = g(\phi_1)\rho(x_0), \tag{A.6}$$

which is the product of two independent functions where

$$\phi = \phi_1 + \gamma g(2\tau - t)x_0, \tag{A.7}$$

$$g(\phi_1) = \frac{1}{\sqrt{2\pi\langle\phi_1^2\rangle}} e^{-\frac{\phi_1^2}{2\langle\phi_1^2\rangle}}. \tag{A.8}$$

The spin density function, $\rho(x_0)$, describes the distribution of chords of a circle, a semi-ellipse, and is zero for a position outside the sample. Noting that the length of a chord at a distance of x_0 from the centre of a circle of radius r is given using the Pythagorean theorem by $2(r^2 - x_0^2)^{1/2}$, we define

$$\rho(x_0) = \begin{cases} \frac{2}{\pi r^2}\left(r^2 - x_0^2\right)^{1/2} & |x_0| \leq r \\ 0 & |x_0| > r. \end{cases} \tag{A.9}$$

and the normalisation factor, $1/\pi r^2$, ensures that

$$\int_{-\infty}^{\infty} \rho(x_0)dx_0 = \int_{-r}^{r} \rho(x_0)dx_0 = 1. \tag{A.10}$$

Thus,

$$\begin{aligned} S(2\tau) &= S(2\tau)_{g=0} \int_{-\infty}^{\infty} P(\phi)\cos\phi\, d\phi \\ &= S(2\tau)_{g=0} \int_{-\infty}^{\infty} g(\phi_1)\, d\phi_1 \int_{-\infty}^{\infty} \rho(x_0)\cos\left(\phi_1 + \gamma g(2\tau - t)x_0\right)dx_0. \end{aligned} \tag{A.11}$$

Starting with the inner integral in Eq. (A.11) and using the trigonometric identity $\cos(A+B) = \cos A \cos B - \sin A \sin B$,

$$\int_{-\infty}^{\infty} \rho(x_0)\cos\left(\phi_1 + \gamma g(2\tau - t)x_0\right)dx_0 =$$

$$\cos\phi_1 \int_{-\infty}^{\infty} \rho(x_0)\cos\left(\gamma g(2\tau - t)x_0\right)dx_0 - \sin\phi_1 \int_{-\infty}^{\infty} \rho(x_0)\sin\left(\gamma g(2\tau - t)x_0\right)dx_0. \tag{A.12}$$

The second integral is an odd function of x_0 (i.e., $-f(x_0)=f(-x_0)$) and so equals 0. Thus, Eq. (A.11) becomes

$$S(2\tau) = S(2\tau)_{g=0}\int_{-\infty}^{\infty} g(\phi_1)\cos\phi_1 d\phi_1 \int_{-\infty}^{\infty} \rho(x_0)\cos(\gamma g(2\tau - t)x_0)dx_0. \quad (A.13)$$

The integral over x_0 in Eq. (A.13) is

$$\int_{-\infty}^{\infty} \rho(x_0)\cos(\gamma g(2\tau - t)x_0)dx_0 = \frac{2}{\pi r^2}\int_{-r}^{r} \left(r^2 - x_0^2\right)^{1/2}\cos(\gamma g(2\tau - t)x_0)dx_0, \quad (A.14)$$

and set $x_0 = rt$ and so $dx_0 = r\,dt$ and thus

$$= \frac{2}{\pi}\int_{-1}^{1} \left(1 - t^2\right)^{1/2}\cos(\gamma g(2\tau - t)rt)dt. \quad (A.15)$$

We use the integral representation of the first order Bessel function (e.g., 8.411.8 in ref. 3)

$$J_v(z) = \frac{\left(\frac{z}{2}\right)^v}{\Gamma\left(v + \frac{1}{2}\right)\Gamma\left(\frac{1}{2}\right)}\int_{-1}^{1} \left(1 - t^2\right)^{v-1/2}\cos(zt)\,dt \qquad \left[\text{Re}\, v > -\frac{1}{2}\right], \quad (A.16)$$

with $v=1$ and $z=\gamma g(\tau - t)r$, $\Gamma(1/2) = \sqrt{\pi}$, $\Gamma(3/2) = \sqrt{\pi}/2$ and so Eq. (A.16) becomes

$$\begin{aligned} J_1(z) &= \frac{\frac{z}{2}}{\Gamma\left(\frac{3}{2}\right)\Gamma\left(\frac{1}{2}\right)}\int_{-1}^{1} \left(1 - t^2\right)^{1/2}\cos(zt)dt \\ &= \frac{z}{\pi}\int_{-1}^{1} \left(1 - t^2\right)^{1/2}\cos(zt)dt. \end{aligned} \quad (A.17)$$

Using Eq. (A.17), Eq. (A.14) becomes

$$\begin{aligned} &\frac{2}{\gamma g(2\tau - t)r}\left[\frac{\gamma g(2\tau - t)r}{\pi}\int_{-1}^{1} \left(1 - t^2\right)^{1/2}\cos(\gamma g(2\tau - t)rt)dt\right] \\ &= \frac{2J_1(\gamma g(2\tau - t)r)}{\gamma g(2\tau - t)r}. \end{aligned} \quad (A.18)$$

We now consider the integral over ϕ_1 in Eq. (A.13),

$$\int_{-\infty}^{\infty} g(\phi_1)\cos\phi_1\, d\phi_1 = \frac{2}{\sqrt{2\pi\langle\phi_1^2\rangle}}\int_{0}^{\infty} e^{-\frac{\phi_1^2}{2\langle\phi_1^2\rangle}}\cos\phi_1\, d\phi_1. \quad (A.19)$$

By noting the standard integral (e.g., 3.896.4 in ref. 3)

$$\int_0^{\infty} e^{-\beta x^2} \cos(bx)dx = \frac{1}{2}\sqrt{\frac{\pi}{\beta}}\exp\left(-\frac{b^2}{4\beta}\right) \quad [\mathrm{Re}\,\beta > 0], \tag{A.20}$$

where in the present case $\beta = 1/(2\langle \phi_1{}^2 \rangle)$ and $b = 1$ and so continuing with Eq. (A.19) we get

$$\int_{-\infty}^{\infty} g(\phi_1)\cos\phi_1 d\phi_1 = \exp\left(-\frac{\langle \phi_1^2 \rangle}{2}\right). \tag{A.21}$$

Using Eq. (A.18) and (A.21) we finally obtain the solution to Eq. (A.11) as[4–6]

$$\begin{aligned} S(2\tau) &= S(2\tau)_{g=0}\int_{-\infty}^{\infty} g(\phi_1)d\phi_1 \int_{-\infty}^{\infty} \rho(x_0)\cos(\phi_1 + \gamma g(2\tau - t)x_0)dx_0 \\ &= \exp\left(-\frac{\langle \phi_1^2 \rangle}{2}\right)\frac{2J_1(\gamma g(2\tau - t)r)}{\gamma g(2\tau - t)r}. \end{aligned} \tag{A.22}$$

(Note: Refs. 5 and 6 contain misprints; the 2 has been omitted in the numerator).

The exponential term in Eq. (A.22) is a constant (it is the attenuation factor due to diffusion and does not affect the echo shape, only its initial amplitude) and to consider the Fourier transform of $\frac{J_1(\gamma g(2\tau - t)r)}{\gamma g(2\tau - t)r}$ and thereby obtain the frequency spectrum

$$S(\omega) = \int_{-\infty}^{\infty} \frac{J_1(\gamma g(2\tau - t)r)}{\gamma g(2\tau - t)r} e^{-i\omega t}dt. \tag{A.23}$$

We set $a = \gamma g r$, $x = a(2\tau - t)$, $t = 2\tau - x/a$ and $dt = -dx/a$ and so Eq. (A.23) becomes

$$\begin{aligned} S(\omega) &= -\int_{-\infty}^{\infty} \frac{J_1(x)}{ax} e^{-i\omega\left(2\tau - \frac{x}{a}\right)}dx \\ &= -\frac{e^{-i\omega 2\tau}}{a}\int_{-\infty}^{\infty} \frac{J_1(x)}{x} e^{i\omega\frac{x}{a}}dx \\ &= -\frac{e^{-i\omega 2\tau}}{a}F(\omega). \end{aligned} \tag{A.24}$$

We set $k = \omega/a$ and consider the integral

$$F(\omega) = \int_{-\infty}^{\infty} \frac{J_1(x)}{x} e^{ikx}dx. \tag{A.25}$$

Using Euler's relation (Eq. (2.1)), this becomes

$$F(\omega) = \int_{-\infty}^{\infty} \frac{J_1(x)}{x} \cos(kx)dx + i\int_{-\infty}^{\infty} \frac{J_1(x)}{x} \sin(kx)dx. \tag{A.26}$$

The first integrand is even and the second integrand is an odd function and therefore goes to zero and so Eq. (A.26) becomes

$$F(\omega) = 2\int_{0}^{\infty} \frac{J_1(x)}{x} \cos(kx)dx. \tag{A.27}$$

Next, noting the standard integral (e.g., 6.693.2 in ref. 3),

$$\int_0^{\infty} \frac{J_\nu(\alpha x)}{x} \cos\beta x\, dx = \begin{cases} \frac{1}{\nu}\cos\left(\arcsin\frac{\beta}{\alpha}\right) & [\beta \le \alpha] \\ \dfrac{\alpha^\nu \cos\frac{\nu\pi}{2}}{\nu\left(\beta + \sqrt{\beta^2 - \alpha^2}\right)^\nu} & [\beta \ge \alpha] \end{cases} \quad [\mathrm{Re}\, \nu > 0], \tag{A.28}$$

in the present case $\nu = 1$, $\alpha = 1$ and $\beta = k$. Thus, Eq. (A.23) becomes

$$S(\omega) = -\frac{e^{-i\omega 2\tau}}{\gamma g r} \cos\left(\arcsin\left(\frac{\omega}{\gamma g r}\right)\right) \quad [|\omega| \le \gamma g r] \tag{A.29}$$

and equals 0 otherwise.

Gradient directed along the cylinder

In a superconducting magnet the gradient direction is usually along the z-axis (i.e., $g = g_z$ in Fig. 6.5). In this case,

$$\rho(z_0) = \begin{cases} \frac{1}{l} & |z_0| \le l \\ 0 & |z_0| > l. \end{cases} \tag{A.30}$$

(i.e., $1/l$ is the normalisation factor) and performing the same procedure as in the previous subsection we have

$$S(2\tau) = S(2\tau)_{g=0} \int_{-\infty}^{\infty} g(\phi_1) \cos\phi_1 d\phi_1 \int_{-\infty}^{\infty} \rho(z_0) \cos\left(\gamma g(2\tau - t) z_0\right) dz_0. \tag{A.31}$$

We first consider the integral over z_0 in Eq. (A.31),

$$\begin{aligned} \int_{-\infty}^{\infty} \rho(z_0) \cos\left(\gamma g(2\tau - t) z_0\right) dz_0 &= \frac{1}{l} \int_0^l \cos(\gamma g(2\tau - t) z_0) dz_0 \\ &= \mathrm{sinc}(\gamma g(2\tau - t) l). \end{aligned} \tag{A.32}$$

The integral in Eq. (A.31) over ϕ_1 is given by Eq. (A.21) and so becomes

$$S(2\tau) = \exp\left(-\frac{\langle\phi_1^2\rangle}{2}\right)\mathrm{sinc}(\gamma g(2\tau - t)l). \tag{A.33}$$

The first term in Eq. (A.33) is a constant and so we perform the Fourier transform of $\mathrm{sinc}(\gamma g(2\tau - t)l)$ to obtain the frequency spectrum

$$S(\omega) = \int_{-\infty}^{\infty} \frac{\sin(\gamma g(2\tau - t)l)}{\gamma g(2\tau - t)l} e^{-i\omega t} dt. \tag{A.34}$$

We set $a = \gamma gl$, $x = a(2\tau - t)$, $t = 2\tau - x/a$ and $dt = -dx/a$ and so Eq. (A.34) becomes

$$\begin{aligned} S(\omega) &= -\int_{-\infty}^{\infty} \frac{\sin x}{ax} e^{-i\omega\left(2\tau - \frac{x}{a}\right)} dx \\ &= -\frac{e^{-i2\tau\omega}}{a} F(\omega), \end{aligned} \tag{A.35}$$

where $k = \omega/a$. We now consider the integral

$$F(\omega) = \int_{-\infty}^{\infty} \frac{\sin x}{x} e^{ikx} dx. \tag{A.36}$$

Using Euler's relation (Eq. (2.1)) to obtain

$$F(\omega) = F_1(\omega) + iF_2(\omega), \tag{A.37}$$

where

$$F_1(\omega) = \int_{-\infty}^{\infty} \frac{\sin x \cos x}{x} dx = 2\int_{0}^{\infty} \frac{\sin x \cos x}{x} dx \tag{A.38}$$

and

$$F_2(\omega) = \int_{-\infty}^{\infty} \frac{\sin x \sin kx}{x} dx = 0, \tag{A.39}$$

because the integrand is an odd function. Next, noting the identity

$$\sin x \cos(kx) = \frac{1}{2}[\sin(x[1 - k]) + \sin(x[1 + k])], \tag{A.40}$$

Eq. (A.38) becomes

$$F_1(\omega) = \int_{0}^{\infty} \frac{\sin([1 - k]x)}{x} dx + \int_{0}^{\infty} \frac{\sin([1 + k]x)}{x} dx. \tag{A.41}$$

Finally, noting the standard integral (e.g., see Eq. 2.5.3.12 in ref. 7)

$$\int_0^\infty \frac{\sin bx}{x} dx = \frac{\pi}{2} \operatorname{sgn} b, \tag{A.42}$$

to obtain

$$\begin{aligned} F(\omega) &= \frac{\pi}{2}[\operatorname{sgn}(1-k) + \operatorname{sgn}(1+k)] \\ &= \frac{\pi}{2}[\operatorname{sgn}(\gamma gl - \omega) + \operatorname{sgn}(\gamma gl + \omega)]. \end{aligned} \tag{A.43}$$

Thus, Eq. (A.35) becomes

$$S(\omega) = -\frac{e^{-i2\tau\omega}}{\gamma gl}\frac{\pi}{2}[\operatorname{sgn}(\gamma gl - \omega) + \operatorname{sgn}(\gamma gl + \omega)]. \tag{A.44}$$

The bracketed term in Eq. (A.44) is non-zero when $|\omega| \leq \gamma gl$ as expected for the transform of a sinc function and thus gives the same results as Eq. (6.7).

References

1. C. W. Gardiner, *Handbook of Stochastic Methods: For Physics, Chemistry and the Natural Sciences*, 2nd edn. (Berlin: Springer-Verlag, 1996).
2. N. G. Van Kampen, *Stochastic Processes in Physics and Chemistry*, 3rd edn. (Amsterdam: North Holland, 2001).
3. I. S. Gradshteyn and I. M. Ryzhik, *Table of Integrals, Series, and Products*, 7th edn. (New York: Academic Press, 2007).
4. H. Y. Carr and E. M. Purcell, Effects of Diffusion on Free Precession in Nuclear Magnetic Resonance Experiments. *Phys. Rev.* **94** (1954), 630–8.
5. D. C. Douglass and D. W. McCall, Diffusion in Paraffin Hydrocarbons. *J. Phys. Chem.* **62** (1958), 1102–7.
6. D. W. McCall, D. C. Douglass, and E. W. Anderson, Self-Diffusion Studies by Means of Nuclear Magnetic Resonance Spin-Echo Techniques. *Ber. Bunsenges. Phys. Chem.* **67** (1963), 336–40.
7. A. P. Prudnikov, Yu. A. Brychkov, and O. I. Marichev, *Elementary Functions*. (New York: Gordon and Breach, 1990).

Index

D

H

For EU product safety concerns, contact us at Calle de José Abascal, 56–1°, 28003 Madrid, Spain or eugpsr@cambridge.org.

www.ingramcontent.com/pod-product-compliance
Lightning Source LLC
LaVergne TN
LVHW062205200926
848484LV00006B/1321

* 9 7 8 0 5 2 1 8 0 6 9 6 1 *